CERAMIC INTERFACES 2

CERAMIC INTERFACES 2

Edited by
H. -I. YOO
and
S. -J. L. KANG

Book 0741
First published in 2001 by
IOM Communications Ltd
1 Carlton House Terrace
London SW1Y 5DB

IOM Communications
is a wholly-owned subsidiary of
The Institute of Materials
1 Carlton House Terrace
London SW1Y 5DB

ISBN 1-86125-118-1

Typeset in India by
Emptek Inc.
Printed and bound in the UK at
The University Press, Cambridge

Contents

Preface

This series of International Workshops on Ceramic Interfaces was first held at the Lucas Heights, Sydney, Australia in February 1993, then moved to Nagano, Japan, in September 1994, and then back to Australia in Kuranda Forest in July 1996. With the later addition of Korea to what had previously been an Australia-Japan bilateral endeavour, the 4th Workshop was held at the Korea Advanced Institute of Science and Technology, Taejon, Korea, from 16 to 19 September, 1998.

By nature, interfaces are the offspring of neighbouring bulks. Following the long tradition, this 4th Workshop thus continued to be concerned with the things happening at and/or across the interfaces and neighbouring bulks in advanced materials including electronic/energy-related materials. All 27 specialists were brought together from 8 countries to discuss, exchange ideas and enjoy intellectual pleasures about the structure, composition and segregation, reaction at/across the interfaces, phase mixtures/ nanostructure/thin films/composites, microstructure evolution and control, defects and electrochemical transport, interface-sensitive properties and applications, and so on. The workshop proceeded in three serial sessions: Microstructural Development, Transport, and Interfacial Phenomena and Kinetics with 4 keynote lectures. In the middle of the Workshop, there was a hopping excursion through the natural interfaces in the nearby Mt. KyeRyong (cover picture), followed by a discussion over whether those interfacial hoppers were intellectual or wise.

The present volume is a precipitate of this workshop in the hope that someday we can say that it has taken too long to recognise the obvious [interfaces]. We wish it to be of some help to those who want to know the interfaces and about the interfaces.

As has been the case in the past, financial support is a main ingredient of today's scientific endeavours. The workshop would have been impossible without generous support from the Korea Science and Engineering Foundation (KOSEF) through the Engineering Research Centers, the Centre for Interface Science and Engineering of Materials (CISEM) at the Korea Advanced Institute of Science and Technology, and the Research Centre for Thin Film Fabrication and Crystal Growing of Advanced Materials (RETCAM) at the Seoul National University. Her financial support is gratefully appreciated.

Finally, but not least, credit should also go to the administrative staff of the CISEM who set the stage for the workshop and greased the wheels up until this point.

S.-J. L. Kang
H.-I. Yoo

SECTION I

Microstructural Development

The Step Growth Hypothesis for Abnormal Grain Growth

DUK YONG YOON, CHAN WOO PARK and JAE BON KOO
Department of Materials Science and Engineering
Korea Advanced Institute of Science and Technology
373-1, Kusong-Dong, Yusong-Gu, Taejon 305-701, Korea

ABSTRACT

The implications of the hypothesis that the abnormal grain growth, with or without liquid matrix, occurs by two dimensional nucleation on the boundaries of defect free grains are examined and compared to the observations. The model of coarsening by the growth on steps, produced by two dimensional nucleation, predicts three types of coarsening behaviour: the stagnant growth where the growth of all the grains with singular boundaries is limited by the production of surface steps; the abnormal grain growth (AGG) where some of the grains with singular boundaries grow rapidly by kinetic roughening under relative by large driving forces; and normal growth where the boundaries are rough. The transition from the stagnant growth to AGG and from AGG to nearly normal growth can occur by increasing the heat-treatment temperature and the driving force (by deformation, composition change, or grain size change) and by decreasing the step free energy. The normal growth, however, is possible only if the boundaries are thermodynamically rough with 0 step free energy. Some experimental observations of the dislocation effects are also consistent with the step growth hypothesis for singular boundaries. More quantitative analysis and observations are needed to test this hypothesis.

1.0 INTRODUCTION

Two extreme types for the equilibrium shapes of crystals in liquid are, first the polyhedra with sharp corners and edges and second the spheres. A crystal of the first type grows by two dimensional nucleation of steps on its flat surface if it is free of defects such as screw dislocations, and in the second case a crystal grows relatively rapidly without any inhibition of the atom attachment to the rough surface. The growth rate is then determined by the heat or solute flux at the surface. When there are many grains dispersed in a liquid matrix and if the grains are spherical, they undergo diffusion controlled normal coarsening in agreement with the modified Lifshitz, Slyozov[1] and Wagner[2] (LSW) theory.[3-5] When the grains are polyhedral the abnormal grain growth (AGG) occurs because large grains may grow rapidly by two dimensional nucleation of steps at their surfaces as proposed by Park, Hwang and Yoon.[6] The two dimensional nucleation theory indicates that there can be three types of grain growth depending on the size distribution of the grains relative to the heat-treatment temperature and the step free energy: slow, stagnant growth of all grains limited by the steps produced, AGG and the normal growth. Such a possibility and some related observations are examined in this report.

The polycrystalline single phase materials also show AGG during heat-treatment after deformation or during sintering. It has been observed that in Ni, AGG occurs when many of the grain boundaries are faceted at relatively low temperatures and normal growth when the grain boundaries become atomically rough at high temperatures close to the melting point.[7] The possibilities of grain boundary transitions were suggested by Hart[8] and Guttmann.[9] Rottman[10] showed for a dislocation model that low angle grain boundaries exceeding a certain misorientation angle could undergo a roughening transition. The observations of the grain boundary roughening have been mostly indirect,[11-16] but for an Al grain embedded in another large Al grain with a relatively large tilt misorientation angle, Westmacott and Dahmen[17] observed reversible roughening of edges and possibly some flat grain boundary segments during temperature cycling. Since at least some of the segments of faceted grain boundary segments are expected to be singular corresponding to the cusp directions of the γ-plot against the inclination angle, the facet-defacet transitions observed with temperature increase[18] and additives[19-21] indicate at least the rounding of corners and edges of the equilibrium shape of a grain embedded in another. It is then possible that the faceted grain boundaries grow by a step mechanism similar to the grains in liquid matrix and the defaceted grain boundaries grow continuously without any inhibition to the atom attachment to another grain across the grain boundary. The experimental results consistent with such a possibility are examined in this report.

2.0 COARSENING IN LIQUID MATRIX

When the grains are spherical with atomically rough surface structure, the grains show normal diffusion controlled coarsening in agreement with the LSW theory modified for the finite liquid volume effect[3] as pointed out earlier.[4,22] For the growth of a polyhedral crystal in a supercooled melt, the rate N of two dimensional nucleation on the surfaces will be,[23]

$$\dot{N} = n_S n^* \nu e^{-[(\Delta G_m + \Delta G^*_{2D})/kT]} \tag{1}$$

where n_s is the number of atoms (or molecules) per unit surface area, n^* the number of atoms of the liquid (or the vapour) close to the critical nucleus, ν the vibrational frequency of the atom in the liquid, and ΔG_m the activation energy for jumping across the interface. The free energy barrier for nucleation ΔG^*_{2D} is given by

$$\Delta G^*_{2D} = \frac{\pi V_m \sigma_s^2(T)}{h \Delta g} \tag{2}$$

where V_m is the molar volume of the solid phase, $\sigma_s(T)$ the step free energy per unit length, h the height of the step and Δg the molar free energy difference between the liquid and the solid in the supercooled state.

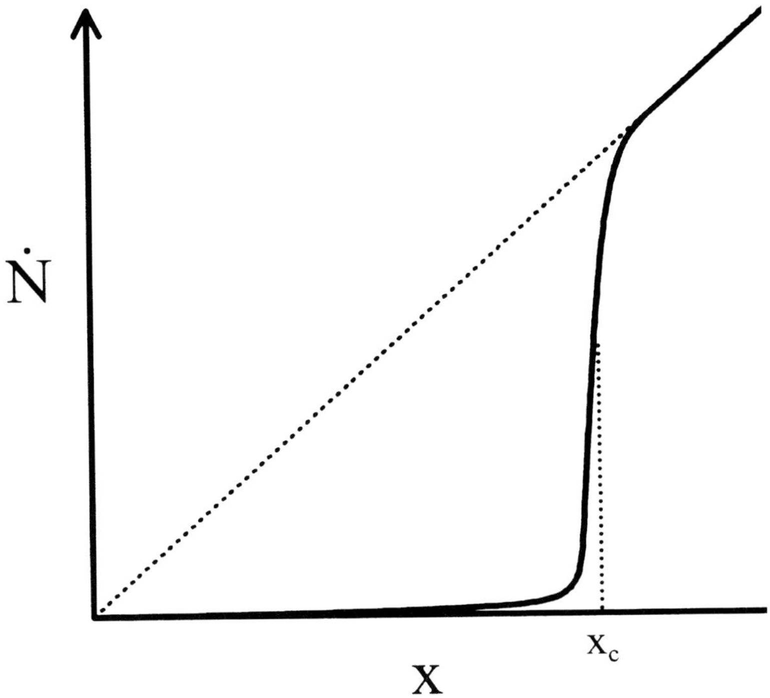

Fig. 1 Schematic variation of Ṅ with X.

Defining x as,

$$x = \frac{kTh\Delta g}{\pi V_m \sigma_s^2 (T)} \tag{3}$$

$$\dot{N} = n_s n^* \nu \, e^{-\frac{\Delta Gm}{kT}} e^{-\frac{1}{x}} \tag{4}$$

The variation of $\dot{N}$ with x is schematically shown in Figure 1. The critical value x_c of x for a substantial nucleation rate may be estimated for $\dot{N}$ = 1 cm^{-2}s^{-1} as,

$$x_c = \frac{1}{ln(n_s n^* \nu) - \Delta G_m / kT} \tag{5}$$

For the coarsening of polyhedral grains dispersed in a liquid matrix, the driving force Δg for a grain of size r can be written as,

$$\Delta g(\bar{r}, r) = 2\sigma V_m \left(\frac{1}{\bar{r}} - \frac{1}{r}\right) \tag{6}$$

where σ is the interfacial energy and $\bar{r}$ is approximately the average size to be determined by the mass balance between the dissolving and the growing grains.

As illustrated in Figure 1, $\dot{N}$ increases abruptly above x_c. In crystal growth, T and $\sigma_s(T)$ (through its T dependence) are assumed to be fixed. Then x_c represents the critical driving force required for substantial nucleation. If Δg is for coarsening as described by Eqn. (6), the AGG behaviour is described by Eqn. (1) as proposed by Park, Hwang and Yoon.[6] There will be a value of x for each growing grain, and if x for a grain is smaller than x_c it will grow slowly by a step nucleation process; if x is larger than x_c, it will grow rapidly by a continuous process. The theoretical analysis of the kinetic roughening during the continuous growth at high driving forces indeed showed that the grain surface structure became disordered resembling that obtained by the thermodynamic roughening transition.[24] Then the number of the abnormally growing grains will be equal to the number of grains with x larger than x_c. If the values of x for all grains are smaller than x_c, the grains will exhibit a slow, stagnant coarsening. In such a stagnant coarsening the grain size distribution may resemble that of a normal growth, but after a prolonged heat-treatment x for the largest grains may exceed x_c giving arise to AGG. In general, if the grain size distribution is such that some of the grains under a heat-treatment condition have values of x larger than x_c, they will grow abnormally while the finer matrix grains will undergo a stagnant growth. A special case of the stagnant growth of all grains arises, as noted by Park et al.,[6] if the average grain size in Eqn. (6) is so large that even for a grain of infinitely large size, the value of x does not exceed x_c. In practice such a condition may be obtained when the initial average grain size or the average size of the abnormally grown grains after their impingement on each other is very large.

Since x_c given by Eqn. (5) may be assumed to be a constant for a material system at a narrow temperature range used for heat-treatment, there will be critical values for any one or combinations of the variables in Eqn. (3) for x which are obtained by equating the Eqns. (3) and (5). As noted earlier, in crystal growth the parameters T and $\sigma_s(T)$ in Eqn.(3) are fixed, so that there is a critical value of Δg. In coarsening, Δg is fixed by the grain size distribution, but T and $\sigma_s(T)$ in Eqn. (3) can be variables with their critical values. If $\sigma_s(T)$ is decreased to a small value close to 0 by an additive, there might be a critical additive content corresponding to x_c. The T term can also be independently varied but only at low temperatures where $\sigma_s(T)$ is nearly independent of T. Since the lowest T is limited by the temperature at which the liquid phase can be maintained, the ranges of the variation of T in most of the solid-liquid systems will be limited compared to those of Δg and $\sigma_s(T)$. For some solid surfaces in equilibrium with liquids, $\sigma_s(T)$ was observed to decrease discontinuously to 0 at T_R,[25] but the solutions of solid-on-solid models predict that $\sigma_s(T)$ will decrease continuously to 0 at T_R[26, 27] in agreement with some experimental observations.[25] In the latter case, with increasing temperature there will be a compound effect of the increase of the T term and the decrease of the $\sigma_s^2(T)$ term in Eqn. (3). Therefore, the critical temperature of each grain for the transition from the step growth to the continuous growth may in practice be determined by the combination of these two effects.

Since x varies as $1/\sigma_s^2(T)$, the number of grains growing continuously will increase with decreasing $\sigma_s(T)$ and the overall growth, although abnormal, may show size distributions resembling the normal growth. But the normal growth with all of the grains growing continuously is possible only when the interface becomes rough with $\sigma_s(T) = 0$. The transitions from stagnant growth to AGG can occur by the variation of the driving force, temperature, and the step free energy. The driving force can be varied by the grain size distribution but additional driving forces might be obtained, for instance, by deformation, composition change, phase transformation of the solid phase, or temperature change.

We will now examine some observations that indicate these growth behaviours. The transition from AGG to normal growth with temperature increase in 70 NbC-30 Fe was previously described.[22,28] With polyhedral grains at a low temperature, AGG was observed, but at high temperatures above the surface roughening transition temperature, normal growth was observed as the grains became nearly spherical.[28] The temperature dependence of the grain coarsening behaviour of this alloy might be, however, more complex, because the early observations of Sarian and Weart[29] indicate that the roughening begins at corners and edges of the grain as shown by the analysis of Rottman and Wortis.[30] The temperatures at which the corners and edges begin to roughen are not known nor predictable, but if they are relatively low, the rough corners and edges may drastically change the growth behaviour because continuous growth can now occur at these surface regions. Rottman and Wortis[30] showed that the surface normal can change either continuously or discontinuously from a rough region to an adjacent flat region. If the surface normal changes continuously, as appears to be the case for the NbC-Fe alloy,[29] it is possible that the growth layers on the rough surface spill over to the flat region, thereby inducing its growth as illustrated in Figure 2. In such a case, the growth will be no longer limited at the flat surface and the grains may undergo normal growth. If, on the other hand, the surface normal changes discontinuously, the step spill-over may not occur from the rapidly growing rough surface to the flat surface. In such a case, the growth at the flat surface will be still limited by the step nucleation process, although there will be now an additional driving force for the growth of the flat surface because of the equilibration tendency of the elongated grains. This latter possibility may be found in the elongated grains growing abnormally with apparently a rough surface in the elongated direction and a flat face often perpendicular to it. As shown by the example of partially roughened grains in NbC-Fe,[29] it is usually difficult to determine from the shape of grains if the slope change from the rough to the flat region is continuous or discontinuous. This ambiguity can probably be resolved more clearly by observing the possible faceting (hill-and-valley) structure of the liquid films between the grains of various average surface orientations in the specimens with relatively low liquid contents.

An interesting example of the additive effect on grain coarsening is found in the alumina-anorthite-MgO system. It is well known that the addition of anorthite (CaO and SiO_2) causes abnormal growth of alumina grains elongated along their basal planes[31,32] as shown in Figure 3a. The microstructures at earlier stages before the impingement among the large grains indicated that the non-basal planes were also flat. The TEM observation of the liquid pockets at the triple junctions of the fine matrix grains exhibited flat surfaces of the grains as shown in Figure 4a. Although

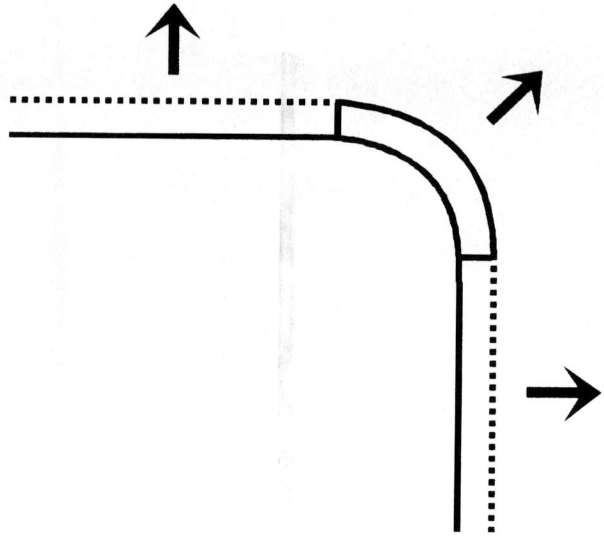

Fig. 2 Illustration of the growth by step spill-over from the rough edge to a flat surface with continuously changing surface slope.

the surface shapes of the grains in contact with their neighbours across the thin liquid layers could not be readily resolved because the grain surfaces were randomly oriented, they are likely to form hill-and-valley faceted structures and grain boundaries at certain regions as observed by Kim et a1.[33] It is not yet understood how these intergranular layers will move during the coarsening, but even though nucleation occurs readily at the reentrant edges, the coarsening will be still controlled by the movement of the triple junctions which can only move, without any reentrant edges, by step nucleation at the flat surfaces. (Nevertheless, it still will be instructive to observe the structure of the grain surfaces at the contact regions).

When 0.1 wt.% MgO was added, AGG still occurred, but the number density of the large grains was much higher and the large grains were equiaxial as shown in Figure 3b. When heat-treated further, the volume fraction of the small grains decreased and after prolonged heat-treatments they probably will disappear completely, leaving only the large grains impinging upon each other and the entire microstructure resembling that obtained by normal growth with a narrow size distribution. At the triple junctions some grains still showed flat surfaces but others now showed curved surfaces as exhibited in Figure 4b. Figure 3b also shows some flat grain surfaces, and in Figure 4b it appears that the surface normal might change discontinuously from the curved to the flat surface sections. It thus appears that the addition of MgO increased the surface energy of the basal plane relative to the others to make the grains equiaxial, reduced the step free energy of some surface planes to make them rough, and also, probably reduced the step free energy of

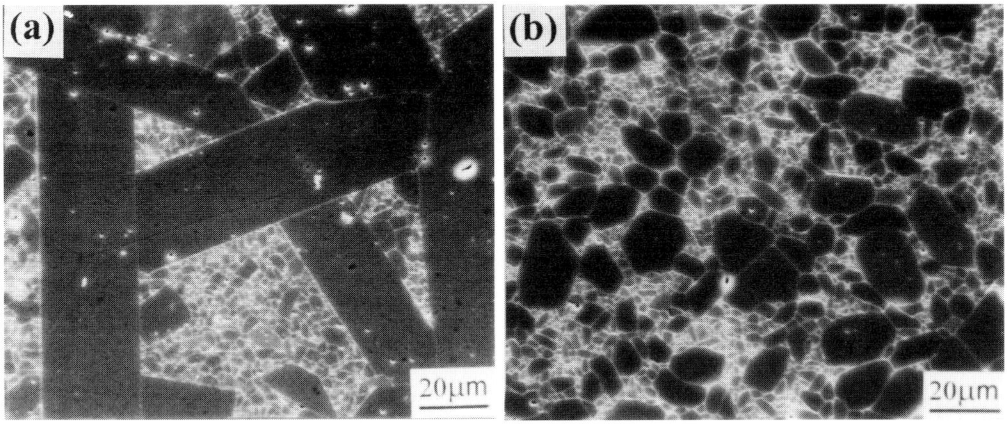

Fig. 3 Microstructures of Al_2O_3 sintered at 1650°C for 6 hrs. dopted with (a) 0.1 wt.% anorthite and (b) 0.1 wt.% anorthite + wt.% MgO.

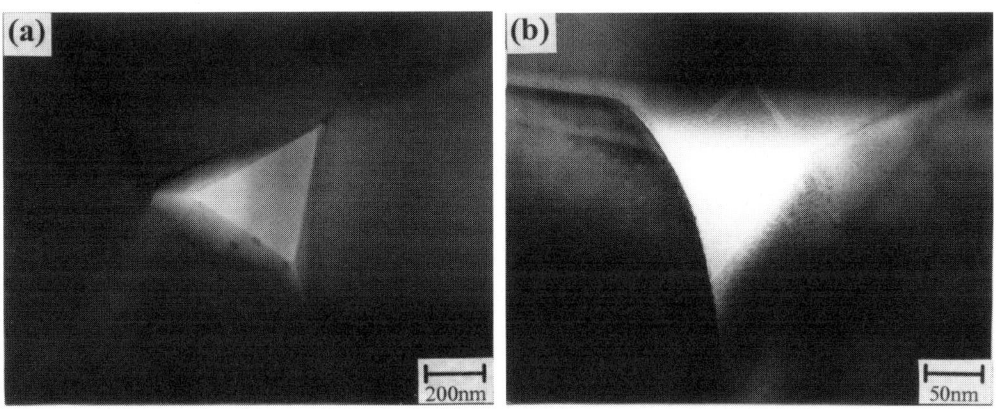

Fig. 4 Grain boundary triple junctions in Al_2O_3 sintered at 1650°C for 1 hr. doped with (a) 0.1 wt.% anorthite and (b) 0.1 wt.% anorthite + 0.1 wt.% MgO.

those flat planes which still remained singular but not sufficiently to make them rough. It also appears that the step spill-over does not occur from the rough surfaces to the neighbouring flat ones like in the case of the continuous slope change illustrated in Figure 2, because AGG still occurs, probably by the step nucleation on the flat surfaces. But now with a reduced step free energy, more grains can grow continuously (because their x values exceed x_c). It has been often suggested that the addition of MgO to alumina (usually with some liquid phase) prevents AGG. But these results show that the MgO addition actually enhances the growth of even relatively small grains such that the number density of the large grains increases with a decrease of their size after the impingement. After the large grains impinge upon each other, they will again undergo step limited stagnant growth with a narrow size distribution resembling that of a normal growth, because their x values would have shifted back to those smaller than x_c in the absence of a large driving force for coarsening. There can, however, be cases where the additive suppresses AGG by increasing the step free energy or by decreasing the surface energy. In such a case, AGG might be only temporarily suppressed to the stagnant growth mode with the possibility that a few grains will grow to very large sizes after prolonged heat-treatments. The effect of adding VC to WC-Co might be such a case.[34]

We have thus far examined the cases where the number of the grains growing continuously was enhanced from that of AGG by either increasing the temperature or by additives. If such effects became extreme, the growth can become normal as observed in the NbC-Fe at high temperatures. We now examine two examples of stagnant growth and the transition to AGG. The first example was shown by Park, Hwang and Yoon,[6] where they observed practically no coarsening when a WC-15 Co alloy made from WC powder of 5.48 μm average size was heat-treated for a long time at the normal sintering temperature, where AGG readily occurs if finer powders are used. When alumina of sub-micron size mixed with 0.05 eqv. wt.% anorthite was sintered and heat-treated further at 1400°C for 12 hrs., the grains grew slightly to an average size of about 1.5 μm with a relatively narrow size distribution. When heat-treated at 1620°C for 12 hrs., AGG was observed. These results appear to show the transition from the stagnant growth to AGG with temperature increase and hence an apparent minimum temperature required for AGG. As discussed earlier, this critical temperature will depend on the grain size distribution and additives through their effect on the step free energy as well as surface free energy. It should be noted that an additive will probably increase or decrease both the step free energy and the surface free energy, thus producing opposing effects on the step nucleation rate, but at temperatures close to the roughening temperature, the dominant effect will probably arise from its effect on the step free energy. In principle, if the solid-liquid equilibrium can be maintained at higher temperatures, another transition from AGG to normal growth will occur with further temperature increase. We thus see that if the temperature is a variable, three growth modes can appear: stagnant growth at low temperatures, AGG at intermediate temperatures and normal growth at high temperatures. In addition to the effects on these transitions arising from the changes in temperature, step free energy, and size distribution, the driving forces arising from composition changes and deformation of the grains may also induce these transitions from the stagnant growth to AGG and from AGG to nearly normal growth. It will be interesting to see if the stagnant growth conditions can be established for the NbC-Fe or TaC-Ni alloy at low temperatures by controlling the average grain size and to test these transitions of the growth behaviour induced by different variables.

3.0 GRAIN GROWTH IN SINGLE PHASE POLYCRYSTALS

Using a dislocation step model, Rottman[10] showed that the roughening transition of low angle grain boundaries was possible. Dahmen and Westmacott[17] observed that the shape of an Al grain embedded in another grain with an orientation close to Σ99 around a <110> tilt axis was nearly polyhedral when heat-treated at 200°C, but at a higher temperature (430°C), the edges and some flat boundaries became rounded, indicating the roughening transition of these apparently singular grain boundary segments. For surfaces and grain boundaries the roughening transition refers to the disordering of the flat singular planes of their equilibrium shapes with their normals corresponding to the cusps in their γ-plots. As for the surface segments, a grain boundary with an average orientation which does not appear in the equilibrium shape of the grain will have a faceted hill-and-valley structure as observed in many polycrystalline systems. These faceted grain boundaries were observed to become curved with continuously varying surface orientations at high temperatures and with additives. An asymmetric Σ3 <111> tilt boundary in Al, for example, was observed to be faceted at 20°C and become straight (defaceted) at 230°C by Hsieh and Balluffi.[18] All grain boundaries in a polycrystalline Ni were observed to be faceted when heat-treated at 0.55 T_m, where T_m is the melting point, and defaceted at high temperatures under a carburizing atmosphere.[7] The reversible faceting-defaceting was also observed when Bi was diffused into Cu grain boundaries.[19]

In order to describe the faceting-defaceting transition of a grain boundary, it is necessary to know the equilibrium shapes (or the γ-plots against the inclination angle) of the grain with a mis-orientation angle with another grain at all temperatures. Such an analysis appears to be a formidable task of modelling and statistical mechanics, particularly for a grain boundary with a general mis-orientation angle. Therefore, presently only quantitative conjectures are possible. There is even no theoretical analysis which shows that the equilibrium shape of a grain with a general mis-orientation relationship with its neighbour should be polyhedral at low temperatures. But if the equilibrium shape is polyhedral at low temperatures, all grain boundaries will be faceted with each facet plane corresponding to one of the flat singular boundaries of the equilibrium shapes or a cusp in the γ-plot. Those which happen to have an orientation of a singular boundary will be flat. It is indeed observed that under certain heat-treatment conditions, all randomly selected grain boundaries are faceted implying polyhedral equilibrium shapes with sharp corners and edges.[7, 21]

If with temperature increase the corners and edges of the equilibrium shape become rough (rounded) and the surface slope changes continuously to the flat faces, all facet planes of a grain boundary will become rough at a same temperature with the entire grain boundary becoming defaceted with a curved or nearly planar shape. If, on the other hand, the slope change from the rounded corners and edges to the flat segments of the equilibrium shape is discontinuous at high temperatures, the facet planes of a grain boundary will be a mixture of singular and rough boundaries. With further temperature increase the area fraction of the rough boundaries will increase with decreasing angle between the singular and rough facet boundaries until the area fraction of the singular boundary segment becomes 0 and the entire grain boundary becomes defaceted to a curved or a nearly planar shape. It is highly unlikely that all facet planes of a grain boundary will be rough, because an equilibrium shape with a curved boundary meeting another curved boundary with a discontinuous surface slope change is unlikely. Thus, if all singular boundaries become rough, any faceting is unlikely. The defaceting transition is thus not directly dependent on the roughening transition of the singular boundaries, but when a faceted boundary becomes defaceted, its structure should be always rough with the characteristic properties analogous to a rough surface.

The equilibrium structure of grain boundaries at low temperatures can be calculated by minimizing the boundary internal energy usually represented by inter-atomic bonds, but most of the studies were limited to those with special mis-orientation angles often with the conclusions that the observed structures could not be explained in terms of the geometrical relations.[35] For the grain boundaries with general mis-orientation angles, only a few structures corresponding to local minimum of the γ-plots have been calculated.[36,37] Experimentally, there have been only a few observations of the structures of the faceted grain boundaries which are commonly found at low temperatures for general mis-orientations. The high resolution TEM observations of the faceted grain boundaries between grains of nearly general mis-orientations in Al,[17] Au,[37]and $BaTiO_3$[38] show that the facet planes are the low index planes of one of the two adjacent grains. The macroscopic observations of the abnormally large alumina grains with an optical microscope also show their basal planes forming grain boundaries with the smaller neighbouring grains.[39]

There is presently no theory which specifically describes the migration mechanism of these faceted grain boundaries, but the observation and the model for the step growth by Gleiter might be applicable.[40] On the other hand, the step growth mechanism is unlikely to be applicable to the rough grain boundaries. Based on the observations of step structures and spiral growth of grain boundaries in an Al-0.39 wt.% Cu alloy, Gleiter proposed that the grain boundaries would migrate by such step mechanisms as spiral growth on screw dislocations or two dimensional nucleation.[40-42] Specifically, Gleiter proposed that the grain boundaries with (111) planes will either form twin layers or new layers of the same orientation by two dimensional nucleation.[42] It is then possible that because the faceted grain boundaries have low index planes, they also migrate by two dimensional nucleation.

If the grains with faceted grain boundaries migrate by two dimensional nucleation, their coarsening behaviour will be similar to the coarsening of the grains dispersed in liquid matrix if it can be assumed, as commonly done, that the driving force for the coarsening of each grain is still described by the deviation from the average size as shown in Eqn. (6). Each grain will interact with its neighbours in the coarsening process, but if the average size of the neighbouring grains of each grain is equal to the average size of all grains, the driving force may be represented in the form identical to the grains in liquid matrix. Indeed a correlation between the grain boundary faceting and AGG was observed by Lee et al. in polycrystalline Ni.[7] Lee, Hwang and Yoon[7] observed that while all of the randomly selected grain boundaries were faceted after the annealing treatments at low temperatures in (low) vacuum, increasing fraction of them became defaceted with the change to a carburizing atmosphere and increase of temperature until all became defaceted close to the melting point.[7] There was also a concurrent change of the grain growth behaviour after recrystallization from AGG with faceted grain boundaries to normal growth with defaceted grain boundaries. The number density of the abnormally large grains increased with the increasing fraction of the defaceted grain boundaries. There was thus a gradual transition from AGG to normal growth behaviour with the increasing fraction of the defaceted grain boundaries. The correlations between the facet-defacet transition and the AGG to normal growth transition were found also in a model Ni-base superalloy,[43] Ag[44], Cu,[45] 316L stainless steel,[46] and alumina with MgO.[39] Lee et al. argued that the grain boundary segregation effect could not explain the observations in Ni and pointed out that the normal growth with defaceted grain boundaries with nearly isotropic properties was consistent with the analysis of Srolovitz et al.[47] and Thompson et al.[48]

The abnormal growth of the grains with faceted grain boundaries thus appears to be consistent with the growth by two dimensional nucleation as of the grains dispersed in liquid matrix. But there are additional complexities for the grain boundaries. Unlike the grains in liquid matrix, the defaceting temperatures vary among the grain boundaries and AGG occurs even when the faceted and defaceted grain boundaries coexist.[7] In such a mixture of grain boundary types, it is likely that the overall coarsening behaviour is still determined by the faceted grain boundaries, thereby giving rise to AGG. But we do not yet understand the dependence of the migration behaviour of faceted grain boundaries on the types of the constituent boundaries which can be all singular or mixtures of singular and rough.

The grains in single phase materials also showed stagnant growth at low temperatures and transition to AGG with temperature increase similar into those with liquid phases. Figure 5 shows, for example, the slow grain growth without any indication of AGG in Cu during annealing at 700°C after 10% deformation. In contrast, Figure 6 shows AGG during annealing at 800°C after 10% deformation, although the average grain size in Figure 6c was larger than or equal to that of Figure 5a. This stagnant growth at low temperatures and the transition to AGG at higher temperatures with an apparent critical temperature for AGG are similar to the coarsening of alumina grains in anorthite liquid matrix and consistent with the two dimensional nucleation theory. The temperature effect is likely to arise from the reciprocal temperature dependence of the nucleation rate arising from the Boltzmann factor as well as from the decrease of the grain boundary step free energy with increasing temperature similarly to the growth in liquid matrix.

A more striking evidence for the step growth mechanism is found in the recent observations of AGG in $BaTiO_3$ by Lee and Kang.[49] When $BaTiO_3$ with excess TiO_2 was sintered and heat-treated in air at a temperature below the eutectic point, AGG initially occurred and the stagnant growth stage was apparently reached after the impingement of the grains. But then double twins were formed in some grains and only those grains grew abnormally to large sizes, elongated parallel to the (111) double twin planes. The grains furthermore showed long (111) faces, which appeared to be flat in contact with the fine grains. The grain boundaries were observed to be

Fig. 5 Microstructures of Cu annealed at 700°C for (a) 1 hr., (b) 48 hrs. and (c) 100 hrs. after 10% compression.

faceted under TEM. When the specimen initially heat-treated in air was heat-treated further in hydrogen atmosphere, the grain boundaries became defaceted, the grains no longer grew fast along the twin direction, and the large grains grew rapidly in the directions perpendicular to the (111) planes to form nearly equiaxial shapes. The initially flat (111) faces of the large grains became randomly orientated in the hydrogen atmosphere. The most plausible explanation of these observations is that in the stagnant growth stage during the heat-treatment in air, the rapid growth occurred by step nucleation at the re-entrant edges of the double twins in contact with the faceted grain boundaries. When the grain boundaries became rough (defaceted) in hydrogen atmosphere, all grain boundaries apparently moved continuously with the re-entrant edges at the end of the double twins no longer providing the preferred sites for rapid growth. The (111) faces of the large grains

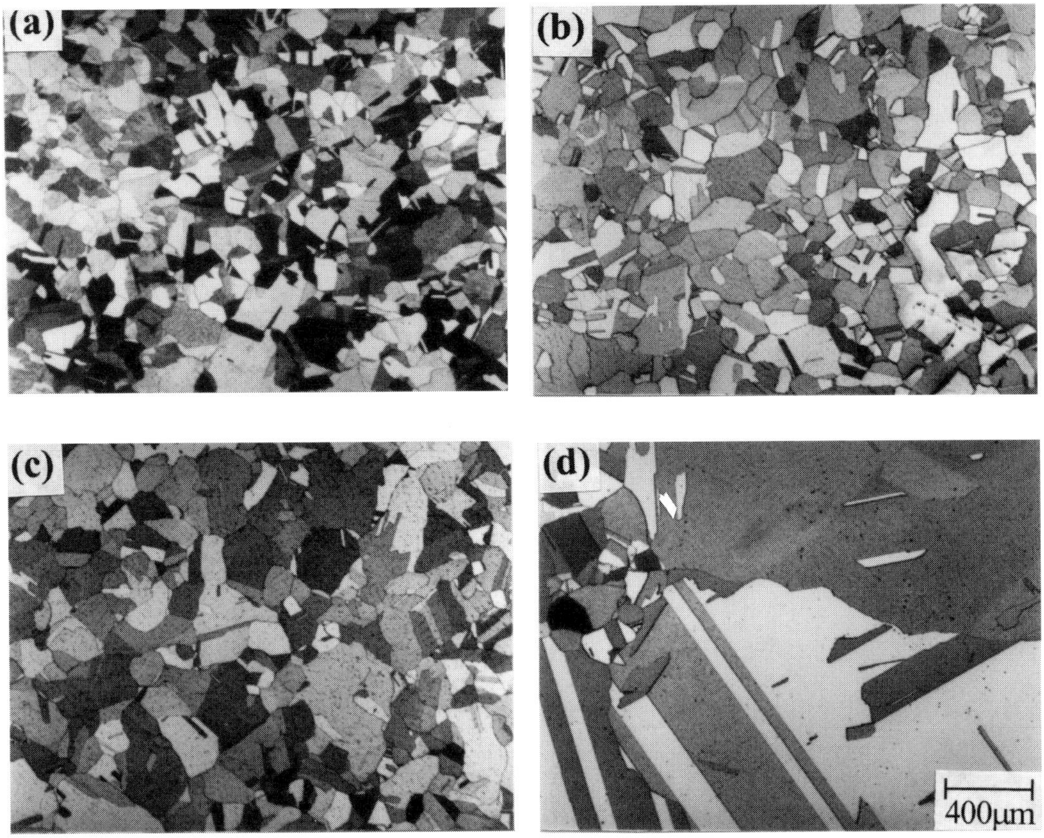

Fig. 6 Microstructures of Cu annealed at 800°C for (a) 15 min., (b) 30 min., (c) 2 hrs. and (d) 24 hrs. after 10% compression.

forming grain boundaries with the fine grains probably represent the low energy singular boundaries when heat-treated in air and they may have become more apparent because of their slow growth by two dimensional nucleation. It will be desirable to observe the structure of these grain boundaries with (111) faces of the large grains by high resolution TEM. When these grain boundaries also became rough in hydrogen atmosphere, the (111) faces of the large grains were no longer maintained and the growth perpendicular to the initially (111) planes of the large grains could readily occur by continuous growth, forming the equiaxial shapes of the large grains. In addition to the driving force from the small adjacent grains, their lateral movements were probably driven also by the tendency of the large grains to attain equiaxial equilibrium shapes. The growth of the fine matrix grains was also much faster in the hydrogen atmosphere than in air because of their rough boundary structure.

If the grains with faceted grain boundaries coarsen by the step mechanism, the transitions from the stagnant growth to AGG and from AGG to nearly normal growth can be induced by the composition changes arising from the diffusion of solute atoms from external sources or dissolving precipitates. Such a phenomenon will be a compounded effect of the coarsening and the diffusion induced grain boundary migration (DIGM). It is thus possible that AGG is quite pronounced when the precipitates are dissolved at temperatures close to their solvus because of the additional driving force for migration arising from the composition change.

4.0 DISLOCATION EFFECTS

The grains in liquid matrix or in single phase solids are likely to have dislocations and it is well known that they have a large effect on the growth of crystals from melts or vapour. The theoretical analysis and the experimental observations of the growth of a single crystal show that the rate of spiral growth at a screw dislocation will be low at low driving forces and increase parabolically to high values at large driving forces.[50,51] It is therefore possible that even with dislocations, AGG can still occur. Although more quantitative analyses are needed to confirm this possibility, the experimental observations indicate the important role of dislocations in grain coarsening. In the liquid phase sintering of NbC and WC powders, ball-milling of the powders was found to produce a large effect on the AGG behaviour during the sintering treatment. Although the powder will become finer by ball-milling, the results appear to be consistent with the effect of dislocations produced by ball-milling.

The controlled experiments also show the important effect of dislocations. Recently, Chung and Kang[52] observed that when a $SrTiO_3$ single crystal with different dislocation densities at two equivalent (100) surfaces was embedded in a fine grain $SrTiO_3$ powder with a liquid forming phase, the surface with higher dislocation density grew faster while the fine grains were showing AGG during the heat-treatment.[52] This result is qualitatively similar to the crystal growth behaviour from the melt.

It is also well known that small deformations of about 2-4% can enhance the AGG rate. It was indeed observed that when Cu specimens in stagnant grain growth conditions were slightly deformed, AGG readily occurred. If AGG was already occurring, the small deformation caused more grains to grow abnormally and sooner. The TEM observations show that the dislocation density within the grains after some growth in the deformed specimens is about same as that in the undeformed specimens, but the dislocation density at the grain boundaries of the deformed specimens remained high during the heat-treatment. Therefore, the main effect of small deformation was to increase the dislocation density at grain boundaries, which increased in effect their mobility at relatively small driving forces. Such a dislocation effect is also consistent with the step growth mechanism of the faceted grain boundaries. If the specimens are deformed to higher levels, recrystallization will occur during the heat-treatment and the grains will be relatively free of dislocations. Therefore, the maximum densities of the dislocations which remain during the heat-treatment can be produced by deforming at low levels, and it is significant that AGG still occurs with relatively large number of dislocations at grain boundaries. It is however possible that under certain conditions, the presence of dislocations at the boundaries may induce the transition from AGG to nearly normal growth behaviour. As the driving force, temperature, and the step free energy, the dislocations may also induce the transitions from stagnant growth to AGG. If the grain boundaries are rough, the dislocations produced by a small deformation are expected to enhance the normal coarsening rate momentarily by increasing the driving force without changing the type of the growth behaviour. Controlled experiments with varying dislocation densities in the grains dispersed in liquid matrix are also desirable.

5.0 CONCLUSIONS

Because we are at the early stage of understanding the relationship between the interface structural change and the grain coarsening behaviour, most of the present explanations and ideas are highly speculative. There are a number of fundamental problems which need to be analyzed quantitatively. The definitive experiments are also needed. For the grains dispersed in a liquid matrix, we do not yet know the effect of roughening at the corners and edges or the effect of continuous or discontinuous surface slope change. In real materials the liquid volume fraction is normally quite limited and the grains will form grain boundaries with re-entrant edges. We do not yet know the role of these re-entrant edges. These are the problems which have not been dealt with in the studies of the growth of single crystals from the melt. For the polycrystals with grain boundaries, we have to know more about the structure of the faceted grain boundaries and the defaceting transition. It is also highly desirable to observe the atomic structure of rough grain boundaries. The dependence of the migration mechanism on the grain boundary structural change has to be examined. The effect of solute segregation on the grain boundary structural transition and migration also has to be studied. Although the dislocation effects have been extensively studied in crystal growth, they must be important also in grain coarsening and need to be studied more quantitatively. It is really unclear if the same principles for the grains in liquid matrix can be applied to the grains with grain boundaries as suggested in this paper, but it may be worthwhile to turn the question around and ask, "What are the implications of the grain boundary roughening transition on their growth behaviour?", because we now know that even the general boundaries can be faceted at normal temperatures with solute atoms and undergo defaceting transitions. The grain boundary roughening must have a large effect on various grain boundary properties and it is possible that it produces a pronounced effect on grain coarsening as proposed here. The key to understanding the grain coarsening behaviour may hinge upon the experiments and analysis of the stagnant growth behaviour.

6.0 ACKNOWLEDGEMENT

This work was supported by the Korea Science and Engineering Foundation through the Centre for Interface Science and Engineering of Materials, by the Korea Ministry of Science and Technology through the Program of the Technology for the Future, by the Corporate Research and Development Centre of the General Electric Company, U.S.A., and by Pohang Iron and Steel Co., Ltd.. We are grateful for discussions with Prof. S. J. L. Kang at KAIST, Prof. D. Y. Kim and Dr. N. M. Hwang at Creative Research Initiative Centre for Microstructure Science of Materials at Seoul National University.

7.0 REFERENCES

1. I. M. Lifshitz and V. V. Slyozov: The Kinetics of Precipitation from Supersaturated Solid Solutions, *Journal of Physics and Chemistry of Solids*, **19**, 1961, pp.35-50.
2. C. Wagner: Theory of Precipitate Change by Redissolution, *Zeitschrift für Elektrochemie*, **65**, 1961, pp.581-591.
3. A. J. Ardell: The Effect of Volume Fraction on Particle Coarsening: Theoretical Consideration, *Acta Metallurgica et Materialia*, **20**, 1972, p.61
4. C. H. Kang and D. N. Yoon: Coarsening of Cobalt Grains Dispersed in Liquid Copper Matrix, *Metallurgical Transactions A*, **12A**, 1981, pp.65-69.
5. S. S. Kang and D. N. Yoon: Kinetics of Grain Coarsening During Sintering of Co-Cu and Fe-Cu Alloys with Low Liquid Contents, *Metallurgical Transactions A*, **13A**, 1982, pp.1405-1411.
6. Y. J. Park, N. M. Hwang and D. Y. Yoon: Abnormal Growth of Faceted (WC) Grains in a (Co) Liquid Matrix, *Metallurgical and Materials Transactions A*, **27A,** 1996, pp.2809-2819.
7. S. B. Lee, N. M. Hwang, D. Y. Yoon and M. F. Henry: Grain Boundary Faceting and Abnormal Grain Growth in Ni, to be published.
8. E. W. Hart: Grain Boundary Phase Transformation, *The Nature and Behaviour of Grain Boundaries*, H. Hu, ed., Plenum Press, New York, 1972.
9. M. Guttmann: Grain Boundary Segregation, Two Dimensional Compound Formation and Precipitation, *Metallurgical Transactions A*, **8A**, 1977, pp.1383-1401.
10. C. Rottman: Roughening of Low-Angle Grain Boundaries, *Physical Review Letters*, **57**, 1986, pp.735-738.
11. H. Sautter, H. Gleiter and G. Baro: The Effect of Solute Atoms on the Energy and Structure of Grain Boundaries, *Acta Metallurgica et Materialia*, **25**, 1977, pp.467-473.
12. U. Erb and H. Gleiter: The Effect of Temperature on the Energy and Structure of Grain Boundaries, *Scripta Metallurgica*, **13**, 1979, pp.61-64.
13. E. M. Fridman, C. V. Kopezky and L. S. Shvindlerman: Effects of Orientation

and Condensation Factors on Migration of Individual Grain Boundaries in Aluminium, *Zeitschrift für Metallkunde*, **66**, 1975, pp.533-539.

14. L. S. Shvindlerman and B. B. Straumal: Regions of Existence of Special and Non-Special Grain Boundaries, *Acta Metallurgica et Materialia*, **33**, 1985, pp.1735-1749.
15. E. L. Maksimova, L. S. Shvindlerman and B. B. Straumal: Transformation of Σ17 Special Tilt Boundaries to General Boundaries in Tin, *Acta Metallurgica et Materialia*, **36**, 1988, pp.1573-1583.
16. E. L. Maksimova, E. I. Rabkin, L. S. Shvindlerman and B. B. Straumal: Phase Transitions at Grain Boundaries in the Presence of Impurities, *Acta Metallurgica et Materialia*, **37**, 1989, pp.1995-1998.
17. U. Dahmen and K. H. Westmacott: Studies of Faceting by High Voltage/High Resolution Microscopy, *Interface: Structure and Properties*, S. Ranganathan, C. S. Pande, B. B. Rath and D. A. Smith, eds., Trans Tech Publications, Switzerland, 1993.
18. T. E. Hsieh and R. W. Balluffi: Observations of Roughening/De-faceting: Phase Transitions in Grain Boundaries, *Acta Metallurgica et Materialia*, **37**, 1989, pp.2133-2139.
19. T. G. Ference and R. W. Balluffi: Observation of a Reversible Grain Boundary Faceting Transition Induced by Changes of Composition, *Scripta Metallurgica et Materialia*, **22**, 1988, pp.1929-1934.
20. A. M. Donald and L. M. Brown: Grain Boundary Faceting in Cu-Bi Alloys, *Acta Metallurgica et Materialia*, **27**, 1979, pp.59-66.
21. G. Henry, J. Plateau, X. Wache, M. Gerber, I. Behar and C. Crussard: Mise en évidence de Phénomenes d'adsorption Intergranulaire et étude de leurs Relations avec la fragilite intergranulaire, *Mém. Science Review Metallurgical*, **56**, 1959, pp.417-426.
22. D. Y. Yoon and C. W. Park: Surface Roughening Transition and Coarsening of Grains in Liquid Matrix, Presented at *Solid-Solid Phase Transformations '99*, Kyoto, Japan, 1999.
23. J. P. Hirth and G. M. Pound: *Condensation and Evaporation, Nucleation and Growth Kinetics; Prog. Materials Science*, B. Chalmers, ed., Pergamon Press, Oxford, **11**, 1963.
24. W. Van Saarloos and G. H. Gilmer: Dynamic Properties of Long-Wavelength Interface Fluctuations During Nucleation-Dominated Crystal Growth, *Physical Review B*, **33**, 1986, pp.4927-4935.
25. X. Y. Liu, P. Van Hoof and P. Bennema: Surface Roughening of Normal Alkane Crystals: Solvent Dependent Critical Behaviour, *Physical Review Letters*, **71**, 1993, pp.109-112.
26. H. Van Beijeren: Exactly Solvable Model for the Roughening Transition of a Crystal Surface, *Physical Review Letters*, **38**, 1977, pp.993-996.
27. J. M. Kosterlitz and D. J. Thouless: Ordering, Metastability and Phase Transitions in Two-Dimensional Systems, *Journal of Physics C*, **6**, 1973, pp.1181-1203.
28. J. Y. Jun: M.S. Thesis, Seoul National University, Korea, 1993.

29. S. Sarian and H. W. Weart: Factors Affecting the Morphology of an Array of Solid Particles in a Liquid Matrix, *Transactions of the Metallurgical Society of AIME*, **233**, 1965, pp.1990-1994.
30. C. Rottman and M.Wortis: Statistical Mechanics of Equilibrium Crystal Shapes: Interfacial Phase Diagrams and Phase Transitions, *Physical Review*, **103**, 1984, pp.59-79.
31. C. A. Bateman, S. J. Bennison and M. P. Harmer: Mechanism for the Role of Magnesia in the Sintering of Alumina Containing Small Amounts of a Liquid Phase, *Journal of American Ceramic Society*, **72**, 1989, pp.1241-1244.
32. W. A. Kaysser, M. Sprissler, C. A. Handwerker and J. E. Blendell: Effect of a Liquid Phase on the Morphology of Grain Growth in Alumina, *Journal of American Ceramic Society*, **70**, 1987, pp.339-343.
33. D. Y. Kim, S. M. Wiederhorn, B. J. Hockey, C. A. Handwerker and J. E. Blendell: Stability and Surface Energies of Wetted Grain Boundaries in Aluminium Oxide, *Journal of American Ceramic Society*, **77**, 1994, pp.444-453.
34. D. Y. Kim, Private Communication.
35. P. J. Goodhew, T. Y. Tan and R. W. Balluffi: Low Energy Planes for Tilt Grain Boundaries in Gold, *Acta Metallurgica et Materialia*, **26**, 1978, pp.557-567.
36. T. Muschik, W. Laub, U. Wolf, M. W. Finnis and W. Gust: Energetic and Kinetic Aspects of the Faceting Transformation of a Σ3 Grain Boundary in Cu, *Acta Metallurgica et Materialia*, **41**, 1993, pp.2163-2171.
37. K. L. Merkle and D. Wolf: Low-Energy Configurations of Symmetric and Asymmetric Tilt Grain Boundaries, *Philosophical Magazine A*, **65**, 1992, pp.513-530.
38. T. Yamaoto, Y. Ikuhara, K. Hayashi and T. Sakuma: Grain Boundary Structure in TiO_2-Excess Barium Titanate, *Journal of Materials Research*, **13**, 1998, pp.3449-3452.
39. C. W. Park, Private Communication.
40. H. Gleiter: The Mechanism of Grain Boundary Migration, *Acta Metallurgica et Materialia*, **17**, 1969, pp.565-573.
41. H. Gleiter: The Theory of Grain Boundary Migration Rate, *Acta Metallurgica et Materialia*, **17**, 1969, pp.853-862.
42. H. Gleiter: The Formation of Annealing Twins, *Acta Metallurgica et Materialia*, **17**, 1969, pp.1421-1428.
43. S. B. Lee: Grain Boundary Faceting and Grain Growth in Ni and Ni alloys, Ph.D. Thesis, Kaist, Korea, 1998.
44. J. B. Koo: The Effects of Annealing Temperature and Oxygen on the Grain Growth of Pure Ag, M. S. Thesis, Kaist, Korea, 1996.
45. J. B. Koo: Private Communication.
46. J. S. Choi: Effects of Grain Boundary Structure on Grain Growth Behaviour in 316 L Stainless Steel, M. S. Thesis, Kaist, Korea, 1997.
47. D. J. Srolovitz, G. S. Grest and M. P. Anderson: Computer Simulation of Grain Growth-V. Abnormal Grain Growth, *Acta Metallurgica et Materialia*, **33**, 1985, pp.2233-2247.
48. C. V. Thompson, H. J. Frost and F. Spaepen: The Relative Rates of Secondary

and Normal Grain Growth, *Acta Metallurgica et Materialia*, **35**, 1987, pp.887-890.

49. B. K. LEE and S. J. L. KANG: Private communication.
50. W. K. Burton, N. Cabrera and F. C. Frank: The Growth of Crystals and the Equilibrium Structure of Their Surfaces, *Philosophical Transactions*, **A243**, 1951, pp.299-358.
51. S. D. PETEVES and R. ABBASCHIAN: Growth Kinetics of Solid-Liquid Ga Interfaces-Part I Experimental, *Metallurgical Transaction A*, **22A**, 1991, pp.1259-1270.
52. S. Y. CHUNG and S. J. L. KANG: Private Communication.

Anisotropic Template Growth in Textured Alumina

MATTHEW SEABAUGH, GARY L. MESSING and BHASKAR BRAHMAROUTU
Materials Research Laboratory and Department of Materials Science and Engineering
Pennsylvania State University, University Park, PA 16802

ABSTRACT

Templated Grain Growth (TGG) relies on the anisotropic growth of template particles to develop highly textured polycrystalline ceramics. Radial growth of alumina particles was much faster than the growth on the basal surface. When compared to growth of $\{11\bar{2}0\}$ and $\{01\bar{1}2\}$ surfaces, radial growth was similar but decreased significantly for the template particles after a short time. A model for TGG using single crystal growth data only fit the early growth period because template grain impingement significantly changed the growth environment for radial growth. In contrast, impingement did not appear to alter growth on the basal surface and thus the 2-D nucleation and growth model described thickening of template grains for the entire process.

1.0 INTRODUCTION

Anisotropic grain growth is an intrinsic phenomenon in many important ceramic systems, and its controlled implementation during ceramic processing has led to improved materials and properties. Commercially important oxides such as alumina,[1,2] mullite,[3] ferrites,[4] and high temperature superconducting oxides all exhibit anisotropic grain growth. Non-oxides such as Si_3N_4 and SiC also demonstrate growth anisotropy.

The whisker - or plate-shaped grains resulting from anisotropic growth also have anisotropic physical properties, (e.g. electrical and thermal conductivity). When the anisotropic grains in a specimen are aligned, a textured polycrystalline material with directionally dependent properties results. Magnetic ferrites are a classic example of materials in which tailored properties are obtained with a textured microstructure.[4] Textured bismuth titanate, which exhibits strongly anisotropic piezoelectric properties,[5,6] and textured alumina ceramics, which demonstrate anisotropic thermal conductivity, dielectric response, and thermal expansion, are other examples[7-10] of textured ceramics possessing anisotropic properties.

Until recently, the utilization of materials with anisotropic grains was limited, because the growth processes in these systems were poorly understood and uncontrolled anisotropic growth resulted in strength-limiting flaws. Recent efforts have shown that anisotropic grain growth can be used favourably to yield textured ceramics by templated grain growth. In this approach texture is obtained by growth of a small number of larger, well-oriented anisotropic-shaped particles distributed in a fine-grained matrix. Because growth is directly related to the oriented, anisotropic particles they serve to template subsequent texture development[8] and consequently, this process is referred to as templated grain growth, or TGG. TGG has been investigated for a variety of materials, and has typically resulted in specimens with high degrees of crystallographic texture.[6-14]

The growth of template particles in TGG systems is typically anisotropic and does not follow normal grain growth kinetics. The sintering conditions, liquid phase content, and initial template loading of the TGG specimens all have a strong effect on template growth and the associated texture development. Sintered TGG microstructures can range from assemblies of well-oriented anisotropic grains, to large, widely-spaced anisotropic grains in a matrix of equiaxed grains. To fully understand the process of TGG and to enable greater tailoring of microstructure and properties, it is essential to better understand anisotropic grain growth.

TGG processing yields a model system for characterizing anisotropic grain growth. The anisotropic grain population is large enough to allow the simultaneous measurement of many grains in a single micrograph, presenting a more statistically valid picture of anisotropic grain growth than isolated single crystals. In particular, liquid phase sintered alumina is a model TGG system for evaluating anisotropic grain growth where template particles retain their morphology by virtue of their anisotropic growth rates and are a distinct population throughout microstructure development. Grain growth in liquid phase sintered alumina is highly anisotropic, and strong textures develop using TGG. We have recently reported that texture measurement techniques can be used to characterize the orientation distribution of anisotropic grains in liquid phase sintered α-Al_2O_3, and that the March-Dollase function provides an extremely good fit to measured distributions.[15]

Anisotropic grain growth has been observed in many ceramic systems, but it has been difficult to quantify. The stereological treatment of disc-shaped grains is difficult if the orientation distribution of the discs is unknown. For some special cases, the stereological correction factors have been derived. In randomly oriented microstructures, the ratio of the average three dimensional template diameter to the average major axis dimension is $\pi/2$, while the ratio between disc thickness and minor axis dimension is $3\,\pi/8$.[16,17] To properly account for the texture of the material, the orientation distribution function of the material should be measured. The measured distributions can be fit with an equation, like the March-Dollase equation, that can be substituted into the derivations by Fullman to produce exact correction factors for the textured material.[18]

Investigations of single crystal alumina growth in liquid phase sintered alumina provide experimental data for crystal growth in different crystallographic directions.[19,20] For comparison to TGG kinetics, single crystals were grown into untemplated matrix material under sintering conditions expressly designed to mimic template growth during TGG.[21] The growth kinetics of $\{11\bar{2}0\}$ and $\{01\bar{1}2\}$ single crystal surfaces suggests that growth in these directions is controlled by diffusion through the intervening liquid phase. In contrast, growth on the $\{0001\}$ basal surface correlated well with a 2-D nucleation and growth model.

Coordinated measurements of grain size, texture development and crystal growth during TGG provide the necessary data for evaluating anisotropic grain growth in liquid phase sintered α-Al_2O_3. In this paper we compare the growth of single crystal surfaces to the growth in the respective directions for templates in the TGG sample. In addition models to describe single crystal growth were modified to predict template growth during TGG.

2.0 EXPERIMENTAL PROCEDURE

Submicron α-Al_2O_3, $CaCO_3$ and SiO_2 were dispersed in water using a polyelectrolyte dispersant. Two compositions were made, the first with 1 wt.% of an equimolar amount of CaO and SiO_2 and a second with 5 wt.% of the same 1:1 CaO:SiO_2 mixture. 1 wt.% of the CaO:SiO_2 mixture resulted in 2.7 vol.% liquid at 1550°C and 3.5 vol.% liquid at 1661°C. Hexagonal platelets of α-Al_2O_3, (aspect ratio ≈ 6, 9 μm major axis, minor axis in c-axis direction) were added to the dispersion. Three template concentrations were investigated, 1, 5 and 20 wt.%. The slurry was sonicated to aid dispersion and allowed to stir overnight. Polyvinyl alcohol and glycerol were added as a binder and plasticizer, respectively. The slurry had a viscosity of 150 MPa·s at a shear rate of 200 s^{-1}, the rate at which the tape was cast. The as-cast tape had a thickness of 500 μm; and was 300 μm thick after drying at 80% relative humidity for two days.

The dried tape was cut into sections, stacked in groups of six layers, and laminated at 30 MPa. Samples were calcined at 1000°C for 3 hours, then sintered at temperatures at 1550 and 1661°C for times ranging between 1 second and 1000 minutes.

After sintering, the texture of the specimens was measured by x-ray diffraction rocking curves.[15] The intensity of the basal peak (000.12) was measured over a rocking angle of -40 to +40°α. The resulting rocking curves were fit with the March-Dollase function, and the appropriate stereological correction factors were determined. After texture analysis, samples were prepared for examination by SEM. Samples were cross-sectioned using a diamond saw, polished to a 1 μm finish with diamond paste, then thermally etched at 100°C below the sintering temperature for 30 minutes. For each specimen, four to six micrographs were taken for image analysis. Using image analysis software, the major axis dimension, minor axis dimension, and the angle of the major axis to the horizontal frame of reference were measured for each anisotropic grain (i.e. with aspect ratio greater than 2). The average major and minor axis dimensions were calculated, as well as the standard deviation (e.g. the width of the measured distribution, not the measurement error). The previously calculated stereological correction factor was applied to the measured dimensions to obtain diameter and thickness values.

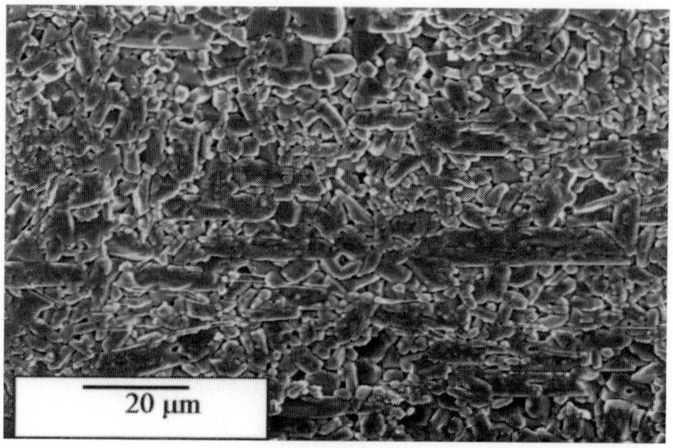

(a)

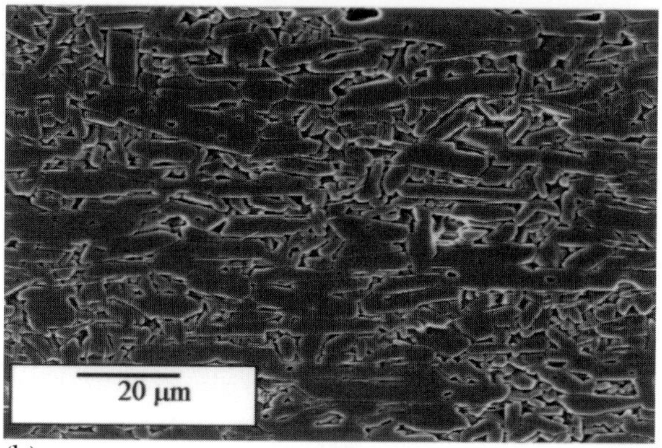

(b)

Fig. 1 Liquid phase sintered Al_2O_3 textured by TGG, 5 wt.% (SiO_2 + CaO) sintered at 1550°C, 100 min. (a) 1 wt.% initial template concentration, (b) 5 wt.% initial template concentration, (c) 5 wt.% (SiO_2 + CaO), 20 wt.% initial template concentration and (d) 1 wt.% (SiO_2 + CaO), 5 wt.% initial template concentration. (Please see Fig. (c, d) in next page)

3.0 RESULTS

The growth of template particles was characterized for four compositions. Three compositions contained 5 wt.% liquid phase formers, with 1, 5 and 20 wt.% initial template concentrations. The fourth composition contained 1 wt.% liquid phase formers and 5 wt.% initial template particles. All samples were sintered for 100 minutes at 1550°C. The micrographs in Figure 1 show the range of microstructures possible by varying the template concentration and liquid phase content during TGG. For example, as the template concentration in the 5% liquid samples increases, (Figure 1a-c) the microstructure changes from one consisting of widely dispersed, large (36.7 x 5.2 μm) anisotropic grains in a relatively equiaxed matrix (Figure 1a), to a highly textured microstructure consisting of smaller (16.9 x 2.1 μm) anisotropic grains with a minority phase of matrix grains (Figure 1c). When the liquid phase

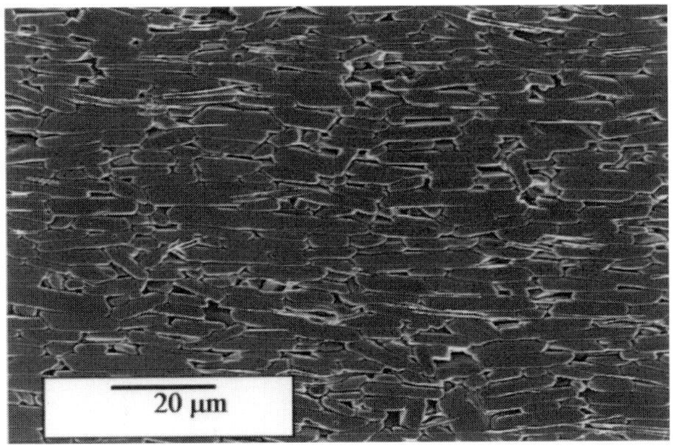

(c)

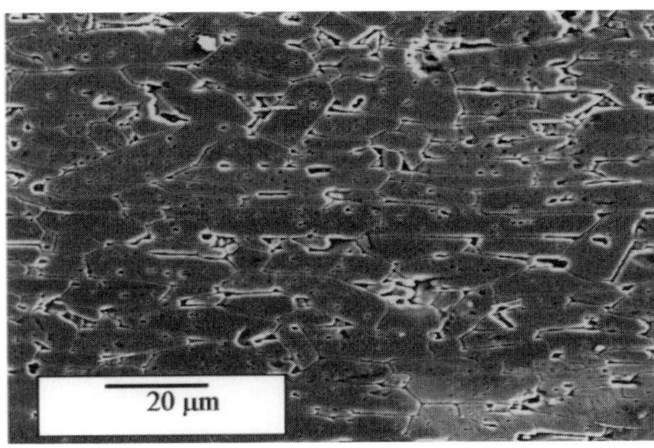

(d)

Fig. 1 Continued.

content is reduced from 5% to 1%, and template concentration is constant at 5%, the grain size is larger, resulting in a greater textured fraction.

In spite of differences in the growth environment, there are many common features between template growth in the samples, and the growth of large single crystals. Most importantly, the growth is anisotropic. Figure 1 provides evidence of growth anisotropy, but the micrographs in Figure 2 show decisively that the growth of templates in a sample with 5% templates and ~13.5 vol.% liquid phase sintered at 1550°C is anisotropic. The microstructures show that the template particles grow from 21.5 to 30.7 μm in 1000 min, while the thickness only increases from 2.4 to 5.1 μm. The growth in the radial and thickness directions for the templated samples (shown in Figure 2) is plotted in Figure 3, along with the amount of single crystal growth on the {11$\bar{2}$0} and surfaces (see Figure 3a) and the {0001} surface (see Figure 3b). As seen in Figure 3, growth anisotropy is also observed in the single crystal growth data. Modelling of single crystal growth[21] indicates two growth mechanisms operate in the alumina system. Growth in non-basal directions for the single crystal correlates

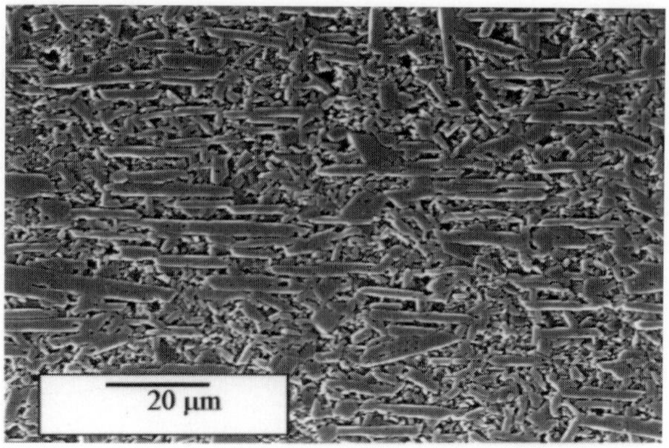

(a)

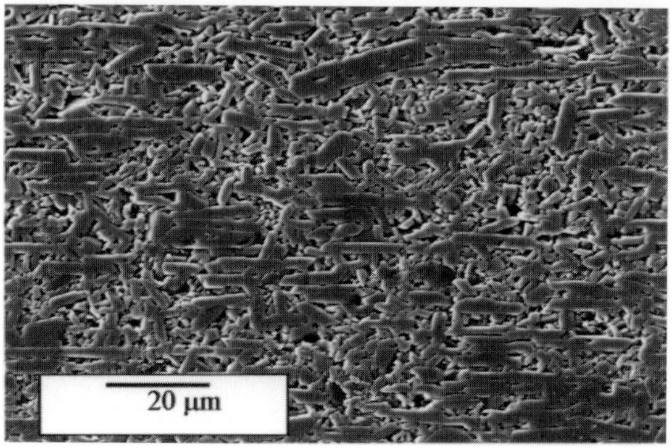

(b)

Fig. 2 Liquid phase sintered Al_2O_3 textured by TGG sintered at 1550°C, 5 wt.% (SiO_2 + CaO), 5 wt.% initial template concentration. (a) 1 min. sintering time, (b) 10 min. sintering time, (c) 30 min. sintering time and (d) 1000 min. Sintering time. (Please see Fig. (c, d) in next page).

well with a diffusion-controlled model, while growth on the basal surface occurs by 2-D nucleation and growth. 2-D nucleation and growth is further supported by the presence of growth islands on the basal surfaces of template particles. Since the radial and thickness growth mechanisms are different they will be considered separately for single crystal growth and template growth.

The difference in the amount of growth between the templates and single crystals is clearly seen in Figure 3a. The templates and the single crystals both show rapid initial growth followed by a much slower rate of growth, but the extent of template growth is significantly less than that of single crystals. A less apparent but distinct difference is the shape of the growth curve. Single crystal growth correlates with a $t^{1/3}$ dependence, which also describes the early growth behaviour of the template

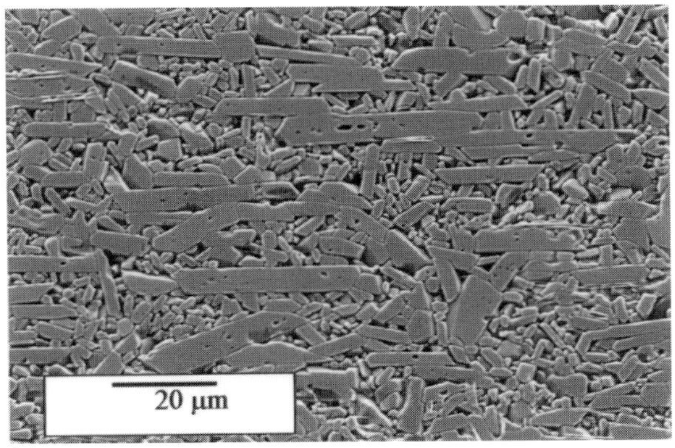

(c)

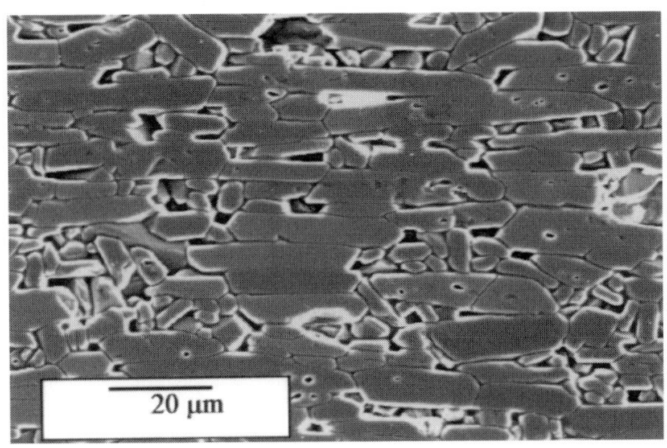

(d)

Fig. 2 Continued.

particles, but at longer times, the template growth rate slows to a much greater degree than that of the single crystals. The most likely cause of the change in radial growth rate of templates is their impingement after a very short period. Evidence of the onset of template impingement after 1 minute at 1550°C in samples containing 5% templates and ~13.5 vol.% liquid can be seen in Figure 2a. The initial growth period is less than one minute in the temperature and template loading ranges of relevance for developing high degrees of texture.

Radial growth of the templates, single crystal growth, and the matrix grain growth are all strongly affected by the liquid content of the specimen. As shown in Figure 4, the radial growth of template particles at 1550 and 1661°C re-emphasizes the results of Figures 1b and d. The amount of growth in the samples with a low liquid content is greater for both temperatures.

As reported earlier,[8] initial template loading was found to have a strong effect on the growth of templates. In Figure 5, the average radial growth of the template particles in samples containing 1, 5 and 20 wt.% template concentrations is shown.

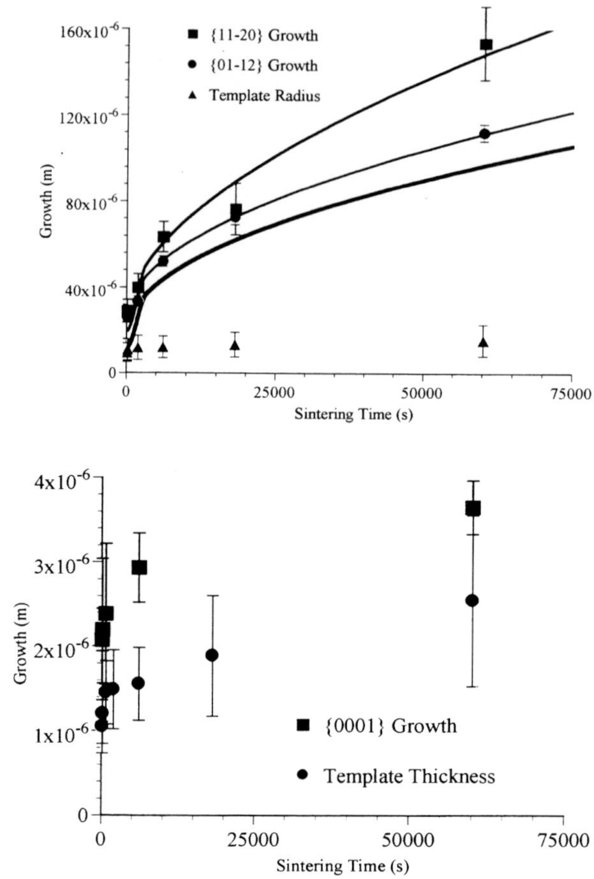

Fig. 3 Comparison of single crystal and template growth, 1550°C 5 wt.% (CaO + SiO_2) and 5 wt.% initial template concentration, (a) Radial growth and (b) Thickness growth.

It is interesting to observe the strong effect template concentration has on the amount of template growth. Samples with 1 wt.% templates grow to a radius of 20 μm whereas samples with 20 wt.% templates grow to 10 μm after 300 min. The decreasing trend in radial growth is a result of the closer spacing of the templates with increased template loading.[22]

Just as growth on the {0001} surface was much different than non-basal surfaces in the single crystal experiments, the growth behaviour of the template thickness is quite different than growth in the radial direction. In Figure 3b, thickening of the templates is compared to growth on the {0001} single crystal surface. While the growth of the templates is less than that of the single crystal, it is much closer to the single crystal growth rate than was observed for radial growth. In addition, the two growth curves share similar shapes, another departure from the observation for radial growth.

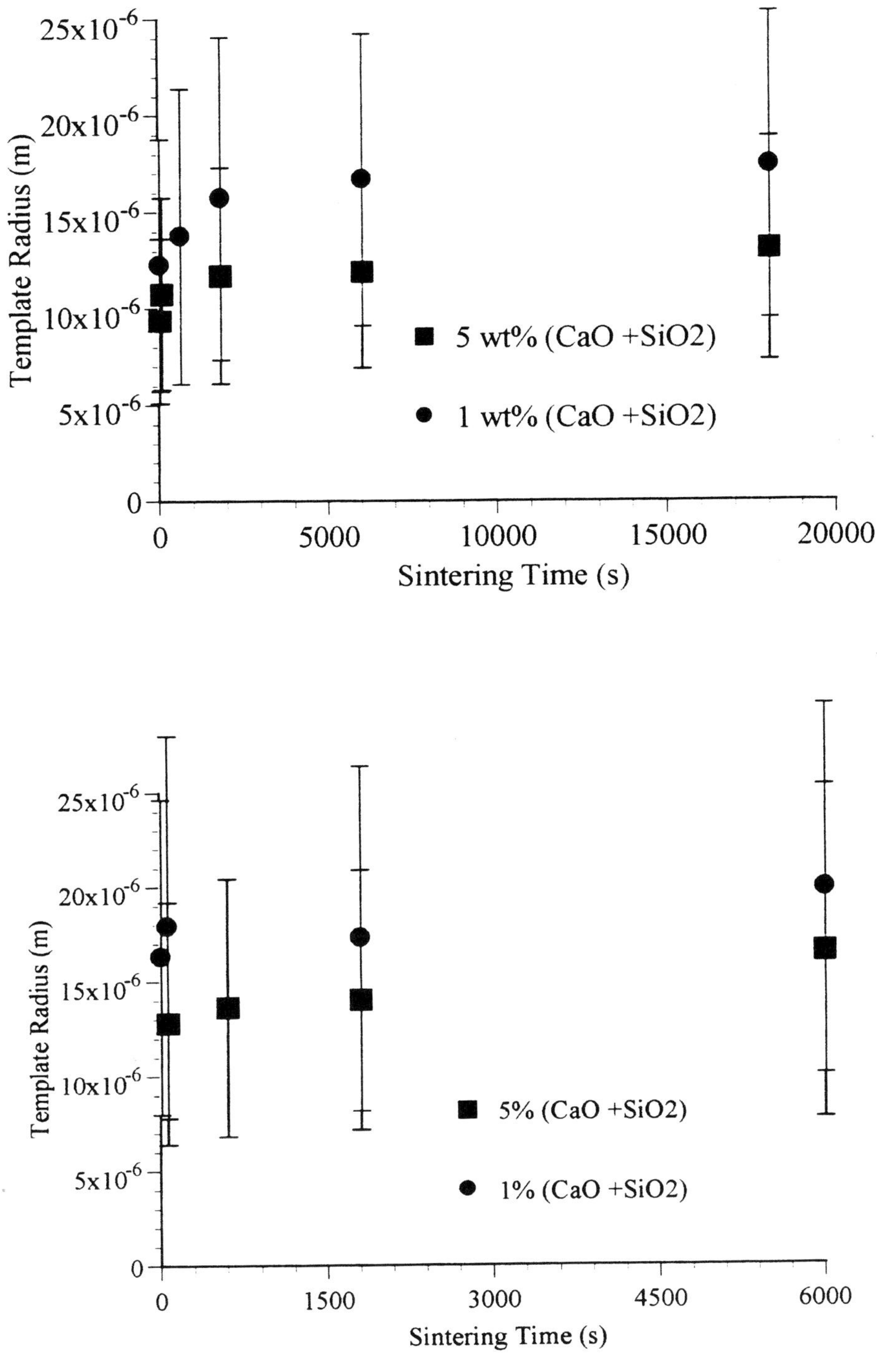

Fig. 4 Comparison of radial template growth for two liquid phase contents 5 wt.% (CaO + SiO_2) and 5 wt.% initial template concentration 1 wt.% (CaO + SiO_2) and 5 wt.% initial template concentration. (a) 1550°C and (b) 1661°C.

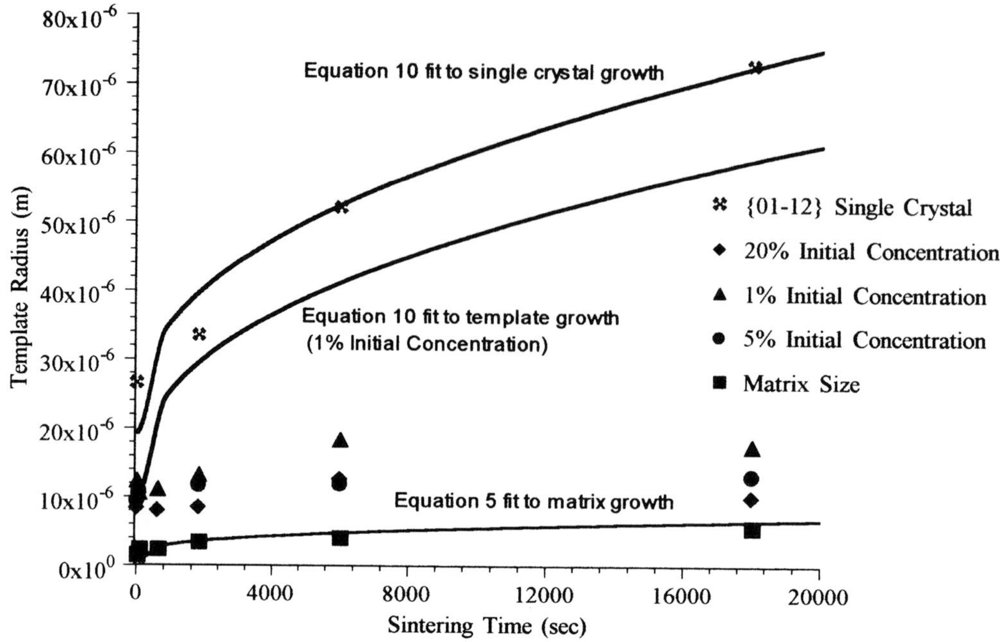

Fig. 5 Comparison of radial template growth for three template concentrations and single crystal growth on the surface at 1550°C, 5 wt.% ($CaO+SiO_2$).

Figure 6 shows the effect of temperature and liquid phase content on template thickening. Again, low liquid content and high temperature increase the growth rate in the thickness direction. In Figure 7, the growth of template particles in specimens with different template loadings is shown. As in the radial growth case, increasing template concentration leads to decreased template growth. These results agree with earlier observations[8] where template growth in both the radial and thickness directions was affected by decreasing template spacing.

4.0 DISCUSSION

The growth of α-Al_2O_3 single crystals in different crystal directions was measured and mechanisms of growth were proposed and modelled.[21] Growth on the {0001} surface was shown to be controlled by 2-D nucleation and growth, while growth on the $\{11\bar{2}0\}$ and $\{01\bar{1}2\}$ surfaces was controlled by diffusion through the liquid phase.

The single crystal experiments were chosen as model experiments, as they are free from two complicating factors of TGG (1) the effect of finite template size and (2) the effect of template/template impingement. As in several other investigations

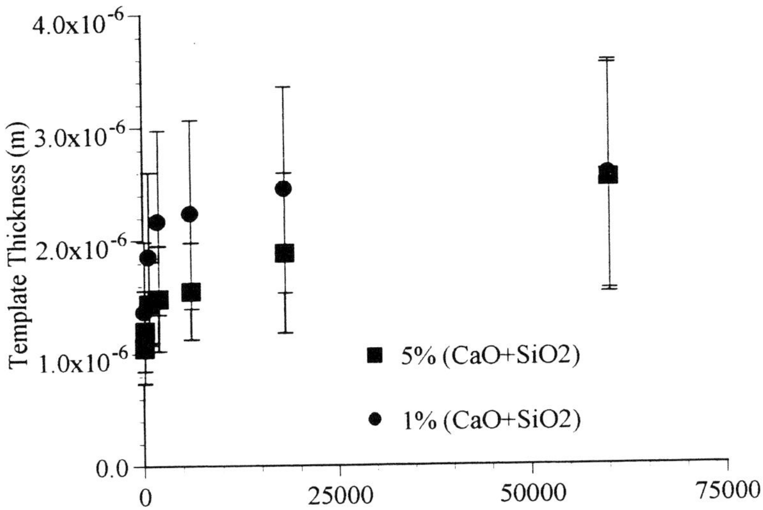

(a)

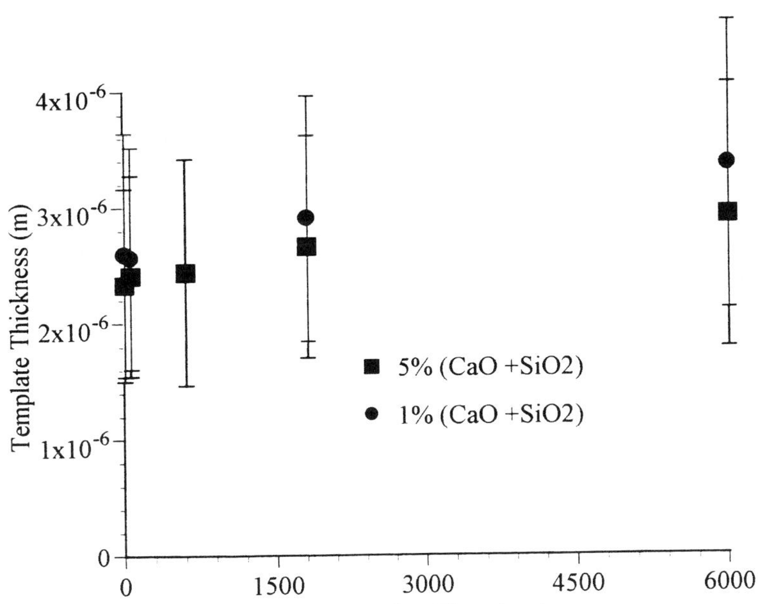

(b)

Fig. 6 Comparison of template thickness growth for two liquid phase contents 5 wt.% (CaO + SiO_2) and 5 wt.% initial template concentration 1 wt.% (CaO + SiO_2) and 5 wt.% initial template concentration. (a) 1550°C and (b) 1661°C.

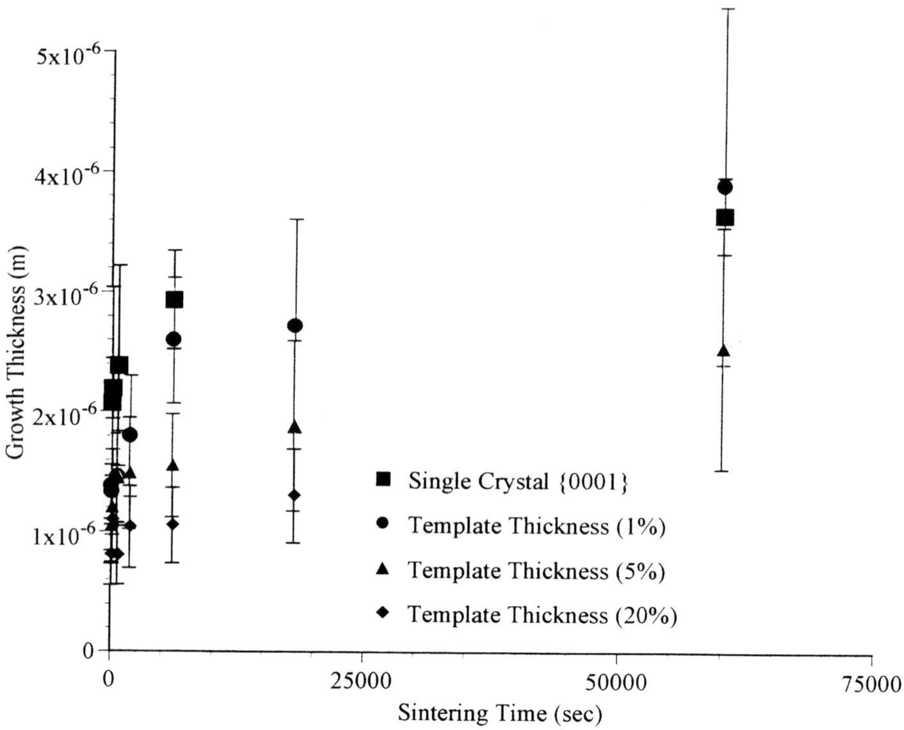

Fig. 7 Comparison of template thickening growth for three template goncentrations and single growth on the basal plane at 1550°C, 5 wt.% (CaO+SiO_2).

of single crystal alumina growth, growth on the basal surface {0006} was found to be the slowest, and to have a fundamentally different dependence on supersaturation compared to the $\{11\bar{2}0\}$ and $\{01\bar{1}2\}$ surfaces.

Having derived expressions for the growth of single crystals[21] it is possible to determine whether these same mechanisms are operative in TGG systems and to evaluate the effect of template size and impingement on template growth.

5.1 Radial Growth

By comparing growth among the TGG samples, it becomes apparent that increased template loading reduces the average growth of template particles in the radial direction. In fact, even for a low template loading (1% templates), growth of the template particles is much less than that of the single crystal samples. The amount of growth on the $\{11\bar{2}0\}$ and $\{01\bar{1}2\}$ single crystals is compared in Figure 3 to the radial growth of template particles in a TGG specimen with an initial template concentration of 5%. As explained above, the extent of template growth is limited by impingement of fast growing templates. Therefore, the similarity in radial growth of the template particles and the growth of a single crystal in non-basal directions occurs only for the earliest stage of template growth.

The models proposed by Lay and Hennings et al.,[23,24] were modified to explain single crystal growth[21] and template growth prior to impingement. The driving force for template growth is provided by the supersaturation gradient between the matrix-liquid interface, S_m, and the template-liquid interface S_R across a liquid layer of thickness δ. The difference in supersaturation ΔS can be obtained by evaluating S_m and S_R in terms of the supersaturation at the interface of the liquid and a very large planar interface, using the Gibbs-Thomson equation. For a matrix of average grain size r,

$$RTln\left(\frac{S_m}{S_o}\right)=\frac{2\sigma_m M}{\rho r} \tag{1}$$

where σ_m is the average surface energy for matrix grains and S_o is the equilibrium solubility at an infinite plane of the matrix material.

From eqn . 1,

$$S_m \approx S_o\left(1+\frac{2\sigma_m M}{\rho RTr}\right) \tag{2}$$

The rate of growth of a template particle,

$$\frac{dR_{disc}}{dt}$$

can be expressed by accounting for diffusion across a thin liquid layer of thickness δ between the matrix grains and the template, driven by a gradient in supersaturation $\Delta S = S_m - S_R$. Hennings et al.[24] showed for exaggerated grain growth that

$$\frac{dR_{disc}}{dt}=\frac{-J}{\rho}=\frac{D\Delta S}{\rho\delta}=\frac{DS_o}{\rho\delta}\left[(1-k_T)+\frac{2\sigma_m M}{r\rho RT}\right] \tag{3}$$

From Lay's treatment[23] of the grain growth in liquid phase sintered alumina, the time dependence of average matrix grain size, r, can be expressed as

$$r^3 = 1.05K_m t \tag{4}$$

$$\text{where } K_m=\frac{DS_oM\sigma_m}{\rho RTc} \tag{5}$$

$$\text{and the liquid layer thickness, } \delta = cr \tag{6}$$

$$\text{where } c=\frac{2}{3}\frac{V_{liquid}}{V_{solid}} \tag{7}$$

The supersaturation of the template-liquid interface does not change significantly with change in R_{disc}. Hence, the initial template radius R_o can be used at all times. Using eqn. 4, 5 and 6, eqn. 3 can be rewritten as,

$$\frac{dR_{disc}}{dt} = 0.98\frac{K_m^{2/3} rRT}{Ms_m}(1-k_t)t^{-1/3} + 1.94K_m^{1/3}t^{-2/3} \tag{8}$$

The growth of template particles can be calculated by integrating eqn. 8,

$$R_t - R_o = 5.81K_m^{1/3}t^{1/3} + 1.48K_m^{2/3}\left[\frac{\rho RT}{M\sigma_m}(1 - k_T)\right]t^{2/3} \tag{9}$$

As seen in Figure 5, this simple model can explain the growth of single crystals and the initial stage of template growth in the radial direction. Impingement of growing templates leads to a change in the growth environment which cannot be explained purely on the basis of a diffusion controlled mechanism as described above. One method of estimating the limiting radial size of templates during TGG is to calculate the average center-to-center distance for neighbouring template particles based on template loading.[24] Models to describe post-impingement radial growth must take into account the modified growth environment of the impinging templates.

5.2 Thickness Growth

In the case of radial growth, the size and concentration of template particles were observed to have an effect on template growth. In the c-axis case, in which growth is controlled by 2-D nucleation and growth, the situation is not the same. If we examine growth of the template particles and the single crystal (Figure 3b), it appears that basal growth is initially faster for the single crystal material.

The relation between supersaturation and growth rate for the 2-D nucleation and growth process can be expressed as follows[25]

$$\frac{dT_T}{dt} = C_1(S_B)^n \, exp(\frac{-C_2}{T^2 S_B}) \tag{10}$$

where T_T is template thickness, S_B is the supersaturation at the basal plane, the exponent n = 5/6 for the birth and spread model, and C_1 and C_2 are constants which depend on the basal surface energy and molar volume of the template material. The similarity of template thickening and growth on the basal plane of a single crystal is borne out in comparisons of growth rate versus supersaturation in which extremely similar growth rates and rate dependence on supersaturation are observed (Figure 8). For the faces of a template, the supersaturation is the same as that of a planar surface, and the mechanism of growth should be the same.

It is possible to see that while the growth of the template particles is less than the single crystal, it retains the same behaviour. While the growth rate in the thickness direction seems to be insensitive to template loading, as suggested by the similar growth rates in the 1, 5 and 20% loading cases, and the good fit with the 2-D nucleation and growth equation for single crystals, it is apparent in Figure 7 that some reduction of growth occurs with increased template loading. This effect may

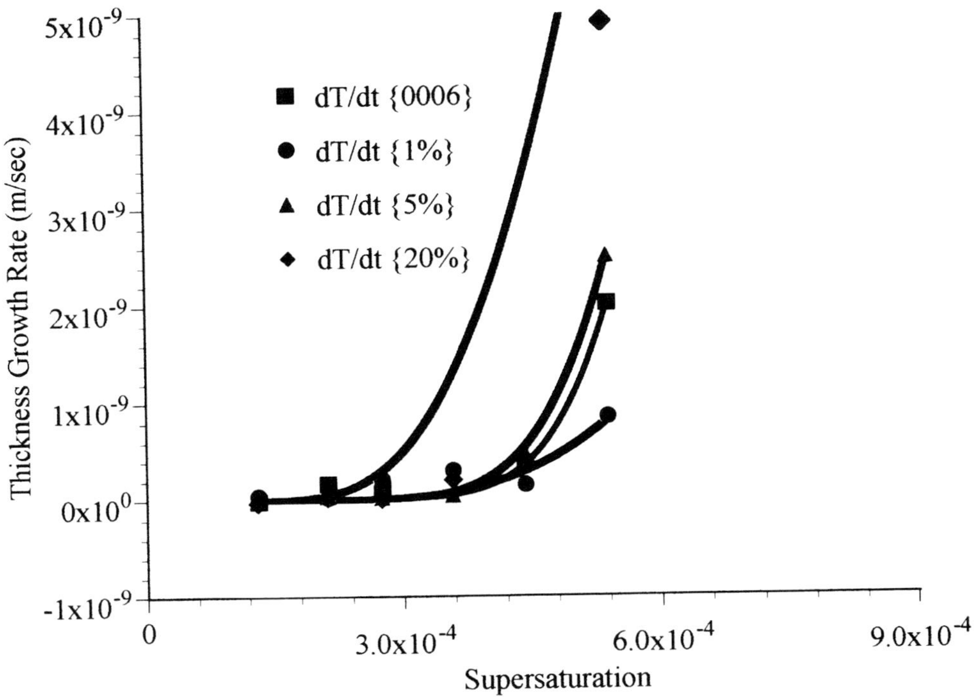

Fig. 8 Correlation of template thickness growth rate to two-dimensional nucleation and growth kinetics as a function of supersaturation.

be due to a decrease in the supersaturation as a result of the high template loading and less of the more soluble matrix material. The difference in the growth rate between the templates and single crystals leads to the conclusion that the template size affects the growth rate on the basal surface, too. Initially, the lower rate of template thickening might be a result of a lower nucleation density on the much smaller area of a template as compared to a large single crystal. This difference in nucleation density could also result from a difference in concentration of surface defects that can act as nucleation sites. However, as the templates grow rapidly in the radial direction, the area available for nucleation of new islands increases, and the nucleation density on the templates is similar to that on single crystals.

6.0 SUMMARY

Similar to the observed behaviour of single crystal samples, growth of template particles appears to be controlled by two different growth mechanisms. Template growth is also strongly affected by the TGG processing environment. As in single crystal samples, the liquid phase content has a strong effect on the growth rate as a result of the direct effect on boundary thickness between the growing templates and the matrix grains. However, template loading appears to have the greatest effect on the extent of radial growth of template grains. Thus, the diffusion-controlled model of template growth can only explain the initial radial growth of templates. The radius of curvature of the template particles is finite, and the equilibrium supersaturation at an infinitely flat template-liquid interface is modified due to the presence of edges and corners on the template particles. Both these factors result in a reduction of the driving force for growth of templates as compared to that of a single crystal. Onset of template impingement leads to significant retardation of template growth rate and limits the final grain size that can be obtained by TGG.

In contrast to radial growth, basal growth seems to be less affected by the difference in size between template particles and the single crystals. Thus, the thickness growth rate retains the same exponential dependence as single crystal growth in all experimental conditions.

7.0 REFERENCES

1. J. Rödel and A. M. Glaeser: 'Anisotropy of Grain Growth in Alumina,' *Journal of American Ceramics Society,* **73**(11), 1990, pp.3292 - 301.
2. D. S. Horn and G. L. Messing: 'Anisotropic Grain Growth in TiO_2-doped Alumina,' *Journal of Materials Science and Engineering,* A195, 1995, pp.169-78.
3. S. H. Hong and G. L. Messing: 'Anisotropic Grain Growth in Diphasic Gel Derived Titania-doped Mullite', *Journal of American Ceramics Society,* **81**(5), 1998, pp.1269 - 1277.
4. G. W. Ratheneau, J. Smit and A. L. Stuyts: 'Ferromagnetic Properties of Hexagonal Iron-Oxide Compounds With and Without a Preferred Orientation', *Z. Phys.,* **133**, 1952, 250.
5. H. Watanabe, T. Kimura and T. Yamaguchi: 'Sintering of Platelike Bismuth Titanate Powder Compacts with Preferred Orientation', *Journal of American Ceramics Society,* **74**(1), 1991, pp.139 - 47.
6. J. A. Horn, S. C. Zhang, U. Selvaraj, G. L. Messing and S. Trolier-mckinstry: 'Templated Grain Growth of Bismuth Titanate', *Journal of American Ceramics Society,* 82(4), 1999, pp.921 - 996.
7. D. Brandon, D. Chen and H. Chan: 'Control of Texture in Monolithic Alumina', *Materials Science and Engineering,* **A195**, 1995, pp.189-96.
8. M. M. Seabaugh, I. H. Kerscht and G. L. Messing: 'Texture Development by Templated Grain Growth in Liquid-Phase-Sintered α-Alumina', *Journal of American Ceramics Society,* **80**(5), 1997, pp.1181-1188.
9. S. M. Sabol, G. L. Messing and R. E. Tressler: 'Textured Alumina Fibres with Elongated Grains', *NASA Hi-Temp Review,* **1**(21), 1992, pp.21-1-21-14.
10. E. Suvaci, M. M. Seabaugh and G. L. Messing: 'Reaction-Based Processing of Textured Alumina by Templated Grain Growth', *Journal of European Ceramics Society.,* **19**, 1999, pp.2465 - 2474.

11. S. H. HONG and G. L. MESSING: 'Development of Textured Mullite by Templated Grain Growth' , *Journal of American Ceramics Society*, **82**(4), 1999, pp.867-872.
12. K. HIRAO, T. NAGAOKA, M. E. BRITO and S. KANZAKI: 'Microstructure Control of Silicon Nitride by Seeding with Rodlike β-Si_3N_4 Particles', *Journal of American Ceramic Society*, **77**(7), 1994, pp.1857-62.
13. M. J. SACKS, G. W. SCHEIFFELE and G. A. STAAB: 'Fabrication of Textured Silicon Carbide via Seeded Anisotropic Grain Growth', *Journal of American Ceramics Society*, **79**, No.6, 1996, pp.1611-1616.
14. B. BRAHMAROUTU, G. L. MESSING, S. TROLIER-MCKINSTRY and U. SELVARAJ: 'Templated Grain Growth of Textured $Sr_2Nb_2O_7$', ISAF '96, *Proceedings of the Tenth IEEE International Symposium on Applications of Ferroelectrics*, B. M. Kulwicki, A. Amin and A. Safari, eds., IEEE, Piscataway, NJ, **2**, 1996, pp.883 - 886.
15. M. M. SEABAUGH, M. D. VAUDIN, J. P. CLINE and G. L. MESSING: 'Comparison of Texture Analysis Techniques for Highly Oriented α-Al_2O_3', *Journal of American Ceramics Society*, **83**(8), 2000, pp.2049-2054.
16. R. L. FULLMAN: 'Measurement of Particle Size in Opaque Bodies', *Journal of Metals*, 1953, pp.447 - 452.
17. B. W. KIBBEL and A. H. HEUER: 'Anisometric Shape Factors for Ceramic Microstructures', *Journal of American Ceramic Society*, **72**(3), 1989, pp.517-519.
18. M. M. SEABAUGH, 'Texture Development in Liquid Phase Sintered Alpha Alumina via Anisotropic Grain Growth, Ph.D. Dissertation, Pennsylvania State University, University Park, PA. 1998.
19. R. D. MONAHAN and J. W. HALLORAN: 'Single -Crystal Boundary Migration in Hot-Pressed Aluminum Oxide,' *Journal of American Ceramics Society*, **62**(11-12), 1979, pp.64-67.
20. C. A. HANDWERKER, P. A. MORRIS and R. L. COBLE: 'Effects of Chemical Inhomogeneities on Grain Growth and Microstructure in Al_2O_3', *Journal of American Ceramic Society*, 72(1), 1989, pp.130 - 36.
21. M. M. SEABAUGH, E. SUVACI, B. BRAHMAROUTU and G. L. MESSING: 'Modeling Anisotropic Single Crystal Growth Kinetics in Liquid Phase Sintered Alpha Alumina', *Interface Science*, **8**, 2000, pp.257-267.
22. M. M. SEABAUGH, M. D. VAUDIN and G. L. MESSING: 'Texture Development and Microstructure Evolution in Liquid Phase Sintered Alpha Alumina Ceramics Prepared by Templated Grain Growth', submitted to *Journal of American Ceramics Society*, 2000.
23. K. W. LAW: 'Grain Growth in UO_2-Al_2O_3 in the Presence of a Liquid Phase', *Journal of American Ceramic Society*, 51(7), 1968, pp.373 - 376.
24. D. F. K. HENNINGS, R. JANSSEN and P. J. L. REYNEN: 'Control of Liquid-Phase-Enhanced Discontinuous Grain Growth in the Barium Titanate', *Journal of American, Ceramics Society*, 70(1), 1987, pp.23-27.
25. Modeling Crystal Growth Rates From Solution, M. Ohara and R.C. Reid, Prentice-Hall, Inc., Englewood Cliffs, NJ. 1973.

Effect of $PbTiO_3$ Additions on Grain Growth and Microstructural Evolution of $BaTiO_3$

Ho-Yong Lee and Jae-Suk Kim
Division of Metallurgical and Materials Engineering
Sunmoon University, Chungnam, Asan, 336-840, Korea

Suk-Joong L. Kang
Centre for Interface Science and Engineering of Materials
Korea Advanced Institute of Science and Technology
Taejon, 305-701 Korea

ABSTRACT

Chemically induced grain boundary migration (CIGM) and its effect on grain growth and microstructure development of $BaTiO_3$ have been studied. When sintered $BaTiO_3$ specimens of two different grain sizes were annealed in contact with $PbTiO_3$ particles at 1250°C, CIGM occurred in the specimen with coarse grains ($\sim$ 200 μm), but fast growth of $BaTiO_3$ grains occurred in the specimen with fine grains ($\sim$ 4 μm). EDS analysis confirmed that the observed microstructural changes were related to CIGM, of which the driving force is the coherent strain energy built in a thin coherent diffusion layer on a receding grain. The coherent strain energy in the $BaTiO_3$-$PbTiO_3$ system was calculated using a previously derived general equation and compared with capillary energy. The calculated strain energy was much higher than the capillary energy for grains larger than a few tenths of microns. It appears that the total driving force for grain growth under CIGM conditions is the sum of the coherent strain energy due to solute diffusion and the capillary energy due to grain boundary curvature. The present result suggests that CIGM may induce abnormal grain growth.

1.0 INTRODUCTION

When a polycrystalline solid comes in contact with a solute source, its equilibrium composition changes and thus the solid becomes unstable. The chemical instability induced in the solid can be eliminated by alloying of solute atoms within the solid. In the 1970s, it was discovered that the equilibration could occur by migration of interfaces forming behind them a new solid solution with the same crystalline structure and orientation as the parent grain, but enriched with solute atoms.[1-8] This interface migration is called the chemically induced interface migration (CIIM): chemically induced grain-boundary migration (CIGM, also called diffusion induced grain-boundary migration, DIGM) for grain boundary or chemically induced liquid film migration (CILFM or LFM) for liquid films.[5,6] Since the CIIM can occur in various processes accompanied by compositional changes in the solid (for example, sintering and heat treatment of systems containing volatile materials, welding of different metals, etc.) so changing the microstructure and several properties of solids, an understanding and control of the phenomenon have practical importance.

For continuous (apparently steady-state) migration of the boundaries, Hillert[9,10] proposed the coherent strain model where the driving force for the CIIM arose from the coherent strain energy built in a thin coherent diffusion layer on a receding grain. When a boundary is migrating at a velocity v, the solution of macroscopic diffusion equations shows that a solute diffusion zone of a thickness of D/v (where D is the lattice diffusivity) exists in front of the boundary. If the thickness is thin, the diffusion zone can remain coherent with the lattice of the solid being consumed.[11] Yoon and co-workers[12,13] showed that, by introducing two solute species of different size into Mo-Ni and cubic zirconia, the migration did not occur when the expected coherent strain in a energy of both positive and negative signs were reduced to very small values. These observations provided definitive evidence that the driving force for a boundary to migrate with a finite velocity arose from the diffusional coherent strain energy in a solute diffusion zone in front of the migrating boundary.

For the initiation of liquid film migration, Yoon et al.[14] proposed that the migration began and continued to the grain with the surface orientation of the higher coherent strain energy, when solute atoms diffused into the surfaces of two adjacent grains in the initial stage, producing diffusion layers coherent with the parent grains. The migration behaviour will thus be determined only by the coherent strain energy at the surface of the retreating grains This theory for the initiation may be applied to a grain boundary, if it is assumed that the coherent strain and stress at the surface of one grain are not transmitted across the grain boundary to the other grain. On the basis of calculation of coherent strain energy, Lee et al.[15,16] have tested the theory for the migration direction in pairs of Al_2O_3 grains of known surface orientations. For all the grain boundaries studied, the CIGM always occurred in the direction of the grain with the surface orientation corresponding to higher coherent strain energy. This critical experiment has provided additional experimental support for the coherent strain model.

The proposed theories and previous experimental results demonstrated that the coherent strain energy was the driving force for grain boundary migration during grain boundary diffusion of solutes. In general, driving force components may act in different directions so that the resulting driving force should be their algebraic sum.[17] When grain growth occurs in a polycrystalline solid during grain boundary diffusion of solutes, the total driving force for grain growth may be the sum of the energies due to the capillary effect and the coherent strain effect. If the coherent strain energy has a value comparable to or higher than the capillary energy, the total driving force for grain growth and thus the rate of grain growth may become much higher than those resulting only from the capillary effect. This implies that the diffusion of solutes along grain boundaries can increase the total driving force for grain growth and thus the grain growth rate. In order to understand this effect of coherent strain energy on grain growth, it is necessary to investigate the grain growth behaviour in a polycrystalline material during grain boundary diffusion of solutes and compare the coherent strain energy with the capillary energy.

In the present work a general equation of coherent strain energy, which is applicable to all crystal systems,[18] is reviewed and then experimental results on CIGM and grain growth of $BaTiO_3$ during diffusion of $PbTiO_3$ along grain boundaries are presented. As a solute compound, $PbTiO_3$ was selected in order to take advantage of its high solid solubility in $BaTiO_3$.[19] After annealing of sintered $BaTiO_3$ in the presence of $PbTiO_3$, grain growth observation and chemical composition analysis were made. The observed growth behaviour was explained in terms of calculated coherent strain energy in the $BaTiO_3$-$PbTiO_3$ system and the energy induced by capillary pressure.

2.0 GENERAL EQUATION OF COHERENT STRAIN ENERGY

The problem of coherency at an interface involves many metallurgical and ceramic processes, such as precipitation, spinodal decomposition, interface migration and thin film growth. If the interface between two different phases or a thin layer on a bulk phase is coherent, elastic deformation of the adjacent phases or in the thin layer is inevitable and therefore coherent strain energy is produced. Recently, Lee and Kang[18] derived a general equation of coherent strain energy applicable to all crystal systems by taking advantage of Hilliard's[20,21] procedure. The derivation of the general equation is as follows. First, consider an infinite crystal containing an arbitrary composition fluctuation which is one-dimensional and parallel to the $<x'>$ axis of a set of orthogonal axes $<x'>$, $<y'>$, $<z'>$, as shown in Figure 1. The prime is used to distinguish the reference axes from the standard axes, x, y and z of each crystal system.[22] The composition will then be uniform across any $(y'\,z')$ plane in the crystal. To maintain coherency under this condition, there must be no lattice parameter variation in the $(y'\,z')$ plane of the composition fluctuation although there can be a variation in spacing along the $<x'>$ direction. The preceding mode of lattice parameter change and lattice deformation, which are required to maintain coherency, can be understood by considering a slab and a slice cut from the slab as in Figure 1, similar to Hiliard's[20,21] procedure.

When solute atoms whose size is different from that of the matrix atoms diffuse into the slice within the solubility limit, uniform internal stresses build up within it and produce stress-free strains ε_{ij}°, if the internal stresses are within the elastic limit. The work needed to attach coherently this slice to the slab will be the coherent strain energy of the slice. The attachment can be made through two separate steps: first, elastic deformation of the slice to the original shape before cutting from the slab under an external stress equal to the internal stress produced by alloying, and second, reversible release of the external stress in the $<x'>$ direction or the direction normal to the face of the slice while the sides of the slice parallel to the $<x'>$ direction are clamped. The elastic strain energy remaining after these two steps is then the coherent strain energy, E_c,

$$E_c = E(I) - E(II), \tag{1}$$

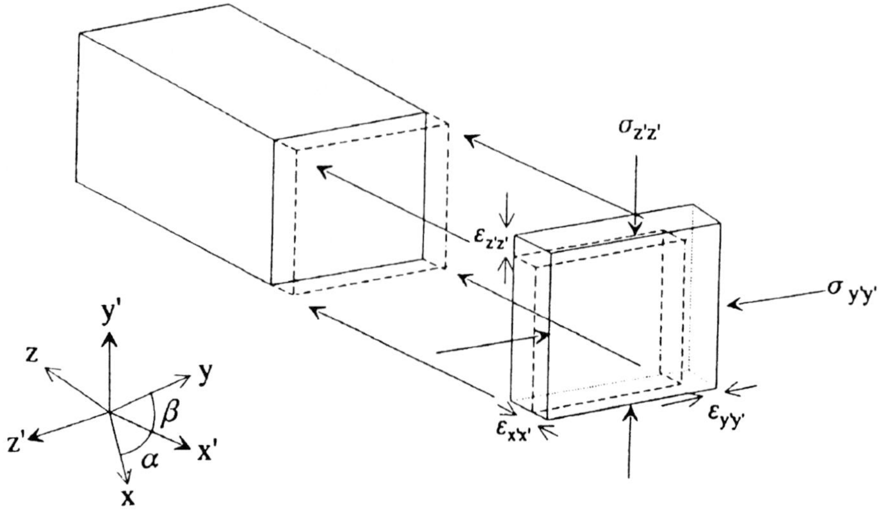

Fig. 1 Schematic showing the procedure for the calculation of coherent strain energy in a thin coherent diffusion layer. If a slice having lattice parameters different from those of slab is coherent with the slab, the slice is subjected to the stress s in the *y*' and *z*' directions and the elastic strain ε is produced in the *x*' direction as well as *y*' and *z*' (only normal components are shown).

where $E(I)$ and $E(II)$ are the elastic work per unit volume of the slice during the first and the second steps, respectively.

Since the elastic work during the first step, $E(I)$, is equal to the elastic strain energy per unit volume required to deform the slice:

$$\begin{aligned} E(I) &= (1/2)\,\sigma_{ij}\,\varepsilon_{ij}^{\circ}\ (i, j = 1,2,3) \\ &= (1/2)\,C_{ijkl}\,\varepsilon_{kl}^{\circ}\,\varepsilon_{ij}^{\circ} \\ &= (1/2)\,E_o. \end{aligned} \qquad (2)$$

Here, σ_{ij} is the elastic stress, ε_{ij} the elastic strain and C_{ijkl} the elastic stiffness. Since $E(I)$ is determined by stress-free strains ε_{ij}° and elastic stiffnesses C_{ijkl}, the value of $E(I)$ is constant regardless of the orientation of the slice. The stress-free strains of materials with orthotropic symmetry can be directly calculated from lattice misfit parameters η_i and solute mole-fraction X:

$$\varepsilon_{11}^{\circ} = \eta_1 X, \varepsilon_{22}^{\circ} = \eta_2 X, \varepsilon_{33}^{\circ} = \eta_3 X. \qquad (3)$$

But crystals with monoclinic or triclinic symmetry require a matrix transformation to obtain stress-free strains ε_{ij}°, as demonstrated by Hay.[23]

During the second step, the state of stresses and strains induced in the first step varies. The strains parallel to the <*x*'> direction are released but the strains in other directions must remain unchanged in order to maintain coherency. The stresses parallel to the <*x*'> direction are completely released, but the stresses in the other directions should vary in accordance with the change of strains. After the stress release, the stress state of the slice is, in fact, two-dimensional.

To obtain the elastic strain energy released per unit volume, *E(II)*, during the second step of the relaxation process, components of the strain and stress in the <*x*'> direction need to be determined. The stress relaxed, $\sigma_{i'j'}$, is[22]:

$$\sigma_{i'j'} = l_{i'i} l_{j'j} \sigma_{ij}. \tag{4}$$

Using the value of relaxed stresses and generalized Hooke's law, the magnitude of released strains can also be determined. Since all strains except $\varepsilon_{1'1'}$, $\varepsilon_{1'3'}$, and $\varepsilon_{1'2'}$ are zero under the state of strain release, Hooke's law is expressed as:

$$\sigma_{1'1'} = C_{1'1'1'1'} \varepsilon_{1'1'} + 2C_{1'1'1'3'} \varepsilon_{1'3'} + 2C_{1'1'1'2'} \varepsilon_{1'2'} \tag{5}$$

$$\sigma_{1'3'} = C_{1'3'1'1'} \varepsilon_{1'1'} + 2C_{1'3'1'3'} \varepsilon_{1'3'} + 2C_{1'3'1'2'} \varepsilon_{1'2'} \tag{6}$$

$$\sigma_{1'2'} = C_{1'2'1'1'} \varepsilon_{1'1'} + 2C_{1'2'1'3'} \varepsilon_{1'3'} + 2C_{1'2'1'2'} \varepsilon_{1'2'}. \tag{7}$$

From these equations, $\varepsilon_{1'1'}$, $\varepsilon_{1'3'}$, and $\varepsilon_{1'2'}$, can be obtained by Cramer's rule.[22] Under the above stress and strain condition, the released strain energy *E(II)* during the second step is expressed as:

$$\begin{aligned} E(II) &= \sigma_{i'j'} \varepsilon_{i'j'} \\ &= (1/2)\,(\sigma_{1'1'} \varepsilon_{1'1'} + 2\sigma_{1'3'} \varepsilon_{1'3'} + 2\sigma_{1'2'} \varepsilon_{1'2'}). \end{aligned} \tag{8}$$

A general equation of coherent strain energy is then expressed as:

$$\begin{aligned} E_c &= E(I) - E(II) \\ &= (1/2)\,[E_o - (\sigma_{1'1'} \varepsilon_{1'1'} + 2\sigma_{1'3'} \varepsilon_{1'3'} + 2\sigma_{1'2'} \varepsilon_{1'2'})]. \end{aligned} \tag{9}$$

In this equation, *E(I)* is constant for a fixed concentration, but *E(II)* is a function of crystallographic orientation. The coherent strain energy is, therefore, determined by the crystallographic direction of stress release, i.e., the direction normal to the interface.

3.0 THE EFFECT OF COHERENT STRAIN ENERGY ON GRAIN GROWTH IN $BaTiO_3$ - $PbTiO_3$

The effect of coherent strain energy on grain growth in $BaTiO_3$ grains with diffusion of $PbTiO_3$ along grain boundaries has been investigated. $PbTiO_3$ particles were scattered on a polished surface of sintered $BaTiO_3$, which consisted of large or small grains and annealed to induce CIGM. In a specimen with large grains (~ 200 μm), conventional CIGM was observed while, in a specimen with small grains (~ 4 μm), fast growth of small $BaTiO_3$ grains occurred. The fast grain growth was explained in terms of calculated coherent strain energy and capillary energy for grain growth.

3.1 Calculation of Coherent Strain Energy for the $BaTiO_3$ - $PbTiO_3$ System

The coherent strain energy of a coherent diffusion layer on a $BaTiO_3$ crystal surface can be calculated by using Eqn. (9). For the calculation of E_c for $BaTiO_3$ - $PbTiO_3$,

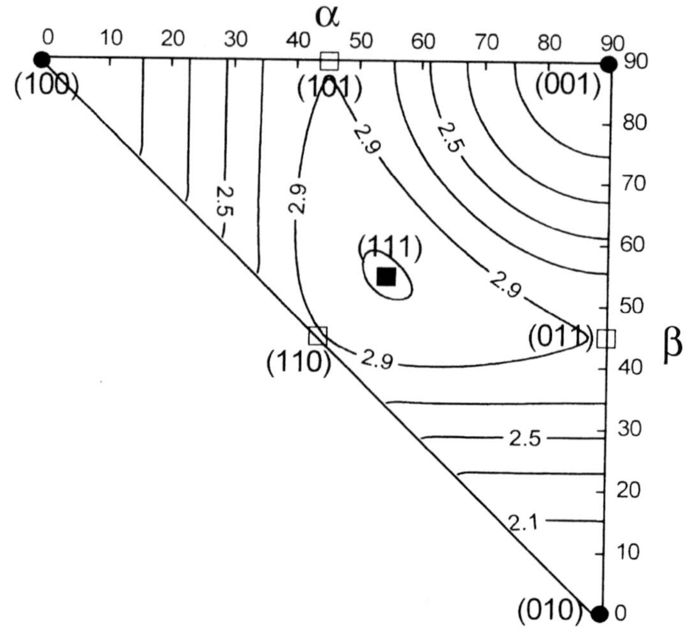

Fig. 2 Calculated coherent strain energy map (CSEM in MJ m^{-3}) of $(Ba_{0.8}\ Pb_{0.2})\ TiO_3$ (with ε of 3.31×10^{-3}) coherent layer on $BaTiO_3$. The axes represent the interplanar angles α and β. The crystallographic directions with minimal and maximal E_c are represented by ● and ■, respectively.

the elastic constants of pure $BaTiO_3$ measured at 423 K were used because of the unavailability of data for (Ba, Pb) TiO_3 solid solution.[24] The stress-free strains ε_{11}° $BaTiO_3$ with cubic symmetry can be calculated directly from the lattice misfit parameter η_1 and solute mole-fraction X:

$$\varepsilon_{11}^{\circ} = \eta_1 X. \tag{10}$$

The lattice parameter of cubic $PbTiO_3$ at 773 K is reported to be ~0.3960 nm, but the lattice parameter of cubic $BaTiO_3$ is known only between 392 and 493 K.[19] The lattice parameter of cubic $BaTiO_3$ at 773 K has been estimated by linear extrapolation from the data obtained between 392 and 493 K. From these lattice parameter data at 773 K, the value of lattice misfit η_1 is calculated to be ~6.64×10^{-4}. Thus the value of elastic strain ε_{11}° of $(Ba_{0.8}Pb_{0.2})\ TiO_3$, which is the composition determined from a EDS (energy dispersive spectroscopy) analysis for migrated regions in the present investigation (see Figures 5 and 8), is estimated to be ~3.31×10^{-3}. The calculated values of E_c with surface normal n (orientation vertical to a surface plane) are graphically represented as a coherent strain energy map

(CSEM), shown in Figure 2, which depicts equi-coherent strain energy lines and some specific crystallographic orientations.[16,18] In the CSEM, the axes represent the interplanar angles between the surface normal of the considered plane and (100) plane (α) and between the surface normal of the considered plane and the (010) plane (β). For $(Ba_{0.8}Pb_{0.2})\ TiO_3$, with $\varepsilon = 3.31 \times 10^{-3}$, the calculation shows that a (111) plane has the maximum E_c ($E_c(111) = 3.11$ [MJ m^{-3}]) and a (100) plane the minimum E_c ($E_c(100) = 1.94$ [MJ m^{-3}]). A (110) plane has an E_c of 2.90 [MJ m^{-3}].

3.2 Experimental Procedure

Hydrothermally prepared $BaTiO_3$ powder (99.95% purity, BT-01, Sakai Chemical Industry Co., Osaka, Japan) and $PbTiO_3$ powder (99% purity, Aldrich Chemical Co., Milwaukee, WI, U.S.A.) were used as starting materials. $BaTiO_3$ powder was ball-milled for 12 hrs. in a Nalgene bottle with ethanol and zirconia (Y-TZP) balls. After drying, the $BaTiO_3$ powder was compacted uniaxially at ~1 MPa into cylindrical compacts of 10 mm diameter and then pressed hydrostatically at 200 MPa. The powder compacts were sintered for 1 hr. in air either at 1350°C or at 1380°C. The heating and cooling rates were 4 K/min. The sintered specimens were cut and polished to 0.25 μm finish. Before annealing of the sintered specimens at 1250°C, $PbTiO_3$ particles, which were obtained by sintering a $PbTiO_3$ powder compact at 850°C for 1 hr. and then crushing, were scattered on polished surfaces of the $BaTiO_3$ specimens. In order to reduce Pb evaporation from the scattered $PbTiO_3$ particles during annealing, fine $PbTiO_3$ powder (atmospheric powder) was placed around the $BaTiO_3$ specimen. After the annealing, the microstructural observation and EDS compositional analysis were made on the surfaces of the $BaTiO_3$ specimens.

3.3 Experimental Results and Discussion

3.3.1 Grain Boundary Migration in $BaTiO_3$ with Large Grains

After sintering of $BaTiO_3$ at 1380°C for 1 hr. the specimens consisted of large grains of ~200 μm average size, as shown in Figure 3. The grains contained numerous fine pores, implying that abnormal grain growth occurred during the sintering. When $PbTiO_3$ particles were scattered on a polished surface of the sintered $BaTiO_3$ and annealed at 1250°C for 5 hrs., grain boundaries migrated, as shown in Figure 4. The lines indicated by arrows with "O" are the original positions of grain boundaries in the sintered specimen before annealing and those indicated by arrows with "M" are the grain boundaries positioned after the annealing. The $PbTiO_3$ particles scattered before the annealing were also observed on the prepolished surface of the annealed specimen, as shown in Figure 4.

Figure 5 plots the measured increase in Pb content in the migrated region and indicates that the boundary migration is accompanied by Pb alloying of ~20 mol% (Pb/(Ba + Pb)) in $BaTiO_3$. EDS analyses also showed that the $PbTiO_3$ particles changed

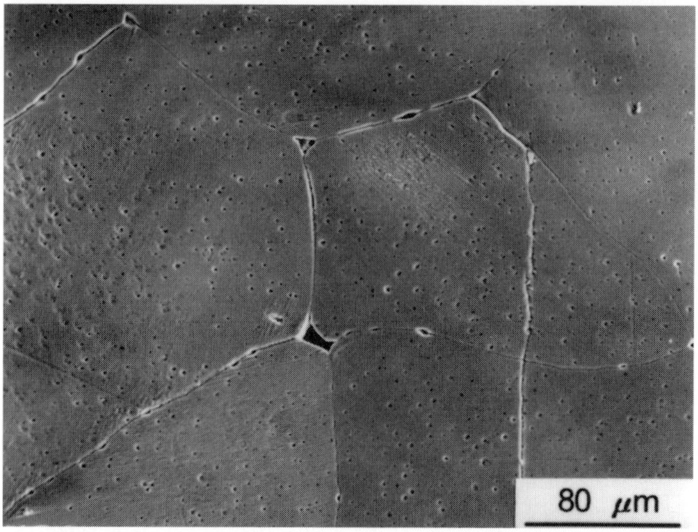

Fig. 3 Microstructure of a polished surface of $BaTiO_3$ specimen thermally etched at 1250°C for 5 hrs. in air after the sintering at 1380°C for 1 hr. in air.

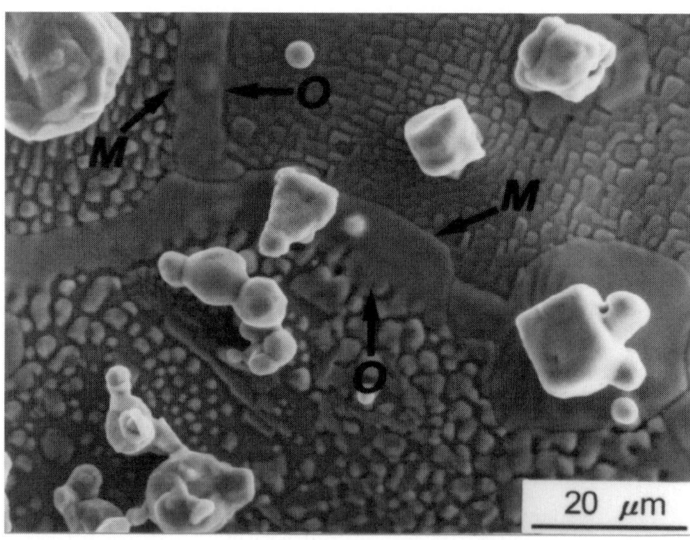

Fig. 4 Surface microstructure of $BaTiO_3$ specimen sintered at 1380°C for 1 hr. in air and then annealed with $PbTiO_3$ particles at 1250°C for 5 hrs. in air. $PbTiO_3$ particles were scattered on the polished surface before the annealing.

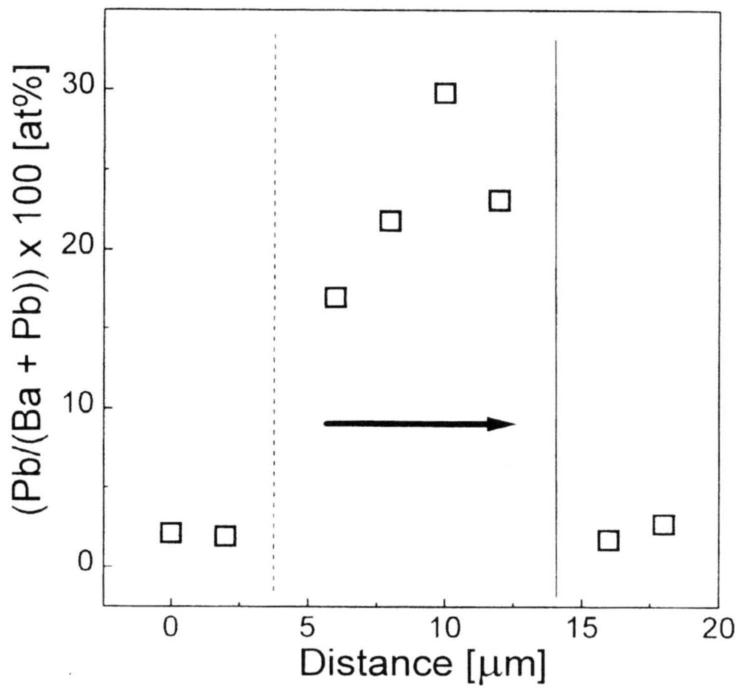

Fig. 5 Measured concentration profile of Pb (in (Pb/(Ba+Pb)) × 100%) along a line perpendicular to a migrating boundary after annealing at 1250°C for 5 h in the presence of scattered $PbTiO_3$ particles. The dotted and solid lines show the initial and final positions of the migrating grain boundary, respectively.

to a solid solution of ~ $(Ba_{0.6}Pb_{0.4})TiO_3$ during the annealing. The solute Pb content in migrated regions is relatively high compared to solute contents in other materials known to induce CIGM.[5-8] For $BaTiO_3$, because of its relatively low stiffness,[25] a high degree of alloying appears to be necessary to build up sufficient coherent strain energy to induce CIGM. On the contrary, for materials with high stiffness, for example Al_2O_3,[25] only 5 mol.% Cr_2O_3 alloying in Al_2O_3 was enough to induce a coherent strain energy of ~25 [MJ m^{-3}],[16] which is a value similar to that induced by Pb alloying of ~20 mol% (Pb/(Ba + Pb)) in $BaTiO_3$.

When a sintered $BaTiO_3$ specimen was annealed without $PbTiO_3$ particles on the surface, the boundary migration did not occur even in the presence of $PbTiO_3$ powder on the specimen. Therefore, the Pb detected in the migrated region was supplied from the scattered $PbTiO_3$ particles via surface diffusion to $BaTiO_3$ grain boundaries during their migration. This boundary migration accompanied by alloying of Pb in $BaTiO_3$ is a chemically induced grain boundary migration, similar to that observed in previous CIGM experiments.[5-8] The boundary migration, however, does not appear to enhance the growth of $BaTiO_3$ grains.

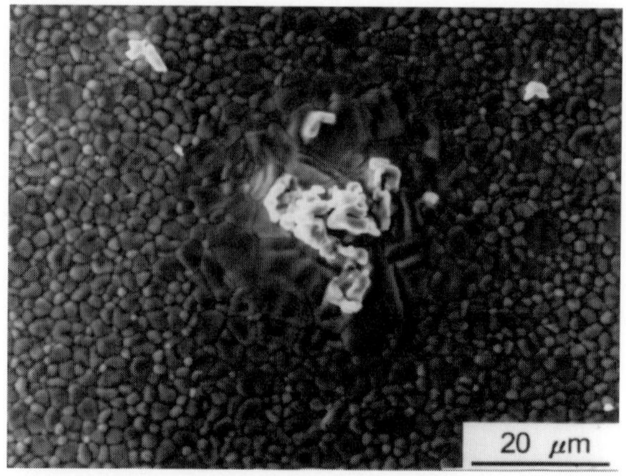

(a)

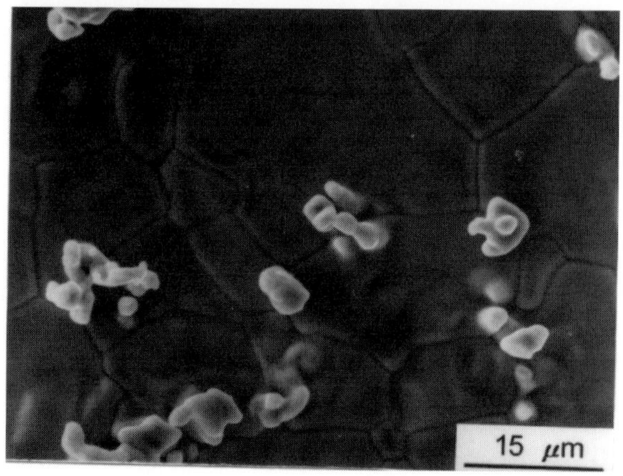

(b)

Fig. 6 Surface microstructures of $BaTiO_3$ specimens sintered at 1350°C for 1 hr. in air and then annealed with $PbTiO_3$ particles at 1250°C for (a) 1 hr. and (b) 5 hr. in air. $PbTiO_3$ particles were scattered on the polished surfaces before the annealing.

3.3.2 Grain Growth in $BaTiO_3$ with Small Grains

When a $BaTiO_3$ specimen was sintered at 1350°C for 1 hr., abnormal grain growth did not occur and the specimen consisted of small grains of ~4 μm average size. When $PbTiO_3$ particles were scattered on a polished surface of this sintered $BaTiO_3$ and annealed at 1250°C for 1 hr., fast growth of small $BaTiO_3$ grains occurred from the interfaces with $PbTiO_3$ particles, as shown in Figure 6a. In Figure 6a, the bright particles are the scattered $PbTiO_3$ particles. EDS analysis showed that, after annealing

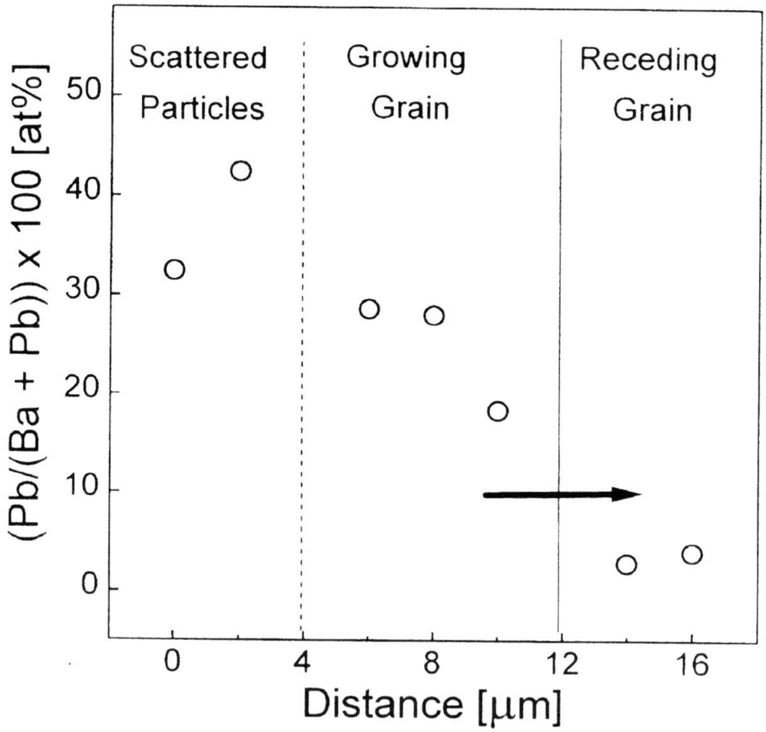

Fig. 7 Measured concentration profile of Pb (in (Pb/(Ba+Pb) × 100%) along a line perpendicular to a migrating boundary after annealing at 1250°C for 1 hr. in the presence of scattered $PbTiO_3$ particles. The dotted and solid lines show the initial and final positions of the migrating grain boundary, respectively.

for 1 hr., the scattered particles changed to a solid solution of ~ $(Ba_{0.6}Pb_{0.4})$ TiO_3 and the growing grains near the $PbTiO_3$ powder to be a solid solution of ~$(Ba_{0.8}Pb_{0.2})$ TiO_3. A discontinuous change in Pb content also was observed across the boundary between a growing $(Ba_{0.8}Pb_{0.2})$ TiO_3 grain and a receding $BaTiO_3$ grain as shown in Figure 7. After annealing for 5 hrs., the grains grew to ~20 μm average size and the scattered particles still remained, as shown in Figure 6b. The Pb content [(Pb/(Ba + Pb)) × 100] in the scattered particles and the grains were ~40 and ~30% respectively. The distribution of Pb in the grains was uniform and a discontinuous change in Pb content across the boundary was not observed.

The fast growth of the $BaTiO_3$ grains around the scattered $PbTiO_3$ in Figure 6 is thought to be due to CIGM. By comparing the microstructures of the $BaTiO_3$ specimens annealed at 1250°C for 5 hrs. without (Figure 8) and with $PbTiO_3$ (Figure 6b), it can be seen that the addition of $PbTiO_3$ particles on the $BaTiO_3$ surface enhanced the growth of $BaTiO_3$ grains.

Fig. 8 Surface microstructure of $BaTiO_3$ specimen sintered at 1350°C for 1 hr. in air and then annealed at 1250°C for 5 hrs. in air.

3.3.3 Driving Force for Grain Growth Under Chemical Inequilibrium

The fast grain growth observed in Figure 7 is obviously a result of chemically induced grain boundary migration. In normal grain growth, the driving force is the capillary energy difference between adjacent grains due to grain boundary curvature. The capillary energy per unit material volume can be expressed as:

$$\Delta G_{CP} = C_{GB}\,\gamma_{GB}, \tag{11}$$

where C_{GB} is the grain boundary curvature and γ_{GB} the grain boundary energy. The curvature is expressed in terms of the principal radii of curvature, and ρ_1 and ρ_2.

$$C_{GB} = 1/\rho_1 + 1/\rho_2 \tag{12}$$

The driving force for CIGM is the coherent strain energy ΔG_{CSE} stored in a thin diffusional layer on the receding grain:

$$\Delta G_{CSE} = E_c, \tag{13}$$

where E_c is the coherent strain energy per unit volume and expressed as in Eqn. (9). Since the driving force components for grain growth may act in different directions, the resulting driving force should be, in general, their algebraic sum.[17] Therefore the total driving force for grain growth under chemical in equilibrium is thought to be the sum of the capillary energy and the coherent strain energy.

$$\Delta G_T = C_{GB}\,\gamma_{GB} + E_c \tag{14}$$

To evaluate the relative contributions of the two components, the coherent strain energy and capillary energy are calculated for the case of $BaTiO_3$ - $PbTiO_3$. For the calculation of coherent strain energy of a coherent diffusion layer on a $BaTiO_3$ crystal surface, the chemical composition of the layer is assumed to be $(Ba_{0.8}Pb_{0.2})\,TiO_3$ because the composition of a coherent diffusion layer on a receding grain is identical to that of the migrated region (Figure 7). The coherent strain energy of a $(Ba_{0.8}Pb_{0.2})$

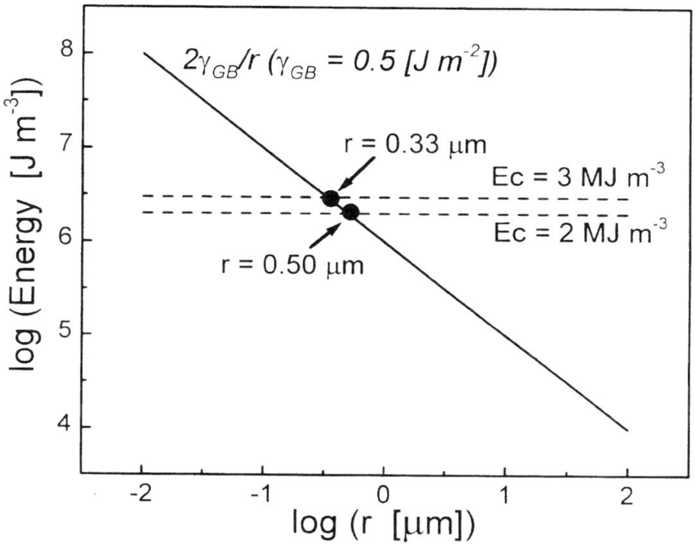

Fig. 9 Comparison between capillary energy ($2\gamma_{GB}/r$, r = the average radius of grain boundary curvature) and calculated coherent strain energies stored in a $(Ba_{0.8}Pb_{0.2})TiO_3$ layer on $BaTiO_3$. Specific grain-boundary energy assumed to be 0.5 J m^{-2}.

TiO_3 layer on $BaTiO_3$ is then read between 2 and 3 MJ m^{-3}, depending on the surface orientation, as shown in Figure 2. For the calculation of capillary energy the grain boundary energy is assumed to be 0.5 J m^{-2}, a typical value of ceramic grain boundary energies.

Figure 9 plots the estimated coherent strain energy and capillary energy as a function of the radius of grain boundary curvature. Since the coherent strain energy is independent of the boundary curvature, the energy is plotted as a horizontal dotted line in the figure. In contrast, the capillary energy is inversely proportional to the radius of boundary curvature. When the curvature radius r is 0.33 μm, the capillary energy is approximately the same as the coherent strain energy of 3 MJ m^{-3}. When r = 0.50 μm, the energy is ~2 MJ m^{-3}. This calculation means that, if the boundary radius is larger than 0.3 – 0.5 μm, the coherent strain energy is the major driving force for grain growth. Since the curvature radius of grain boundaries is much larger than the grain radius, the grain size must be much smaller than a few tenths of micrometers for the capillary energy to become the major driving force for grain growth. Since the average radius of $BaTiO_3$ grains in Figure 8 is ~2 μm, the coherent strain energy should dominate the driving force for grain growth. The fast growth of $BaTiO_3$ grain adjacent to $PbTiO_3$ particles must be due to the coherent strain energy.

The grain boundary diffusion of solute atoms in polycrystalline materials is known to change the composition, structure and mobility of grain boundaries. When the grain growth was enhanced by the addition of solute atoms into a polycrystalline material, the grain growth enhancement was, in general, thought to be due to an increase in grain boundary mobility. The present investigation shows that fast grain growth under chemical inequilibrium is due to chemically induced grain-boundary migration. For many other cases of grain growth under chemical inequilibrium, the coherent strain energy should dominate the driving force for the grain growth.

4.0 SUMMARY

The coherent strain theory and critical experiments with Mo-Ni, cubic zirconia, and Al_2O_3-Fe_2O_3 have shown that the driving force for CIGM is the coherent strain energy in a thin coherent diffusion layer formed at the surface of the receding grain during the migration.[12,13,15,16] These results indicate that the coherent strain energy acts as an additional driving force for grain growth under chemical inequilibrium in the $BaTiO_3$ - $PbTiO_3$ system. The effect of coherent strain energy on grain growth has been found to be considerable. When $PbTiO_3$ particles were scattered on a polished surface of sintered $BaTiO_3$ and annealed at 1250°C, fast growth of $BaTiO_3$ grains occurred at the interface with $PbTiO_3$ particles. A discontinuous change in Pb content was also observed across the boundary between a growing $(Ba_{0.8}Pb_{0.2})TiO_3$ grain and a receding $BaTiO_3$ grain. For the $BaTiO_3$-$PbTiO_3$ system, the coherent strain energy of $(Ba_{0.8}Pb_{0.2})$ TiO_3 on $BaTiO_3$ was calculated to be between 2 and 3 MJ m^{-3}, depending on the surface orientation of $BaTiO_3$. The calculated coherent strain energies were much higher than the capillary energy produced by $BaTiO_3$ grains of 2 μm average radius. The present calculation and observation show that the coherent strain energy can induce fast growth of grains and is the major driving force for the grain growth under chemical inequilibrium.

5.0 ACKNOWLEDGEMENT

This work was supported by the Ministry of Science and Technology, Korea, through the Nanostructure Technology Project.

6.0 REFERENCES

1. F. N. Rhines and A. M. Montgomery: A New Type of Structure in the α-Copper-Zinc Alloy, *Nature*, **141**, 1938, p.413.
2. F. J. A. den Broeder: Interface Reaction and a Special Form of Grain Boundary Diffusion in the Cr-W System, *Acta Metallurgica et Materialia*, **20,** 1972, p.319.

3. K. N. Tu: Kinetics of Thin-Film Reactions between Pb and AgPd Alloys, *Journal of Applied Physics*, **48,** 1977, p.3400.
4. M. Hillert and G. R. Purdy: Chemically Induced Grain Boundary Migration, *Acta Metallurgica et Materialia,* **26,** 1978, p.333.
5. D. N. Yoon: Chemically Induced Interface Migration in Solids, *Annual Review of Materials Science*, **19**, 1989, p.43.
6. D. Y. Yoon: Theories and Observations of Chemically Induced Interface Migration, *International Materials Review*, **40**, 1995, p.149.
7. A. H. King: Diffusion Induced Grain Boundary Migration, *International Materials Review*, **32**, 1987, p.173.
8. C. A. Handwerker: Diffusion-Induced Grain Boundary Migration in Thin Films, *Diffusion Phenomena in Thin Films and Microelectronic Materials*, D. Gupta and P. S. Ho, eds., Noyes Publications, Park Ridge, NJ, 1988.
9. M. Hillert: On the Theories of Growth During Discontinuous Precipitation, *Metallurgical Transactions*, **3A**, 1972, p.2729.
10. M. Hillert: On the Driving Force for Diffusion Induced Grain Boundary Migration, *Scripta Metallurgica*, **17,** 1983, p.237.
11. W. A. Tiller, K. A. Jackson, J. W. Rutter and B. Chalmers: *Acta Metallurgica et Materialia*, **1**, 1953, p.428.
12. W. H. Rhee and D. N. Yoon: The Grain Boundary Migration Induced by Diffusional Coherent Strain in Mo-Ni Alloy: *Acta Metallurgica et Materialia*, **37**, 1989, p.221.
13. J. W. Jeong, D. N. Yoon and D. Y. Kim: Chemically Induced Instability at Interfaces of Cubic ZrO_2-Y_2O_3 Grains in a Liquid Matrix, *Acta Metallurgica et Materialia*, **39**, 1991, p.1275.
14. D. N. Yoon, J. W. Cahn, C. A. Handwerker, J. E. Blendell and Y. J. Baik: Coherent Strain Induced Migration of Liquid Films Through Solids, *Interface Migration and Control of Microstructure,* C. S. Pande, et al., eds., ASM International, Metals Park, OH., 1985.
15. H. Y. Lee, S. J. L. Kang and D. Y. Yoon: The Effect of Elastic Anisotropy on the Direction and Faceting of Chemically Induced Grain Boundary Migration in Al_2O_3, *Acta Metallurgica et Materialia,* **41**, 1993, p.2497.
16. H. Y. Lee, S. J. L. Kang and D. Y. Yoon: Coherent Strain Energy and the Direction of Chemically Induced Grain Boundary Migration in $Al_2O_3 - Cr_2O_3$ and $Al_2O_3 - Fe_2O_3$, *Journal of American Ceramics Society*, **77**, 1994, p.1301.
17. V. Novikov: Grain Growth and Control of Microstructure and Texture in Polycrystalline Materials, CRC Press, Boca Raton, 1997.
18. H. Y. Lee and S. J. L. Kang: A General Equation of Coherent Strain Energy and Its Application, *Z. Metallkd.*, **85**,1994, p.426.
19. K. H. Hellwege and A. M. Hellwege: *Ferroelectrics and Related Substances (Subvolume a: Oxides)*, Berlin, 1981.
20. J. E. Hillard: Nucleation within Crystalline Phases, *Ind. Eng. Chem.*, **58**, 1966, p.19.
21. J. E. Hilliard: Spinodal Decomposition, *Phase Transformations*, H. I. Aaronson, ed., ASM International, Metals Park, OH., 1968.

22. J. F. Nye: *Physical Properties of Crystals*, Clarendon Press, Oxford, 1967.
23. R. S. Hay: Coherent Strain Energy and Thermal Strain Energy of Thin Films in Any Crystal System, *Scripta Metallurgica*, **26**, 1992, p.535.
24. D. Berlincourt and H. Jaffe: Elastic and Piezoelectric Coefficients of Single-Crystal Barium Titanate, *Physical Review*, **111,** 1958, p.143.
25. G. Simmons and H. Wang: *Single Crystal Elastic Constants and Calculated Aggregated Properties: A Handbook*, The MIT Press, Cambridge, U.S.A., 1971.

The Influence of Ceramic Interphase on the Creep Behaviour of Silicon Nitride

S. M. Wiederhorn
National Institute of Standards and Technology
Gaithersburg, MD, U.S.A

ABSTRACT

In this paper we discuss the role played by the intergranular phases in the creep of silicon nitride, which is completely controlled by the amount and effective viscosity of the intergranular phases. After briefly reviewing the structure and creep behaviour of silicon nitride, we discuss two theories developed to rationalize known facts on the creep behaviour of this material. One of the theories is based on grain boundary sliding, the other on cavity formation and the flow of the sintering aid away from the cavitation site. A detailed evaluation of each model with regard to transport properties of the sintering aid, supports the cavitation model as the explanation for high temperature creep of silicon nitride.

1.0 INTRODUCTION

Structural silicon nitride is made by the process of liquid phase sintering. It is used commercially as ball bearings in a variety of applications, as cutting tools for cast iron, as cam followers in diesel engines and as turbine nozzles in auxiliary power engines for aircraft. It is also being tested as valves in automobile engines and as blades and nozzles in stationary turbines. At low temperatures, commercial grades of silicon nitride have a toughness as high as 8 MPa • $m^{1/2}$ and a creep resistance far superior to that of the best grades of super-alloys. At the same applied stress and lifetime, silicon nitride is capable of sustained temperatures about 300°C higher than the super-alloys. These favourable properties and its potential use as a high temperature replacement for the super-alloys are the reasons that materials scientists are interested in characterizing and improving the mechanical behaviour of silicon nitride.

2.0 PROCESSING AND STRUCTURE

Silicon nitride is made by liquid phase sintering with sintering aids such as MgO, Y_2O_3, Yb_2O_3 and Al_2O_3. This method of processing must be used because of the very high temperatures and pressures needed to sinter silicon nitride in the absence of a sintering aid. Sintering aids promote fast sintering at temperatures of 1825 to 2080°C.[1] High pressure nitrogen (1 to 8 MPa), is also needed to prevent decomposition of the silicon nitride. Because of the way silicon nitride is made, its grain boundaries are completely wetted by the grain boundary phase, which results from a reaction between the sintering aid and silica normally found on the surface of silicon nitride powder. The grains of silicon nitride are separated by a layer of amorphous silicate phase, ≈0.5 to ≈1 nm thick. The thickness of the amorphous layer depends on the composition of the sintering aid.[2, 3] At temperatures between 1200 and 1400°C, the silicate phase in the multi-grain junctions partially devitrifies, forming a variety of crystalline phases.[3-5] Devitrification is never total; a thin layer of an amorphous phase (2-3 nm), always separates the crystalline silicate phases from the silicon nitride grains.[2, 3] The thickness of the amorphous phases can be estimated by the theory proposed by Clarke.[2]

3.0 PROPERTIES

3.1 Strength and Toughness

The mechanical properties of the silicon nitride at high temperatures depend on the amount and composition of the sintering aid.[6] Because the sintering aid softens at a lower temperature than the silicon nitride grains, a refractory sintering aid is required for high-temperature applications. Rare earth oxides such as yttrium oxide, which form high temperature eutectics are normally used as sintering aids. For lower temperature applications, toughness and strength are enhanced by *in situ* toughening, in which sintering aids such as Al_2O_3, MgO and CaO cause the grains of silicon nitride to grow in length to increase their aspect ratio.[7-10] High toughness is achieved by debonding along the silicon nitride grains during fracture, thus promoting bridging across the propagating crack.[11, 12] Commercial grades of silicon nitride made in this way have critical stress intensity factors as high as $K_{Ic} = 8$ MPa • $m^{1/2}$[13, 14] and bending strengths, σ_b, greater than 1000 MPa.[14, 15]

3.2 Creep Behaviour

Based on recent investigations of the creep behaviour of a number of grades of silicon nitride, these materials have the following characteristics:

a. For much of the range of applied stress, creep occurs faster in tension than in compression,[16-24]

b. In contrast to the creep behaviour of most metals, creep in tension is not a power function of the applied stress,[16-18, 25-27]

c. Cavity formation accompanies creep in tension,[17, 18, 28-30]

d. The volume fraction of cavities in tension is linearly proportional to the axial creep strain,[17, 18, 25]

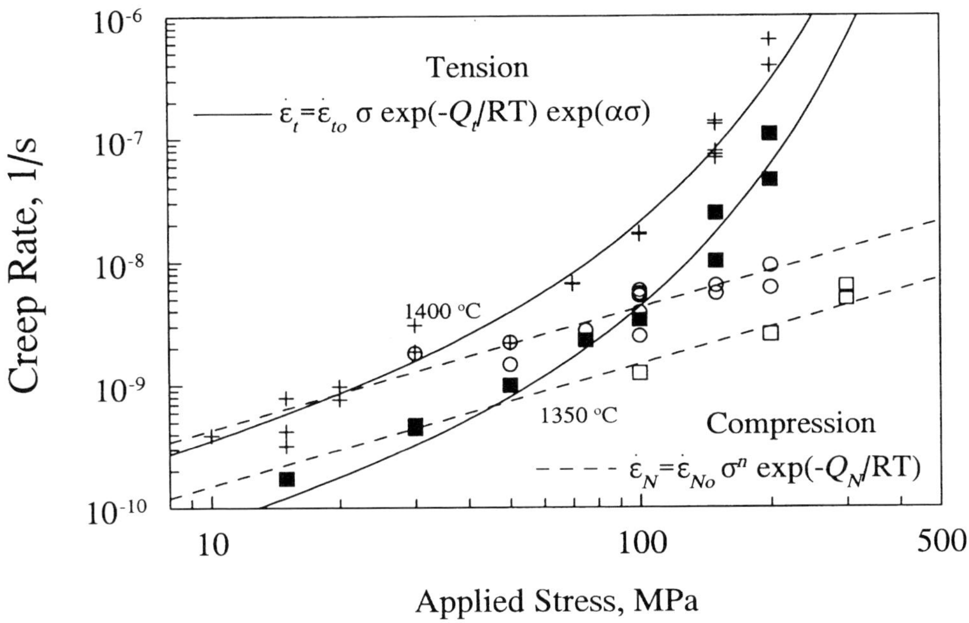

Fig. 1 A comparison of tensile and compressive creep for SN88.[31] The tensile data is fitted to an equation for cavitation creep, equation 5, while the compression data is fitted to a Norton equation.

e. cavitation in compression is much less than that in tension.[17, 20]

Figure 1 is a log-log plot of the tensile and compressive creep rates of SN88 (NGK Insulator, Inc., Nagoya, Japan)* as a function of applied stress.[31] The creep data in compression fit a power law function of applied stress with a stress exponent of $n = 0.998 \pm 0.112$ and an activation energy of $\Delta H = (491.6 \pm 54.9)$ kJ/mol. The intercept on Figure 1 is $ln\dot{\epsilon}_0 = (11.52 \pm 3.69)\ s^{-1}$. The uncertainty presented is the standard error of the constants. The stress exponent and activation energy are both within the range of values obtained in other studies of the compressive creep of silicon nitride.[32] In tension, the creep data are curved over most of the data range, but approach a straight line with a slope of 1 at low stresses.

Microstructural examination of the gauge sections of crept specimens of many different grades of silicon nitride show that cavities form primarily at multi-grain junctions within pockets of the residual sintering aid.[17, 26, 28-30] The shapes of these cavities are the same as the silicate pockets in the undeformed material. Cavity growth involves the flow of the sintering aid away from the cavitation site, along triple junction channels to other silicate pockets in the microstructure. Simultaneously, a slight

* The use of commercial designations is for purposes of identification only and does not imply endorsement by the National Institute of Standards and Technology.

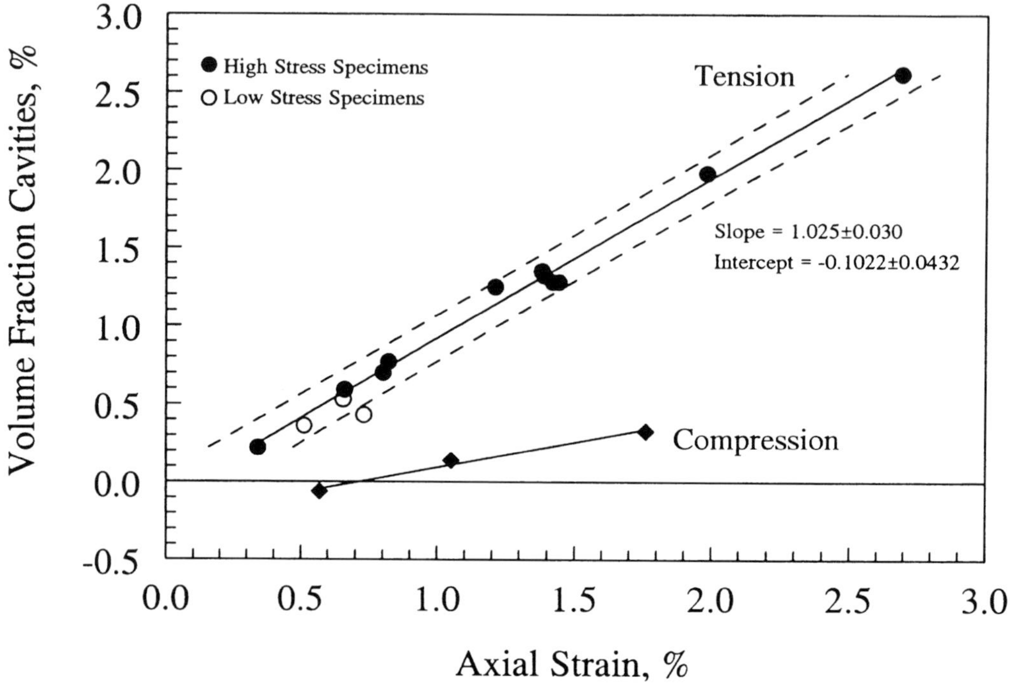

Fig. 2 Cavitation during tensile and compressive creep.[31] Three sets of data are represented: cavitation data obtained at high tensile stresses, >100 MPa,[16] cavitation data obtained at low tensile stresses, ≤ 30 MPa,[31] cavitation data obtained in compression.[31] The dashed curves represent 95% confidence limits for the location of the individual data points.

dilatation of the structure occurs to accommodate the sintering aid previously located in the pocket. A similar creep asymmetry[17, 20, 21] and nonlinear behaviour[17, 18, 25] have also been observed on other grades of silicon nitride.

The relationship between the volume fraction of cavities and the total axial strain is shown in Figure 2 for SN88.[16, 31] Cavities form in tension at all applied stresses used in Figure 1, even at stresses as low as 15 MPa. The volume fraction of cavities increases linearly with the axial strain for each material tested.[17] This linear relationship occurs in most silicon nitrides,[17, 18, 25, 30, 31] usually with a slope close to unity.[16] The volume fraction of cavities generated in compression was less than 18% of the volume fraction in tension,[31] Figure 2, suggesting that compressive stresses retard cavity formation.

4.0 MECHANISMS OF CREEP

In a recent analysis, Luecke and Wiederhorn[16] considered two possible mechanisms that account for the creep of silicon nitride. Both mechanisms were associated with

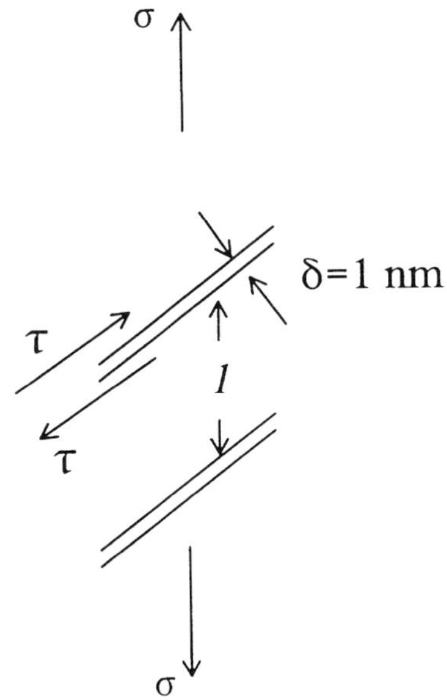

Fig. 3 Model for sliding-limited creep.[16] The model assumes that the creep rate is limited by the rate at which the grains can slide over one another as the material elongates. Sliding is assumed to occur on planes 45° to the applied stress. The planes on which skidding occurs are spaced a distance l apart.

cavity formation during the creep process. One assumes that creep was limited by grain boundary sliding of the silicon nitride in the vicinity of the cavity. The other assumes that creep was limited by the flow of intergranular phase away from the sight of cavitation. Both processes depended on the effective viscosity of the intergranular phase. In this section we discuss the two models and show that they both give excellent fits to the creep data. In the next section we present arguments to differentiate between the two, and show that the flow process is the more reasonable of the two mechanisms.

In the grain boundary sliding model the rate of deformation of glass at the grain boundaries is assumed to limit the rate of deformation.[16] As a cavity forms, the material around the cavity has to deform to make room for the cavity. The volume of material around the cavity must get a bit larger during the cavitation process. One can imagine the cube of silicon nitride to be crossed by many slip planes along which slip is occurring,[16] Figure 3. The size of the cube, L, is equal to the distance between cavities. The distance between slip planes is assumed to be l. If all of the grains slide then l will be equal to the grain size; if clusters of grains slide then l will be larger than the grain size. In either case, the main resistance to sliding will be the amorphous silicate phase at the grain boundaries, which has a thickness, δ, and a viscosity η. Assuming that on the average the sliding elements are oriented at 45° to the applied stress, the creep rate is determined by the following equation:[16]

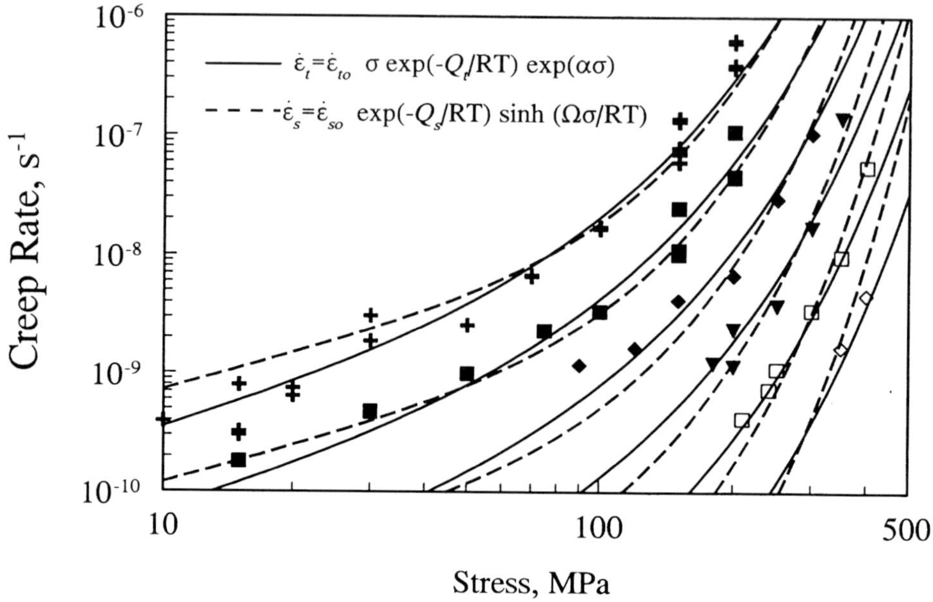

Fig. 4 Tensile creep data compared with the grain boundary sliding model, equation 2, and the flow model, equation 5, for creep. Both equations fit the data very well.

$$\dot{\varepsilon} = \frac{1}{4}\frac{\delta}{l}\frac{\sigma}{\eta} \tag{1}$$

Equation 1 predicts that the creep rate will increase linearly with the applied stress, which is inconsistent with experimental data for silicon nitride. Gasdaska[18] suggested that the viscosity of the amorphous phase was non-linear with stress as it is for very high strain rates and high stresses.[33, 34] Gasdaska used the Eyring model of viscosity[35] to model the viscous flow of the amorphous phase between the grains. Following this procedure, Luecke and Wiederhorn[16] derived the following equation for the creep rate:

$$\dot{\varepsilon}_s = \frac{A}{2}\frac{\delta}{l}\, exp\left(-\frac{Q_s}{RT}\right) sinh\left(-\frac{\sigma\Omega}{RT}\right), \tag{2}$$

where A is a constant, Q_s is the apparent activation energy for the shear process and Ω is related to the shear volume for the flow process.[16] Equation 2 has the correct form to fit the creep data, and as can be seen from Figure 4, fits the data very well.

The flow mechanism for creep of silicon nitride assumes that flow of silicate phase from the cavitation site to other multi-grain junctions in the solid controls the creep rate. The distance over which the silicate phase moves during the formation of a cavity is equal to about one-half the distance between active cavities,* $L/2$ in

* Active cavities are the cavities that are in the process of forming. Once the pocket in which the cavity forms is completely drained of silicate phase, cavity growth stops.

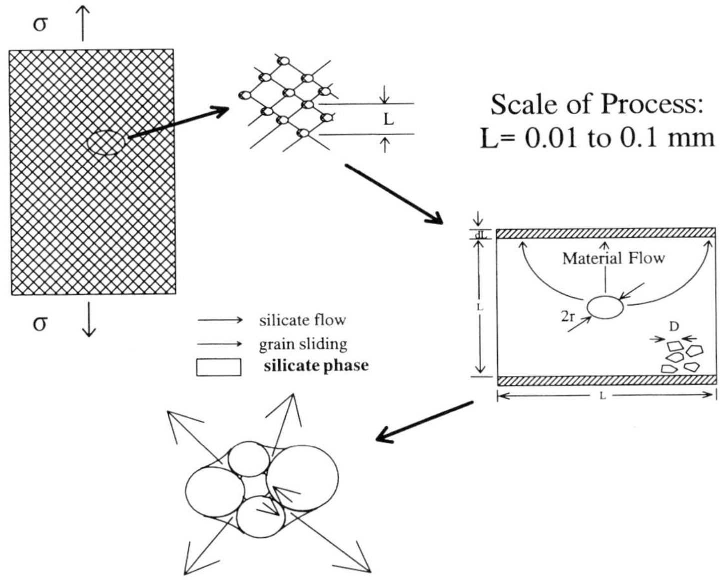

Fig. 5 Model for flow limited creep.[16] The model assumes that the flow of silicate phase away from the cavitation sight is the rate limiting step for the deformation. As the silicate phase leaves the cavitated pocket, the surrounding structure of silicon nitride grains must dilate to accommodate the silicate phase. The net effect of the dilation is a slight increase in length of the material surrounding the cavity.[16]

Figure 5. Because the silicon nitride is fully dense before the silicate pockets cavitate, a slight dilation of the microstructure is required to make room for the silicate phase that was in the pockets. Based on experimental observation, Luecke and Wiederhorn[16] showed that most of the dilatation occurred in the axial direction, i.e. the creep direction, and in fact was the main component of creep. The importance of dilatation to the axial strain was emphasized recently by Lofaj et al.[30] in a discussion of the role of cavition on the creep of silicon nitride.

The analysis used by Luecke and Wiederhorn[16] required two steps to obtain the final equation. The first step dealt with the problem of silicate flow from the cavity to the surrounding material; the second dealt with the number of cavities nucleated as a function of applied stress. Clearly both are important, since the creep rate will depend on the number of cavities being emptied of silicate phase and the rate at which they are emptied.

The rate of emptying was treated by the Carman-Kozeny equation[36, 37] which is a generalization of the Hagen-Poiseuille equation. The Carman-Kozeny equation is commonly used to describe the flow of fluids through packed beds.[37] From the Carman-Kozeny equation, an equation that describes creep in terms of microstructural variables can be derived. The equation relates creep rate, $\dot{\epsilon}_t$, to the applied stress, σ, the effective viscosity of the silicate phase, η, the grain size, D, the distance between cavities, L, and the volume fraction of silicate phase, V_f:

$$\dot{\varepsilon}_t = \frac{\sigma}{3\,k_o \eta}\left(\frac{D}{L}\right)^2 \frac{V_f^3}{(1 - V_f)^2}, \tag{3}$$

where k_o is a geometric term with a value of about 150. With a well-defined microstructure it is possible to use equation 3 to make order of magnitude estimates of the creep rate.

Equation 3 also predicts a linear dependence of the creep rate on applied stress, σ, in contradiction to experimental observations. This contradiction is solved by determining the number of cavities actively emptying during creep. A negative pressure within the silicate phase located at multi-grain junctions is assumed to result from the deformation process.* The average pressure within the multi-grain junctions is assumed to be equal to the mean hydraulic pressure, which is $-(1/3)\sigma$. A critical pressure is also assumed to exist for cavity formation. At any time, a small fraction of multi-grain junctions will have pressures that are super-critical. This fraction of junctions define the density of active cavities, which is the other parameter needed to determine the creep rate. Following this analysis, the following equation was obtained by Luecke and Wiederhorn[16] from equation 3:

$$\dot{\varepsilon}_t = \frac{\dot{\varepsilon}_0}{\eta} \frac{V_f^3}{(1 - V_f)^2}\, \sigma \exp(\alpha\sigma) \tag{4}$$

The creep rate in equation 4 increases exponentially with the applied stress at high values of the stress and linearly with the applied stress at low values of the stress. Since V_f is normally small, < 0.03, the creep rate increases as V_f^3. Since viscosity depends on temperature, $\eta \approx \eta_o \exp(-Q_t/RT)$, equation 3 can be rewritten as follows:

$$\dot{\varepsilon}_t = \frac{\dot{\varepsilon}_0}{\eta_o} \frac{V_f^3}{(1 - V_f)^2}\, \sigma \exp(\alpha\sigma) \exp\left(\frac{-Q_t}{RT}\right) \tag{5}$$

In Figure 4, equation 5 has been fitted to the same set of data as equation 2. As can be seen, equation 5 fits the data equally well, in fact, statistically, there is no difference between the two fits.[16] Therefore, the fits of the equation to the creep data cannot be used to distinguish between the two creep models. Other means of deciding between the two models must be made. In the next section, the absolute creep rate, determined from equations 1 and 3, is used to distinguish between the two theories. The calculation clearly shows that the cavitation and flow model of creep gives a better estimate of the creep rate than the grain boundary sliding process, suggesting that it also is a better description on the creep process.

---

* Cavitation can only occur if the pressure is negative, which is equivalent to a tensile stress. The pressure will not be the same within each multi-grain junction. As the material is deformed the grains will move away from some multi-grain junctions, decreasing the pressure, in others the grains move towards the multi-grain junction increasing the pressure.

5.0 ESTIMATE OF CREEP RATE FROM PROPERTIES OF INTERGRANULAR PHASE

5.1 Grain Boundary Sliding

The applicability of this model to the creep of Si_3N_4 can be tested by substituting values of δ, σ, η and l into equation 1. Some of these values are well known for a given material and experiment; others are less well known. Extensive microscopy on silicon nitride, for example, indicate the thickness of these boundaries to be approximately 1 nm.[38-40] Recent studies on the viscosity of nitrogen containing glasses similar in composition to those found at grain boundaries in silicon nitride can be used to obtain a reasonable idea of the value of η. Data by Lemercier et al.[41] and Hampshire et al.[42] are particularly pertinent since their data were for glasses containing alumina, yttria and nitrogen, and so are similar in composition to those found in some of the grades of silicon nitride tested in our own laboratory. The separation between slip planes is not known, but should be larger than the grain size, and less than the cube size L. A value of l somewhere between 10 and 100 μm is probably reasonable. To test equation 1, we use values for the viscosity determined on nitrogen glasses containing yttria and alumina:[42] $\eta = 1 \times 10^6$ Pa · s at 1250°C. This value of viscosity is used to estimate the creep rate of a commercial grade of silicon nitride (AY6 made by GTE) containing 6% Y_2O_3 and 2% Al_2O_3 by mass (atomic Y/Al ratio of about 1.76). We feel justified in using this data because the glass from it was close in composition to the ternary eutectic determined from Si_3N_4-$Y_2Si_2O_7$-$Y_3Al_5O_{12}$.[43] Substituting estimated values for l, η and δ into equation 1, the creep rate ranges from 2.5×10^{-3} s^{-1} to 2.5×10^{-4} s^{-1}, as l varies from 10 to 100 μm. These values are from 4 to 5 orders of magnitude greater than the creep rate measured on AY6 at 100 MPa and 1250°C, i.e. 8×10^{-8} s^{-1}.[44]

From the calculation just given, the intergranular frictional model of creep does not seem to provide sufficient deformation resistance to explain the creep data. However, a number of factors have been ignored, which in principal, could bring the measured and calculated values closer to the measured values. First, the slip planes were treated as if they were smooth. In fact, if the slip plane followed some rough intergranular path, resistance to creep would be larger than that quoted here. The effect of roughness on slip is important in the science of tribology. Friction can be increased by rough surfaces "because of the need to lift one surface over asperities on the other".[45] If asperities have an inclination of θ to the direction of sliding, then roughness adds a term equal to $\tan\theta$ to the coefficient of friction. For an inclination of 45°, $\tan\theta$ is equal to 1, which is only a factor of two larger than the expected coefficient of friction of a normal unlubricated ceramic surface. If the effect of roughness on creep is the same as that for friction, roughness will have little effect on the creep rate.

A second factor that could alter our conclusion is based on the possibility of errors in estimating the composition of the glass. Since glass viscosity depends strongly on composition, selecting the correct composition is important for an accurate estimate of the viscosity. Recent studies of SiAlON glasses as a function of nitrogen and aluminium content give some indication of how viscosity changes as a function of composition. A study by Hampshire et al.[42] showed that the viscosity of glass can increase by two orders of magnitude as nitrogen is added to the glass. Glass containing 17 equivalent % nitrogen, for example, has a viscosity that is about two orders of magnitude greater than nitrogen free silica. The alumina to yttria ratio also has a significant effect on the viscosity of nitrogen glasses. Lemercier et al.[41] showed that as the Y/Al atomic ratio varied from 10 to 1, the viscosity decreased by about 1.5 orders of magnitude. The viscosity used in our paper was taken from a glass that was close in composition to ternary eutectic reported in reference 46. The Y/Al ratio of the glass was also identical to the Y/Al ratio, 1.75, in the silicate phase used to make AY6. Based on these two considerations, we feel that the actual glass in the AY6 was close in composition to the glasses on which viscosity measurements were made. Therefore we feel confident that the estimated viscosity was close to that of the actual glass found at grain boundaries of the silicon nitride.

Finally, creep rate may also be influenced by the thickness of the glass layer between the silicon nitride grains. The glass located at grain boundaries of silicon nitride are in thin layers, ≈1 nm thick, and their behaviour as a viscous fluid might not be the same as that of the bulk glass used for our calculation. Experiments on organic fluids between sheets of mica using the surface force apparatus[47-49] show that when the liquid is of the order of two molecular layers thick, confined layers of organic liquids or polymers can reach values as much as approximately 10^7 greater than the bulk velocity. If silicate glasses behave in the same way as the inorganic fluids, then substantially higher viscosities might be expected for silicate glasses at grain boundaries.

Recent experiments by Ferber et al.[50] on two commercial grades of silicon nitride, NT154 and NT164 (St. Gobain/Norton Industrial Ceramics Corp, Worcester MA), suggest that the enhancement of viscosity as a consequence of grain boundary thinning does not occur in silicon nitride. NT154 and NT164 were both made with approximately 4% Y_2O_3 by mass as the silicate phase. Both were annealed to devitrify the remaining silicate phase after sintering. Yttrium disilicate was the secondary phase at multigrain junctions in both materials. NT164 had measurable amounts of silicon oxynitride indicating a higher oxygen content in the residual silicate phase. From a microstructural view, the grain-boundary thickness of the amorphous phase was substantially less in the NT164 than in the NT154: < 0.3 nm for the NT164[51] versus 1 nm for the NT154. If the results on organic fluids are applicable to silicon nitride, then this factor of approximately 3 in grain boundary thickness should have decreased the creep rate of NT164 by a factor of about 1000. Instead, the creep rate of the NT164 decreased by a factor of 2, which is consistent with eqn. 2. This modest effect of grain boundary thickness on creep rate suggests that the viscosity of the grain boundary phase does not depend on the grain boundary thickness.

Use of the bulk viscosity to model the creep rate is also supported by recent studies of the internal friction of silicon nitride by Pezzotti et al.[52] These authors were able to model the relaxation peaks they detected by an ensemble of sliding grain boundaries. The viscosity of the glass could be determined from the relaxation peak. Pezzotti et al. showed that viscosity determined from relaxation peaks was the same as that of the bulk glass of the same composition. Thus, two experiments indicate that the kind of increase in viscosity needed to yield a high creep resistance by sliding friction, does not occur as a function of grain boundary thickness in silicon nitride.

5.2 Flow Model

The applicability of the flow model can be tested by substituting values for σ, η, D, L and V_f into equation 3 to calculate the creep rate, $\dot{\epsilon}$. Again we used data for AY6 for comparison. At 1250°C, the viscosity of the residual glass is approximately 1×10^6 Pa • s.[42] As the multigrain junctions of AY6 will be partially devitrified[44] by the 20 h anneal at the creep temperature prior to testing, the fact that the cross-sectional area is only partially amorphous should be taken into account. Microscopic observations[44] show that the amorphous boundary between the crystalline silicate phase, and the silicon nitride grains is approximately 3 nm thick, and that the intergranular channels have a cross-sectional diameter of 200 to 400 nm. If the channels can be treated as devitrified tubes with a diameter of 300 nm each coated by an amorphous layer of 3 nm, then about 4% of the cross-sectional area of the channels consists of amorphous phase. Hence the flow rate (and the creep rate) from the channel will only be 4% of that given by estimating the entire channel as amorphous.

Based on an examination of the microstructure by transmission electron microscopy,[44] the grain size, D, in silicon nitride is about 1 μm,* and the volume fraction of silicate phase is about 10%. Substituting these values into equation 7, and taking into account the fact that only 4% of each channel is amorphous, a creep rate of 3.3×10^{-7} s^{-1} is obtained for $L = 0.01$ mm and 3.3×10^{-9} s^{-1} for $L = 0.1$ mm. These estimates of the creep rate bracket the measured value of 8×10^{-8} s^{-1}. Thus, the estimate of creep rate obtained from the viscous flow model is closer to the experimentally measured creep rates than the estimate obtained from the grain boundary sliding model.

6.0 CONCLUSION

In this paper we discussed the role of the intergranular phase in the creep behaviour of silicon nitride. Two models for creep were discussed, one based of grain boundary sliding as the rate limiting step for deformation, the other based on the flow of the bonding phase away from cavities. Both models fit the creep data equally well, so that the creep data can not be used to distinguish between the two models of deformation. By substituting experimental values for the materials parameters in each of the models, we were able to show that the flow model yielded predictions of

*AY6 actually has a bimodal microstructure, elongated grains 1 to 3 μm wide and up to 10 μm long, embedded in a matrix of submicron size grains.[41] The 1 μm selected for the grain size fits within the middle of the bimodel distribution. An error of a factor of three in the grain size will have little effect on our conclusions.

the creep rate that agreed with experimental values. By contrast, the grain boundary sliding model yielded predictions of the creep rate that were four to five orders of magnitude greater than that obtained experimentally. We conclude from this calculation that silicon nitride deforms by the nucleation of cavities within the bonding phase and the flow of that phase away from the site of cavitation.

7.0 REFERENCES

1. F. L. RILEY: "Silicon Nitride," *Concise Encyclopedia of Advanced Ceramic Materials*, R. J. Brook, ed., Pergamon Press, Oxford, 1991, pp.434-437.
2. D. R. CLARKE: "On the Equilibrium Thickness of Intergranular Glass Phases in Ceramic Materials", *Journal of American Ceramic Society*, **70**(1), 1987, pp.15-22.
3. H. J. KLEEBE: "Structure and Chemistry of Interfaces in Si_3N_4 Ceramics Studied by Transmission Electron Microscopy", *Journal of Ceramics Society Jpn.*, **105**(6), 1997, pp.453-475.
4. D. A. BONNELL: "Structure of Grain Boundary Phases in Silicon Nitride", *Mater. Sci. Forum*, **47**, 1989, pp.132-142.
5. L. K. L. FALK and G. DUNLOP: "Crystallization of the Glassy Phase in an Si_3N_4 Material by Post Sintering Heat Treatment", *Journal of Materials Science*, **22**, 1987, pp.4369-4376.
6. G. WOETTING and E. GUGEL: "Influence of the Grain-Boundary Phase Amount on Properties of Dense Si_3N_4", *CFI-Ceramic Forum International*, **74**(5), 1997, pp.239-244.
7. A. J. PYZIK, D. F. CARROLL, C. J. HWANG and A. R. PRUNIER: "Self-Reinforced Silicon Nitride - A New Microengineered Ceramic", *Fourth International Symposium on Ceramic Materials and Components for Engines*, R. Carlsson, T. Johansson and L. Kahlman, eds., Elsevier Applied Science, London, 1992, pp.584-593.
8. A. J. PYZIK and D. R. BEAMAN: "Microstructure and Properties of Self-Reinforced Silicon Nitride", *Journal of American Ceramic Society*, **76**(1), 1993, pp.2737-2744.
9. K. R. SELKREGG, K. L. MORE, S. G. SESHADRI and C. H. MCMURTY: "Microstructural Characterisation of Silicon Nitride Ceramics Processed by Pressureless Sintering, Overpressure Sintering and Sinter HIP", *Ceramics Engineering Sci. Proc.*, **11**(7-8), 1990, pp.603-615.
10. G. WÖTTING, B. KANKA and G. ZIEGLER: "Microstructural Development, Microstructural Characterization and Relation to Mechanical Properties of Dense Silicon Nitride", *Non-Oxide Technical and Engineering Ceramics*, S. Hampshire, ed., Elsevier, London, 1986, pp.83-96.
11. P. F. BECHER, S. L. HWANG and C. H. HSUEH: "Using Microstructure to Attack the Brittle Nature of Silicon Nitride Ceramics", *MRS Bull.*, **20**(2), 1995, pp.23-27.
12. P. F. BECHER: "Microstructural Design of Toughened Ceramics", *Journal of American Ceramic Society*, **74**(2), 1991, pp.255-269.

13. G. PEZZOTTI and M. SAKAI: "Effect of Silicon Carbide "Nano Dispersion" on the Mechanical Properties of Silicon Nitride", *Journal of American Ceramic Society*, **77**(11), 1994, pp.3039-3041.
14. G. WOETTING, H. FEUER, L. FRASSEK, L. SCHOENFELDER and G. LEIMER: "Si_3N_4 Materials: Their Properties and Technical Applications", *CFI-Ceram. Forum Int.*, **75**(7), 1998, pp.25-30.
15. Data Sheet SP-00244 on GS-44CL, GS-44SG and AS-800 from Allied Signal Ceramic Components, Torrance, CA.
16. W. E. LUECKE and S. M. WIEDERHORN: "A New Model for Tensile Creep of Silicon Nitride", *Journal of American Ceramic Society*, **82**(10), 1999, pp.2769-2778.
17. W. E. LUECKE, S. M. WIEDERHORN, B. J. HOCKEY, R. KRAUSE, JR. and G. G. LONG: "Cavitation Contributes Substantially to Tensile Creep in Silicon Nitride", *Journal of American Ceramic Society*, **78**(8), 1995, pp.2085-2096.
18. C. J. GASDASKA: "Tensile Creep in an in Situ Reinforced Silicon Nitride", *Journal of American Ceramic Society*, **77**(9), 1994, pp.2408-2418.
19. R. KOSSOWSKY, D. G. MILLER and E. S. DIAZ: "Tensile and Creep Strengths of Hot Pressed Si_3N_4", *Journal of Materials Science*, **10**, 1975, pp.983-997.
20. B. J. HOCKEY, S. M. WIEDERHORN, W. LIU, J. G. BALDONI and S. T. BULJAN: "Tensile Creep of Whisker Reinforced Silicon Nitride", *Journal of Materials Science*, **26**, 1991, pp.3931-3930.
21. M. K. FERBER, M. G. JENKINS and V. J. TENNERY: "Comparison of Tension, Compression, and Flexure Creep for Alumina and Silicon Nitride Ceramics", *Ceram. Eng. Sci. Proc.*, **11**(7), 1990, pp.1028-1045.
22. K. C. LIU, C. O. STEVENS, C. R. BRINKMAN and N. E. HOLSHAUSER: "A Technique to Achieve Uniform Stress Distribution in Compressive Creep Testing of Advanced Ceramics at High Temperatures", *Journal of Engineirng Gas Turbines Power*, **119**, 1988, pp.500-505.
23. A. A. WERESZCZAK, T. P. KIRKLAND, H. T. LIN and M. K. FERBER: "Tensile Creep Performance of a Developmental, In-situ Reinforced Silicon Nitride", *Ceram. Eng. Sci. Proc.*, **18**(4), 1997, pp.45-55.
24. F. LOFAJ, H. USAMI, A. OKADA and H. KAWAMOTO: "Long-Term Creep Damage Development in a Self-Reinforced Silicon Nitride", *Engineering Ceramics '96: Higher Reliability Through Processing*, G. N. Babini et al., eds., Kluwer Academic Publishers, 1997, pp.337-352.
25. H. MAUPAS, B. J. HOCKEY and S. M. WIEDERHORN: "Tensile Creep of a Commercial Grade of Silicon Nitride: Limitation of Norton's Relationship?" To be published.
26. R. F. KRAUSE, JR., W. E. LUECKE, J. D. FRENCH, B. J. HOCKEY and S. M. WIEDERHORN: "Tensile Creep and Rupture of Silicon Nitride", *Journal of American Ceramic Society*, **82**(5), pp.1233-1241.
27. M. GÜRTLER and G. GRATHWOHL: "Tensile Creep Testing of Sintered Silicon Nitride", *Proceedings of the Fourth International Conference on Creep and Fracture of Engineering Materials and Structures*, Institute of Metals, 1 Carlton House Terrace, London, SW1Y 5DB, 1990, pp.399-408.

28. M. N. Menon, H. T. Fang, D. C. Wu, M. G. Jenkins, M. K. Ferber, K. L. More, C. R. Hubbard and T. A. Nolan: "Creep and Stress Rupture Behaviour of an Advanced Silicon Nitride: Part I - Experimental Observation", *Journal of American Ceramic Society*, **77**(5), 1994, pp.1217-1227.
29. Tatstuki Ohji and Yukihiko Yamauchi: "Tensile Creep and Creep Rupture Behavior of Monolithic and SiC Whisker-Reinforced Silicon Nitride Ceramics", *Journal of American Ceramic Society*, **76**, 1993, pp.3105-3112.
30. F. Lofaj, A. Okada and H. Kawamoto: "Cavitation Strain Contribution to Tensile Creep in Vitreous Bonded Ceramics", *Journal of American Ceramic Society*, **80**(6), 1997, pp.1619-1623.
31. K. J. Yoon, S. M. Wiederhorn and W. E. Luecke: A Comparison of Tensile and Compressive Creep Behavior in Silicon Nitride", *Journal of American Ceramic Society*, **83**(8), 2000, pp.2017-2022.
32. W. E. Luecke: "Review: Creep of Si_3N_4", in preparation.
33. J. H. Li and D. R. Uhlmann: "The Flow of Glass at High Stress Levels, I Non-Newtonian Behavior of Homogeneous 0.08 Rb_2O-0.92 SiO_2 Glasse", *Journal of Non-Crystalline Solids*, **3**, 1970, pp.127-147 .
34. S. Webb: "Silicate Melts: Relaxation, Rheology and the Glass Transition", *Reviews of Geophysics*, **35**(2), 1997, pp.191-218.
35. H. Eyring: "Viscosity, Plasticity and Diffusion as Examples of Absolute Reaction Rates", *Journal Chem. Phys.*, **4**, 1936, pp.286-291.
36. P. C. Carman: "Fluid Flow through Granular Beds", *Trans. Inst. Chem. Eng.*, **15**, 1937, pp.150-161.
37. R. B. Bird, W. E. Stewart and E. N. Lightfoot: *Transport Phenomena*, John Wiley & Sons, Inc. New York, 1966.
38. D. R. Clarke and G. Thomas: "Boundary Phases in a Hot-Pressed MgO Fluxed Silicon Nitride", *Journal of American Ceramic Society*, **60**(11-12), 1977, pp.491-495.
39. L. K. V. Lou, T. E. Mitchell, A. H. Heuer: "Impurity Phases in Hot-Pressed Si_3N_4", *Journal of American Ceramic Society*, **61**(9-10), 1978, pp.392-396.
40. H. J. Kleebe, M. K. Cinibulk, R. M. Cannon and M. Rühle: "Statistical Analysis of the Intergranular Film Thickness in Silicon Nitride Ceramics" *Journal of American Ceramic Society* ,**76**(8), 1993, pp.1969-1977.
41. H. Lemercier, T. Rouxel, D. Fargeot, J. L. Besson and B. Piriou: "Yttrium Sialon Glasses: Structure and Mechanical Properties - Elasticity and Viscosity", *Journal of Non-Crystalline Solids*, in press.
42. S. Hampshire, R. Flynn, V. Morrissey, M. Pomeroy, T. Rouxel, J. L. Besson and P. Goursat: "Effects of Temperature on Viscosities and Elastic Modulus of Oxynitride Glasses", *Silicon Nitride 93 Stuttgart*, International Conference on Silicon Nitride-Based Ceramics, Stuttgart, Germany, Key Engineering Materials, Trans. Tech. Publications, Switzerland, 1994, pp.351-356.
43. Z. K. Huang and T. Y. Tien: "Solid-Liquid Reaction in the System Si_3N_4-$Y_2Si_2O_7$-$Y_3Al_5O_{12}$ under 1 MPa Nitrogen", *Journal of American Ceramic Society*, **77**(10), 1994, pp.2763-2766.
44. B. J. Hockey, S. M. Wiederhorn, W. Liu, J. G. Baldoni and S. T. Buljan: "Tensile

Creep of Whisker-Reinforced Silicon Nitride", *Journal of Materials Science*, **26**, 1991, pp.3931-3930.

45. ERNEST RABINOWICZ: *Friction and Wear of Materials*, John Wiley & Sons, Inc. New York, 1965, p.61.
46. L. J. GAUKLER, H. HOHNKE and T. Y. TIEN: "The System Si_3N_4-SiO_2-Y_2O_3", *Journal of American Ceramic Society*, **63**(1-2), 1980, pp.35-37.
47. M. L. GEE, P. M. MCGUIGGAN and J. N. ISRAELACHVILI: "Liquid to Solid Like Transitions of Molecularly Thin Films under Shear", *J. Chem. Phys.*, **93**(3), 1990, pp.1895-1906.
48. J. N. ISRAELACHVILI, P. M. MCGUIGGAN and A. M. HOMOLA: "Dynamic Properties of Molecularly Thin Liquid Films" *Science*, **240**, 1988, pp.189-191.
49. J. V. ALSTEN and S. GRANICK: "Shear Rheology in a Confined Geometry: Polysiloxane Melts", *Macromolecules*, **23**, 1990, pp. 4856-4862.
50. M. K. FERBER, M. G. JENKINS and T. A. NOLAN: "Comparison of the Creep and Creep Rupture Performance of Two HIPed Silicon Nitride Ceramics", *Journal of American Ceramic Society*, **77**(3), 1994, pp.657-665.
51. M. K. FERBER, E. LARA-CURZIO and A. A. WERESZCZAK: "Time -Dependent Failure of Structural Ceramics and Ceramic -Matrix Composites at Elevated Temperatures", Paper No. SXVI-048-97, *99th Annual Meeting of the American Ceramic Society*, 1997 .
52. G. PEZZOTTI, K. OTA and H. J. KLEEBE: "Grain-Boundary Relaxation in High-Purity Silicon Nitride", *Journal of American Ceramic Society*, **79**(9), 1996, pp.2237-2246.

Growth of Solid Grains Dispersed in a Liquid Matrix: Numerical Approach

MYUNG-KOO KANG, NONG MOON HWANG* and DOH-YEON KIM
School of Materials Science and Engineering and Centre for Microstructure Science of Materials
College of Engineering, Seoul National University
Seoul 151-742, Korea

ABSTRACT

The coarsening process of solid grains dispersed in a liquid matrix has been determined numerically in relation to the grain shape: spherical and angular grains were assumed to grow by a diffusion process and a 2-D controlled process, respectively. The coarsening of spherical grains by a diffusion-controlled process resulted in a uniform microstructure as analytically predicted by LSW theory. In contrast, the coarsening of angular grains by 2-D nucleation resulted in abnormal grain growth (AGG). The effects of initial average particle size, particle size distribution, surface energy and edge energy on grain coarsening by 2-D nucleation were also determined.

1.0 INTRODUCTION

The microstructure of liquid-phase sintered materials can be characterized by solid grains dispersed in a liquid matrix. If there is sufficient amount of liquid, deformation of the grains due to impingement between them will not occur. In such a case, the grain shape will be mainly determined by the requirement to minimize the solid-liquid interfacial energy (Wulff's theorem),[1] and the equilibrium shape of solid grains is either angular (polyhedral) or spherical. For instance, W grains in Ni matrix are spherical,[2] while Si_3N_4 grains in silicate liquid are angular.[3] Although the spherical grains are common in metallic alloys and the angular grains are frequent in ceramics, there are a few exceptions. MgO grains dispersed in $CaMgSiO_4$ are spherical.[4] On the other hand, angular ZnO grains are observed to become spherical at high temperature.[5]

Regardless of its shape, the average size of grains dispersed in a liquid matrix increases by dissolution and reprecipitation (Ostwald ripening). Grains smaller than the average go into solution and larger ones grow during heat-treatment. Greenwood,[6] has analysed the growth or dissolution rate, as a function of grain size

* Permanently with Korea Research Institute of Standards and Science, Daedok 305-600, Korea

during Ostwald ripening and predicted that the dissolution rate, can be very large, as grain size decreases, but the growth rate of large grains is limited. Therefore, a rather uniform grain size distribution was suggested[6, 7] to maintain once the quasi-equilibrium state is reached, which is in agreement with the experimental results[8]. In some cases, however, abnormal or exaggerated grain growth (AGG or EGG) is observed to occur during heat-treatment: only a limited number of grains grow to an exceptional size at the expense of small grains.

The control of AGG is quite important since many properties of liquid-phase sintered materials are critically dependent on it. On the other hand, we have observed that AGG occurs only for the angular grains with apparently straight solid-liquid interfaces. From this observation, AGG has been suggested to be the consequence of a growth process controlled by 2-dimensional nucleation.[9, 10] In fact, Wynblatt et al.[11, 12] have shown that the "nucleation inhibited" growth of faceted particles can account satisfactorily for the coarsening of Pt particles on flat alumina substrates. Although they did not explicitly mention AGG during liquid-phase sintering, their works are believed to lay a firm base for the 2-D nucleation controlled coarsening process.

In this investigation, the results of numerical calculation of the grain coarsening process are reported. For spherical and angular grains, grain growth controlled by diffusion and 2-D nucleation processes, respectively, are assumed. A personal computer is used for calculation and the results are illustrated schematically. Effects of various processing parameters on the grain coarsening are treated and AGG is examined in detail.

2.0 MODELS AND CALCULATION

Depending on the atomic configuration, the structure of the solid-liquid interface is classified into a rough interface or a smooth interface.[13] The interface energy for the system with a rough interface is isotropic while that of the smooth interface is anisotropic. As a result, the equilibrium shape of the solid grain with the rough interface is spherical while that with the smooth interface is angular. Since the atomic attachment to the rough interface has negligible energy barrier, the material transfer to the rough interface is diffusion controlled. On the other hand, atomic attachment to the singular interface needs ledge-generating sources such as screw dislocations or 2-D nuclei. Therefore, grain growth behaviour depends naturally on the solid-liquid interface structure and in turn on the grain shape.

The driving force for grain coarsening is mainly provided by the difference in capillary pressure and can be written as

$$\Delta G_d = 2\gamma_{sl} V_m \left(\frac{1}{G^*} - \frac{1}{G} \right) \tag{1}$$

where V_m and γ_{sl} are the molar volume and the solid-liquid interfacial energy, respectively. G^* represents the size of grains which are neither growing nor dissolving. The model for coarsening of spherical grains with the rough solid-liquid interface assumes that all interface sites are available for atomic attachment. The net growth for a grain of size G is then the difference between the arrival and departure rates of atoms at the interface and proportional to the driving force for grain growth, ΔG_d. Such behaviour can be referred to as a linear or continuous growth. The growth rate for a grain of size G has been deduced to be[6]

$$\frac{dG}{dt} = \frac{D_f C_0 \gamma_{sl} V_m^2}{kTG}\left(\frac{1}{G^*} - \frac{1}{G}\right) \tag{2}$$

where D_f and C_o represent diffusivity and solubility, respectively. Although all grains larger than G^* will grow, their growth rate is known to be maximum when $G = 2G^*$.[6] The distribution of normalized grain size, G/G^*, for instance, is thus uniform during coarsening.

In contrast, the growth of angular grains with an atomically smooth interface will proceed by a process similar to the 2-D nucleation. Assuming that the 2-D nucleus on the terrace surface is a circular disk of atomic height h and radius r, the free energy change associated with the nucleus formation of such a disk of perimeter $2\pi r$ is given by

$$\Delta G_t = -\pi r^2 h \Delta G_d + 2\pi r \varepsilon \tag{3}$$

where ε is the edge energy. From this equation, the free energy of formation of a critical nucleus is

$$\Delta G_c = \frac{\pi \varepsilon^2}{h \Delta G_d} \tag{4}$$

Furthermore, the nucleation rate is known to have essentially the following form;[9]

$$I = A \exp\left(-\frac{\Delta G_c}{kT}\right) \tag{5}$$

where A is a constant determined by the atomic jumping frequency and kT has the usual meaning. In practice, it is found that the constant A does not differ greatly from system to system and is commonly of the order of $e^{50} \sim e^{65}$.[14, 15] The factor that exerts the largest influence on I is ΔG_c which is again mainly determined by ΔG_d and ε from Eqn. 4.

The growth rate for a grain of size G can be obtained by simply multiplying atomic height h to I. Therefore,

$$\frac{dG}{dt} = hA\exp\left(-\frac{\Delta G_c}{kT}\right) = h\exp\left(lnA - \frac{\pi\varepsilon^2}{2kTh\gamma_{sl}\left(\frac{1}{G^*} - \frac{1}{G}\right)}\right) \tag{6}$$

The main feature of Eqn. (6) is that the growth rate is negligible when ΔG_c is larger than a certain critical value. This means by Eqn. (4) that the growth rate will

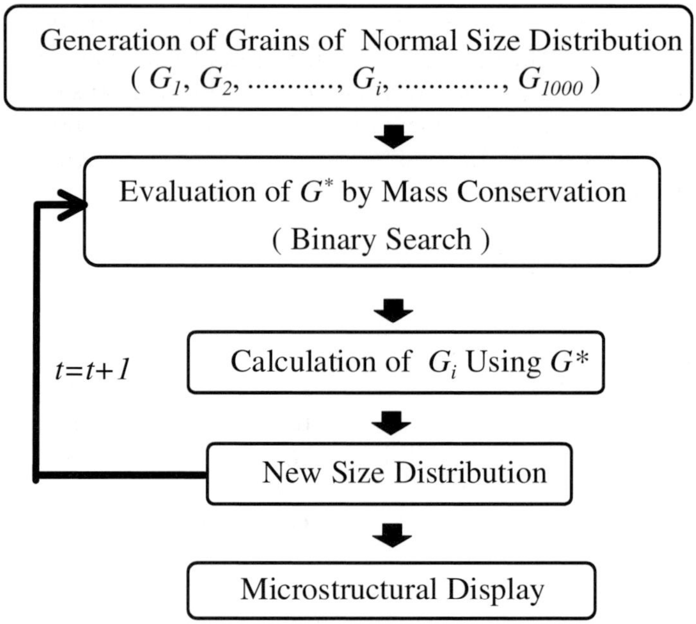

Fig. 1 Flow chart of the calculation procedure.

be very small (approximately zero) when ΔG_d is small and rise abruptly as ΔG_d exceeds a certain critical limit. Therefore, it can be noted that the grains of size larger than G^* is not the criterion for growth. The grains which are much larger than G^* will have a sufficient driving force for the onset of 2-D nucleation, and only they can grow.

In order to calculate the coarsening behaviour of solid grains, which depends on the solid-liquid interface structure, three basic equations were used including Eqns. (2) and (6). The third one is the mass conservation equation, i.e. the dissolving mass from small grains should be identical to the mass gained in the growing grains. In 2-dimension, this is given as;

$$-\sum_{i=1}^{k} G_i \left(\frac{dG_i}{dt} \right)_{dissolve} = \sum_{i=k+1}^{1000} G_i \left(\frac{dG_i}{dt} \right)_{grow} \tag{7}$$

For the computation, the computer program was coded in Turbo-C and run on personal computer. The overall computational procedure is shown in Figure 1. 1000 random numbers were first generated, and they were regarded as the sizes of each grain. They had a normal distribution with a given average value. From Eqn. (7), G_k was determined by using the binary search method, which is equivalent to G^* in Eqns. (2) and (6). For grain coarsening by a diffusion process, both $(dG_i/dt)_{dissolve}$ and $(dG_i/dt)_{grow}$ were calculated by Eqn. (2). For grain coarsening by 2-D nucleation, on the other hand, $(dG_i/dt)_{dissolve}$ was calculated by Eqn. (2), while $(dG_i/dt)_{grow}$ was calculated by Eqn. (6). Even for the faceted grains, it is reasonable to assume a zero activation energy barrier for dissolution because the detachment of atoms is expected to occur at the corners of the grain where the interfacial energy does not change.

Once G^* was determined, the changes in size of each grain were again calculated using the following equation:

$$G_i^{new} = G_i^{old} + \left(\frac{dG_i}{dt}\right)dt \tag{8}$$

This step was considered to increase heat-treatment time from $t = t$ to $t = t + 1$. The same process was repeated until a notable change in grain size occurred. The results obtained were illustrated schematically by using a computer graphic technique. Circles were assumed for grains growing by a diffusion process and the diameter was regarded as the grain size. On the other hand, squares were adopted to represent the grains growing by 2-D nucleation with the edge length regarded as the grain size.

The microstructures theoretically predicted in this study will deviate from that observed in experiment. This is mostly due to the assumption of a uniform and identical grain shape regardless of size. In a real system, the grain shape can deviate considerably from a uniform one because they impinge upon each other, particularly when the liquid volume fraction is not very high. Furthermore, the microstructural observation is usually made with the polished plane from a 3-dimensional specimen. As a result, a wider grain size distribution and a smaller average size are observed compared with the real 3-dimensional values. In contrast, the microstructures illustrated in this study show the grains which exhibit their maximum and real size.

3.0 RESULTS AND DISCUSSION

Figures 2a and b show the results of calculation for the growth of spherical and angular grains, respectively. As we did not take into account the effect of liquid volume fraction, these correspond to the cases where the grains are highly dispersed. For calculations, the initial average grain size, the standard deviation of size distribution and temperature are taken to be 0.3 μm, 0.2 μm and 1300 K, respectively. In addition, $D_f = 10^{-3}$ m^2 sec^{-1}, $C_o = 1$ at.%, $\gamma_{sl} = 0.1$ Jm^{-2} and $V_m = 10^{-5}$ m^{3}mol.$^{-1}$ are used, which are typically reported values in liquid-phase sintered materials, except D_f. The excessive D_f value is chosen on purpose to minimize the computation time. The value of ε has not yet been reported. According to Beijeren,[16, 17] however, the values of ε/h and γ_{sl} are the same at 0 K but both of them decrease with temperature. The decrease of ε/h has been predicted to be rather significant compared with that of γ_{sl}. In this respect, $\varepsilon/h = 0.2\ \gamma_{sl}$ at high temperatures of heat-treatment has been assumed.

Figure 2a shows the coarsening of spherical grains which exhibit a uniform coarsening behaviour. These results are in agreement with those already reported

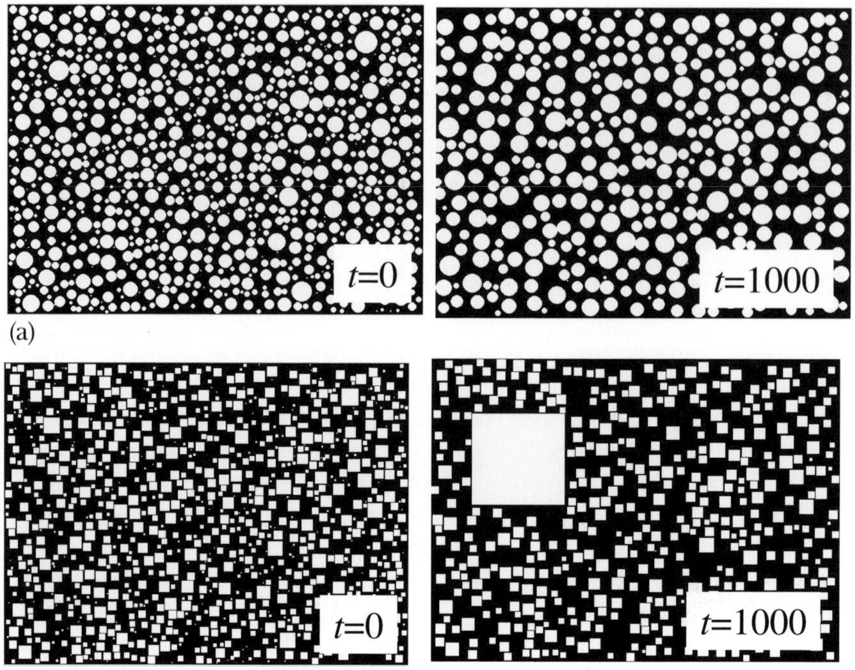

Fig. 2 Coarsening of grains (initial grain size: 0.3 μm) by (a) diffusion controlled and (b) 2-D nucleation controlled processes.

in the theoretical analysis of Ostwald ripening. The uniform coarsening has also been experimentally observed in many systems consisting of spherical grains.[4, 8] Coarsening by a diffusion process therefore results in a "normal" microstructure. The sizes of individual grains are relatively uniform and this uniformity is maintained during heat-treatment. On the other hand, Figure 2b shows the coarsening of angular grains controlled by a 2-D nucleation mechanism. In this case, the microstructure becomes "abnormal" as only a few grains grow very rapidly. Thus, a typical AGG occurs for the angular grains. In the experiments, indeed, the materials that are reported to exhibit AGG have the angular shape of the solid grains without exception. Examples are WC-Co,[9] $BaTiO_3$,[10, 18] Si_3N_4,[3] Ferrite,[19] SiC,[20] Y-Ba-Cu-O superconductors,[21] Al_2O_3[22, 23] and B_4C.[24] From these results, the model of 2-D nucleation is believed to be appropriate for the exaggerated growth of grains dispersed in a liquid.

In Figure 3, the effect of initial average grain size on grain coarsening by 2-D nucleation is shown. As previously shown in Figure 2b, AGG is a conspicuous coarsening characteristic when the initial grain size is 0.3 μm. The use of initial powders with a fine particle size was reported to result in the AGG in liquid phase-sintered alumina[22] and WC-Co alloys.[9] When grain size becomes 0.6 μm, however, the tendency of

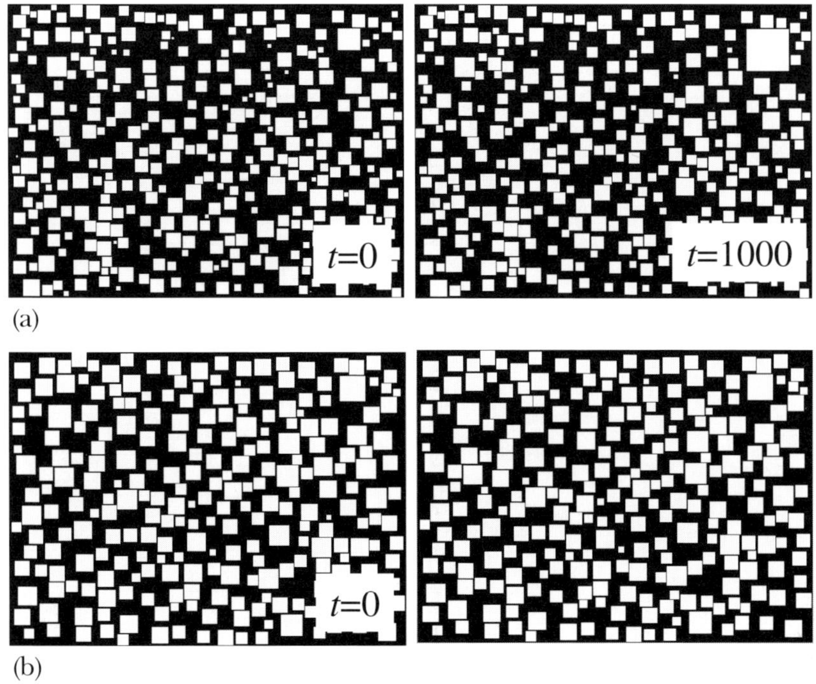

Fig. 3 Coarsening of grains by 2-D nucleation, initial average particle size (a) 0.6 μm and (b) 0.8 μm.

AGG decreases as illustrated in Figure 3a. As the initial average grain size reaches 0.8 μm, practically no grain coarsening occurs (Figure 3b). This can be regarded as a critical grain size above which the grain coarsening is negligible. Such a dependence of grain coarsening on initial grain size can be easily explained in terms of ΔG_c in Eqn. (6). The increase of initial average grain size results in a decrease of ΔG_d hence an increase of ΔG_c, which results in a negligible (dG/dt). In this respect, the parameters that determine ΔG_d and ΔG_c such as ε and γ_{sl}, are important. The higher the ε or the lower the γ_{sl}, the smaller the critical grain size expected.

During sintering of ceramics, the large distribution of initial powder particle size has also been attributed to be the main cause of AGG.[25, 26] Figure 4 shows the results of the calculation when the initial powders have a larger distribution compared with the case of Figure 3a. For a given average initial grain size of 0.6 μm, three different standard deviations are assumed: 0.2, 0.4 and 0.6 μm for Figures 3a, 4a and 4b, respectively. From the results, it can be noted that the larger grain size distribution results in a faster grain coarsening as well as a higher extent of AGG. On the other hand, Figure 5 shows the effect of γ_{sl} value on the grain coarsening. For these, the initial average grain size and the standard deviation are assumed to be 0.3 μm and 0.2 μm, respectively. Therefore, the results should be compared with

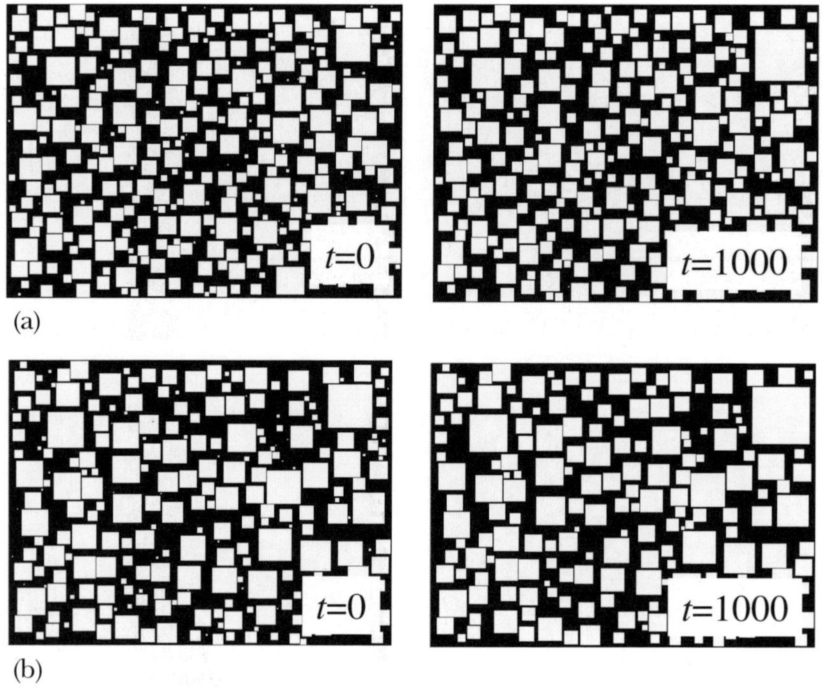

Fig. 4 Coarsening of grains (initial grain size: 0.6 μm) by 2-D nucleation, standard deviation in size distribution (a) 0.4 μm and (b) 0.6 μm.

Figure 2b, which describes the case when $\gamma_{sl} = 0.3$ Jm^{-2}. When $\gamma_{sl} = 0.1$ Jm^{-2}, as shown in Figure 5a, the kinetics of grain coarsening becomes very slow. Consequently, AGG is observed to occur only after the excessive heat-treatment time. With an increase of $\gamma_{sl} = 0.2$ Jm^{-2}, a notable grain coarsening as well as AGG are observed to occur. These results are also due to ΔG_d which increases with the width of the grain size distribution and γ_{sl} value.

The last parameter that influences grain coarsening by 2-D nucleation is the edge energy. As can be expected from Eqn. (4), ΔG_c is proportional to ε^2 so that a small variation in edge energy may result in a notable change in coarsening behaviour. Instead of $\varepsilon/h = 0.2\ \gamma_{sl}$ which was assumed for the calculation of Figures 2b, 6a and b show the cases where $\varepsilon/h = 0.19\ \gamma_{sl}$ and $\varepsilon/h = 0.3\ \gamma_{sl}$, respectively. The low edge energy means that the critical nucleus can be formed easily. As a result, AGG by 2-D nucleation is accelerated. As observed in Figure 6b, when the edge energy is higher than a certain limit, the grain coarsening becomes negligible. During sintering of a WC-Co alloy, for instance, the addition of TaC or VC has been reported to strongly restrict grain coarsening. Although such an inhibition has been related to impurity segregation[27] and change in γ_{sl},[28] the relevant mechanism has not yet been

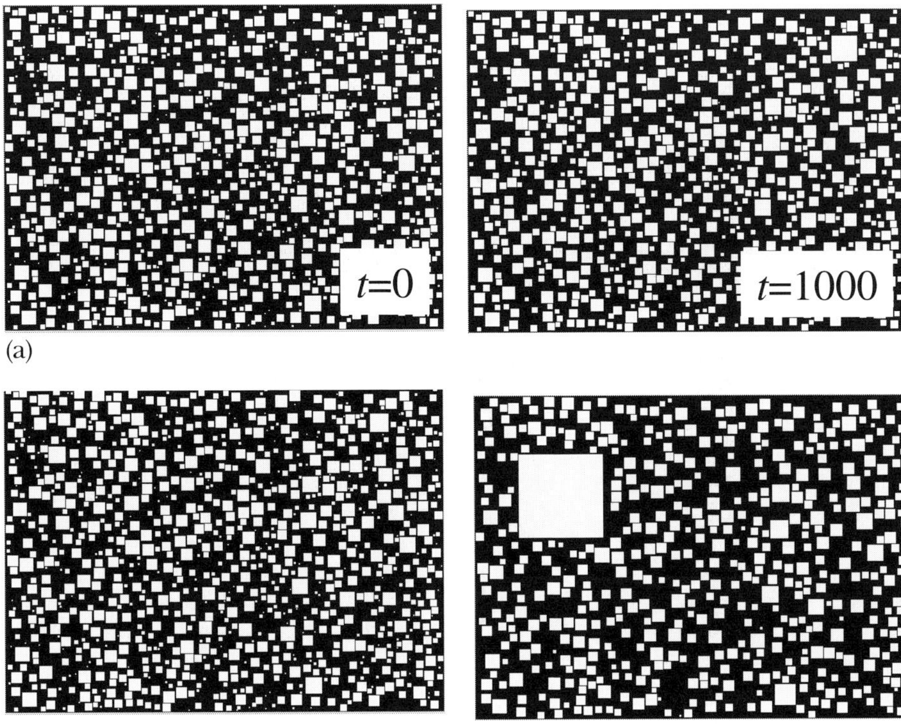

(b)

Fig. 5 Coarsening of grains (initial grain size: 0.3 μm) by 2-D nucleation, surface energy (a) 0.1 and (b) 0.2 Jm^{-2}.

determined thoroughly. Our analysis predicts that a possible change in edge energy by a small amount of additive during liquid-phase sintering may also cause an inhibition of grain coarsening.

4.0 CONCLUSIONS

The coarsening behaviour of solid grains dispersed in a liquid matrix was analysed and a theoretical model that depends on the grain shape was developed. When grains are spherical and the solid-liquid interface structure is atomically rough, grain coarsening is predicted to occur by the well-known diffusion controlled Ostwald ripening process. In this case, the grains grow uniformly so that a normal microstructure is maintained. When grains are angular and the solid-liquid interface structure is atomically smooth, on the other hand, 2-D nucleation and growth are suggested to control grain coarsening. Therefore, the large angular grains tend to grow abnormally because only the large grains have enough driving force for the

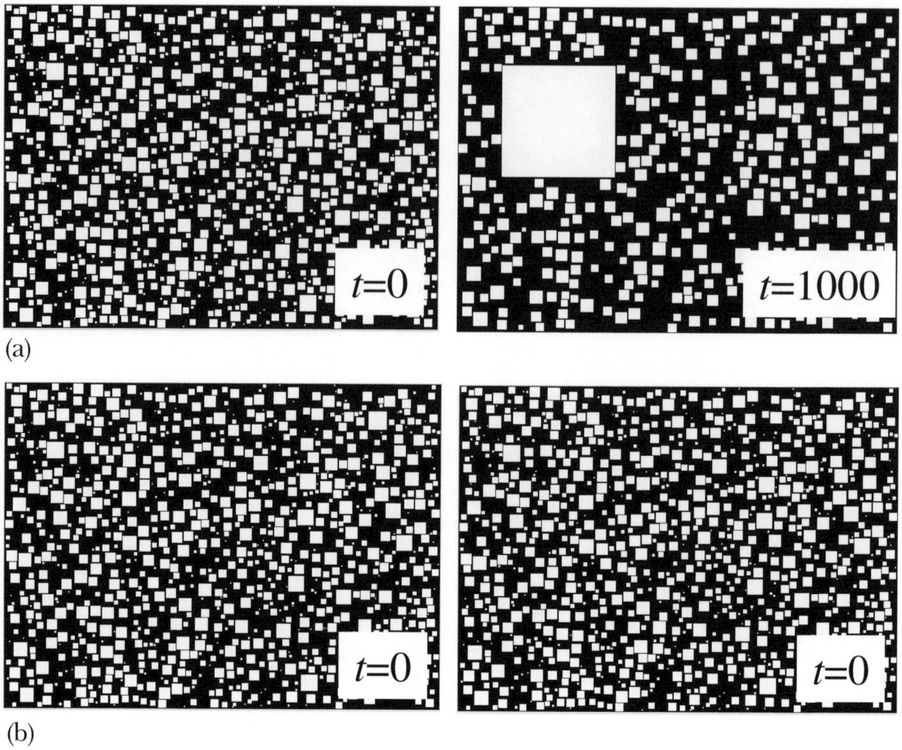

Fig. 6 Coarsening of grains (initial grain size: 0.3 μm) by 2-D nucleation, edge energy (a) 0.19 $h\gamma_{sl}$ and (b) 0.3 $h\gamma_{sl}$ Jm^{-1}.

formation of stable 2-D nuclei. The results were in good agreement with various experimental observations. In particular, the abnormal grain growth behaviour observed in the angular grains could be described adequately. We determined also the effect of important parameters on the coarsening of angular grains, such as initial particle size, size distribution and edge energy.

5.0 REFERENCES

1. C. HERRING: Some Theorems on the Free Energy of Crystal Surfaces, *Physics Review*, **82**, 1951, p.87.
2. D. N. YOON and W. J. HUPPMANN: Grain Growth and Densification During Liquid Phase Sintering of W-Ni', *Acta Metallurgica et Materialia*, **27**, 1979, p.693.
3. S. H. RHEE, J. D. LEE and D. Y. KIM: Effect of Heating Rate on the Exaggerated Grain Growth Behaviour of β-Si_3N_4, *Materials Letters*, **32**, 1997, p.115.

4. U. C. OH, Y. S. CHUNG, D. Y. KIM and D. N. YOON: Effect of Grain Growth on Pore Coalescence During the Liquid-Phase Sintering of MgO-$CaMgSiO_4$ Systems, *Journal of American Ceramics Society*, **71**(10), 1988, p.854.
5. J. H. HAN, Y. K. CHUNG, D. Y. KIM, S. H. CHO and D. N. YOON: Temperature Dependence of the Shape of ZnO Grains in a Liquid Matrix, *Acta Metallurgica et Materialia*, 1989, **37**, p.2705.
6. G. W. GREENWOOD: The Growth of Dispersed Precipitates in Solutions, *Acta Metallurgica et Materialia*, **4**(5), 1956, p.243.
7. I. M. LIFSHITZ and V. V. SLYOZOV: The Kinetics of Precipitation from Supersaturated Solid Solutions, *Journal of Physics Chemical Solids*, **19**(1/2), 1961, p.35.
8. T. K. KANG and D. N. YOON: Coarsening of Tungsten Grains in Liquid Nickel-Tungsten Matrix, *Metallurgical Transactions*, **9A**, 1978, p.433.
9. Y. J. PARK, N. M. HWANG and D. Y. YOON: Abnormal Growth of Faceted (WC) Grains in a (Co) Liquid Matrix, *Metallurgical and Material Transactions*, **27A**, 1996, p.2809.
10. M. K. KANG, Y. S. YOO, D. Y. KIM and N. M. HWANG: Growth of $BaToO_3$ Grains by the Twin-Plane Rentrant Edge Mechanism, *Journal of American Ceramics Society*, **83**(2), 2000.
11. P. WYNBLATT and N. A. GJOSTEIN: Particle Growth in Model Supported Metal Catalysts – I. Theory, *Acta Metallurgica et Materialia,* **24**, 1976, p.1165.
12. P. WYNBLATT: Particle Growth in Model Supported Metal Catalysts – I. Comparison with of Experiment with Theory, *Acta Metallurgica et Materialia,* 1976, **24**, p.1175.
13. D. P. WOODRUFF: *The Solid-Liquid Interface*, Cambridge University Press, Cambridge, U.K., 1973.
14. I. V. MARKOV: *Crystal Growth for Beginners*, World Scientific Publication Co. Pte. Ltd., Singapore, 1994.
15. J. P. HIRTH and G. M. POUND: Condensation and Evaporation, *Progress in Materials Science*, 11, Pergamon Press Ltd., London, 1963.
16. H. VAN BEIJEREN: Exactly Solvable Model for the Roughening Transition of a Crystal Surface, *Physics Review Letter,* **38**, 1977, p.993.
17. E. D. WILLIAMS and N. C. BARTELT: Thermodynamics and Statistical Mechanics of Surfaces, *Handbook of Surface Science,* W. N. Unertl, ed., Elsevier Science B. V., Amsterdam, 1996.
18. Y. S. YOO, H. KIM and D. Y. KIM: Effect of SiO_2 and TiO_2 Addition on the Exaggerated Grain Growth of $BaTiO_3$, *Journal of European Ceramic Society,* **17**, 1997, p.805.
19. H. H. YOON and D. Y. KIM: Effect of Heating Rate on the Exaggerated Grain Growth During the Sintering of Sr-Hexaferrite, *Materials Letters*, **20**, 1994, 293.
20. N. P. PADTURE: In Situ-Toughened Silicon Carbide, *Journal of American Ceramics Society*, **77**(2), 1994, p.519.
21. I. K. JEONG, D. Y. KIM, Z. G. KHIM and S. J. KWON: Exaggerated Grain Growth During the Sintering of Y-Ba-Cu-O Superconducting Ceramics, *Materials Letters*, **8**(3,4), 1989, p.91.

22. H. S. Song and R. L. Coble: Origin and Growth Kinetics of Platelike Abnormal Grains in Liquid-Phase-Sintered Alumina, *Journal of American Ceramics Society*, **73**(7), 1990, p.2077.
23. H. S. Song and R. L. Coble: Morphology of Platelike Abnormal Grains in Liquid-Phase-Sintered Alumina, *Journal of American Ceramics Society*, **73**(7), 1990, p.2086.
24. S. Prochazka, S. L. Dole and C. I. Hejna: Abnormal Grain Growth and Microcracking in Boron Carbide, *Journal of American Ceramics Society*, 1985, **68**(9), p.C-235.
25. M. Hillert: On the Theory of Normal and Abnormal Grain Growth, *Metallurgical and Material Transactions*, **13**, 1965, p.227.
26. G. R. Chol: Influence of Milled Powder Particle Size Distribution on the Microstructure and Electrical Properties of Sintered Mn-Zn Ferrites, *Journal of American Ceramics Society*, **54**(1), 1971, p.34.
27. A. Jaroenworaluck, T. Yamamoto, Y. Ikuhara and T. Sakuma: Segregation of Vanadium at the WC/Co Interface in VC-doped WC-Co, *Journal of Material Research*, **13**(9), 1998, p.2450.
28. D. Y. Kim and A. Accary: Mechanisms of Grain Growth Inhibition During Sintering of WC-Co Based Hard Metals, *Sintering Processes, Material Science Research*, G.C. Kuczynski, ed., Plenum Press, New York, **13**, 1979.

New Mechanism of Thin Film Growth: Charged Cluster Model

Nong-Moon Hwang,* In-Deok Jeon and Doh-Yeon Kim
Centre for Microstructure Science of Materials
School of Materials Science and Engineering, Seoul National University
Seoul 151-742, Korea

1.0 INTRODUCTION

It is generally believed that the crystals grow by the atomic or the molecular unit. The atom first adsorbs on the terrace, diffuses to the ledge and finally becomes the crystal at the kink.[1, 2] A drastically different way of crystal growth was suggested by Glasner et al.[3-6] during their study on the crystal growth of alkali halides from the supersaturated aqueous solution. In this case, the nanometer-size nuclei are formed in the solution and they become the growth unit. They showed that an almost perfect crystal grew by orderly packing of these nuclei with perfection of a crystal increasing with decreasing size of nuclei.

This new mechanism suggested by Glasner et al.[7] was so revolutionary that it was severely criticized by Botsaris and Reid.[8] Later, most of their important results on the new mechanism were published in the Israeli domestic journal.[4-6] Their important findings have been neglected in the community of crystal growth. To our knowledge, only Sunagawa[9] referred their results and suggested a similar mechanism in the diamond synthesis by the high and the low pressure processes.

The charged cluster model suggested by Hwang et al.[10, 11] is practically the same concept as the new mechanism suggested by Glasner et al. The only difference is that the charged cluster model deals with the thin film growth. The model is also similar to the mechanism suggested by Sunagawa[9, 12] in that the growth unit of the synthetic diamond is not an individual carbon atom but is much larger than the atom.

In the charged cluster model,[10, 11] the charged clusters, which might be induced by ions or by photo-excited molecules existing in the thin film reactor, exist in the gas phase. The charged clusters contain hundreds to thousands of atoms, being suspended in the gas phase like colloid suspension; the Brownian coagulation is inhibited by the presence of charge. The charged nanometer size clusters, being invisible to an unaided eye, are the growth unit of thin films. The thin film grows by the three-dimensional self-assembly of the charged clusters. Initially we developed the model

On leave from Korea Research Institute of Standards and Science, P. O. Box 102, Taedok Science Town, Taejon 305-600, Korea

in the diamond CVD process but it turned out to be a general growth mechanism in many thin film processes. The purpose of this paper is to describe the overall aspect of the charged cluster model in the thin film process. We will show how the charged cluster model can approach two important issues: the growth mechanism of the CVD diamond and the mechanism of selective deposition of silicon.

2.0 THERMODYNAMIC AND KINETIC APPROACH TO THE DOMINANT FORMATION OF DIAMOND IN THE CVD PROCESS

In the diamond CVD process, diamond is formed dominantly over graphite under the pressure where graphite is more stable. In this section, it will be dealt with how the metastable diamond can be formed dominantly over the stable graphite. Although this problem is not an essence of the charged cluster model, it is an important part of the model specific to the diamond CVD.

The CVD process is the precipitation of the solid phase from the gaseous phase. The deposition takes place only when a driving force for precipitation exists. Also in the CVD diamond process, there should exist a driving force for precipitation of diamond from the chemical reactions of gases. Since graphite is more stable than diamond, the driving force for precipitation of graphite is larger than that of diamond. Therefore, the precipitations of graphite and diamond from the gaseous phase are kinetically two parallel paths. The point is that in these two paths, a larger driving force does not guarantee the dominant formation. The kinetic barrier should be considered.

Since diamond and graphite compete in the precipitation kinetics from the gas phase, the nucleation barrier would be the most important factor in determining the dominant nucleation of the one phase over the other. If the nucleation barrier for graphite is higher than that of diamond, diamond will nucleate dominantly over graphite. Although there are other carbon allotropes, only diamond and graphite will be compared for simplicity.

The free energy change for the formation of diamond from the gaseous phase is expressed as

$$\Delta G_{dia} = -\Delta\mu_{dia}\, n + \eta_{dia}\, \sigma_{dia}\, n^{2/3} \tag{1}$$

and for the formation of graphite as

$$\Delta G_{gra} = -\Delta\mu_{gra}\, n + \eta_{gra}\, \sigma_{gra}\, n^{2/3} \tag{2}$$

where $\Delta\mu$ is the bulk free energy change per atom associated with the transfer of an atom from the vapour to the solid phase, n is the number of atoms in the particle, σ is the specific surface energy and η is the shape factor, which is $(4\pi)^{1/3}(3\Omega)^{2/3}$ for sphere, where Ω is an atomic volume.

Figure 1 shows the free energy curves for nucleation between diamond and graphite. These curves are based on the assumption of the spherical shape and the isotropic surface energy of the nucleus. The surface energies used for diamond and graphite are 3.7 J/m^2 and 3.1 J/m^2,[13] respectively.

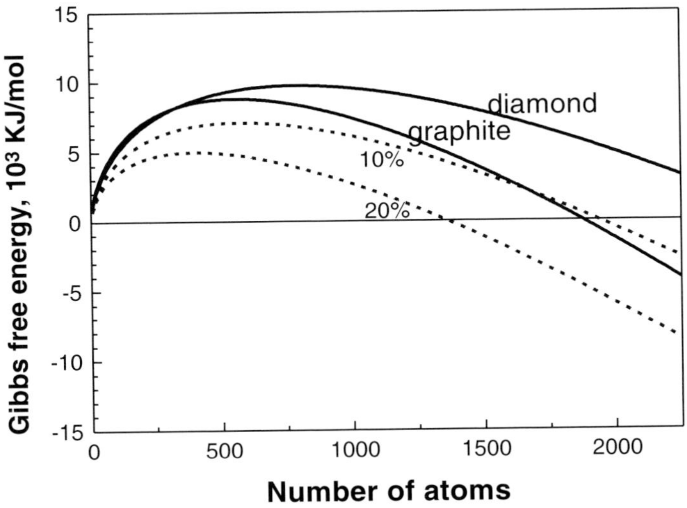

Fig. 1 Dependence of the Gibbs free energy for diamond and graphite on the size of the cluster. Two dashed lines are, respectively, for 10% and 20% reduction of the surface energy of diamond.

It should be noted that for the sufficiently small size, diamond becomes more stable than graphite. This stability change takes place at the intersection of the two free energy curves. The intersection can be obtained by equating eqns. (1) and (2) and solving for n as

$$n^* = 36\pi\left(\frac{\sigma_{dia}(\Omega_{dia})^{2/3} - \sigma_{gra}(\Omega_{gra})^{2/3}}{\Delta\mu_{dia\rightarrow gra}}\right)^3 \qquad (3)$$

$\Delta\mu_{dia\rightarrow gra}$ for the gas mixture of 1% CH_4-99%H_2 at 2700 Pa and 1200 K is - 1.2101 × 10^{-20} J/atom. Substituting the related data, the number of atoms at the intersection is ~350. This means that if the cluster contains less than ~350 atoms, diamond is more stable than graphite. The stability of diamond comes from the high capillary pressure built up inside a small nucleus.

However, Figure 1 shows that the nucleation barrier of diamond is larger than that of graphite. In this condition, the nucleation of graphite will be dominant over that of diamond. For the dominant nucleation of diamond, n^* should be increased. In eqn. (3), the atomic volume (Ω) and the free energy difference between diamond and graphite ($\Delta\mu_{dia\rightarrow gra}$) are not affected by the processing condition and the surface energy (σ) is the only factor that can be modified by the processing condition. And as shown by the dashed line in Figure 1, if the surface energy of diamond is assumed to decrease by 10%, n^* becomes ~1780. In this case, the nucleation barrier of diamond is lower than that of graphite and diamond would nucleate dominantly. For the decrease by 20%, n^* becomes ~5090 and the nucleation barrier of diamond decreases further.

These analyses indicate that the dominant formation of diamond over graphite in the CVD diamond process is identical to the well-established phenomena of the metastable nucleation.[14] For example, the metastable tetragonal zirconia nucleates dominantly over the stable monoclinic zirconia when zirconia is precipitated from the gas phase.[15] The other famous example is that when water vapour is supercooled into a freezing temperature, the metastable water droplet is almost always precipitated first instead of the stable ice crystal.[16]

The difference between the CVD diamond and the other metastable phases is the number of atoms at which the stability changes. In the CVD diamond process, the stability change between diamond and graphite takes place at the number of atoms close to the size of the critical nucleus. Therefore, the dominance of nucleation can change from graphite to diamond by the processing condition. In the case of the other metastable phases, the stability change between the stable and the metastable phases takes place at the size much larger than the critical nucleus. In such a case, the dominant nucleation of the metastable phase is a rule rather than exception.

Phenomenologically, the gas activation is necessary for making diamond. As explained in Figure 1, the slight decrease in the surface energy of diamond can result in the dominant nucleation of diamond over graphite. The gas activation seems to stabilize diamond by increasing n^* in eqn. (3). As mentioned earlier, the surface energy is the only parameter that can be varied by the processing conditions. One possibility is that the unbalanced reduction of the surface energy between diamond and graphite can be achieved by the adsorption of atomic hydrogen or hydrocarbon radicals. Another possibility, which was suggested to be a more dominant factor in the charged cluster model, is as follows.

Once the carbon clusters are charged, the dielectric diamond has the ion-induced dipole interaction at the surface while the conducting graphite does not. Jang and Hwang[17] calculated the decrease of the free energy of the diamond cluster arising from the ion-induced dipole interaction and showed that the nucleation barrier of the charged diamond cluster could be far smaller than that of the graphite cluster.

Here, we will make a simple treatment in order to show how the charged diamond cluster can reduce its surface energy compared with the charged graphite cluster. Figure 2 shows the schematic diagram of the charged diamond cluster. Here, we assume that the diamond cluster is negatively charged. Figure 2 shows that the negative charge is distributed on the surface and the dipole is induced inside the cluster, making an electric double layer at the surface. The relation between the surface energy and the surface charge is described by the Lippmann equation as

$$\frac{d\sigma}{dV} = -q, \tag{4}$$

where σ is the surface energy, V is the electric potential difference across the interface and q is the amount of charge built up at the surface.[18] In order to solve the differential equation of eqn. (4), we will approximate the double layer as a parallel plate condenser, which is called the Helmholtz-Perrin model in the electrochemistry. Then,

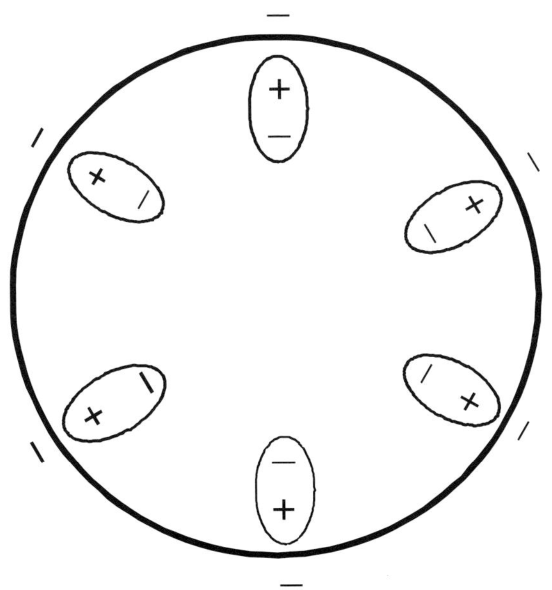

Fig. 2 Schematic diagram of the charged diamond cluster showing the dipole induced by the negative charge on the surface.

$$V = \frac{4\pi d}{\varepsilon} q, \tag{5}$$

where d is the distance between the double layer and ε the dielectric constant of the material between the double layer. By substituting eqn. (5) into eqn. (4), the following solutions are obtained.

$$\sigma = \sigma_{max} \frac{4\pi d}{\varepsilon} \frac{q^2}{2} \tag{6}$$

or

$$\sigma = \sigma_{max} - \frac{\varepsilon}{4\pi d} \frac{V^2}{2} \tag{7}$$

Eqns. (6) and (7) show that if the interface is charged, the interface energy decreases when the interface makes an electric double layer. If the surface energy of diamond decreases only by 10%, Figure 1 shows that the nucleation barrier of diamond becomes smaller than that of graphite, leading to the dominant nucleation of diamond.

In this section, we explained how nucleation of diamond can be dominant over that of graphite especially in the presence of abundant charges provided by the gas activation. This part is specific to the diamond CVD process, where the charge plays an additional role to stabilize diamond. In the following sections, it will be shown that the role of the charge is not limited to the diamond CVD process but is general in the other thin film processes.

3.0 GAS PHASE NUCLEATION IN THE CVD PROCESS

The starting point of the charged cluster model is the gas phase nucleation. Normally, once the nuclei are formed, they grow instantly into the visible size by Brownian coagulation. But in the presence of appreciable charges, the nuclei should be charged from the energetic point of view. Then, the growth of the charged nuclei is significantly inhibited and the nanometer size can be maintained during their residence time in the CVD reactor. The charged nuclei would be invisible and unnoticed during the CVD process. Special tools should be used to detect their presence in the gas phase as will be explained in a later section.

The gas phase nucleation can be triggered in many ways. Especially, ions and photo-excited species are known to trigger the gas phase nucleation and the phenomena are called the ion-induced and the photoinduced nucleation, respectively. These two phenomena will be explained briefly.

In early experiments on the homogeneous nucleation by the Wilson cloud chamber, Wilson[19] found that when an x-ray tube or some uranium or other radioactive body is brought close to the chamber, the few drops that are always produced with expansion ratios between 1.25-1.37 are replaced by a dense cloud. Thomson[20] showed that the cloudy condensation that took place in the presence of an x-ray beam have ions as nuclei. By applying a strong electric field between the parallel plate placed in the chamber, he could show that the dense cloud that was previously produced when the chamber was traversed by x-rays, disappeared in the presence of the electric field. Wilson[21-23] took advantage of these phenomena in the development of his cloud-chamber technique for photographing the tracks of α-rays, fast electrons, x-rays, and γ-rays. Later, it was realized that in the normal condition, the formation of the charged-cluster was unavoidable because the charges were continuously produced in the carrier gas by cosmic rays and natural radioactivity.[24, 25] The rate in air is several ion pairs per cubic centimetre per second, leading to a steady-state concentration of about 1000 ion pairs per cc.

These facts indicate that even if the concentration of ions in the room temperature air is so negligibly small, the ion-induced nucleation was unavoidable. In order to estimate the amount of ions in the CVD reactor, we measured the electric current in the plasma, the hot-filament and the thermal CVD reactor, where the currents of ~mA, ~μA and ~nA were measured, respectively, with the probe of one square centimetre near the substrate. Since one ampere represents the flux of $\sim 10^{19}$ charges per second, one nanoampere corresponds to the flux of $\sim 10^{10}$ charges per second, which is an enormous amount compared to the ion concentration in a room temperature air.

Although such a high current is expected in the plasma and the hot-filament reactors, it is not expected in the thermal CVD reactor. We observed the abrupt increase in the current in the thermal CVD reactor when the air or the inert gas is replaced by the carrier gas containing the precursors. This abrupt increase takes place when the temperature is high enough to decompose the precursors. Tentatively, we concluded that the ions should be generated by the chemionization during decomposition of the precursors.

The ion-induced nucleation was first treated theoretically by Thomson.[26] The Monte Carlo simulation for the enhanced nucleation in the presence of abundant charges was made by Shevkunov.[27, 28] Especially focused on the diamond CVD process, the enhanced nucleation in the presence of abundant charges was theoretically treated by Jang and Hwang.[29]

On the other hand, the photo-induced nucleation has been well established since its discovery by Katz et al.[30] The phenomenon refers to the nucleation triggered by the species, photo-excited by a specific wavelength. The photoinduced nucleation is much more powerful than the ion-induced nucleation in triggering nucleation in the gas phase.[31, 32] We expect that in the plasma, the flame and the hot-filament CVD processes where the photo-excited species are abundant, the photoinduced nucleation might be dominant and that in the thermal CVD process, the ion-induced nucleation might be dominant.

Photoinduced nucleation will produce the neutral clusters. In the ionizing environment, however, the neutral clusters will be easily charged because of their low ionization potential and the high electron affinity whose values approach the work function of the bulk. Therefore, regardless of whether nucleation is induced by ions or by the photo-excited species, the resultant clusters would be charged.

Considering that the formation of charged clusters is unavoidable even in the room temperature air, where the amount of ions is negligibly small, the gas phase nucleation triggered either by ions or by photo-excited species should be a general phenomenon in the thin film process. The exceptional case would be the very low pressure process (less than ~mtorr) such as the low pressure sputtering where the mean free path is large enough to suppress any appreciable collisions among the gas species during their residence time in the reactor.

Although the ion-induced and the photoinduced nucleation is phenomenologically well-established and the ions or the photo-excited species are abundantly produced in many thin film processes, the possibility of the gas phase nucleation in the thin film process has long been neglected. Recently, however, great interests have been shown in the gas phase nucleation or in the particle formation in the gas phase since the particles can fail the semiconductor device.[33]

4.0 CONFIRMATION OF CLUSTERS IN THE GAS PHASE OF THE CVD REACTOR

The experimental confirmation of the hypothetical charged clusters is essential to the validity of the charged cluster model. As to their experimental confirmation, we not only performed our own experiments but also made an extensive literature survey. First we will review our literature survey then report our experimental results.

The gas phase nucleation or the particle formation in the CVD reactor has been a big issue in the microelectronics, where the attention shifted from the clean room to the clean reactor for the processing of the giga bite memory chip.[33] In this case, the particle formation may originate from the reaction gases or from the reactor wall. Especially in the plasma CVD reactor, the particle formation is so general as to be called the dusty plasma.[34-36] In the dusty plasma, the negatively-charged clusters are suspended in the positively-charged plasma. Currently, the particle formation in the gas phase is regarded as being almost unavoidable in the plasma CVD process.

On the other hand, in pursuit of the good quality film, Adachi et al.[37-40] tried to find the SiO_2 CVD condition where the gas phase nucleation is prevented. The process is the atmospheric-pressure thermal CVD using tetraethylorthosilicate (TEOS). They used the aerosol sizing technique such as differential mobility analyser (DMA) and the particle number counter to detect the particles nucleated in the gas phase. They observed that when the particles were not detected in the gas phase, the film did not grow either and further that the growth rate of the film increased with increasing number density of clusters. These behaviours were consistent for three different gas mixtures of the precursor and the carrier gas.

Okuyama et al.[41] also performed similar experiments. They observed that for the small particles of a few nanometers, the film surface was smooth while for the coarse ones larger than 10 nanometers, the film surface was irregular. Based on these observations, they made an important suggestion that the molecules and the particles co-deposit in the film growth. Although they did not consider the role of charge, their suggestion is one of the very rare ones that the growth unit of the film can be the particle formed in the gas phase. But it is generally believed that the particles generated during the CVD process do not contribute to the film deposition.[42] The incorporation of the particles into the film is believed to impair the film quality significantly.

It should be noted that the terminology of the particle used in the dusty plasma or in the aerosol science and technology is similar to the cluster in this paper. However, the particle and the cluster typically refer to the size, respectively, larger and less than tens of nanometers. The size dependence of clusters on the microstructure will be explained in a later section.

The typical size of clusters referred in the charged cluster model is less than ~10 nm, which is beyond the detection limit of the laser scattering method. The charged clusters in the CVD diamond process are even smaller, estimated to be ~2 nm. This size range is difficult to measure because it is much larger than the widely used quadruple or the time-of-flight (TOF) mass analyser and too small to be measured by the conventional DMA. The TOF mass analyser designed for the large mass or the nano-DMA should be used. When the clusters are captured on the metal mesh grid with the carbon or the SiO_2 membrane, they can be observed by transmission electron microscopy (TEM).

The presence of the charged clusters in the gas phase can be confirmed by an energy analyser, a Wien filter and TOF. In all cases, the gas is extracted through one or two orifices by molecular beam sampling using differential pumping. The gas species together with the charged clusters go from the reactor chamber into the measuring chamber through the orifice. In the case of the energy analyser, the repelling voltage is applied on the grid in the measuring chamber, where the mean free path is long enough. The charged clusters which have the kinetic energy less than that given by the repelling voltage will be repelled and would not be measured on the Faraday cup detector. Therefore, with increasing repelling voltage, the detected current will decrease. Differentiation of the current with respect to the voltage will give the energy distribution of the charged clusters. If the dependence of the mass of clusters on the velocity is known, the mass distribution can be derived from the energy distribution.

In the case of a Wien filter, only the charged clusters with the same velocity go to the detector. Therefore, if the energy distribution is determined as described above, the mass distribution of the clusters with the specific velocity is determined. Summation of the data for the various velocities will give the mass distribution of the charged clusters in the reactor. In the case of TOF, the beam of the charged clusters is deflected to the vertical direction by the applied voltage so that it may have an initial velocity of zero before going into the TOF chamber.

Homann et al.[43-47] used these three methods to measure the size distribution of the charged carbon clusters during their study on combustion of the hydrocarbon-oxygen flame. The composition they used covers the ratio of C/O from 0.8 to 1.2. In all ratios, they measured the presence of the charged clusters though the cluster size increased with the ratio. It should be noted that the diamond films are made by the combustion flame when the ratio of C/O is close to unity.[48]

After the method of Homann group, we measured the distribution of charged clusters in typical conditions of the hot-filament diamond CVD process by both energy analyser and Wien filter and confirmed the presence of the charged clusters containing hundreds to thousand of carbon atoms.

Figure 3 shows the energy distribution of the negatively charged carbon clusters determined by the energy analyser. The processing condition was at the filament temperature of 2000°C under 8.5 Torr for the gas mixture of 1% CH_4-99%H_2. The distance between the orifice and the filament was 10 mm. The kinetic energy of 12 eV for the cluster corresponds to ~3000 atomic mass unit (amu), which is equivalent to ~250 carbon atoms. The details will be published elsewhere.[49]

As to the TEM observation of the clusters after capturing them onto the grid for TEM, Adachi et al. observed the SiO_2 clusters of 10~20 nm, which had existed in the gas phase of the CVD reactor. We observed the silicon clusters of 2~10 nm in the hot-filament CVD process.[50] Figures 4a and b are the TEM microstructures showing these clusters captured on the carbon membrane of the TEM grid for 20 seconds. At the filament temperature of 1800°C (Figure 4a), the clusters had almost monosize distribution with the size of ~2 nm while at 1600°C (Figure 4b), they were larger and deviated a little from the monosize distribution with the size of 5~10 nm. We also observed the clusters of relatively uniform size less than 2 nm on the grid prepared in the diamond synthesis by the acetylene flame.[51] We also made TEM observations on tungsten clusters during thermal evaporation coating of tungsten[52] and on ZrO_2 clusters during thermal CVD of ZrO_2.[53]

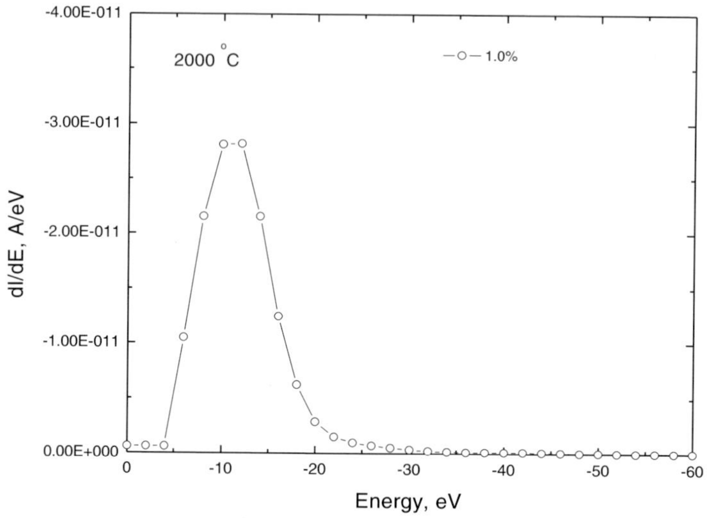

Fig. 3 Energy distribution of clusters extracted from the hot-filament diamond CVD reactor at the filament temperature of 2000°C under 8.5 Torr for the gas mixture of 1% CH_4-99% H_2. The distance between the orifice and the filament was 10 mm.

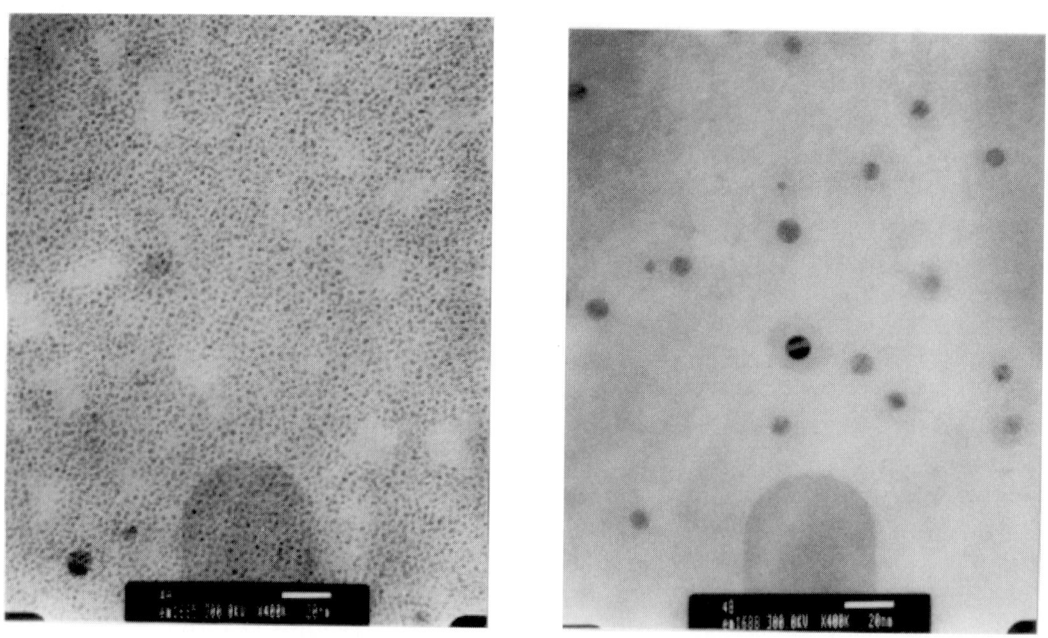

(a) (b)

Fig. 4 TEM microstructures showing the silicon clusters formed in the gas phase of the CVD reactor under 10 Torr for the gas mixture of 5% SiH_4-1% HCl-94% H_2 at the filament temperatures (a) of 1800°C and (b) of 1600°C. The exposure time of the TEM grid was 20 seconds. The scale bar is 20 nm.

It can be argued that the clusters of Figures 4a and b were not formed in the gas phase but nucleated directly on the carbon membrane. However, the microstructural features are against the possibility of the direct nucleation on the substrate surface in several respects. One is the highly uniform size of the clusters. If the clusters were nucleated directly on the carbon membrane, such a high monosize character is not expected; the newly nucleated clusters should be small and the old ones should be large. However, the clusters prepared at 1800°C and at 1600°C belong to two distinctly different size classes of nearly monosize. Another fact is the dependence of the exposure time on the size of the cluster. With increasing exposure time before coalescence, only the number density of the cluster increases with the size of the cluster being constant.

5.0 EFFECT OF CHARGE DENSITY ON THE CLUSTER SIZE

Neutral and charged clusters markedly differ in their coagulation rate. In the aerosol science, the half-life is defined as the time elapsed for the number density of clusters to be reduced by half. In the typical CVD condition, the half-life is of the order of 1 μs and 1 ms for the clusters of 1 nm and 20 nm in diameter, respectively.[54] Considering the residence time of the clusters in the CVD reactor, neutral clusters are expected to grow instantly by Brownian coagulation into visible macro particles.

If clusters are charged, however, their coagulation rate would be retarded markedly. This retardation would depend on the charge density. The theoretical treatment for the dependence of the charge density on the cluster size was made by Jang and Hwang[17] for the case where an equal amount of opposite charges exists and thus the total charge neutrality is maintained. Here, the charged clusters were treated like colloid particles. The calculation showed that with increasing charge density, the cluster size decreased. The calculation further showed that there exists a free energy well for the cluster of specific size, implying that the charged clusters should tend to have a monosize distribution. These aspects of decreasing cluster size with increasing charge density and the tendency for the monosize distribution are experimentally confirmed in the silicon CVD process.[50]

The retardation would be favourable if one sign of charge is dominant over the other sign of charge because the chance for neutralization of charge by collisions between clusters of an opposite sign will decrease. This situation occurs in the hot filament CVD, where the negative charge is dominant over the positive one since the main source of charge is the electron emission. This fact might be the reason that even if the overall charge density in the hot-filament diamond CVD is low compared to that in the plasma diamond CVD, the high quality diamond films can be grown by the hot-filament method.

The clusters in the rf plasma reactor are expected to be larger than those in the microwave plasma reactor because in the former, the negatively-charged clusters tend to be suspended by the positively-charged plasma and thus, the residence time in the gas phase increases. This might be the reason why the quality of diamond

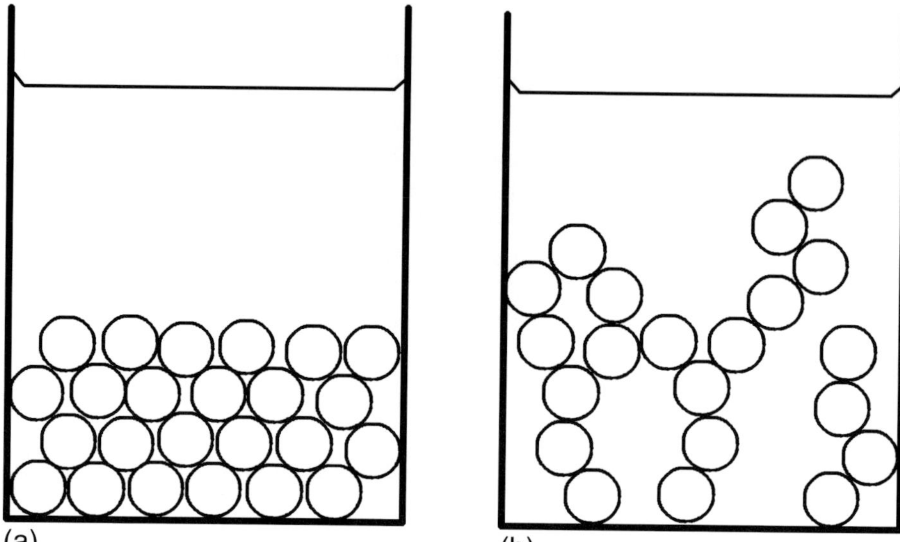

Fig. 5 Sedimentation behaviours for (a) deflocculation and (b) flocculation of colloidal particles.[56]

films prepared by the rf plasma process is worse compared to those made by the microwave plasma process.

6.0 DEPOSITION BEHAVIOUR OF THE CHARGED CLUSTERS

In the previous section, it was analysed that the charge plays a critical role in suppressing the growth of clusters. The charge also plays a critical role in the evolution of the dense film. The neutral clusters undergo the Brownian coagulation in the gas phase. It is well established that the cluster-cluster aggregation by Brownian coagulation makes the fractal aggregates of a highly porous skeletal structure.[55] And the resulting macroparticles of a fractal shape would settle down on the substrate surface. The phenomenon is called "snowing" in the CVD process and the resulting deposited layer is very porous. If the clusters are charged, however, they behave like colloid suspension. Interactions between charged clusters are the Coulomb repulsion and the van der Waals attraction. Especially, when the Coulomb repulsion is dominant, their landing on the surface is highly ordered and the behaviour is known as deflocculation in the colloid chemistry, leading to close packing. Figures 5a and b show the schematic diagram comparing the sedimentation behaviour of the repulsion-dominant and attraction-dominant colloidal particles.[56] In Figure 5a, the charged particles land selectively on the macro kink sites. The phenomenon is identical to

(a) (b)

Fig. 6 SEM microstructures showing (a) diamond deposited on the silicon substrate and (b) soot deposited on the iron substrate for the gas mixture of 1% CH_4-99% H_2 for 2 hrs. at a substrate temperature of 990°C under 20 Torr for the filament temperature of 2100°C. The scale bars in (a) and (b) are 704 and 532 nm, respectively.

the self-assembly of the nano-clusters treated in the organic chemistry.[57] Such close packing structure would be favourable for the evolution of the dense film if it is combined with the high diffusivity of the atoms in the small clusters.[58]

Even if the clusters are charged, the charge can be lost to the substrate just before they touch on the growing surface. This property depends on the charge transfer rate of the substrate materials.[10] Substrate materials in the order of decreasing charge transfer rate are Pd, Pt, Rh, Ir, Ni, Fe, Au, W, Ag, Nb, Mo, Cu, Ta, Bi, Al and Ti.[18] The substrate with a high charge transfer rate will make the porous skeletal structure in the initial stage of deposition until the deposits cover their surface. These aspects were revealed in the diamond CVD process[10] and other systems.[59] In the CVD diamond process, the porous skeletal structure of soot is obtained on the substrates with a high charge transfer rate while the diamond crystals grow on the substrates with a low charge transfer rate. We also observed the similar tendency in the thermal CVD of silicon and zirconia and in the evaporation coating of tungsten.

Figures 6a and b show that the diamond crystal grows on the silicon substrate while the porous graphitic soot grows on the iron substrate. In the case of the CVD diamond process, the charge plays an additional role in stabilizing the diamond cluster over the graphite cluster as mentioned in the section II. Therefore, the high charge transfer rate on the iron substrate not only transforms the diamond cluster into the graphite cluster but also produces the porous fractal structure of soot.

7.0 EFFECT OF CLUSTER SIZE ON THE MICROSTRUCTURE EVOLUTION

In the charged cluster model, the microstructural evolution depends on the cluster size. As the cluster size increases, the growth mode changes from the crystal with well-defined facet to the cauliflower structure. We realized that the study by Fujita[58] on the magic size of the clusters is closely related to this problem. His study will be briefly described.

According to Fujita,[58, 60] the cluster property changes abruptly at a specific size, which was called the magic size. The magic size corresponds to the number of atoms in the cluster from 10^3 to 10^5, depending on the bonding mode and the bonding strength. In the order of the increasing cluster size, the particle coalescence, epitaxial recrystallization, and the diffusional sintering were suggested to take place.

Fujita[58] estimated the magic size for several materials. The magic size of isolated clusters was estimated to be 15-20 nm ∅ for fcc metals and alloys, 10-15 nm ∅ for bcc metals and alloys, 15-20 nm ∅ for Al_2O_3, and 3-5 nm ∅ for TiC and SiC, where ∅ represents the diameter of the cluster. For embedded clusters, it is ~2 nm ∅ for Si and ~1 nm ∅ for diamond. The magic size for an isolated diamond cluster is expected to be ~2 nm ∅. The diamond cluster of 2 nm ∅ contains a few hundreds of atoms.

The cluster smaller than the magic size deforms and diffuses like a liquid phase and the rapid coalescence takes place when the clusters are in contact. As a result, the epitaxial film can grow by these clusters. If the initial nuclei on the substrate have different orientations the continued epitaxial growth of each nucleus will lead to evolution of the columnar structure. As the cluster size becomes larger than the magic size, the epitaxial coalescence will begin to fail. Then, the twin or grain boundaries will form and the polycrystalline film will grow. As the frequency for the formation of the boundary increases, the grain size will decrease. The grain size will be minimal when all the clusters have their individual orientations, leading to a cauliflower structure or a nanostructure. The similar aspect was observed by Glasner et al.[5] in their growth of KCl and KBr from the solution.

Figures 7a and b are the microstructures of the diamond film with low and high methane concentrations, respectively. The crystals with well-defined facets are evolved in Figure 7a while the cauliflower structure is evolved in Figure 7b. Figure 7c is the magnified view of one particle in Figure 7b. Numerous nodules appear on the surface in Figure 7c. Note that each faceted particle in Figure 7a is a single crystal while each spherical particle in Figures 7b and c appears to consist of numerous nanocrystals. The cauliflower structure tends to be evolved with increasing methane concentration and decreasing substrate temperature. Figure 8 shows the cauliflower structure evolved in the thermal CVD of silicon in the Si-Cl-H system. These cauliflower structures are generally observed in the thin film process. Especially, the evolution of the cauliflower structure is pronounced in the deposition by laser ablation.[61-63]

According to our analysis, the cauliflower structure cannot be evolved by the atomic unit deposition. If the deposition is by the atomic unit, each bulge of the cauliflower structure should be formed by three-dimensional nucleation on the growing surface. If so, the nucleation rate should be extremely high, for which a

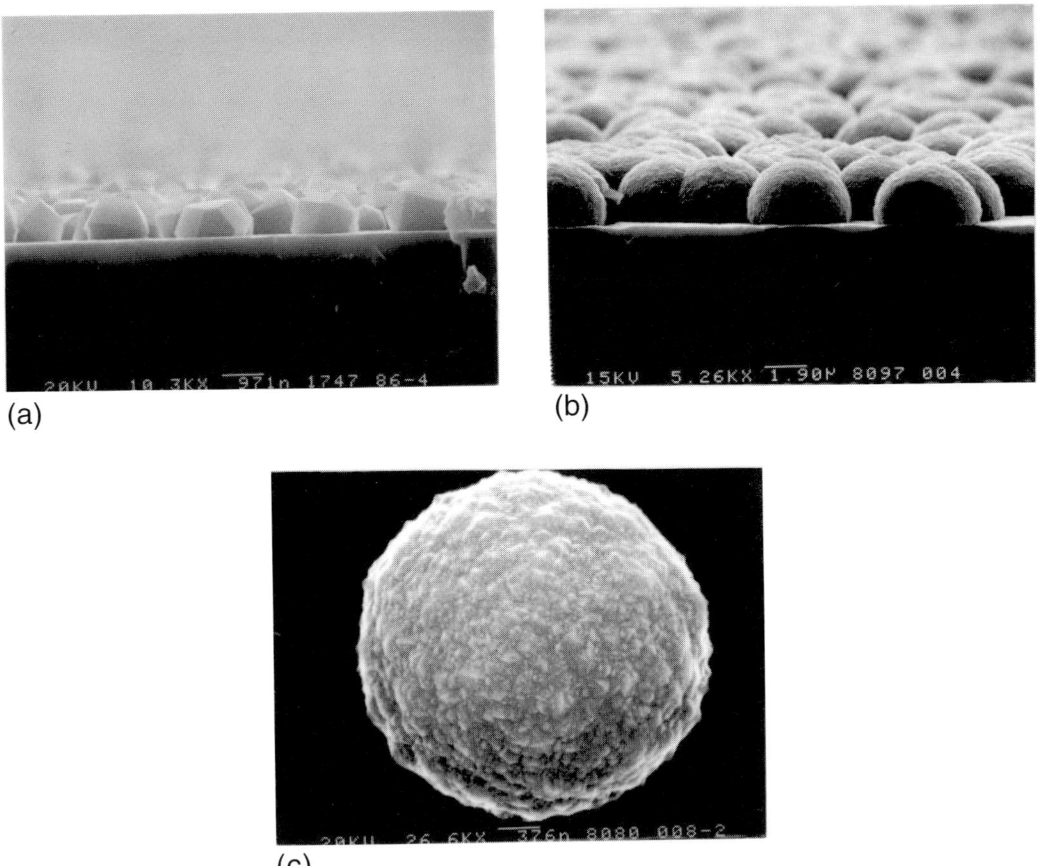

Fig. 7 SEM microstructure showing the profiles of (a) diamond with well-defined facet for the gas mixture of 1% CH_4-99% H_2 and of cauliflower-shaped diamond structures (b) at low magnification and (c) at high magnification for the gas mixture of 3.5% CH_4-96.5% H_2. Other processing conditions are the same as Figure 6. The scale bars in (a), (b) and (c) are 971, 1900 and 376 nm, respectively.

very high supersaturation is required. However, it is well established in the crystal growth that the kinetic roughening of the growing surface takes place at the supersaturation just higher than that for onset of two-dimensional nucleation. If the kinetic roughening takes place, the atoms from the gas phase easily find the kink sites and the supersaturation cannot be built up higher for the three-dimensional nucleation. In such a case, the growing surface acts like a rough interface and the growth process changes from interface-controlled growth to diffusion-controlled dendritic or spherulic growth as well established in the crystal growth.[9] For this reason, the grain size has a lower limit in the deposition by the atomic unit. And the cauliflower structure, which is a nanostructure, implies that the growth unit is clusters

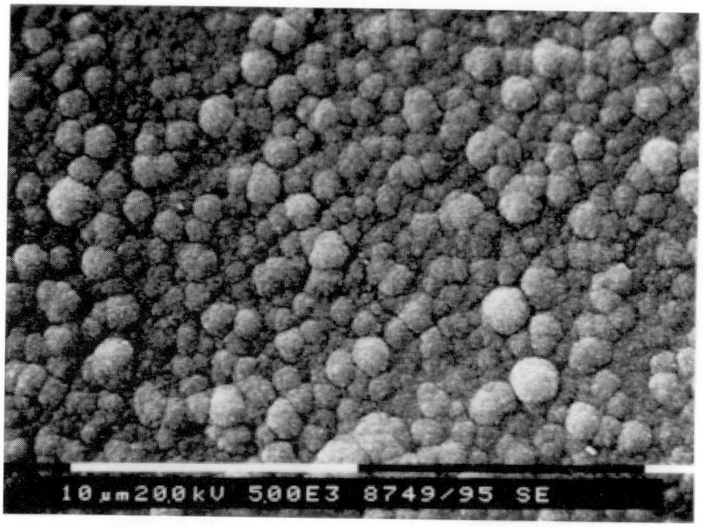

Fig. 8 Cauliflower structure of silicon evolved during thermal CVD at the substrate temperature of 850°C under 100 Torr with the gas mixture of 1% SiH_4-1% HCl-98% H_2. The deposition time was 10 minutes.

formed in the gas phase. In this respect, the cauliflower structure can be the evidence that the film grows by the mechanism of the charged cluster model.

The effect of adsorbates on the cluster will be briefly mentioned. Since lots of gas species exist in the deposition chamber, the gas species can adsorb on the clusters. For example, the carbon or the silicon clusters can be hydrogenated. In the case of carbon clusters, Homann et al.[43] reported that the negatively-charged cluster consists of almost pure carbon but the positively-charged cluster is hydrogenated, which is called "the polyaromatic hydrocarbon ion (PAH^+)".

In case of the deposition of clusters adsorbed by gas species, the crystalline film cannot grow and the amorphous phase will grow unless the gas species are desorbed at a sufficiently high rate. Desorption kinetics will be low at low temperature. If the adsorbed clusters are small as not to maintain the crystalline lattice, the amorphous phase will deposit. If the clusters are large enough to maintain the crystalline lattice but adsorbed by the gas species, the mixture of crystalline and the amorphous phases will deposit. Many micro or nano crystalline structures contain some portion of the amorphous phase and would fall into this category.

8.0 EFFECT OF CLUSTER SIZE ON THE DEPOSITION RATE

We expect most of the charged clusters to be singly charged. The cluster of the same kind tends to have dominantly one sign of charge. For example, in the CVD diamond process, the diamond cluster is negatively charged and the hydrogenated carbon cluster is positively charged. If the thin film material is not electrically conducting, the growing surface will be charged and will exert the Coulomb repulsion to the incoming charged clusters of the same kind. As the cluster size decreases, Coulomb repulsion tends to be dominant over van der Waals attraction. Therefore, small clusters will decrease the growth rate even though they will improve the crystal quality. These aspects are clearly revealed in the CVD diamond process as well as in the experiments by Glasner et al.[5,6]

In the MOCVD process, it is well known that the growth rate of the film first increases with the substrate temperature then decreases above a certain temperature. But the films prepared above the temperature where the growth rate begins to decrease, show better quality of crystal with well-defined facets. This decrease in the growth rate has been attributed to the gas phase nucleation. From the viewpoint of the charged cluster model, however, the decrease in the growth rate is due to the small size of the clusters arising from the increase in charge density, which again comes from increase in the decomposition temperature of the precursors.

We observed such temperature dependence in the ZrO_2 CVD using solid $ZrCl_4$ precursor. At the precursor temperature of 250°C the cauliflower structure was evolved, while at 450°C, the crystalline structure with well-developed facets was evolved, implying that the cluster size is larger at the precursor temperature of 250°C than at 450°C, which is experimentally verified.[64] Besides, the growth rate of the film was about 10 times higher at the precursor temperature of 250°C than at 450°C, in spite of the fact that the evaporating flux at 250°C is a few orders of magnitude lower than that at 450°C. These puzzling behaviours could be explained by the temperature dependence of the electric current generated during precursor decomposition. The electric current in the chamber was about 10 times higher at the precursor temperature of 450°C than at 250°C.

9.0 SELECTIVE DEPOSITION

The selective deposition is extensively utilized in microelectronics. In this process, a given element, alloy or compound is made to deposit selectively on a specific pattern. The process is based on the assumption that there can be a substantial difference in the nucleation rate on different substrates.[65] In reality, the nucleation tends to be much easier on conductors or semiconductors than on insulating or dielectric materials. It is not clearly understood why the conducting material is favourable for nucleation over the insulating one. Quite often, the selectivity is lost before the desirable thickness of the selective growth is achieved. Selective growth can be changed to nonselective growth by introducing the plasma CVD. Also, the reason for the loss of selectivity is not clearly understood.

The charged cluster model can also approach the selective deposition.[59] In addition to its effect on the cluster size, the charge plays another important role in the selective deposition. Considering the electrostatic interaction between the charged cluster and the substrate surface, the conducting substrate will be favourable for the deposition of the charged clusters over the insulating substrate.

The electrostatic interaction would be minimal if both the substrate and the depositing material are conducting. Even if both are insulating, the continuous deposition would be possible if the gas phase is conducting. The conductivity in the gas phase is high in the plasma and the hot filament reactors while it is low in the thermal CVD reactor. The selective deposition would be favoured in the thermal CVD process. At the same time, the high conductivity in the gas phase in the plasma and the hot filament reactors would be unfavourable for the selective deposition. These behaviours are in agreement with the well-known phenomenon that the selectivity is lost when the plasma is used.[65]

10.0 DEPOSITION AND SIMULTANEOUS ETCHING IN THE DIAMOND AND SILICON CVD PROCESSES

The highlight of the charged cluster model is the successful explanation of the puzzling phenomena of deposition and simultaneous etching taking place in the diamond and the silicon CVD processes.[59, 66-68] To our knowledge, only the charged cluster model can explain these phenomena on a sound basis of thermodynamics. And their detailed and rigid analysis is critical to the validity of the model. Before approaching these phenomena by the charged cluster model, the phenomena will be described in detail.

It was experimentally well established in the diamond CVD process that when graphite is used as a substrate, diamond deposits on the graphite substrate, which etches away simultaneously. Although the dominant formation of diamond over graphite can be approached by the kinetic barrier in the nucleation stage, the deposition of the less stable diamond with the simultaneous etching of the stable graphite cannot be explained by the kinetic barrier and appears to be against the concept of thermodynamics.

Let's assume that the deposition unit is the atom and then apply the thermodynamic concept to the phenomenon. The universal criterion for the transfer of atoms between phases is the chemical potential difference. The diamond deposition indicates that the chemical potential of carbon in the gas phase (μ_c^{gas}) is higher than the chemical potential of carbon in diamond (μ_c^{dia}). The graphite etching indicates that the chemical potential of carbon in graphite (μ_c^{gra}) is lower than the chemical potential of carbon in the gas phase (μ_c^{gas}). Therefore the following inequalities in chemical potential of carbon among the three phases results.

$$\mu_c^{gra} > \mu_c^{gas} > \mu_c^{dia} \tag{8}$$

From the well-established phase diagram in the carbon system, we know that graphite is more stable than diamond and the inequality of chemical potential of carbon becomes

$$\mu_C^{gra} < \mu_C^{dia} \tag{9}$$

It should be noted that the graphite substrate does not have any appreciable capillary effect. Therefore, the stability reversal between diamond and graphite arising from the capillary effect, which was described in the previous section, cannot take place.

Eqn. (8), which is derived from the experimental data, is against Eqn. (9), which is derived from our knowledge of thermodynamics in the carbon system. This is why Yarbrough[69] called this phenomenon "thermodynamic paradox". This problem had been a big issue in the diamond community in the year of 1992. Since at that time the majority of the community thought that the experimental fact is more persuasive than the thermodynamic argument, the conclusion was that the phenomenon should not be thermodynamic paradox. Since then, the scientists in the community have taken this paradoxical phenomenon for granted and have tried to explain this phenomenon based on kinetics. In my opinion, however, the pursuit of kinetics violating thermodynamics would lead to a wrong way. Thermodynamic laws are the necessary condition for any phenomena to take place. Kinetics should be pursued within the framework of thermodynamic laws.

Similar phenomenon takes place in the silicon CVD process. In the selective epitaxial growth of silicon, the silicon tends to grow selectively on the conducting surface patterned on the insulating substrate.[65] Kumomi et al.[70-72] prepared the slightly conducting SiN_x patterned on the insulating SiO_2 substrate and deposited the silicon selectively on the SiN_x patterns using the dichlorosilane and hydrochlorine. According to them, the silicon deposited initially on the SiN_x pattern on the SiO_2 substrate and then the deposited particles etched away while one of the particles grew into a large particle during selective epitaxial growth by the thermal CVD process.[70-72] As a result, the initially formed numerous silicon particles etched away except one, which grew into a large particle. In some SiN_x patterns, all the silicon particles etch away.

We did our own experiments for this phenomenon and observed a similar result. The gas 1% SiH_4-2%HCl-97% H_2 was used with the substrate temperature of 950°C and the reactor pressure of 100 Torr. The deposition behaviours were compared between the conducting and the insulating substrates. The silicon continued to deposit on the conducting substrate while numerous silicon particles deposited initially on the insulating SiO_2 and SiN_4 substrates and almost all of them etched away with increasing processing time.[59]

In the case of the silicon CVD, two irreversible phenomena of deposition and etching, which are in opposite direction, take place simultaneously. The irreversible deposition means that the chemical potential of silicon in the gas phase (μ_{Si}^{gas}) is higher than that of the solid (μ_{Si}^{solid}). And the irreversible etching means that μ_{Si}^{gas} is lower than that of μ_{Si}^{solid}. Therefore, if we assume that two irreversible phenomena take place simultaneously in opposite directions, the inequality in chemical potentials leads to the thermodynamic contradiction.

The phenomena should not be confused with the fact that deposition and etching takes place simultaneously in terms of the principle of microscopic reversibility. Let's take an example of water at 60°C. When the water vapour adjacent to the water is lower than the equilibrium vapour pressure of water at 60°C, the irreversible evaporation will take place. Even if the net flux is for evaporation, there exists a flux of condensation by the principle of microscopic reversibility. It should be noted, however, that when the net flux is for evaporation, the irreversible process is the evaporation and the condensation is the reversible flux. Always the reversible flux in one direction is cancelled out by another reversible flux in an opposite direction. The irreversible process refers to the case only when the net flux exists.

On the other hand, the deposition and the simultaneous etching in diamond and silicon CVD processes are in terms of net flux and indicate definitely that two irreversible processes in opposite directions take place simultaneously. The resultant thermodynamic contradiction comes from the implicit assumption that the deposition unit should be the atom. If we assume that the deposition unit is the clusters formed in the gas phase as claimed by the charged cluster model, the phenomena can be explained without any thermodynamic contradiction. In this case, the deposition unit is the cluster and the etching unit is the atom. That is, the thermodynamic driving force at the surface is for etching while the deposition takes place by landing of the clusters nucleated in the gas phase on the substrate.

Then, we need to explain how the gas phase nucleation in the diamond CVD and in the silicon CVD processes makes the driving force at the substrate for etching. For this, we should examine the CVD phase diagram of the C-H and the Si-Cl-H systems. The phase boundary in the CVD phase diagram, which demarcates the gas phase region and the gas plus solid region, represents the solubility of the depositing species in the gas phase. Normally, this solubility decreases with decreasing temperature. In the C-H and the Si-Cl-H systems,[66,67,73] however, the solubility is minimal at the intermediate temperature, which is above the substrate temperature in the typical processing condition. In the C-H and the Si-Cl-H systems, the minimums are at ~1150°C under 20 Torr and at ~1200°C under 150 Torr, respectively. The amount of precipitation is maximal at these temperatures. The carbon cluster precipitated in its maximum amount at ~1150°C will etch either above or below ~1150°C. This means that except at the temperature of ~1150°C, the driving force at the substrate can be for etching. Similarly, the silicon clusters precipitated in its maximum amount at ~1200°C will etch either above or below ~1200°C.

These analyses indicate that the gas phase nucleation in the C-H and the Si-Cl-H systems can make the driving force for etching at the substrate temperature. In this case, all the deposition flux comes from the clusters and the atomic flux is for etching. This picture can explain the net flux of deposition and the simultaneous net flux of etching without the thermodynamic contradiction. According to this picture, diamond as well as graphite etches away at the substrate by the atomic unit in the diamond CVD process. However, the deposition flux for diamond from the clusters is dominant over its etching flux by the atomic unit while graphite continues to etch at a high rate. What we observe is the deposition of the less stable diamond with simultaneous etching of the stable graphite.

In the case of the silicon CVD process, the deposition seems to be dominant in the initial stage and the etching takes place in the later stage on the insulating or the semi-insulating substrate, while the deposition continued on the conducting substrate. As the charged clusters deposit, the charges build up on the substrate. These built-up charges exert Coulomb repulsion inhibiting further attachment of the charged clusters while the driving force at the substrate is for etching by the atomic unit. In the initial stage before saturation of charge, the numerous silicon particles can be formed on the substrate because the deposition flux from the charged clusters are dominant over the etching flux by the atomic unit. In the later stage after saturation of charge, further landing of the charged clusters will be inhibited and then the etching flux will be dominant. What we observe is the formation of numerous silicon particles in the initial stage and their etching in the later stage.

11.0 CONCLUSIONS

The invisible nanometer charged clusters exist in the gas phase in many thin film processes. These clusters are the major flux for the growth of the film. This mechanism can explain some unusual phenomena such as evolution of a cauliflower structure, selective deposition and simultaneous deposition and etching, which cannot be explained by the atomic unit deposition.

12.0 ACKNOWLEDGEMENTS

This work was supported by Creative Research Initiatives Program of the Korean Ministry of Science and Technology.

13.0 REFERENCES

1. P. Bennema and G. H. Gilmer: *Kinetics of Crystal Growth*, North-Holland Publ. Co., Amsterdam, 1973.
2. J. P. Van Der Eerden: *Crystal Growth Mechanism*, North-Holland, Amsterdam, 1993.
3. A. Glasner and J. Kenat: The Crystallisation of KCI from Aqueous Solutions in the Presence of Lead Ions. I. A Calorimetric Study, *Journal of Crystal Growth*, **2**, 1968, pp.119-127.
4. A. Glasner and S. Skurnik: A New Mechanism for the Crystallisation and Growth of Ionic Crystals, With Special Reference To KCl in the Presence of Pb^{2+} Ions, *Israel Journal of Chem.* **6**, 1968, pp.69-72.
5. A. Glasner and M. Tassa: The Thermal Effects of Nucleation and Crystallization of KBr and KCl Solutions II, The Heat of Crystallisation and the Co-Precipitation of Lead Ions, *Israel Journal of Chem.*, **12**(4), 1974, pp.817-826.
6. A. Glasner and M. Tassa: The Thermal Effects of Nucleation and Crystallisation of KBr and KCl Solutions II, The Heat of Nucleation and the Supersaturated Solution, *Israel Journal of Chem.* **12**(4), 1974, pp.799-816.
7. A. Glasner and S. Skurnik: Growth of Potassium Chloride Crystals from

Aqueous Solutions I, The Effect of Lead Chloride, *Journal of Chem. Physics*, **47**, 1967, pp.3687-3688.
8. G. D. Botsaris and R. C. Reid: Comments on the Letter By Glasner and Skurnik Entitled "Growth of Potassium Chloride Crystals from Aqueous Solutions I, The Effect of Lead Chloride, *Journal of Chem. Physics*, **47**, 1967, pp.3689-3690.
9. I. Sunagawa: Morphology of Crystals, Part B, Terra Science Pub., Tokyo, Japan, 1987.
10. N. M. Hwang, J. H. Hahn and D. Y. Yoon: Charged Cluster Model in the Low Pressure Synthesis of Diamond, *Journal of Crystal Growth*, **162**, 1996, pp.55-68.
11. N. M. Hwang: Crystal Growth by Charged Cluster: Focused on CVD Diamond, *Journal of Crystal Growth*, **198/199**, 1999, pp.945-950.
12. I. Sunagawa: Growth and Morphology of Diamond Crystals Under Stable and Metastable Conditions, *Journal of Crystal Growth*, **99**, 1990, pp.1156-1161.
13. D. V. Fedosayev, B. V. Deryaguin and I. G. Varasavskaja: *Surface Coatings Technology*, **38**, 1989, p.9.
14. I. N. Stranski and D. Totomanow: Keimbildungsgeschwindigkeit und Ostwaldsche Stufenregel, *Z. Physics Chem.*, **33**, 1933, pp.399-408.
15. G. Skandan: Processing of Nanostructured Zirconia Ceramics, *Nanostructured Materials*, **5**(2), 1995, pp.111-126.
16. L. A. Madonna, C. M. Sciulli and L. N. Canjar et al.: Low Temperature Cloud Chamber Studies on Water Vapour, *Proceedings of Physics Society*, London, U.K., **78**, 1961, pp.1218-1222.
17. H. M. Jang and N. M. Hwang: Theory of the Charged Cluster Formation in the Low Pressure Synthesis of Diamond: Part II, Free Energy Function and Thermodynamic Stability, *Journal of Materials Research*, **13**(12), 1998, pp.3536-3549.
18. J. O. M. Bockris and A. K. N. Reedy: *Modern Electrochemistry*, Plenum, New York, 1970.
19. C. T. R. Wilson: On the Condensation Nuclei Produced in Gases by the Action of Roentgen Rays, Uranium Rays, Ultraviolet Light and Other Agents, *Phil. Transactions Royal Society*, **192**, 1899, p.403.
20. J. J. Thomson: Charge Carried by Roentgen Ions, *Phil. Mag.*, **46**, 1898, p.528.
21. C. T. R. Wilson: Cloud-Chamber Technique, Proceedings of Royal Society, **85**, 1911, p.285.
22. C. T. R. Wilson: Cloud-Chamber Technique, Proceedings of Royal Society, **87**, 1912, p.277.
23. C. T. R. Wilson: Investigation on X-Rays and Beta-Rays by the Cloud Method, Proceedings of Royal Society, **104**(1), 1923, p.192.
24. W. E. Hazen: Formation of Charged Clusters by Measuring the Mobility, Physics Review, **65**, 1944, p.259.
25. M. Pollermann and I. N. Stranski: Diskussionstagung Der Bunsengesellschaft, Berlin, 1952.
26. J. J. Thomson: Conduction of Electricity Through Gases, Cambridge, U.K., 1906.
27. S. V. Shevkunov and P. N. Vorontsov-Velyaminov: Thermodynamic Properties

of Ion Clusters and their Effect on the Mechanism of Homogeneous Nucleation in an Ion Plasma, Calculation by the Monte Carlo Method in an NpT Ensemble, *High Temperature*, **20**(6), 1983, pp.803-809.

28. S. V. Shevkunov: Mechanism for Retarding the Nucleation Rate in a Dense Nonneutral Ion Plasma, *JETP*, **77**(3), 1993, p.413.
29. H. M. Jang and N. M. Hwang: Theory of the Charged Cluster Formation in the Low Pressure Synthesis of Diamond: Part I, Charge-Induced Nucleation, *Journal of Materials Research*, **13**(12), 1998, pp.3527-3535.
30. J. L. Katz, F. C. Wen and T. McLaughlin et al.: Nucleation on Photoexcited Molecules, *Science*, **196**(10), 1977, pp.1203-1205.
31. F. C. Wen, T. McLaughlin and J. L. Katz: Photoinduced Nucleation of Water Vapour, *Science*, **200**(19), 1978, pp.769-771.
32. J. L. Katz, T. McLaughlin and F. C. Wen: Condensation of a Supersaturated Vapour V, The Nucleating Effects of Ultraviolet Light on Vapours Containing Very Low Concentration of O-Tolualdehyde, *Journal of Chem. Physics*, **75**(3), 1981, pp.1459-1474.
33. B. Y. H. Liu: Aerosol Science and Microelectronics, *Journal of Aerosol Science*, **26**(4), 1995, pp.523-528.
34. L. Boufendi, A. Plain and J. PH. Blondeau et al.: Measurements of Particle Size Kinetics from Nanometer to Micrometer Scale in a Low-Pressure Argon-Silane Radio-Frequency Discharge, *Applied Physics Letters*, **60**(2), 1992, pp.169-171.
35. A. A. Howling, J. L. Dorier and CH. Hollenstein et al.: Negative Ion Mass Spectra and Particulate Formation in Radio Frequency Silane Plasma Deposition Experiments, *Applied Physics Letters*, **62**(12), 1993, pp.1341-1343.
36. A. Garscadden, B. N. Ganguly and P. D. Haaland et al.: Overview of Growth and Behaviour of Clusters and Particles in Plasma, *Plasma Sources Science Technology*, **3**, 1994, pp.239-245.
37. M. Adachi, K. Okuyama and N. Tohge et al.: Gas-Phase Nucleation in an Atmospheric Pressure Chemical Vapour Deposition Process for SiO_2 Films Using Tetraethylorthosilicate (TEOS), *Japan Journal of Applied Physics*, **31**(10a), 1992, pp.L1439-L1442.
38. M. Adachi, K. Okuyama and N. Tohge et al.: Precursors in Atmospheric-Pressure Chemical Vapour Deposition of Silica Films from Tetraethylorthosilicate/ Ozone System, *Japan Journal of Applied Physics*, **33**(3b), 1994, pp.L447-L450.
39. M. Adachi, K. Okuyama and N. Tohge: Particle Generation and Film Formation in an Atmospheric-Pressure Chemical Vapour Deposition Process Using Tetraethylorthosilicate, *Journal of Materials Science*, **30**, 1995, pp.932-937.
40. M. Adachi, K. Okuyama and T. Fujimoto et al.: Morphology Control of Films Formed by Atmospheric-Pressure Chemical Vapour Deposition Using Tetraethylorthosilicate / Ozone System, *Japan Journal of Applied Physics*, **35**(8), 1996, pp.4438-4443.
41. K. Okuyama, T. Fujimoto and T. Hayashi: Gas-Phase Nucleation in the Tetraethylorthosilicate (TEOS)/O3 APCVD Process, *AiChE Journal*, **43**(11a), 1997, pp.2688-2697.

42. N. P. Rao, S. Nijhawan and T. Kim et al.: Investigation of Particle Generation During the Low-Pressure Chemical Vapour Deposition of Borophosphosilicate Glass Films, *Journal of Electrochemical Society*, **145**(6), 1998, pp.2051-2057.
43. K. H. Homann and H. Wolf: Charged Soot in Low-Pressure Acetylene/Oxygen Flames, Presented at the Twenty-First Symposium (International) on Combustion, 1986.
44. K. H. Homann and J. Traube: Use of a Wien Filter for Mass and Velocity Analysis of Charged Soot Particles in a Nozzle Beam, *Ber. Bunsenges. Physics Chem.* **91**, 1987, pp.828-833.
45. Ph. Gerhardt and K. H. Homann, Ions and Charged Soot Particles in Hydrocarbon Flames I. Nozzle Beam Sampling: Velocity, Energy and Mass Analysis, Total Ion Concentrations, *Combust. Flame*, **81**, 1990, pp.289-303.
46. Ph. Gerhardt and K. H. Homann: Ions and Charged Soot Particles in Hydrocarbon Flames 2, Positive Aliphatic and Aromatic Ions in Ethyne/Oxygen Flames, *Journal of Physics Chem.*, **94**, 1990, pp.5381-5391.
47. R. Wegert, W. Wiese and K. H. Homann, Molecular Beam/Wien Filter Application to the Study of Charged Soot in Flames: Methodology and Mass Distributions of Particles in Butadiene Flames, *Combust. Flame*, **95**, 1993, pp.61-75.
48. P. K. Bachmann, D. Leers and H. Lydtin: Towards a General Concept Of Diamond Chemical Vapour Deposition, *Diamond and Related Materials*, **1**, 1991, pp.1-12.
49. I. D. Jeon, C. J. Park and N. M. Hwang et al.: Experimental Confirmation of Charged Carbon Clusters in the Hot Filament Diamond Reactor, *Journal of Crystal Growth*, **213**, 2000, pp.79-82.
50. W. S. Cheong, N. M. Hwang and D. Y. Yoon: Observation of Nanometer Silicon Clusters in the Hot-Filament CVD Process, *Journal of Crystal Growth*, **204**, 1999, pp.52-61.
51. H. S. Ahn, D. Y. Kim and N. M. Hwang, In Preparation.
52. K. S. Seo: Study on the Mechanism of Tungsten Thin Film Prepared by Evaporation, M.S. Thesis, Chunbuk University, 1998.
53. I. D. Jeon, L. Gueroudji and N. M. Hwang: Deposition of Yttria Stabilized Zirconia by the Thermal CVD Process, *Korean Journal of Ceramics*, **5**, 1999, pp.131-136.
54. O. Preining: The Physical Nature of Very, Very Small Particles and its Impact on their Behaviour, *Journal Aerosol Science*, **29**(5/6), 1998, pp.481-495.
55. P. Meakin: *Computer Simulation of Growth and Aggregation Process*, Martinus Nijhoff Publishers, Dordrecht, The Netherlands, 1986.
56. D. J. Shaw: Introduction to Colloid and Surface Chemistry, 2nd ed. Butterworths, London, U.K., 1970.
57. A. Terfort, N. Bowden and G. M. Whitesides: Three-Dimensional Self-Assembly of Millimetre-Scale Components, *Nature*, **386**, 1997, pp.162-164.
58. H. Fujita: Atom Clusters - New Applications of High-Voltage Electron Microscopy, Micro-Laboratory to Materials Science, *Ultramicroscopy*, **39**, 1991, pp.369-381.

59. N. M. Hwang, W. S. Cheong and D. Y. Yoon: Deposition Behaviour of Si on Insulating and Conducting Substrates in the CVD Process: Approach by Charged Cluster Model, *Journal of Crystal Growth*, **206**, 1999, pp.177-186.
60. H. Fujita: Studies on Atom Clusters by Ultra-High Voltage Electron Microscopy, *Materials Transactions*, JIM., **35**(9), 1994, pp.563-575.
61. H. Park, Y. K. Hong, J. S. Kim et al.: Surface Morphology of Laser Deposited Diamond-Like Films by Atomic Force Microscopy Imaging, *Applied Physics Letters*, **69**(6), 1996, pp.779-781.
62. R. Dat, O. Auciello and A. I. Kingon: Surface Roughness of $PbZr_xTi_{1-x}O_3$ Thin Films Produced by Pulsed Laser Ablation-Deposition, *Thin Solid Films*, **283**, 1996, pp.45-48.
63. K. D. Develos, M. Kusunoki and S. Ohshima: Studies on the Surface Morphology and Orientation of CeO_2 Films Deposited by Pulsed Laser Ablation, *Japan Journal of Applied Physics*, **37**, 1998, pp.6161-6169.
64. I. D. Jeon, L. Gueroudji, D. Y. Kim and N. M. Hwang: In Preparation.
65. T. T. Kodas and M. J. Hampden-Smith, *The Chemistry of Metal CVD*, VCH., Weinheim, 1994.
66. N. M. Hwang, J. H. Hahn and D. Y. Yoon: Chemical Potential of Carbon in the Low Pressure Synthesis of Diamond, *Journal of Crystal Growth*, **160**, 1996, pp.87-97.
67. N. M. Hwang and D. Y. Yoon: Thermodynamic Approach to the Paradox of Diamond Formation with Simultaneous Graphite Etching in the Low Pressure Synthesis of Diamond, *Journal of Crystal Growth*, **160**, 1996, pp.98-103.
68. N. M. Hwang: Deposition and Simultaneous Etching of Si in the Chemical Vapour Deposition (CVD) Process: Approach by the Charged Cluster Model, *Journal of Crystal Growth*, **205**, 1999, pp.59-63.
69. W. A. Yarbrough: Vapour-Phase-Deposited Diamond - Problems and Potential, *Journal of American Ceramics Society*, **75**(12), 1992, pp.3179-3200.
70. H. Kumomi, T. Yonehara and Y. Nishigaki et al.: Selective Nucleation Based Epitaxy (Sentaxy): Investigation of Initial Nucleation Stages, *Applied Surface Science*, **41/42**, 1989, pp.638-642.
71. H. Kumomi and T. Yonehara: Coarsening Phenomenon of Si Clusters, *Proceedings of Materials Research Society Symposium*, **202**, 1991, pp.83-88.
72. H. Kumomi and T. Yonehara: Selective Nucleation-Based Epitaxy (Sentaxy): A Novel Approach for Thin Film Formation, *Japan Journal of Applied Physics*, **36**(3b), 1997, pp.1383-1388.
73. N. M. Hwang and D. Y. Yoon: Thermodynamic Approach to the Chemical Vapour Deposition Process, *Journal of Crystal Growth*, **143**, 1994, pp.103-109.

Morphological Stability of Interfaces Driven by External Electric Fields

M. MARTIN
Institute of Physical Chemistry
Darmstadt University of Technology, Petersenstr. 20, 64287 Darmstadt
Germany

ABSTRACT

We have studied the stability and morphology of moving interfaces between two different cation-conductors AX and BX theoretically and experimentally. For the theoretical analysis we have performed Monte Carlo simulations showing that the morphological stability of the interface is determined by the difference in the ionic conductivities of the two crystals. In the unstable regime interface morphologies can be calculated ranging from fingers to fractals. Experiments have been performed with electrochemical cells containing diffusion couples composed of AgCl and KCl, CuCl or NaCl, respectively, which are either immiscible, partially miscible or totally miscible. As predicted by the theoretical calculations we observe morphological instabilities of the originally planar interface, if only migration takes place, while a critical electric field is necessary if migration and interdiffusion take place. The observed morphologies are discussed and compared with the morphologies obtained by Monte Carlo simulations.

1.0 INTRODUCTION

During a diffusion controlled solid state reaction, matter is transported towards (or away from) an interface which separates the reactants and the product. The progress of the reaction is strongly determined by the morphology of the interface, because the morphology determines the boundary conditions for the transport problem. One-dimensional diffusion with a planar interface is only a special case, since often morphological instabilities result in non-planar interfaces with complicated structures, which can roughly be classified as cellular, dendritic or fractal. The morphological stability of interfaces belongs to the wide class of, self-organisation or pattern-formation problems arising in biology, physics, chemistry and geology, in which a non-equilibrium system forms new structures. Examples are complex biological systems, dendrites during the growth of snow flakes, the Bénard-instability or spatio-temporal structures during the formation of the so-called Liesegang rings.[1-3] In all these different systems new structures develop through instabilities from stationary structures as a result of changed external parameters (control parameters).

In this paper we will review two types of solid state reactions where we have found complicated morphological instabilities. Our first example concerns the migration of a solid/solid interface between two ionic conductors AX and BX which is driven by an applied electric field. Our second example is interdiffusion of ionic conductors in an electric field. Because of the complexity of the processes, even in very simple systems, an accurate mathematical solution is impossible in most cases. A good example for this problem is the attempt to predict morphologies of the solid/gas phase boundary during the chemical reduction of CoO.[4] Linear stability analysis allows to predict stability of the planar phase boundary against infinitesimal disturbances, but cannot explain the experimentally observed more complex morphologies. At this point Monte Carlo (MC) simulations are helpful; they investigate processes on a microscopic scale, performing changes in systems exclusively on an atomic scale with a well-considered probability. Since the first paper by Metropolis et al.[5] MC simulation has become a more and more powerful tool,[6,7] to determine phase diagrams, interatomic interaction energies (inverse MC method), diffusion coefficients and correlation factors in binary solutions or short range order parameters of alloys. Also, a wide class of pattern formation problems has been studied using MC simulations, e.g. the diffusion limited aggregation model (DLA)[8] and improvements of it.[9, 10]

It is the aim of this paper to review theoretical and experimental results on the morphological stability of interfaces between ionic conductors AX and BX driven by external electric fields. The subsequent sections are organised as follows. In section 2 Monte Carlo simulations for the considered problem will be described, starting with the definition of a simple model.[11] Then, simulation results will be discussed.[12] In section 3 experiments with ionic conductors (AgCl, KCl, CuCl and NaCl) will be described and the results will be compared with the Monte Carlo simulations.

2.0 MONTE CARLO SIMULATION

2.1 The Model

Our two-dimensional model consists of the ionic crystals AX and BX, A and B representing the monovalent cations and X the monovalent anion which is identical in both crystals. Both crystals possess the same cubic structure, in which anions and cations form two interpenetrating cubic lattices. The lattice constants of both crystals are identical resulting in a coherent and initially planar phase boundary. The intrinsic disorder is assumed to be exclusively of the Schottky-type, i.e. vacancies occur in both sublattices. However, the anion vacancies are assumed to be immobile, which is a good approximation for the considered halides. The interaction energy of a cation with its ionic surrounding in the ionic crystal is given by the electrostatic energy E^C_{ij}, and the short range interaction energy by ε_{ij}, both of which are assumed to be pairwise additive. In MC simulations the short range interactions are often assumed to be nearest neighbour interactions while the determination of the coulombic energy is generally difficult because of its comparatively longer range. However, the following assumptions for the ions in the model system AX/BX simplify the calculations:

a. Concerning the coulombic interaction the ions are considered to be point charges located exactly at the above described lattice sites. Therefore, in a perfect crystal cations and anions possess the same coulombic potentials, respectively.
b. The short range interaction energies ε_{iX} (i = A, B) with the anions *X* are the same both for A and B.
c. Merely the short range interaction energies between the cations are dependent on the cation nature. This means that the values $E_{i,min}$ of the potential minima of the cations (equilibrium position) differ from each other by their nearest neighbour interaction energies ε_{ij} (i, j = A,B), simply denoted by ε_{AA}, ε_{AB} and ε_{BB}, and are given by

$$E_{i,min} = E^0_{i,min} + \sum_{\mathrm{j}} \varepsilon_{\mathrm{ij}}$$

where $E^0_{i,min}$ is the sum of all interaction energies apart from ε_{ij}.
d. The potential maximum of a cation jumping into a neighbouring vacancy lies exactly halfway between two cation sites where the repulsive forces of the large anions should be largest. Its value E_{max} is assumed to be constant as well as independent of the cation nature.
e. The interaction of cations with vacancies is set to zero, i.e. $\varepsilon_{AV} = \varepsilon_{BV} = 0$.

The jump frequency of a cation i into a neighbouring vacancy is given by a simple Boltzmann approach where the activation energy ΔE_i is determined by the difference of $E_{i,max}$ and $E_{i,min}$ (sum over the three nearest neighbours, since $\varepsilon_{iV} = 0$):

$$\Delta E_i = E_{i,max} - E_{i,min} = \Delta E^0_i - \sum_{j=1}^{3} \varepsilon_{ij} \tag{1}$$

The absolute jump frequency, Γ_i, for a cation with three nearest cation neighbours can be described as

$$\Gamma_i = \Gamma^0_i \cdot \exp(-\Delta E_i/kT) \tag{2}$$

where k is the Boltzmann constant, T the absolute temperature and Γ^0_i the vibrational frequency of the cation i (A or B) in its equilibrium position. This approach is identical to the one of Kikuchi and Sato[13] and often used in simulations.[7] For the sake of simplicity, the vibrational frequencies Γ^0_i and the activation energies without short range interactions, ΔE^0_i, are assumed to be identical for both cations, $\Gamma^0_{\mathrm{A}} = \Gamma^0_{\mathrm{B}} = \Gamma^0$ and $\Delta E^0_A = \Delta E^0_B = \Delta E^0$. For one of the four cations neighbouring the vacancy to exchange sites with it, the so-called relative jump frequency $\Gamma_{i,rel}$ is defined as follows:

$$\Gamma_{\mathrm{i,rel}} = \frac{\Gamma_{\mathrm{i}}}{\sum_{\mathrm{i=1}}^{4} \Gamma_{\mathrm{i}}} \quad \text{with} \quad \sum_{\mathrm{i=1}}^{4} \Gamma_{\mathrm{i,rel}} = 1 \tag{3}$$

The relative jump frequency represents a probability. In the MC simulations the relative jump frequencies of the four cations around the vacancy are determined and mapped to the interval (0,1). Then, a random number Q between 0 and 1 is determined, and the cation with Q in its interval is chosen for the jump.

For the following simulations the jump activation energy is fixed to $\Delta E^0 = 1.0$ eV. Because it is well known that in a regular solution all thermodynamic properties depend only on $\varepsilon = \varepsilon_{AB} - 0.5(\varepsilon_{AA} + \varepsilon_{BB})$, ε is fixed to $\varepsilon = 0.04$ eV, the interaction energy ε_{AB} between A and B is fixed to $\varepsilon_{AB} = 0.2$ eV, and only $\Delta\varepsilon = \varepsilon_{AA} - \varepsilon_{BB}$ is varied in such a way that ε remains constant. The jump rate of atoms in the pure crystal, e.g. Γ_A in AX, increases with increasing ε_{AA} (see Eqn. 1). Thus, $\Delta\varepsilon > 0$ corresponds to a higher ionic conductivity in AX while $\Delta\varepsilon < 0$ corresponds to a higher ionic conductivity in BX.

2.2 The Phase Diagram

The determination of the phase diagram of the quasi binary system (A,B)X with demixing below a critical temperature T_C is possible by MC simulation, as shown by Schweika[14] using the grand canonical ensemble. Starting from a lattice of any desired configuration and composition a site is chosen per random and a particle exchange A↔B or B↔A is performed if the transition probability $W = \exp(-\Delta H/kT)$ is larger than a random number Q between 0 and 1. Here, ΔH is the difference $H^{after} - H^{before}$ of the system energy after and before the exchange. For a given temperature this simulation leads to the minimum of the Gibbs energy, i.e. to the mole fractions $x_A = x_B = 0.5$ for temperatures $T > T_C$ and to the boundaries of the miscibility gap for $T < T_C$. For these simulations periodic boundary conditions were assumed, like in most of the simulations investigating bulk properties.[6] The results of the simulations, with $\varepsilon = 0.04$ eV are shown in Figure 1 and compared to calculations based on the Bragg-Williams approximation for regular solutions. The simulations lead to a critical temperature $T_C = 530$ K while the Bragg-Williams approximation (where the configurational entropy is approximated by the ideal mixing entropy) yields $T_C = 2\varepsilon/k = 923$ K. The miscibility gap is symmetric and independent of $\Delta\varepsilon$, as expected.

2.3 Interdiffusion Simulations

Before we start to simulate transference experiments in an electric field we will test our model by simulating a simple interdiffusion process without external electric field. These simulations have been carried out for several temperatures between 300 and 700 K to observe the morphological development of the phase boundary in the absence of a vacancy flux and to compare the resulting solubilities of AX in BX and vice versa with the miscibility gap of the phase diagram. For each temperature a simulation matrix of the width $\Delta X = 250$a and the height $\Delta Y = 180$a (a = lattice constant) was chosen (see Figure 2), consisting of two initially pure crystals AX and

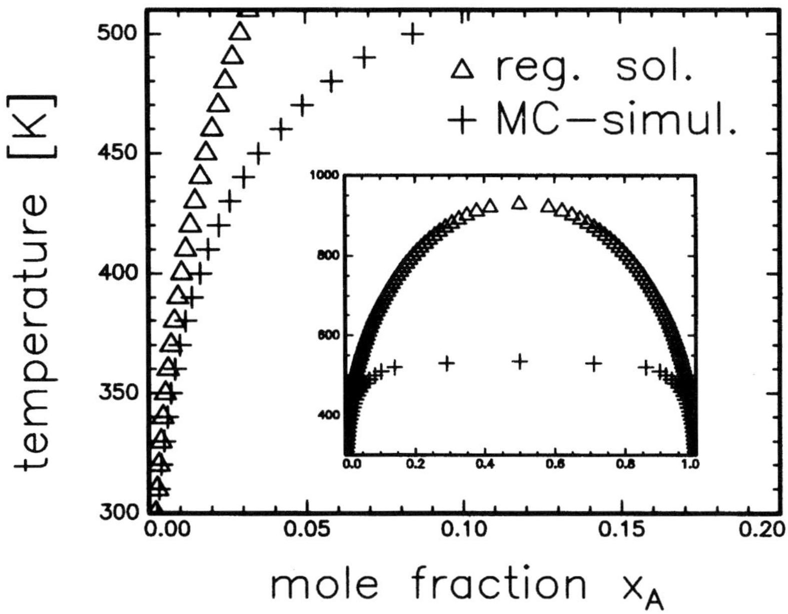

Fig. 1 Phase diagram of the quasi-binary system (A,B)X, determined with the MC-simulation (+) and calculated with the Bragg-Williams-approximation (Δ) for a regular solution.

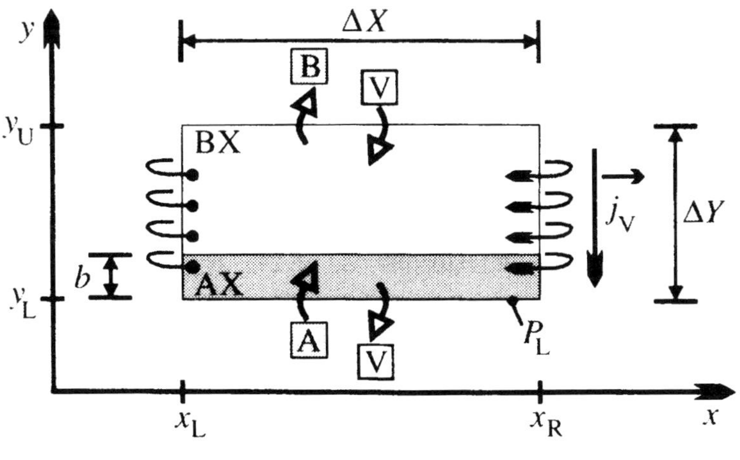

Fig. 2 Simulation matrix, characterised by extensions in x and y directions, ΔX and ΔY, respectively, and the initial position b of the phase boundary. The vacancy flux is directed from top to bottom. During interdiffusion the vacancy is reflected at the upper and lower boundary.

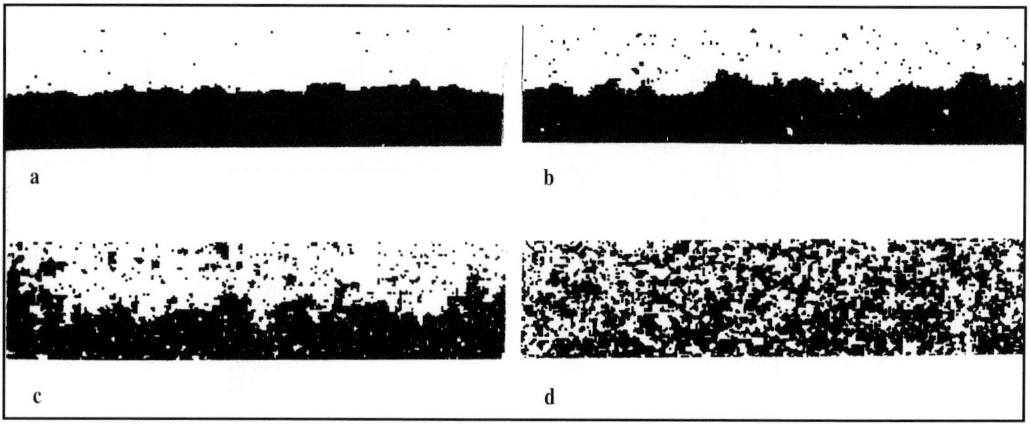

Fig. 3 The area around the initially flat phase boundary after 8.89×10^5 mcs for (a) $T = 300$ K, (b) 400 K, (c) 500 K and (d) 700 K. Initially the simulation matrix consisted of two pure crystals AX (black) and BX (white) of the same size.

BX with a flat phase boundary at $y = y_L + \text{b}$ with $\text{b} = 90\text{a}$. Periodic boundary conditions were assumed at the left and the right matrix boundaries at $x = x_L$ and $x = x_R$ respectively and the vacancy was reflected at the other boundaries. At the beginning of each simulation the vacancy was set to a randomly chosen position directly on the phase boundary. The configurations of the simulation matrices were analysed and stored after 10^8 Monte Carlo steps (mcs), i.e. after 10^8 vacancy jumps. Figure 3 shows the interface between AX (black) and BX (white) for temperatures of 300, 400, 500 and 700 K after 4×10^9 mcs, i.e. after 8.9×10^5 mcs per matrix site. The solubilities of AX in BX and vice versa were determined, i.e the number of dissolved A(B) particles in BX(AX). Comparing these results with the results of the phase diagram simulation, one observes lower solubilities in the case of the interdiffusion for all temperatures. These deviations become smaller with increasing temperature and amount to 8% at 300 K but only 4% at 450 K. This is due to the presence of the phase boundary which is missing in the simulations of the phase diagram. It consists of A-B bonds with relatively high energy and results in an important contribution to the total system energy in the case of relatively small size of simulation matrices. The influence of the phase boundary energy to the total system energy decreases with increasing temperature due to the increasing solubility of AX in BX and vice versa at the same time. In addition to this observation the phase boundary becomes more irregular and rougher with increasing temperature. For $T > 550$ K the phase boundary disappears during the interdiffusion, as expected.

The definition of the phase boundary follows Sapoval et al.[15] According to these authors, the interface is described as the line dividing the set of nearest neighbour A particles and the set of nearest neighbour B particles. This definition may be applied to the system (A,B)X since the phase boundary is solely determined by the cations due to identical anion sub-lattices of both crystals AX and BX.

2.4 Simulation of Transference Experiments

In this section the transference of A-ions through the system resulting in growth of AX at the expense of BX will be simulated. As before periodic boundary conditions are assumed at the boundaries at $x = x_L$ and $x = x_R$. However, to generate a vacancy flux through the diffusion couple from BX to AX, the following boundary conditions were defined for the upper and lower boundaries, respectively (see Figure 2):

a. Reaching the upper boundary of the simulation matrix, the vacancy is always reflected.
b. Reaching the lower boundary the vacancy is reflected with the probability $0 \leq P_L \leq 1$ or taken out of the matrix with the corresponding probability $1 - P_L$.
c. In the latter case the "old" vacancy is replaced by an A-cation, and a "new" vacancy is placed at a random position at the upper boundary.
d. At any time only one cation vacancy is in the matrix, the corresponding anion vacancy is not considered, because it does not influence the motion of the cation.

For $P_L = 1$ the previously discussed interdiffusion takes place, while for $P_L < 1$ a vacancy flux from BX to AX results. In addition to these asymmetric boundary conditions causing a directed vacancy flux, a constant, one-dimensional electric field $\nabla\varphi$ with $\partial\varphi/\partial x = 0$ is superimposed. As a consequence the jump frequencies of the cations parallel to the electric field change to

$$\Gamma_i = \Gamma_i^{\,0} \cdot exp\left(-\left(\Delta E_i^0 - \sum_{j=1}^{3} \varepsilon_{ij} - \Delta E_\varphi \right) / kT \right) \tag{4}$$

resulting in a favoured motion of the vacancy to the negative y-direction (from BX to AX). $\Delta E_\varphi = \frac{1}{2}\, a \cdot e \cdot \nabla\varphi$, where a is the distance between two neighbouring cation sites (lattice constant) and e the electric charge. For all the following simulations the parameter b was set to $b = 20$a, both crystals AX and BX were initially pure crystals and the temperature was set to $T = 300$ K (which means low mutual solubilities, see Figure 1).

2.4.1 *Variation of* $\Delta\varepsilon$

For $\Delta E_\varphi < 5 \cdot 10^{-5}$ eV the simulations gave the same results, independent of the magnitude of ΔE_φ, so for these simulations ΔE_φ was set zero. $\Delta\varepsilon$ was varied systematically in the range 0.04 eV $\leq \Delta\varepsilon \leq 0.06$ eV. P_L was set to zero (vacancy flux from top to bottom). It was observed that for $\Delta\varepsilon \leq 0.004$ eV the phase boundary roughened but remained stable during its motion through the considered matrix. For $\Delta\varepsilon \geq 0.0046$ eV on the other hand the phase boundary became morphologically unstable. For small $\Delta\varepsilon$ ($\Delta\varepsilon = 0.006$ eV) the disturbances can be described as wave-

a b

Fig. 4 The morphology of the phase boundary from transference simulations: (a) $\Delta\varepsilon = 0.01$eV (after 3.56×10^5 mcs) and (b) $\Delta\varepsilon = 0.04$eV (after 4.0×10^5 mcs), respectively, $\Delta E_\varphi = 0$ eV, Matrix size: 250×180 a^2 (width × height), Vacancy flux from the top to the bottom.

like, while with increasing $\Delta\varepsilon$ more finger-like structures appear (Figure 4a). With further increasing $\Delta\varepsilon$ these fingers became wider at their top and sometimes split, as shown in Figure 4b. The number of fingers increased while their width decreased. For $\Delta\varepsilon = 0.06$ eV the thickness of the fingers had decreased to a few lattice constants.

2.4.2 Variation of $\Delta\varepsilon$ and ΔE_φ

In the above described range of $\Delta\varepsilon$ the electrical energy difference ΔE_φ was systematically varied between 10^{-4} eV and 5×10^{-2} eV. For a = 0.5 nm this corresponds to a variation of $\Delta\varphi$ between 4×10^5 Vm^{-1} and 10^8 Vm^{-1}, which can be compared with real systems. Figure 5 shows some representative results from simulations with a given set of $\Delta\varepsilon$ and ΔE_φ. The following observations were made:

a. For $\Delta\varepsilon < 0$, no instabilities of the moving phase boundary could be detected, even for relatively large values of $\Delta\varphi$ (10^8 Vm^{-1}). This means that the interface is always stable if the slower phase (in this case AX) grows.
b. For $\Delta\varepsilon = 0$ eV ($\varepsilon_{AA} = \varepsilon_{BB}$) and $\Delta E_\varphi = 0$ the phase boundary remains planar, while the formation of a smooth wave-like phase boundary could be observed for the first time for $\Delta E_\varphi > 2.5 \times 10^{-4}$ eV. This means that a critical electric field is necessary for instabilities.
c. Generally, the number of the growing disturbances (fingers) increases with increasing ΔE_φ. The same applies to the tendency of the disturbances to split and branch. At the same time the width of these branches became smaller.
d. Sometimes branches appear which are disconnected from the rest of the AX crystal resulting in nearly round precipitations in BX, especially for $\Delta\varepsilon \geq 0.04$ eV.

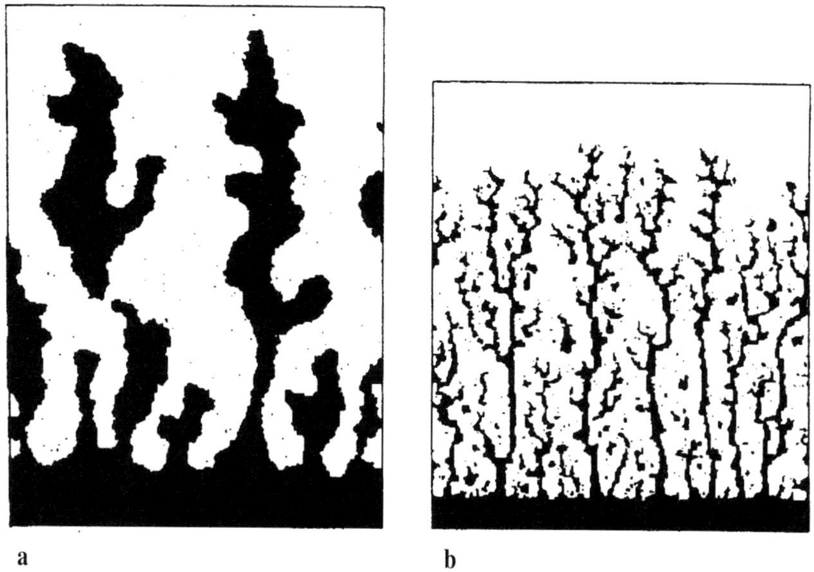

Fig. 5 The morphology of the phase boundary from transference simulations: (a) $\Delta\varepsilon = 0.02$ eV and $\Delta E_\varphi = 10^{-3}$ eV after 5.1×10^4 mcs and (b) $\nabla\varepsilon = 0.04$ eV and $\Delta E_\varphi = 5 \times 10^{-3}$ eV after 2.6×10^4 mcs. Vacancy flux from the top to the bottom.

e. With increasing ΔE_φ, the growth of the disturbances, especially of the split and tree-like shapes, became nearly parallel to the potential gradient $\nabla\varphi$. The same applies for decreasing $\Delta\varepsilon$ at constant ΔE_φ.

2.5 Simulations with Initially Disturbed Phase Boundaries

In these simulations the planar phase boundary was replaced by a periodically disturbed phase boundary. These simulations correspond to a finite amplitude stability analysis in contrast to the previous sections where small fluctuations of the phase boundary caused the instabilities (linear stability analysis). The morphology of such a disturbed phase boundary is described by

$$y(x) = Z\ sin\left(\frac{2\pi}{\lambda}x\right) + b \tag{5}$$

Z is the amplitude, λ the wavelength and b the mean distance of the phase boundary from the lower boundary of the simulation matrix. The parameter b was set to 40a at the beginning of each simulation of this kind. The amplitude Z and the wavelength λ were varied systematically in the range $2a \leq Z \leq 20a$ and $5a \leq \lambda \leq 300a$, respectively. Due to the periodic boundary conditions the width ΔX of the matrices was always a multiple of the wavelength λ of the sine shaped disturbance, at least twice the value of λ. P_L was always set to zero, the temperature remained constant at $T = 300$ K.

2.5.1 Stability Diagrams

The configurations of the simulation matrices were analysed and stored after 10^9 mcs and the shape of the phase boundary was fitted to the sine-function in Eqn. 5 with *b* and Z as fit parameters to observe the development of the amplitude with continuing vacancy flux. The wavelength λ was found to remain constant for a certain time. To compare simulations with different matrix sizes, the dependence of the amplitude on the fraction x'_A of A-particles in the whole simulation matrix was investigated since this fraction is a measure for the real time. Figure 6 shows the development of the shape of the phase boundary obtained from a simulation with λ = 93.3a, Z(start) =10a and $\Delta\varepsilon$ = 0.004 eV after 1, 4, 7, 12 and 17 × 10^8 mcs respectively. For comparison, the fitted sine function is also shown (dotted line). The deviations of the phase boundary from the ideal sine shape increase with increasing simulation time. Nevertheless, the goodness of the fit remains well up to $x'_A = 0.6$ for most of the simulations. Simulations have been carried out in the range of $5a \leq \lambda \leq 300a$ for four different values of $\Delta\varepsilon$ ($\Delta\varepsilon$ = 0.002 eV, 0.0026 eV, 0.003eV and 0.004 eV) at which the initially planar phase boundaries remained stable during the simulations described in section 2. 4.1. The starting amplitudes have been varied from Z = 2a to Z = 20a. For a given starting amplitude Z(start) a critical wavelength λ_{crit} was found below which the initially given disturbance disappeared during the simulation. λ_{crit} is dependent on Z(start) as well as on $\Delta\varepsilon$. For wavelengths $\lambda > \lambda_{crit}$ the amplitudes increased with continuing vacancy flux. Figure 7 shows for different $\Delta\varepsilon$, a stability diagram with stable and unstable regions which are separated by the critical-wavelength curve, λ_{crit}(Z(start)). Generally, λ_{crit} decreases with increasing Z(start). The same applies for increasing $\Delta\varepsilon$ at constant Z(start). The "error bars" in Figure 7 indicate the interval of λ in which the amplitude either disappeared or grew as a simulation is repeated several times with a particular set of parameters. Setting Z(start) < 4a the disturbances always disappeared during the simulation, independent of their wavelengths, i.e. the phase boundary is totally stable against disturbances of finite amplitude Z < 4a. Setting Z(start) > 15a and $\lambda \leq 50a$ the phase boundaries lost their sine-wave like shape after relatively short simulation times. A similar stability diagram is found when the electric field, i.e. ΔE_φ, is varied.[12]

2.5.2 The Growth Velocity of the Disturbances

Setting the starting amplitude Z(start)=10a and $\Delta\varepsilon$ = 0.004 eV the time development of the amplitude Z of the disturbances has been investigated for different wavelengths. An almost linear dependence of the amplitude Z on x'_A i.e. on time, was observed throughout the whole range of the considered wavelengths. The amplitude decreases right from the beginning for wavelengths λ < 60a. Thus the growth velocity

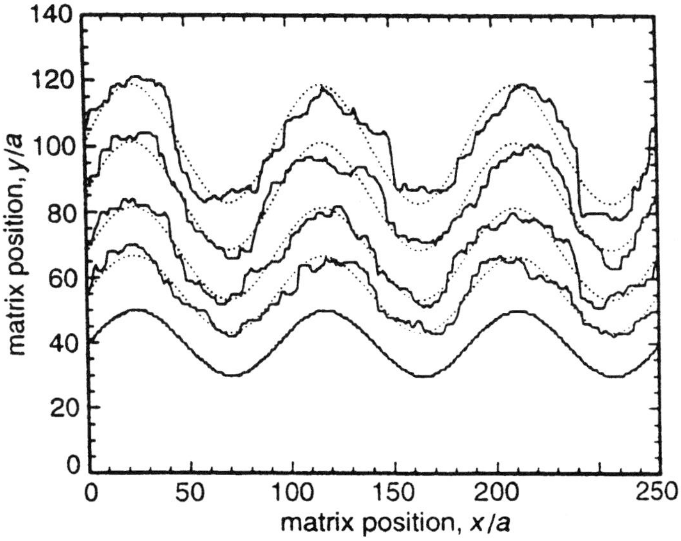

Fig. 6 The morphology of the phase boundary (solid lines) from disturbance simulations with λ= 93.3a, Z(Start) = 10a and $\Delta\varepsilon$ = 0.004 eV after 1, 4, 7, 12 and 17 $\times$ 10^8 mcs, respectively, with the corresponding fits (dotted lines). Matrix size: 280 $\times$ 180 a^2.

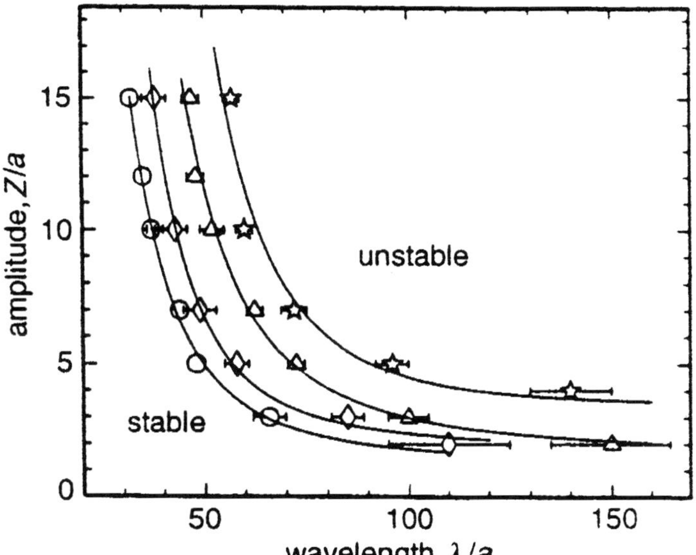

Fig. 7 Stability diagram with the critical wavelength λ_{crit} in dependence of the starting amplitude of the disturbance for different $\Delta\varepsilon$ (0.002 eV (o), 0.0026 eV (Δ), 0.003 eV ($\Diamond$), 0.004 eV (*)) at ΔE_φ = 0 eV. The solid lines have no physical relevance.

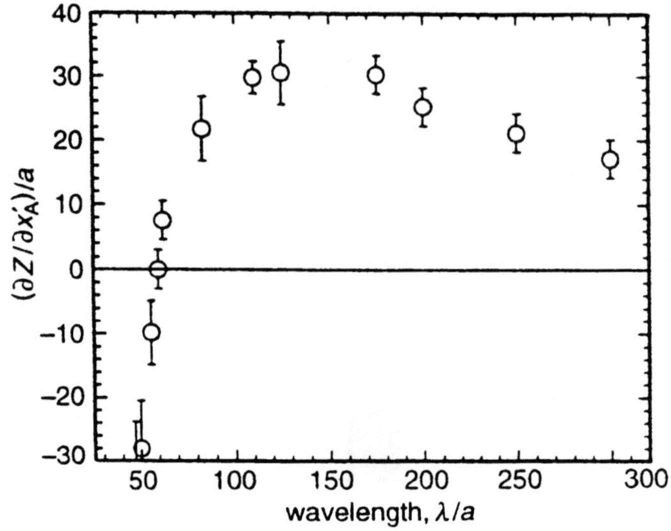

Fig. 8 Growth velocity, $v_z = \partial Z/\partial x'_A$, of a finite amplitude disturbance for Z(start) = 10a and $\Delta\varepsilon$ = 0.004 eV.

$v_Z = \partial Z/x'_A$ is negative, i.e. the disturbances disappear with time. For λ = 60a, Z remains approximately constant, i.e. $v_z \approx 0$. For λ > 60a the amplitude growths exclusively, $v_Z > 0$, i.e. the surface is unstable against finite amplitude disturbances. In Figure 8, v_Z is plotted versus the investigated wavelengths. v_Z shows a maximum at $\lambda \approx$ 125a and decreases afterwards but remains positive. At the wavelength λ_{crit}, here 60a, a transition from positive to negative values of v_Z occurs, i.e. for $\lambda < \lambda_c$ the amplitudes decrease in time showing that the interface is stable against finite amplitude disturbances.

2.5.3 Variation of P_L

For two particular wavelengths, λ = 70a and λ =125a, respectively, setting Z(start) = 10a, $\Delta\varepsilon$ = 0.004 eV and ΔE_φ = 0 eV the reflection probability P_L of the vacancy at the lower boundary of the simulation matrix was varied. The wavelength λ = 70a is only slightly larger than the critical wavelength $\lambda_{crit} \approx$ 60a, while λ = 125a corresponds to the maximum of v_Z. The objective was to investigate whether an increase of P_L, i.e. a decrease of the vacancy flux, would lead to stable phase boundaries, i.e. to a decrease of the disturbances in comparison to the case of $P_L = 0$ where the considered disturbances grew for both λ = 70a and λ = 125a. P_L has been varied in the range of $0.9 \leq P_L \leq 0.99999$. For λ = 70a and for λ = 125a values of P_L above which the disturbances disappeared were found during the simulations. In the case of λ = 70a this occurred at first for P_L = 0.9, for λ = 125a a relatively sharp transition from

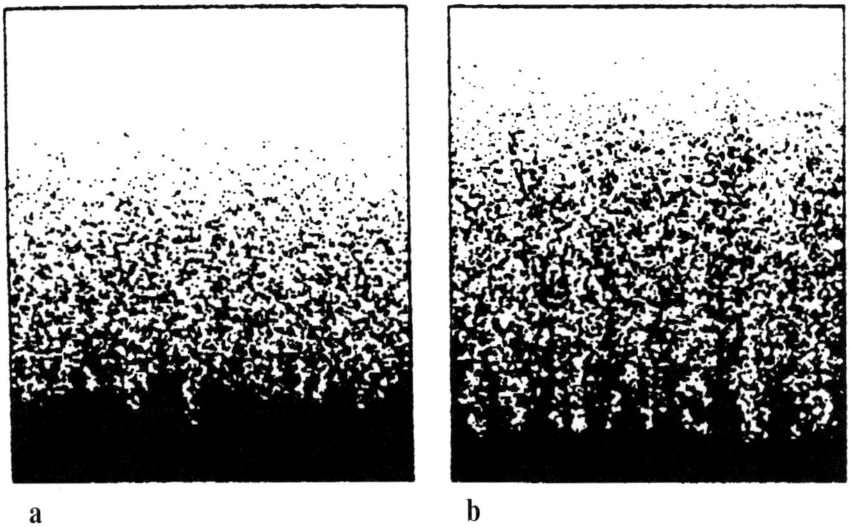

Fig. 9 Interdiffusion simulations above T_C (T = 900 K) with $\Delta E_\varphi = 10^{-2}$ eV, $\Delta\varepsilon = 0.06$ eV, 4.4×10^3 mcs (a) and $\Delta\varepsilon = 0.1$ eV, 8.8×10^3 mcs (b) respectively. Starting conditions: pure crystals AX and BX with a planar phase boundary.

positive to negative values of v_Z was found between $P_L = 0.999$ and $P_L = 0.9995$. As a consequence the stability curve in Figure 8 moves to the right and downwards at the same time with increasing reflection probability of the vacancy at the lower matrix boundary, i.e. with a decreasing vacancy flux. At sufficiently large values of P_L corresponding to sufficiently small vacancy fluxes the phase boundary becomes stable independent of λ.

2.6 Simulations above T_C

For particular values of $\Delta\varepsilon$ and ΔE_φ simulations have been carried out for temperatures above the critical temperatures (900 and 1100 K) to check whether inter-diffusion profiles might become unstable. Throughout these simulations a distinct interdiffusion took place as expected for temperatures above the critical temperature of $T_C \approx 530$ K. It was observed that, for $\Delta\varepsilon \leq 0.08$ eV and $\Delta E_\varphi \leq 5 \times 10^{-3}$ eV the moving inter-diffusion front remained stable, i.e. a relatively homogeneous distribution of A particles in x-direction was found, while a gradient of the mole fraction x_A was detected only in y-direction, comparable to results of inter-diffusion simulations in the absence of a vacancy flux in section 2.3. Increasing $\Delta\varepsilon$ and/or ΔE_φ the mole fraction profile $x_A(x)$ of the A particles became inhomogeneous in special parameter regions. Figure 9 shows the results of simulations with $\Delta E_\varphi = 10^{-2}$ eV and $\Delta\varepsilon = 0.06$ eV and 0.1 eV, respectively. While for $\Delta\varepsilon = 0.06$ eV the inter-diffusion front remains stable, i.e. x_A is constant in x-direction, x_A remains no more constant in x-direction for $\Delta\varepsilon = 0.1$ eV. This means that interdiffusion in a sufficiently strong external electric field results in an instability of the diffusion profiles.

2. 7 The Migration of AX Precipitations in an Electric Field

Placing some round AX precipitations of different diameters into a pure BX crystal one observes a migration of these precipitations in a direction opposite to the vacancy flux. The morphological development of these initially round particles depends on their diameter. Figure 10 shows a) the initial state of the simulation matrix of the size of $600 \times 300a^2$ with precipitations of the diameters $d = 20, 30, 40, 50, 60, 70$ and 80a, respectively, centred at the same y-position and b) the resulting morphologies after sufficient simulation time. The precipitations with diameters $d = 20$a and $d = 30$a keep their more or less round shape during the migration. For precipitations with diameters $d > 50$a one observes the formation of finger-like structures at their upper side which branch in the case of the largest precipitations. It is quite obvious that the lower boundaries of the larger precipitations develop into a smooth shape perpendicular to the direction of the electric potential gradient. Starting with squares instead of round precipitations the same phenomena were observed independent of their initial orientation. In every case the upper side of the precipitations became morphologically unstable while the lower side developed into a flat and stable phase boundary.

2.8 Summary of the Monte Carlo Simulations

Using a simple model of a cation-conducting ionic system AX/BX with identical, perfect anion sublattices and a Boltzmann approach for the cation jump frequencies transference experiments with varying vacancy fluxes and interdiffusion experiments above and below the critical temperature T_C were simulated. The starting point of each simulation of a transference or interdiffusion experiment was a system of two pure crystals AX and BX with an initially planar phase boundary. Concerning the transference simulations the following observations have been made. Below the critical temperature T_C (limited miscibility between AX and BX) the phase boundary roughens but remains morphologically stable if the less mobile phase (with the smaller value of ε_{ii}) is the growing one, independent of temperature and almost independent of the external driving force. In comparison to that one observes a critical parameter $\Delta\varepsilon_c = \varepsilon_{AA}$-$\varepsilon_{BB}$ above which the phase boundary becomes morphologically unstable if the more mobile phase (with the larger value of ε_{ii}) is the growing one. In this case $\Delta\varepsilon_c$ is dependent of the temperature, the boundary conditions (especially the vacancy reflection probability at the lower side of the simulation matrix, i.e. the vacancy flux) and the external driving force. With increasing $\Delta\varepsilon$ and ΔE_φ one observes a transition from finger-like to branched structures. In some cases the latter ones can

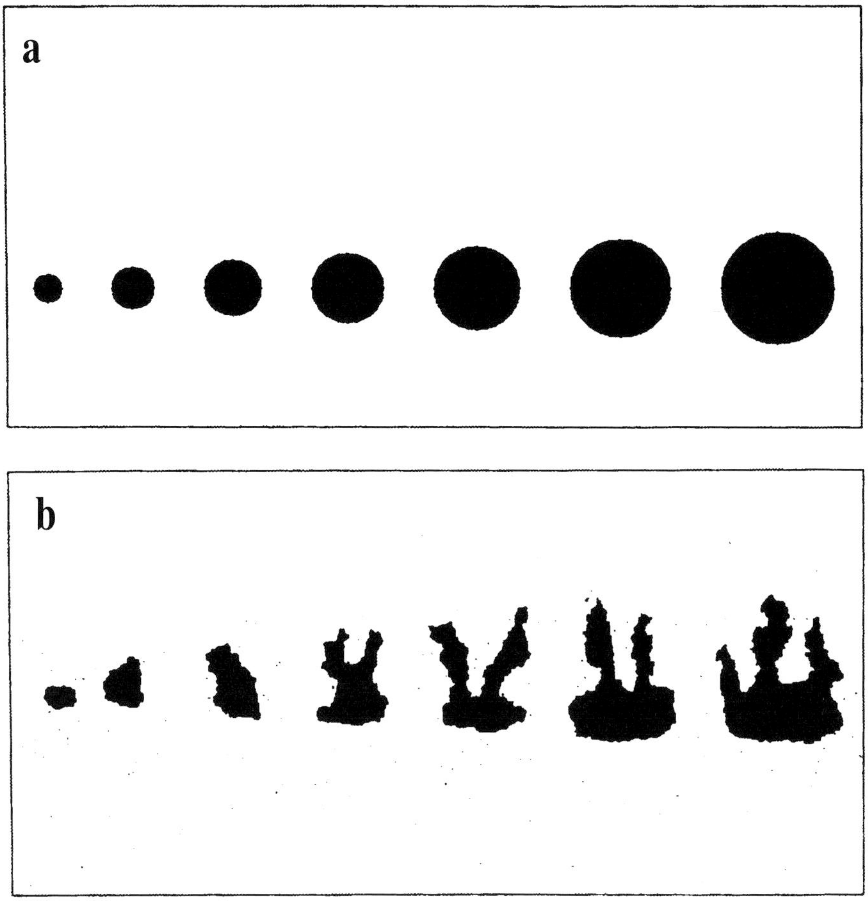

Fig.10 Migration of initially round AX precipitations of different diameters (d = 20, 30, 40, 50, 60, 70 and 80a) in BX with $\Delta\varepsilon = 0.02$ eV in an electric field ($\Delta E_\varphi = 10^{-3}$ eV). Vacancy flux from the top to the bottom. (a) initial matrix and (b) after 7.22×10^3 mcs.

be described as fractals. For temperatures above the critical temperature T_C instabilities of diffusion fronts could be observed by increasing the applied electric field.

In addition to these simulations the stability of periodic sine shaped disturbances was investigated varying the wavelength, the amplitude, the vacancy flux and the external driving force (finite amplitude stability analysis). The growth velocity of the disturbances found in the simulations shows a similar behaviour as known from the analytical linear stability analysis of the so-called Mullins-Sekerka instability.[16] Decreasing the vacancy flux the transition from stable to unstable migration of the disturbed phase boundary no longer occurred, i.e. the phase boundary became stable during its migration, independent of the wavelength of the disturbances.

3.0 EXPERIMENTS

3.1 Interface between Two Immiscible Ionic Conductors

To study the morphology of an interface between two crystalline solids we consider a diffusion couple consisting of two ionic conductors and apply an external electric field as driving force for the motion of the solid/solid interface. First, we study a phase boundary between two immiscible ionic conductors (corresponding to $T << T_C$ in the MC simulations). The experiments were performed with the quasi-binary system AgCl-KCl.[17] Below the eutectic temperature, $T_e = 306°C$, there is no mutual solubility, although AgCl as well as KCl crystallise in the rocksalt-structure. However, the lattice constants of both compounds are very different, resulting in a mismatch of about 12%. Both compounds are practically pure cation conductors. In AgCl (cation Frenkel disorder) silver-ions are mobile in the interstitial sublattice, while in KCl (Schottky disorder) potassium-ions are mobile via cation vacancies. As a result of the different disorder types the ionic conductivity in AgCl is about 6 orders of magnitude larger than that of KCl. Experiments were done with AgCl- and KCl-single crystals, which were used in an electrochemical cell +/Ag/AgCl/KCl/Ag/- . Silver-ions, Ag^+, are driven by the applied electric potential difference from the Ag-anode to the boundary AgCl/KCl. Since Ag is not soluble in KCl, the following exchange reaction takes place:

$$Ag^+ + KCl \rightarrow AgCl + K^+ \quad (6)$$

This exchange reaction has two consequences:

i. It allows further charge transport via K^+-ions

ii. The AgCl-phase grows at the expense of the KCl-phase.

Using the results of the MC simulations in the previous sections we expect the following results: since the ionic conductivity in AgCl is much larger than in KCl the planar solid/solid phase boundary between AgCl and KCl should be morphologically unstable, if we apply the potential difference in such a way that AgCl grows. If we reverse the applied voltage, we expect the boundary to be stable. A typical experimental result, demonstrating that the solid/solid interface is indeed morphologically unstable is shown in Figure 11. We note that AgCl grows into KCl in the form of "trees", which are typical of fractal growth phenomena.[9] Closer inspection shows that these trees are two-dimensional objects, with certain orientations in the KCl-matrix, which are probably the result of elastic effects due to the different lattice constants of both materials. If the polarisation of the cell is reversed, i.e. –Ag/AgCl/KCl/Ag/+, the plane interface between AgCl and KCl is morphologically stable as expected. A more detailed discussion can be found in Ref. 17.

3.2 Interface between Two Partially Miscible Ionic Conductors

To study the morphology of a moving interface between two partially miscible ionic conductors, corresponding to $T \leq T_c$ in the MC simulations, we use the quasi-binary

Fig. 11 Cross section of the unstable interface AgCl/KCl in an electric field.

system AgCl-CuCl. In contrast to the previously discussed case, there is now mutual solubility in the percent range below the eutectic temperature, $T_e = 258°C$. Again, both compounds are practically pure cation conductors, however, in contrast to AgCl/KCl, they have similar ionic conductivities. Both halides crystallise in cubic structures with a small lattice mismatch of about 2%. A typical experimental result[18] is depicted in Figure 12 where the unstable interface AgCl/CuCl is shown after the CuCl-crystal has been dissolved. Now the three-dimensional structure of the unstable interface and the existence of fingers are obvious. At the tip, the fingers exhibit well defined crystal planes, typical for crystal growth. Whether these planes represent the equilibrium shape of the crystal or non-equilibrium forms is not yet clear. Since now interdiffusion and migration cause the interface motion, a critical driving force is necessary for instabilities, as predicted by the Monte Carlo simulations. A detailed description of these experiments including an analytical stability analysis will be published in due course.[19]

3.3 Instability of Diffusion Fronts

In the previous two sections we have demonstrated that the moving interface between two immiscible or partially miscible ionic conductors, which is driven by an external electric field, is morphologically unstable when the "faster" of both phases grows, in agreement with the MC simulations. In this section we will analyse what happens if we go to the other extreme, i.e. if the two ionic conductors are totally miscible. The

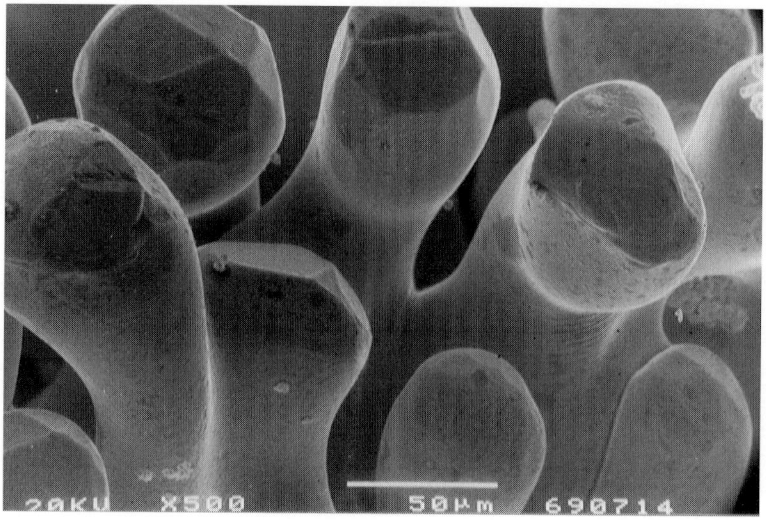

Fig. 12 Three-dimensional view of the unstable interface AgCl/CuCl in an electric field.

MC simulations in section 2.6 predicted instabilities of the diffusion fronts above a critical electric field strength. A good example for two totally miscible ionic conductors is provided by the system AgCl-NaCl. Between the critical temperature T_c = 198°C (below which there is a miscibility gap with mutual solubilities) and the melting temperature of AgCl, both compounds form a complete solid solution. As in AgCl/KCl, Ag^+-ions in AgCl are mobile in the interstitial sublattice, Na^+-ions in NaCl are mobile via cation vacancies, and the ionic conductivity, σ, in AgCl is orders of magnitude larger than in NaCl. In the solid solution, $(Ag_{1-x}Na_x)Cl$, σ decreases approximately exponentially with the composition x.[20] Experiments were performed in the same way as before, i.e. with an electrochemical cell +/Ag/AgCl/NaCl/Ag/-.[21] As in the last experiment, we have now a combination of two effects, interdiffusion of Ag^+ and Na^+, and drift in the externally applied electric potential gradient. Since the diffusion coefficients of Ag^+ and Na^+ depend exponentially on composition, x, pure interdiffusion would result in very steep interdiffusion profiles. If we apply in addition an electric field in such a way that the faster "phase" (in this case the Ag-rich part of the solid solution) grows, we expect morphological instabilities, as predicted by the MC simulations in section 2.6. It has to be emphasised, however, that these expected morphological instabilities are instabilities of diffusion fronts in a single-phase system without any phase boundary.

Figure 13 shows a typical result of an interdiffusion and drift experiment. The picture was obtained in an electron microscope using backscattered electrons, which indicate a phase contrast. In the picture, the steep concentration gradient in the interdiffusion zone appears as sharp phase contrast, but microprobe analysis confirms a steep concentration gradient in a single phase system. While the interdiffusion

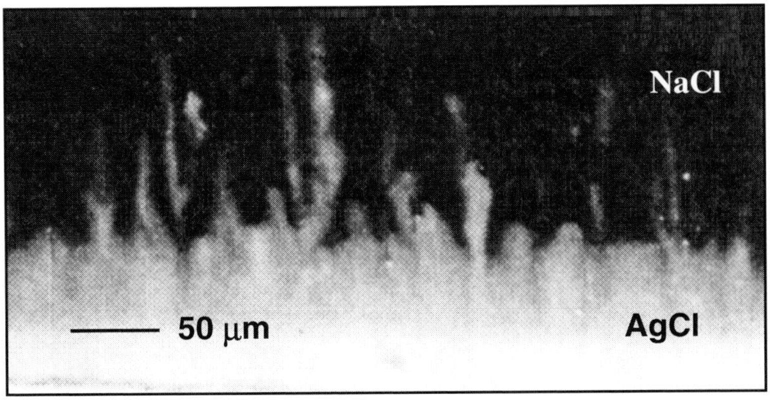

Fig. 13 Cross section of the interdiffusion zone AgCl-NaCl in an electric field.

profile without an external electric field is one-dimensional, i.e. planar as expected, it is by no means one-dimensional with an applied electric field. Instead one can observe instabilities of the diffusion fronts in the form of 'fingers'.

These experimental results are in agreement with the predictions of the Monte Carlo simulations in section 2.6. We have also performed a numerical stability analysis of the interdiffusion and drift problem. Starting from a periodically disturbed initial concentration step (Figure 14a) we have solved the two dimensional transport problem for Ag- and Na-ions using a finite element method.[22] We have modelled the transport problem using two assumptions:

i. The diffusion coefficients of both interdiffusing cations are identical and
ii. They are exponentially dependent on composition x, $D \propto \exp(b \times x)$.

For the special case of AgCl-NaCl the second assumption is a good approximation (both diffusion coefficients show the same exponential dependence on composition), while the absolute values differ by a factor 10. However, this simplification in the calculations should not influence the overall picture, as long as we consider the previously mentioned principle, that the faster phase should grow to obtain instabilities, which is assured by the exponential concentration dependence of the diffusion coefficient(s). The results of the numerical integration for one special parameter set are shown in Figure 14. One can see that the amplitude of the initial disturbance grows in time, on the other hand the width of the concentration step spreads in time, as a consequence of interdiffusion. By varying the strength of the non-linearity, b, of the diffusion coefficient(s), and the applied voltage, $\Delta\phi$, we find a critical voltage, $\Delta\phi_c$, which has to be applied to obtain instabilities. For smaller voltages the diffusion fronts become again planar, while for larger voltages the initial disturbance grows in time. $\Delta\phi_c$ decreases with increasing b, i.e. the stronger the diffusion coefficient varies with composition the weaker the critical voltage for the onset of instabilities may be. It should be emphasised that we have found

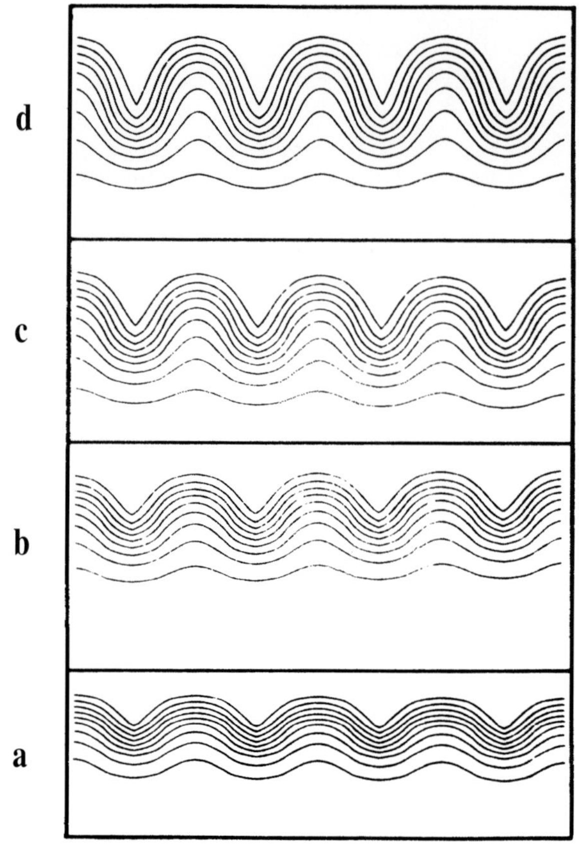

Fig. 14 Contour plots of the diffusion fronts during interdiffusion in an electric field (a) initial condition, (b), (c) and (d) increasing time.

morphological instabilities of diffusion fronts using very simple transport equations for the moving cations where the fluxes of both cations consist of a usual diffusion term ($\propto D(c) \times \nabla c$) and a drift term ($\propto \nabla \phi$). The non-linear concentration dependence of the diffusion coefficient(s) is the real cause for the instabilities. If b vanishes, i.e. the diffusion coefficient is constant, we obtain no instabilities, even for very large applied voltages.

4.0 SUMMARY

To analyse the morphological stability of solid/solid interfaces and diffusion fronts in cation conducting systems we have performed Monte Carlo simulations and experiments with electrochemical cells containing diffusion couples of AgCl and KCl, CuCl and NaCl, which are either immiscible, partially miscible or totally miscible. The MC simulations show that a plane interface between two cation conductors is morphologically unstable if the phase with the higher ionic conductivity grows and a critical electric field strength is exceeded. Varying the difference in the ionic conductivities of both crystals and the electric field strength, finger- and fractal-like morphologies are obtained. In addition, the simulations show that diffusion fronts also can become unstable as a result of an external electric field.

The experiments with the model systems AgCl/KCl, AgCl/CuCl and AgCl-NaCl confirm the theoretical predictions. In the immiscible system AgCl/KCl the ionic conductivity of AgCl is orders of magnitude higher than in KCl. Thus, when AgCl grows the interface is always unstable with a critical field strength of zero. The morphology of the interface can be described as fractal-like, in qualitative agreement with the simulation results. In the partially miscible system AgCl/CuCl the ionic conductivities are of similar magnitude and the interface is only unstable if a critical electric field strength is exceeded. The morphology of the interface can be described as finger-like, in qualitative agreement with the simulation results. Finally, we have performed interdiffusion experiments in an electric field with the totally miscible system AgCl-NaCl. As predicted by the simulations the diffusion fronts exhibit finger-like morphologies if a critical electric field is exceeded. In summary, we have shown that there is qualitative agreement between the MC simulation results (using a simple nearest neighbour interaction model) and experimental results. However, it must be emphasised that the simulations predict morphologies only on an atomistic scale. Whether these results really describe the experimentally found morphologies in the μm-range is still an open question.

5.0 ACKNOWLEDGEMENT

Financial support by the Deutsche Forschungsgemeinschaft and the Fonds der Chemischen Industrie is acknowledged.

6.0 REFERENCES

1. H. HAKEN: *Synergetik*, Springer Verlag, Berlin, 1982.
2. J. S. LANGER: 'Instabilities and Pattern Formation in Crystal Growth', *Review of Modern Physics*, 1980, **52**, pp.1-28.
3. H. E. STANLEY and N. OSTROWSKY: *On Growth and Form*, Martinus Nijhoff Publishers, Boston, 1986.
4. M. MARTIN and H. SCHMALZRIED: 'Cobaltous Oxide in an Oxygen Potential Gradient: Morphological Stability of the Phase Boundaries', *Ber. Bunsenges. Phys. Chem.*, 1985, **89**, pp.124-130.
5. N. METROPOLIS, A. W. ROSENBLUTH, M. N. ROSENBLUTH, A. H. TELLER and

E. TELLER: 'Equation of State Calculations by Fast Computing Machines', *Journal of Chem. Phys.*, 1953, **21**, pp.1087-1092.

6. K. BINDER and D. W. HEERMANN: *Monte Carlo Simulation in Statistical Physics*, Springer Verlag, Berlin Heidelberg, 1989.
7. G. E. MURCH: 'Simulation of Diffusion Kinetics with the Monte Carlo Method', *Diffusion in Crystalline Solids*, G. E. Murch and A. S. Nowick, eds., Academic press Inc., Orlando, 1984.
8. T. A. WITTEN and L. M. SANDER: 'Diffusion-Limited Aggregation, a Kinetic Critical Phenomenon', *Physics Review Letters*, 1981, **47**, pp.1400-1403.
9. P. MEAKIN: 'Fractal Structures', *Prog. Solid St. Chem.*, 1990, **20**, pp.135-233.
10. P. TIGELMANN and M. MARTIN: 'Monte Carlo Simulation of Surface Structures During Oxide Reduction', *Physica A*, 1992, **191**, pp.240-247.
11. G. SCHULZ and M. MARTIN: 'Computer Simulations of Pattern Formation in Ion-Conducting Systems', *Solid State Ionics*, 1997, **101-103**, pp.417-423.
12. G. SCHULZ and M. MARTIN: 'Monte Carlo Simulations of Pattern Formation at Solid/Solid Interfaces', *Faraday Discuss.*, 1997, **106**, pp.291-306.
13. R. KIKUCHI and H. SATO: 'Diffusion Coefficient in an Ordered Binary Alloy', *Journal of Chem. Phys.*, 1972, **57**, pp.4962-4979.
14. W. SCHWEIKA: 'Monte Carlo Simulation of Order-Disorder Phenomena in Binary Alloys', *Structural and Phase Stability of Alloys*, J. L. Morán-López, et al., eds., Plenum Press, New York, 1992.
15. B. SAPOVAL, M. ROSSO and J. F. GOUYET: 'Fractal Interfaces in Diffusion, Invasion and Corrosion', *The Fractal Approach to Heterogeneous Chemistry*, D. Avnir, ed., John Wiley & Sons Ltd., 1989.
16. W. W. MULLINS and R. F. SEKERKA: 'Stability of a Planar Interface During Solidification of a Dilute Binary Alloy', *Journal of Applied Physics*, 1964, **35**, pp.444-451.
17. S. SCHIMSCHAL-THÖLKE, H. SCHMALZRIED and M. MARTIN: 'Instability of Moving Interfaces Between Ionic Crystals KCl/AgCl', *Ber. Bunsenges. Phys. Chem.*, 1995, **99**, pp.1-6.
18. H. TEUBER: Thesis, Darmstadt University of Technology, 1998.
19. H. TEUBER and M. MARTIN: to be published.
20. A. MAGISTRIS and A. SCHIRALDI: 'Thermoelectric Power and Electric Conductivity of the (Ag,Na)Cl and (Ag,Na)Br Solid Solutions', *Z. Naturforsch.*, 1974, **29a**, pp.1330-1334.
21. S. SCHIMSCHAL-THÖLKE, H. SCHMALZRIED and M. MARTIN: 'Stability of Diffusion Profiles in Quasi-Binary Solid Solutions (Ag,Na)Cl' , *Ber. Bunsenges. Phys. Chem.*, 1995, **99**, pp.7-13.
22. M. MARTIN: 'Interface Stability During Reactions', *Materials Science Forum*, 1994, **155-156**, pp.429-443.

SECTION II

Transport

Material Transport via the Grain Boundary of Lanthanum Chromites

N. Sakai, T. Tsunoda, I. Kojima, K. Yamaji, T. Horita and H. Yokokawa
National Institute of Materials and Chemical Research
Tsukuba, Japan

T. Kawada
Institute for Scientific Measurements, Tohoku University, Japan

M. Dokiya
Institute of Environmental Science and Technology
Yokohama National University, Japan

ABSTRACT

Material transport in alkaline-earth substituted lanthanum chromites is discussed. Diffusion coefficients of oxide ion and cations in alkaline-earths doped lanthanum chromites were determined by secondary ion mass spectrometry (SIMS). The contribution of grain boundary as a fast diffusion path was quantitatively evaluated by analysing the depth profiles. The grain boundary diffusion coefficients were 10^4 times higher than bulk diffusion coefficients for oxide ion, and 10^5 times higher for cations. For Sr-Ca interdiffusion, the diffusion mechanism at grain boundary has been found to be closely related to the valence state of chromium ion, which will play a main role of determining the vacancy concentration at A-site in the vicinity of grain boundaries. The oxygen partial pressure dependence of oxide ion diffusion coefficient was determined by annealing in Ar-$^{18}O_2$ and Ar-H_2-$C^{18}O_2$, and the results revealed good agreement with the data of oxygen permeation fluxes which were measured in a steady state under a large gradient of oxygen partial pressure. The effect of grain boundary diffusion was observed only in the oxidizing atmosphere, and bulk diffusion is dominant in the reducing atmosphere.

1.0 INTRODUCTION

Alkaline-earth substituted lanthanum chromites ($La_{1-x}M_xCrO_3$, M = Ca or Sr, $LaCr_{1-y}MgO_3$) have high chemical stability and electrical conductivity. This perovskite - type oxides have high potential as electrodes for magnetohydrodynamic generator (MHD),[1] or interconnects for solid oxide fuel cell (SOFC).[2] According to the point defect model, the substitution of alkaline earth results in the formation of tetravalent chromium ion (Cr^{4+}) to compensate the charge neutrality. For example in $La_{1-x}Ca_xCrO_3$, the following reaction occurs,

$$(2CaO) + 2Cr^{x}_{Cr} + 2La^{x}_{La} + \tfrac{1}{2}O_2(g) \leftrightarrow 2Ca'_{La} + 2Cr^{\bullet}_{Cr} + (La_2O_3) \quad (1)$$

Here, Cr^{x}_{Cr} is a trivalent chromium ion (Cr^{3+}) in a perovskite lattice, La^{x}_{La} is a trivalent lanthanum ion (La^{3+}) in the lattice, Ca'_{La} is a calcium ion (Ca^{2+}) in lanthanum (La^{3+}) site in lattice, $Cr^{\bullet}_{Cr}$ is a tetravalent chromium ion in chromium (Cr^{3+}) site, and O^{x}_{O} is an oxide ion in the lattice. $Cr^{\bullet}_{Cr}$ acts as a hole carrier and results in high electronic (hole) conductivity. However, in an SOFC operating condition, the interconnect plate should be placed under a large gradient of oxygen partial pressure ($p(O_2) = 10^5 \sim 10^{-13}$ Pa) and a high current density at high temperature, ca. 1273 K. In a reducing atmosphere, the tetravalent chromium ion is not stable, so that it is reduced to trivalent ion with liberating gaseous oxygen and forming an oxygen vacancy ($V^{\bullet\bullet}_{O}$),

$$2Cr^{\bullet}_{Cr} + O^{x}_{O} \leftrightarrow 2Cr^{x}_{Cr} + V^{\bullet\bullet}_{O} + \tfrac{1}{2}O_2(g) \quad (2)$$

In SOFC operating condition, the formed oxygen vacancy may diffuse from the low $p(O_2)$ to the high $p(O_2)$ side, which can be observed as the oxide ion conductivity or oxygen permeability in a dense polycrystalline material.

On the other hand, the diffusion of cations and oxide ion in this material is greatly related to the sinterability of powder compact, and also to the reactivity with other ceramic components in the energy converters written above, so that it should be intensively investigated. Since various valence states of chromium ion may affect the vacancy concentration of oxide ions or cations which is directly related to diffusion, it is very interesting to clarify the role of defects in the diffusion mechanism.

In spite of the importance of diffusion properties, the experimental methods for the determination of diffusion coefficient are limited for oxide ceramics. Oxygen chemical diffusion coefficients were determined from the relaxation behaviours of electrical conductivity for $La_{1-x}M_xCrO_{3-\delta}$.[3] Concerning the cation diffusion, very little data is available because the diffusion coefficients are so small that it takes tremendous amount of time to obtain the measurable diffusion lengths. Furthermore, very few papers are found about the quantitative treatment of, the effect of grain boundary on the diffusion.

Recently, secondary ion mass spectrometry (SIMS) has been regarded as a new powerful technique to obtain diffusion profiles of isotopes or tracers. SIMS depth profiling has very high depth resolution of ca. 10 nm, so that very small diffusion lengths, ca. 0.5 ~ 1 μm are enough to obtain good depth profiles to be analysed. This does shorten the time required for annealing. Since the diffusion length is smaller than the grain size of matrix polycrystalline, the grain boundary effect can be quantitatively considered under simple assumptions. We have tried to adopt this experimental technique to determine the diffusion coefficients of oxide ion and cations in calcium substituted lanthanum chromites ($La_{1-x}Ca_xCrO_{3-\delta}$) by using stable isotopes or tracers.[4-9] The effects of temperature, oxygen partial pressure and grain boundary as a fast diffusion path are reviewed in this paper and the diffusion mechanism and its relation with defect chemistry is all-inclusively discussed.

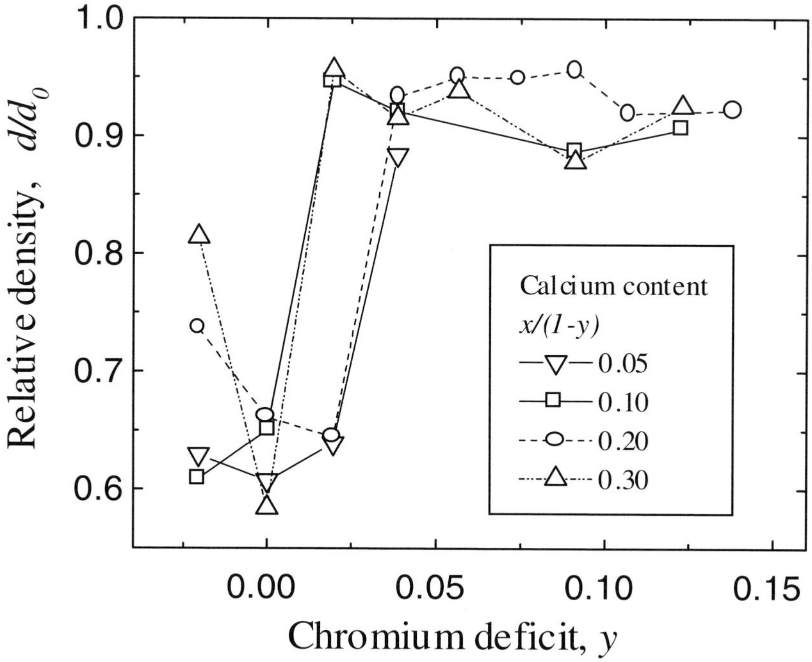

Fig. 1 Effect of nominal chromium deficit (y) on sinterability of $La_{1-x}Ca_xCr_{1-y}O_3$ at T = 1873 K in air for 5 h.

2.0 GRAIN BOUNDARY EFFECT ON CATION DIFFUSION IN $La_{1-x}Ca_xCrO_3$

2.1 Grain Boundary Characteristics of Sinterable $La_{1-x}Ca_xCrO_3$

Cation diffusion in ceramics is the main cause for densification of powder compact by sintering. Generally, lanthanum chromites are regarded as poorly-sinterable material because of their high melting points (~ 2600 K), and of high volatility of the chromium component. The material transport of lanthanum chromites is dominated by the vaporization-condensation mechanism with gaseous CrO_3(g) in an oxidizing atmosphere at high temperature, which prohibits the densification of powder compacts.

The sinterability of $La_{1-x}Ca_xCrO_3$ is drastically improved by preparing the samples with a slight amount of chromium depletion, in other words, calcium excess.[10] Figure 1 shows the relative density of $La_{1-x}Ca_xCr_{1-y}O_3$ sintered at T = 1873 K for 5 h in air. Only 2 mol.% of nominal chromium depletion is enough to obtain a dense polycrystalline with relative density up to 95%. The chromium depletion results in the formation of La_2O_3 or CaO which lowers the equilibrated vapour pressure of

CrO_3(g), and the material transport is governed by bulk diffusion process leading to particle joining and grain growth.

By optimizing the preparation condition, $La_{0.70}Ca_{0.32}CrO_3$ can be sintered up to 94% of theoretical density at T = 1573 K at which temperature the bulk diffusion is not fast enough to result in densification.[11] A slight amount of chromium depletion (or calcium excess) in $La_{1-x}Ca_xCrO_3$ results in the formation of $CaCrO_4$ as a secondary phase at rather low temperatures. This compound, $CaCrO_4$, has higher valence state of chromium (VI) than those in $La_{1-x}Ca_xCrO_{3-\delta}$ (III ~ IV), and it melts and becomes liquid calcium oxychromates, $Ca_m(CrO_4)_n$ ($m > n$) above T = 1333 K. The densification of $La_{1-x}Ca_xCrO_{3-\delta}$ powder compact by fast material transport with particle joining and pore elimination is via $Ca_m(CrO_4)_n$ liquid phase.[11]

From thermodynamic calculations, it was confirmed that $La_{1-x}Ca_xCrO_3$ can co-exist with $Ca_m(CrO_4)_n$ when the calcium content is larger than that corresponding to x = 0.5 at 1573 K.[11] Hence, if $Ca_m(CrO_4)_n$ exists between $La_{1-x}Ca_xCrO_3$ particles in sintering, the interface of each grain can have calcium rich composition. Figure 2a shows a TEM image of a grain boundary of $La_{0.70}Ca_{0.32}CrO_3$ sintered at T = 1573 K in air ($La_{0.70}Ca_{0.30}CrO_3$ with 2 mol.% calcium excess, d/d_0 = 95%).[12] The grain boundary was very thin, at most 1 nm in width, and very little lattice distortion was observed. EDS spectra were collected along the line perpendicular to the grain boundary and the compositional distributions of lanthanum, calcium and chromium are plotted as a function of distance in Figure 2b. Obvious enrichment of calcium, and depletion of lanthanum were observed around the grain boundary (distance = 0). The width of the Ca-enriched region is at most ±2 nm. This experimental results indicate the $Ca_m(CrO_4)_n$ existed between $La_{1-x}Ca_xCrO_3$ particles during the sintering treatment.

The existence of $Ca_m(CrO_4)_n$ and its interaction with $La_{1-x}Ca_xCrO_3$ can be also confirmed by analysing the valence state of chromium in the vicinity of grain boundary. A Cr_{2p} spectrum of an as-fractured surface of $La_{0.70}Ca_{0.32}CrO_3$ is shown in solid line in Figure 3a.[12] The peak in a higher energy region around E = 579.7 eV was assigned to $Cr^{6+}(d^0)$, because the same peak was observed in $CaCrO_4$. The main peak was observed at around E = 577 eV with complex broadening which was similar to that of Cr_2O_3. This peak can be deconvoluted by using a non-linear fitting program to Gauss-Lorenz functions which was developed by Kojima et al.,[13] and an example was shown in Figure 3a. The background was subtracted by using Shirley's method.[14] The existence of $Cr^{3+}(d^3)$ and $Cr^{4+}(d^2)$ is reasonable because the nominal valence number of chromium in $La_{0.70}Ca_{0.30}CrO_3$ is 3.3 to maintain the charge balance with A-site cations, La^{3+} and Ca^{2+}. However, the existence of Cr^{5+} is questioned, so that these components probably indicates the effect of multiplet term or the charge transfer spectrum of chromium as reported by Howng *et al.*[15]

The intensity of the high energy peak decreased with Ar^+ sputtering, whereas the intensity of the main peak was kept almost unchanged as shown in Figure 3b. The deconvoluted intensities of various valence states of chromium were plotted in Figure 4 as a function of etched depth which was calculated assuming the etching rate = 12.5 pm/s. The Cr^{2+} state was formed by the damages with substantial irradiation of high energy ions. The Cr^{6+} peak was observed only in the vicinity of fracture surface, around 1.2 nm in depth, which is in good agreement with the width of Ca-enriched region observed by TEM/EDS.

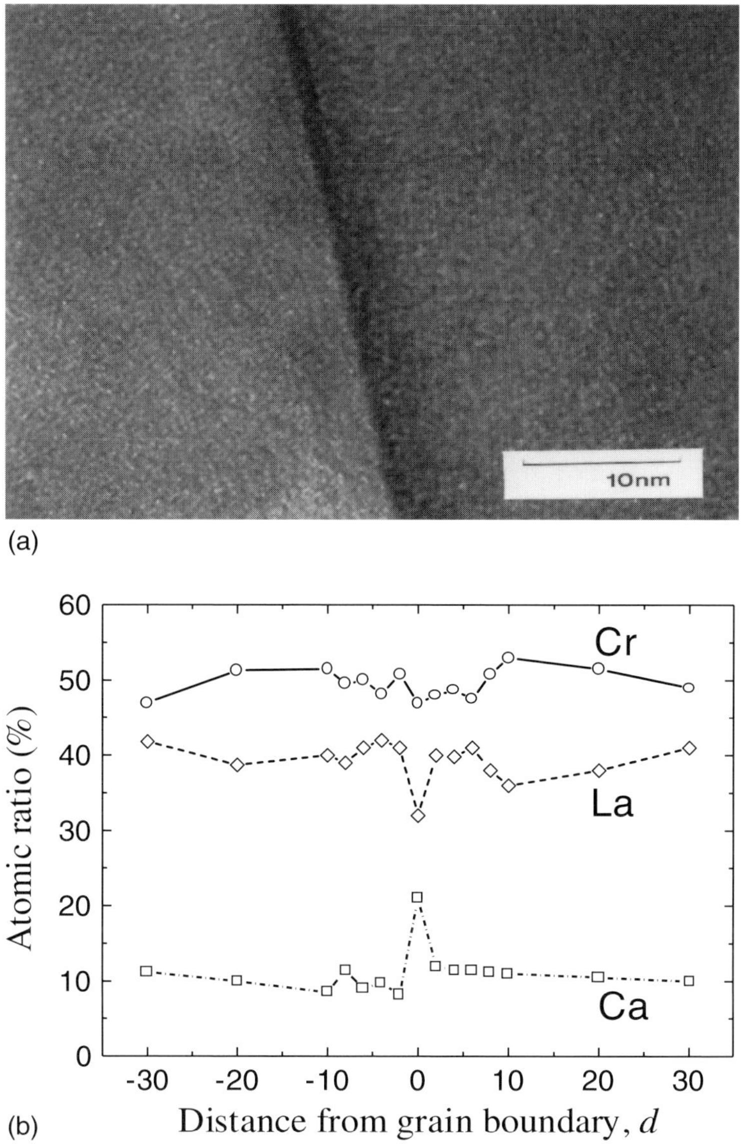

Fig. 2 (a) TEM image of a grain boundary of $La_{0.70}Ca_{0.32}CrO_3$ and (b) Compositional distribution around a grain boundary of $La_{0.70}Ca_{0.32}CrO_3$.

It is easily supposed that such a calcium enriched region at triple points or grain boundaries are not stable in the sintered polycrystalline, and may migrate very fast as liquid $Ca_m(CrO_4)_n$ along grain boundaries in a long term heat treatment. This material transport is enhanced in an SOFC operating condition, at a high

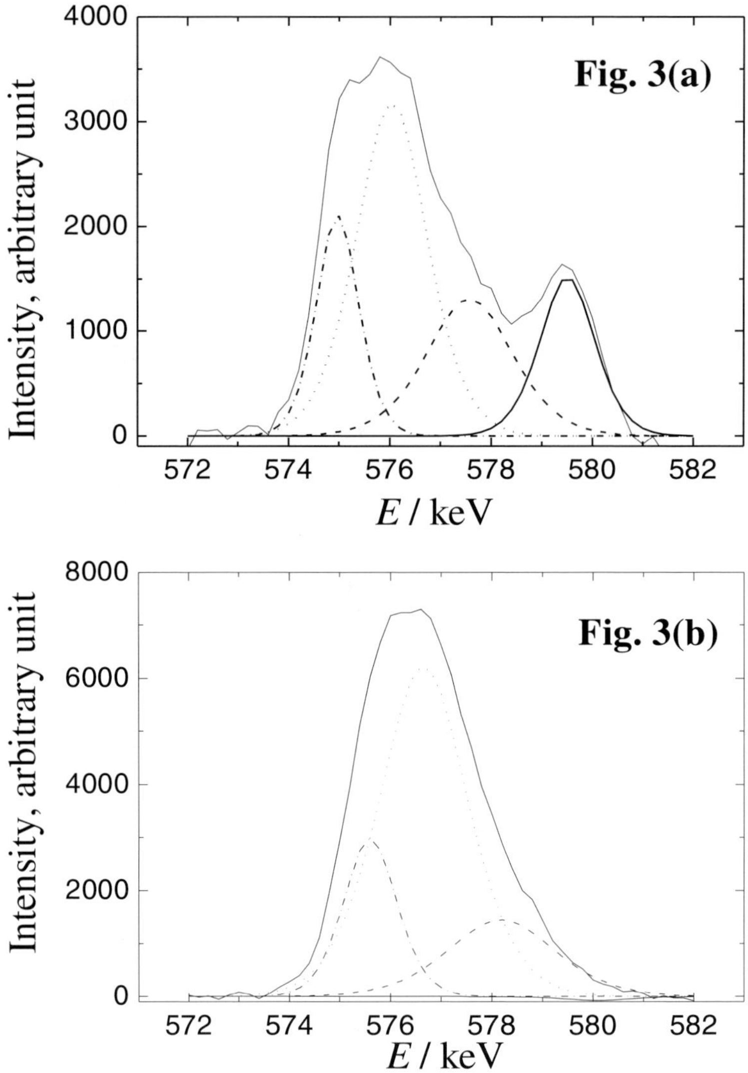

Fig. 3 (a) Intensity of Cr^{2p} peak for as-fractured and (b) sputtered $La_{0.70}Ca_{0.32}CrO_3$ —— raw data. The components of different chromium valence state: —— Cr^{6+}, ----- Cr^{5+}, ... Cr^{4+}, -.-.-.-.- Cr^{3+}.

temperature (T = 1273 K) and under a gradient of oxygen partial pressure ($p(O_2) = 10^4 - 10^{-13}$ Pa) and electrical current (350 mA cm^{-2}). It will play an important role in undesirable degradation or reactions in SOFC components, and also may result in the reaction with other SOFC components in a long term operation as will be discussed in the next section.

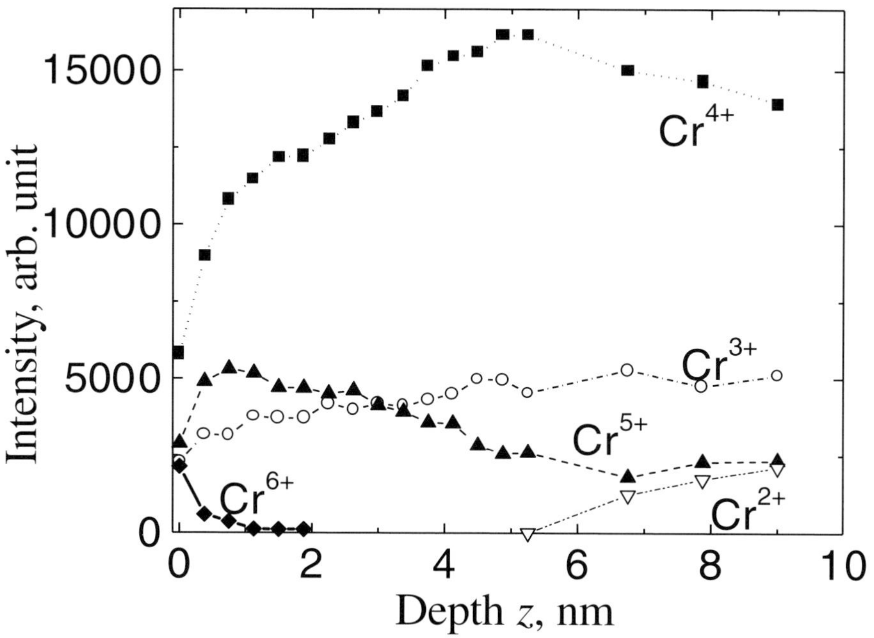

Fig. 4 Chromium valence state in the vicinity of a fractured surface.

2.2 Determination of Sr-Ca Interdiffusion Coefficients

Horita et al. have investigated the diffusion of alkaline -earth in $La_{1-x}Ca_xCrO_{3-\delta}$ via bulk and via grain boundary for the case of Sr^{2+} - $La_{1-x}Ca_xCrO_{3-\delta}$.[6,7] A dilute aqueous solution of strontium nitrate, $Sr(NO_3)_2$, was applied on the $La_{0.75}Ca_{0.25}CrO_{3-\delta}$ and then annealed at a selected temperature and oxygen partial pressure. The intensities of secondary ions, $^{88}Sr^+$, $^{44}Ca^+$, $^{139}La^+$ and $^{52}Cr^+$ were measured in SIMS (CAMECA ims5f, primary ion O_2^+, I = 500 nA, accelerating voltage 12.5 kV) as a function of distance from the surface. The diffusion coefficient of strontium - calcium interdiffusion was determined from the depth profile of relative intensity of strontium ion. Figure 5 shows an example of depth profile of normalized strontium concentration which was calculated from a SIMS depth profile (Sample: $Sr(NO_3)_2$-$La_{0.75}Ca_{0.25}CrO_3$, T = 1273 K, in air, annealing time 72 h.) The steep slope, which was observed in the shallow region, represents the contribution of bulk diffusion, and a long tale in the deeper region represents the contribution of grain boundary as a fast diffusion path. The diffusion coefficients of bulk (D_{bulk}) and of grain boundary ($D_{g.b.}$) are calculated by fitting the depth profile to the Fick's second law, solved assuming a semi-infinite medium,[16] or the equation which was suggested by Le Claire.[17] A similar method was also applied for the determination of the isotope diffusion coefficient of chromium by using $Ca^{50}CrO_4$-HNO_3(aq.).[9]

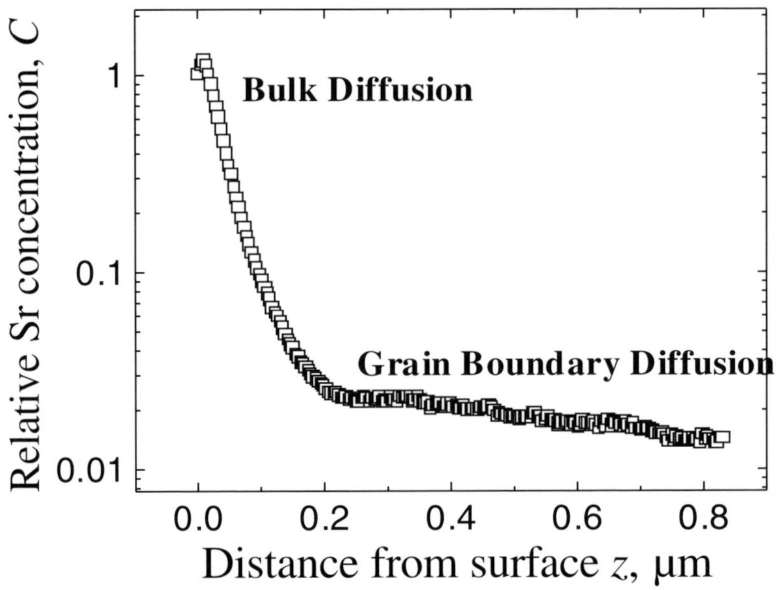

Fig. 5 Depth profile of relative strontium concentration taken by SIMS. The concentration was assumed to be unity at $z = 0$.

Figure 6 summarizes the obtained cation diffusion coefficients, via bulk (D_{bulk}) and via grain boundaries ($D_{g.b.}$), which are plotted as a function of inverse temperature. The activation energy was 318 kJ mol^{-1} for D_{bulk} and 261 kJmol^{-1} for $D_{g.b.}$. It is noteworthy that the strontium - calcium interdiffusion coefficients in bulk are in good agreement with the extrapolated literature data of diffusion coefficients of La^{3+} in $LaCrO_3$.[18] These facts indicate that the diffusion mechanisms are very similar for La^{3+} diffusion and Sr-Ca interdiffusion, i.e. replacing the cations via A-site vacancies. The vacancy diffusion coefficient of A-site is evaluated as:

$$D_M C_M = f D_V C_V \tag{3}$$

Here D_M is the diffusion coefficient of a cation, C_M is the concentration of the cation, f is correlation factor, D_V is the vacancy diffusion coefficient, and C_V is the concentration of vacancy. The correlation factor was assumed to be unity in the following discussion. According to the temperature dependence of $C_V \approx [V_{La}''']$ reported by Akashi et al., the vacancy concentration of A-site at $T = 1273$ K is evaluated to 4.2×10^{-5} mole per 1 mole of $LaCrO_3$.[19] From the present data for Sr^{2+}-$La_{0.75}Ca_{0.25}CrO_3$, together with this estimated vacancy concentration, D_V can be estimated to be 1.1×10^{-14} m^{2}s^{-1}, at $T = 1273$ K in air.

The obtained values of bulk diffusion coefficient of calcium - strontium interdiffusion was ca. 2.38×10^{-19} m^{2}s^{-1} at $T = 1273$ K. This value is sufficiently small and the bulk diffusion of alkaline-earth generally does not affect the chemical stability

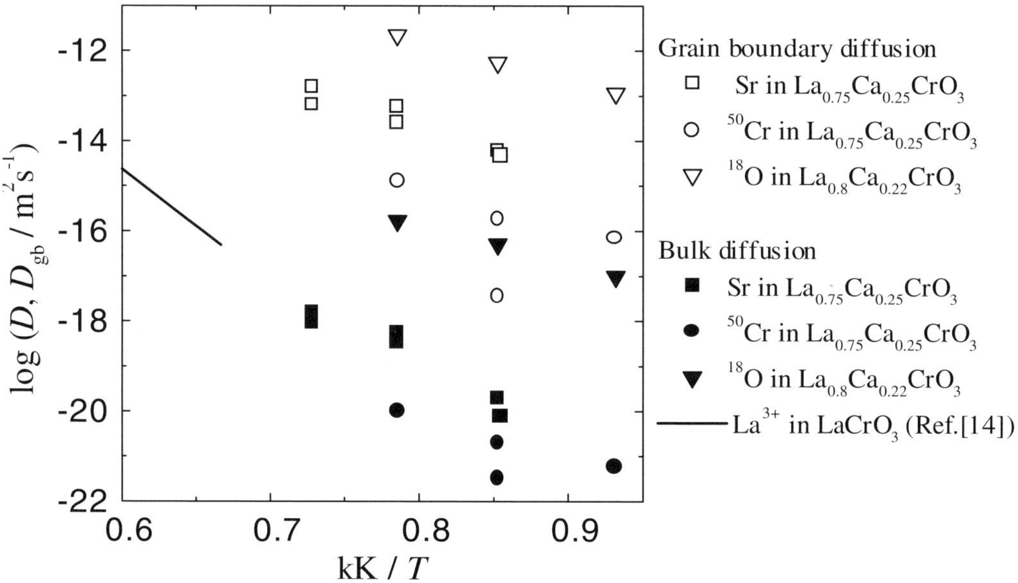

Fig. 6 Temperature dependence of diffusion coefficients via bulk and via grain boundary. The width of grain boundary was assumed to be 1 nm.

of $La_{1-x}Ca_xCrO_3$ polycrystalline in general SOFC operating condition and time (10^4 – 10^5 hours.) However, It should be noted that the grain boundary diffusion coefficients were $10^4 - 10^5$ times higher than bulk diffusion coefficient, which indicates that the material transport in $La_{1-x}Ca_xCrO_{3-\delta}$ is dominated by the fast diffusion via grain boundary around 1273 K.

Since the annealed sample was easily fractured along grain boundary, the compositional distribution around grain boundary can be analysed by obtaining secondary ion images of each cations with a focused primary ion beam (I < 100 pA, beam diameter < 0.3 μm). Several images were taken by etching the fractured surface, resulting in the three dimensional map of strontium content as shown in Figure 7.

The strontium component has higher concentrations at fractured surface (= grain boundary) and strontium migrated deeply inside the $La_{0.90}Ca_{0.11}CrO_3$ polycrystalline. This feature should be compared with the diffusion depth, about a few nm from the fractured surface towards inside polycrystalline.

The width of the strontium-enriched region obtained by present SIMS analyses was larger (4 nm) than those obtained in TEM/EDS or XPS analyses. However in this case, we must take into account the atomic mixing effect which was due to the bombardment on the sample surface by focused primary ion with high energy density. Considering the acceleration voltage (12.5.kV) of O_2^+ primary ion, all data shown in Figure 7. represents the averaged information ca. 10 nm around each measuring spot, which may be the reason why the larger width was obtained.

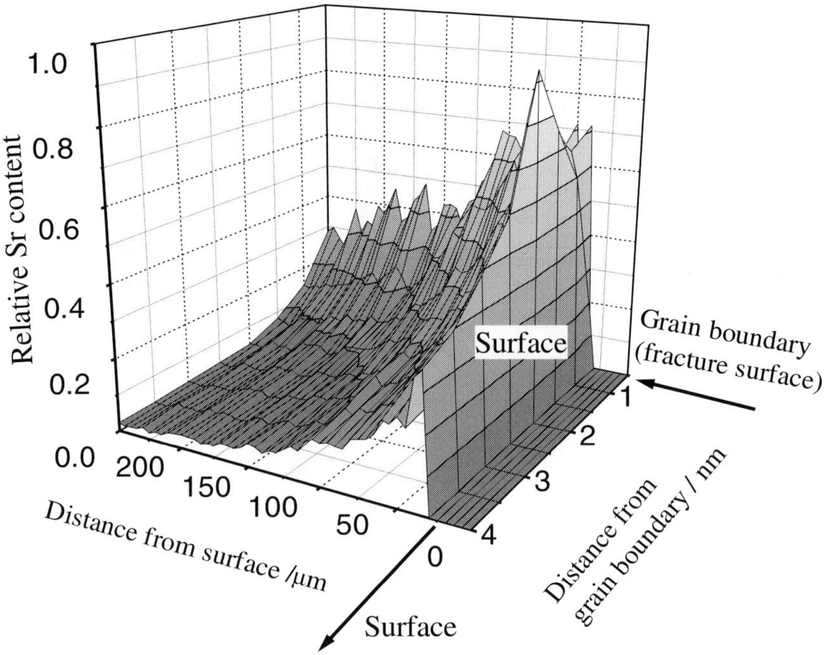

Fig. 7 Relative strontium concentration in the vicinity of grain boundary of Sr – $La_{0.90}Ca_{0.10}CrO_3$, annealed at T = 1273 K for 72 h.

The material transport through a $La_{1-x}Ca_xCrO_3$ interconnect plate can be calculated as molar flux (J_{oxide}) of alkaline -earth oxide in the SOFC operating condition for the case of 100 μm thin layer of $La_{0.75}Ca_{0.25}CrO_3$ as given in Table 1. To calculate the flux via grain boundary, the relative surface area of grain boundary was estimated as 3.51×10^{-4}, which was derived from the length per unit area (3.52×10^{-5} m^{-1}) and width (10^{-9} m) of grain boundary.

This fact is in good agreement with the experimental results on chemical stability of $La_{0.7}Ca_{0.32}CrO_3$ polycrystalline, in which the samples exhibited a significant migration and precipitation of $Ca_m(CrO_4)_n$ in 200 hours treatment in SOFC operating conditions as shown in Figure 8. In the air side, the precipitation of $Ca_m(CrO_4)_n$ was observed as a dark, glassy phase in Figure 8b. Much more $Ca_m(CrO_4)_n$ was also precipitated, and decomposed in the fuel side, and then the surface was completely covered with $CaCr_2O_4$ and CaO (Figures 8c and d).

2.3 Effect of Chromium Valence State on Grain Boundary Diffusion

The grain boundary diffusion coefficient has $p(O_2)$ dependence as shown in Figure 9. The grain boundary diffusion coefficient ($D_{g.b.}$) is proportional to $p(O_2)^{0.66}$,

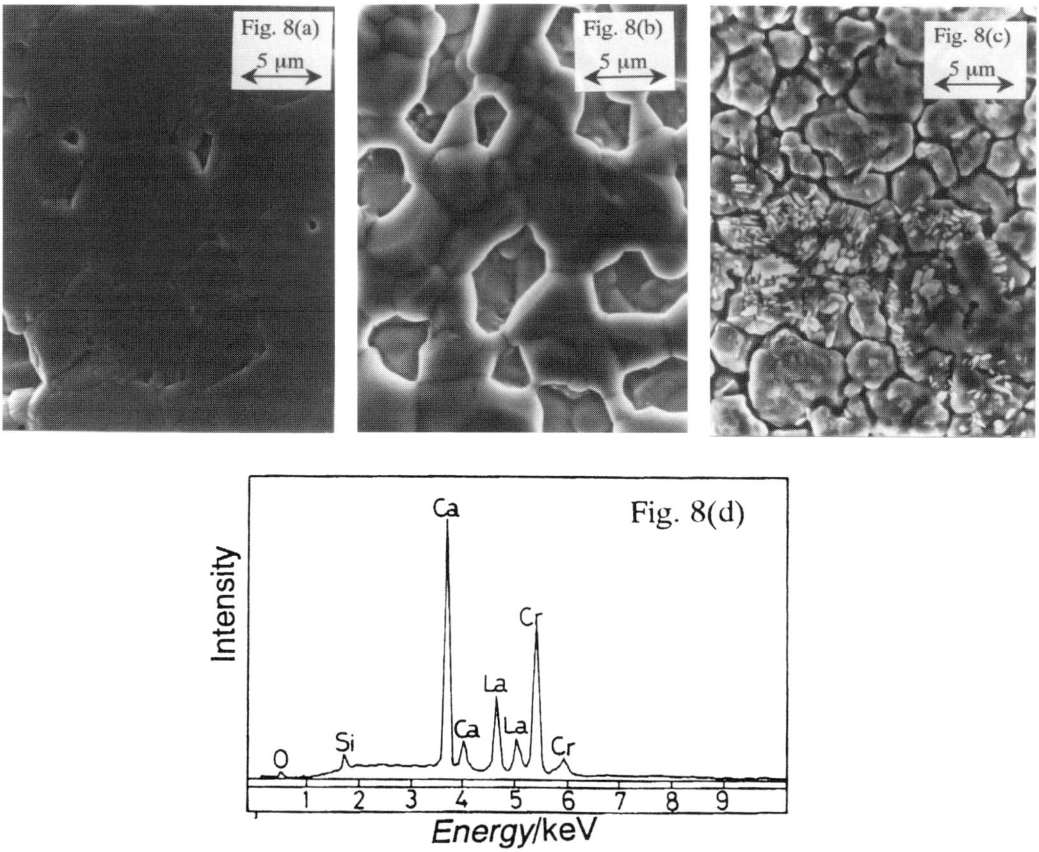

Fig. 8 SEM images of $La_{0.70}Ca_{0.32}CrO_3$ surface. (a) Polished surface after the 200 hour in simulated SOFC condition, T = 1273 K, j = 0.35 A cm^{-2}, (b) Air side, $p(O_2)$ = 2.13 × 10^4 Pa, (c) Fuel side, (H_2 + H_2O, $p(O_2)$ = 10^{-13} Pa) and (d) EDS spectrum in Figure 8c, Ca and Cr components were enriched.

Table 1 Diffusion coefficients in $La_{0.75}Ca_{0.25}CrO_3$ in air and fluxes of cation transport as oxides in SOFC operating condition*

	Temp. (K)	D_0 (m^2s^{-1})	E_A ($kJmol^{-1}$)	D at 1273 K (m^2s^{-1})	J_{oxide} (mol $m^{-2}s^{-1}$)
	Bulk Diffusion Coefficient			Thickness = 100 μm	
Sr-Ca	1173-1673	2.66 × 10^{-6}	318	2.38 × 10^{-19}	3.70 × 10^{-9}
^{50}Cr	1073-1273	2.04 × 10^{-14}	157	7.37 × 10^{-21}	8.47 × 10^{-11}
Grain Boundary Diffusion Coefficient (J_{oxide} is normalized by the relative surface area of grain boundary, $S_{g.b.}/S_{bulk}$ = 3.52 × 10^{-4})					
Sr-Ca	1173-1373	1.01 × 10^{-3}	261	1.97 × 10^{-14}	1.07 × 10^{-7}
^{50}Cr	1073-1273	4.20 × 10^{-9}	161	1.01 × 10^{-15}	4.05 × 10^{-9}

* T = 1273 K, $p(O_2)$ = 10^5 - 10^{-13} Pa, open circuit condition.

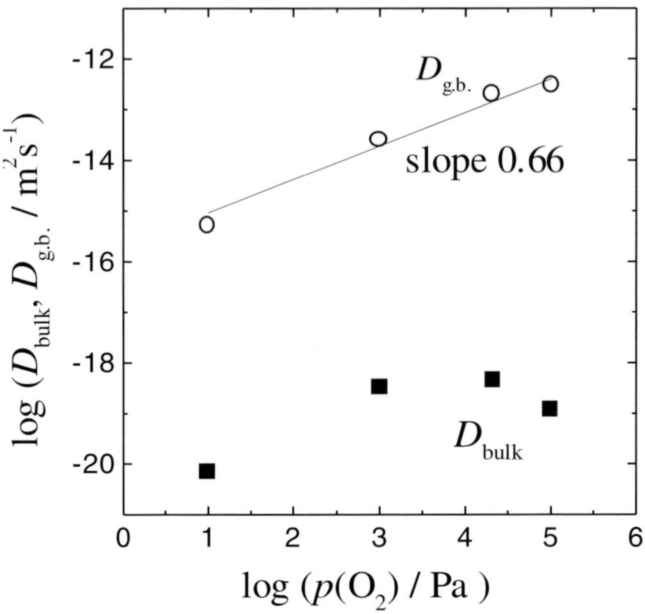

Fig. 9 Oxygen partial pressure dependence of Sr-Ca interdiffusion coefficient via bulk D_{bulk} and via grain boundary $D_{g.b.}$ for $La_{0.75}Ca_{0.25}CrO_3$. T = 1273 K.

whereas the bulk diffusion coefficient (D_{bulk}) is apparently independent of oxygen partial pressure. This $p(O_2)$ dependence suggests that some reduction/oxidation reactions are considered to play important roles in diffusion process.

As we mentioned in the previous chapter, the presently observed TEM/EDS and XPS results show the presence of hexavalent chromium in the vicinity of thin grain boundaries and no lattice distortion, which suggest the presence of A-site vacancy around grain boundary. According to the Kröger-Vink notation, the vacancy formation can be expressed as follows:

$$Ca_{La}' + Cr_{Cr}^{\bullet} + \tfrac{1}{2}O_2 \leftrightarrow V_{La}''' + Cr_{Cr}^{\bullet\bullet\bullet} + CaO \qquad (4)$$

$$K = \frac{[V_{La}'''][Cr_{Cr}^{\bullet\bullet\bullet}][CaO]}{[Ca_{La}'][Cr_{Cr}^{\bullet}]p(O_2)^{\frac{1}{2}}} \qquad (5)$$

Here, Ca_{La}' is a Ca^{2+} at La^{3+} site (A-site), $Cr_{Cr}^{\bullet}$ is a Cr^{4+} at Cr^{3+} site (B-site), $Cr_{Cr}^{\bullet\bullet\bullet}$ is a Cr^{6+} at B-site, V_{La}''' is an A-site vacancy, and K is the equilibrium constant.

Most of the atoms in the vicinity of grain boundary are thought to be not so mobile as atoms in liquid phase, although the grain boundary has much more structural incompleteness than in bulk. Therefore it is reasonable that the strontium diffusion at grain boundary is made by a similar mechanism as that in bulk, that is, cation exchange of Ca^{2+} and Sr^{2+} at A-sites in perovskite lattice. The activation energy of grain boundary diffusion coefficient is 261 kJ/mol,[9] and this value is almost the same as the activation energy of bulk diffusion coefficient (318 kJ/mol), which may support the assumption of the similar diffusion mechanism. As shown in eqn. 4, the formation of A-site vacancy is inevitably accompanied with the oxidation of chromium, or in other words, the formation of Cr^{6+}. Thus the A-site vacancy concentration in the grain boundary vicinity is probably much higher than that in the bulk, which greatly affects the diffusion coefficient at grain boundary.

Although we confirmed the existence of Cr^{6+} at room temperature XPS, there was no proof for the existence of Cr^{6+} at higher temperatures. There still remains the possibility that the Cr^{6+} might form in a cooling procedure, because the oxide ions also diffuse as fast as that of cations at grain boundaries (see next chapter). However, the $p(O_2)$ dependence of Sr -Ca interdiffusion coefficient at grain boundary indicates that the Cr^{6+} indeed exists at annealing temperatures in the diffusion experiment, because the concentration of $Cr_{Cr}^{\bullet\bullet\bullet}$ directly depends on $p(O_2)^{0.5}$ as derived from eqn. 5.

The derived $p(O_2)$ dependence of diffusion coefficient should be compared with the experimental results, which apparently showed the linear dependence on $p(O_2)^{0.66}$ as shown in Figure 9. Although the $p(O_2)^{0.5}$ dependence is expected when the oxidation from Cr^{4+} to Cr^{6+} is assumed as above, the experimental results indicate the possibility of more drastic change in valence state, for example from Cr^{3+} to Cr^{6+} ($p(O_2)^{0.75}$). It is possible that the deviation of the experimental data ($p(O_2)^{0.66}$) is also due to the $p(O_2)$ dependence of other defect contents, such as $[V_{La}''']$ or $[Cr_{Cr}^{\bullet}]$. The actual chemical state at grain boundary is of course much more complicated than that expected in the discussion above. In view of this correlation, the diffusion of chromium ions themselves becomes of great interest.

The tracer diffusion coefficients of chromium in $La_{1-x}Ca_xCrO_{3-\delta}$ were preliminarily determined in the temperature range from 800 to 1000°C, and obtained values were relatively small, around 1/10 - 1/100 of calcium - strontium interdiffusion coefficients as shown in Figure 6 and Table 1. The activation energy of D_{bulk} in ^{50}Cr- $La_{0.75}Ca_{0.25}CrO_3$ was 157 kJ/mol, which was almost half of that of Sr^{2+}- $La_{0.75}Ca_{0.25}CrO_3$ (318 kJ/mol). The low absolute values are probably due to the difference of binding energy between A - site and B - site cations in perovskite lattice. Since the diffusion coefficients are very small, we obtained very shallow depth- profiles below 0.1 μm. Therefore, the accuracy of the value at T = 800°C is somewhat questioned, because the excess amount of defect may be enriched in the vicinity of surface which enhances the cation diffusion. It probably results in relatively higher diffusion coefficients and apparently small activation energy, so that more experimental study is required to discuss about the diffusion mechanism.

3.0 OXIDE ION DIFFUSION IN $La_{1-x}Ca_xCrO_{3-\delta}$

3.1 Determination of Oxide Ion Diffusion Coefficient by SIMS Analyses

The experimental facts in the cation diffusion of $La_{1-x}Ca_xCrO_{3-\delta}$ revealed that the diffusion coefficients are drastically influenced by the vacancy concentration. In this section, the same treatment is adopted to the oxide ion diffusion in $La_{1-x}Ca_xCrO_{3-\delta}$.

As mentioned in introduction, $La_{1-x}Ca_xCrO_{3-\delta}$ has a significant amount of oxygen vacancy ($V_O^{\bullet\bullet}$, $[V_O^{\bullet\bullet}] \approx \delta$ in $La_{1-x}Ca_xCrO_{3-\delta}$) in a reducing atmosphere. The oxygen vacancy content (δ) has been determined with good precision for various compositions by thermogravimetry in different temperatures and oxygen partial pressures.[20-22] On contrary, it should be noted that there is little data for oxide ion (oxygen) diffusion coefficients measured by different experimental methods, such as electrical conductivity relaxation,[3] gas analysis,[23] or oxygen isotope exchanging and SIMS analysis.[5, 8, 23] By using oxygen isotope ($^{18}O_2$) exchange and SIMS depth profiling, the relative concentration of oxygen isotope ion ($^{18}O^{2-}$) in solids can be determined as a function of the distance from surface, which is considered to be a more direct way than gas analysis or electrical conductivity relaxation method.

The tracer (oxygen isotope) exchange and diffusion procedures were carried out by annealing a sample in a closed system, in which a desirable amount of $^{18}O_2$ was introduced. Since the diffusion of oxide ion is much faster than that of cation, the short time annealing (ca. 300 s) was carried out by using an infrared furnace in which sample was rapidly heated or cooled. The heating rate was ca. 100 K/s for heating and 40 K/s for cooling. After quenching the sample, the intensities of secondary ions $^{16}O^-$ and $^{18}O^-$ were collected as a function of spattering time by SIMS (primary ion Cs^+, accelerating voltage 10 kV, I = 100 nA. Secondary polarity: Negative). The depth profile was analysed by using the same method as the case of cations. However, the terms representing surface reactions were included in fitting eqn. 17 for determination of bulk diffusion coefficient (D_{bulk}) and surface reaction constant.

Temperature dependence of the tracer diffusion coefficient of oxide ion (D_{bulk}) of $La_{0.8}Ca_{0.2}CrO_{3-\delta}$ in air is shown in Figure 6. The oxide ion diffusion is much faster than cation diffusion, because the mobility of oxide ion is higher than those of cations in the perovskite lattice, and the concentration of oxygen vacancies formed according to the reaction (2) is expected to be large. The diffusion coefficient increased with calcium content (x in $La_{1-x}Ca_xCrO_{3-\delta}$) as shown in Figure 10, because the oxygen vacancy content is also increased with x. Grain boundary diffusion coefficients are ca. 10^4 times higher than bulk diffusion coefficients, which is very similar to the fact observed for cation diffusion.

3.2 Effect of Oxygen Partial Pressure on Diffusion Coefficient

To clarify the relationship of diffusion coefficient and oxygen vacancy in detail, the tracer diffusion coefficient of oxide ion in $La_{1-x}Ca_xCrO_{3-\delta}$ was determined over a wide range of oxygen partial pressure at T = 1273 K by using oxygen isotopes, Ar + $^{18}O_2$ ($p(^{18}O_2)$ = 10^3 - 10^5 Pa) or Ar + H_2 + $C^{18}O_2$ ($p(^{18}O_2)$ = 10^{-9} - 1 Pa). The oxygen partial pressure in the closed system was calculated by Gibbs energy minimization method using GEM/MALT2 software.[24, 25] To eliminate the effect of thermodynamic factor, each sample was pre-annealed in the Ar + O_2 or the Ar + H_2 + CO_2 gas mixture

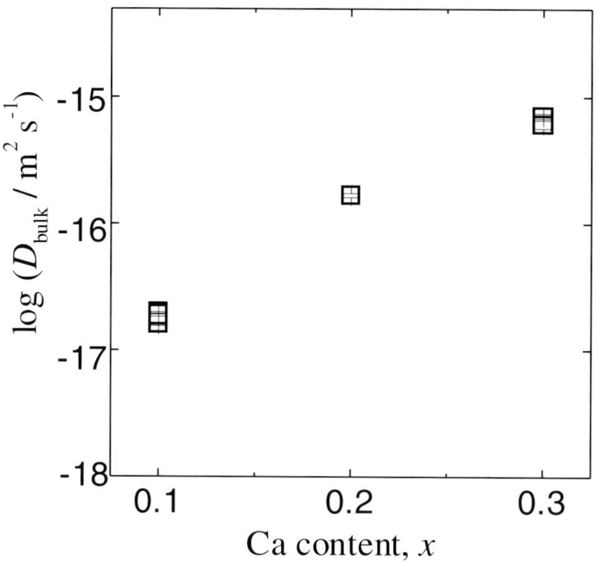

Fig. 10 Tracer diffusion coefficient of oxide ion as a function of calcium content, x. T = 1273 K, $p(^{18}O_2)$ = 2.13 × 10^4 Pa.

which had the same oxygen partial pressure in the subsequent ^{18}O diffusion experiments.

Figure 11 shows the diffusion coefficients of $La_{1-x}Ca_xCrO_{3-\delta}$ (x = 0.1, 0.2 and 0.3) via bulk (D_{bulk}) and via grain boundary ($D_{g.b.}$) plotted as a function of oxygen partial pressure. Grain boundary diffusion coefficient ($D_{g.b.}$) is 10^3 -10^4 times larger than bulk coefficient (D_{bulk}) at $p(O_2)$ region from 10^3 - 10^5 Pa, in which condition the effect of grain boundary as a fast diffusion path should be concerned. A good secondary ion image of $^{18}O^{2-}$ could be obtained on the surface of $La_{0.7}Ca_{0.3}CrO_3$ as shown in Figure 12, which exhibited that the $^{18}O^{2-}$ ions dominantly diffuse via grain boundary.

The logarithmic bulk diffusion coefficients linearly increases with decreasing log $p(O_2)$, in the $p(O_2)$ region from 10^2 to 10^{-10} Pa, which is due to the increase of oxygen vacancy content (δ). As a result, the bulk diffusion is dominant and the effect of grain boundary diffusion is negligible in the reducing atmosphere. Assuming that the diffusion of oxide ion is governed by vacancy mechanism, the following relation is expected between oxide ion diffusion coefficient (D_{bulk}) and oxygen vacancy diffusion coefficient (D_V).

$$C_O D_{bulk} = f C_V D_V \tag{6}$$

$$D_{bulk}(3-\delta) = f D_V \delta \tag{7}$$

$$\log D_{bulk} = \log\left(\frac{\delta}{3-\delta}\right) + \log f + \log D_V \tag{8}$$

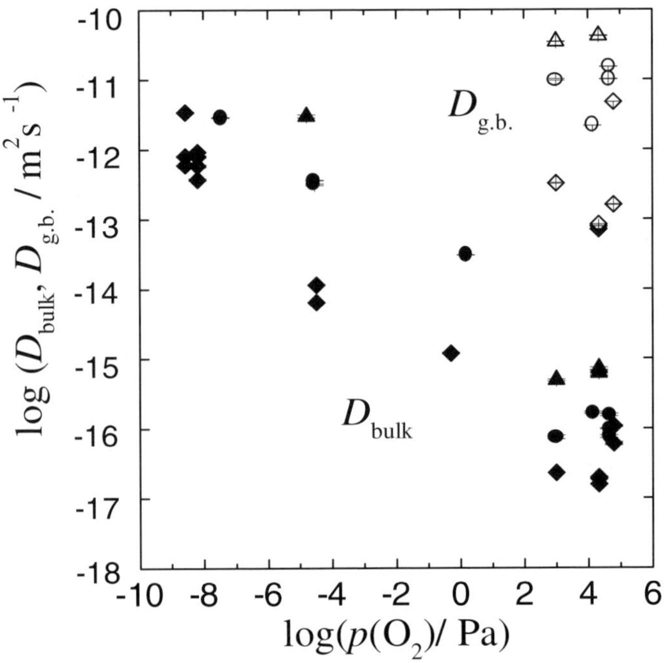

Fig. 11 Tracer diffusion coefficients of oxide ion via bulk (D_{bulk}, closed symbols) and via grain boundary ($D_{g.b.}$, open symbols) in $La_{1-x}Ca_xCrO_{3-\delta}$ as a function of oxygen partial pressure. T = 1273 K. ◆, ◇ x = 0.1, ●, ○ x = 0.2, ▲, Δ x = 0.3.

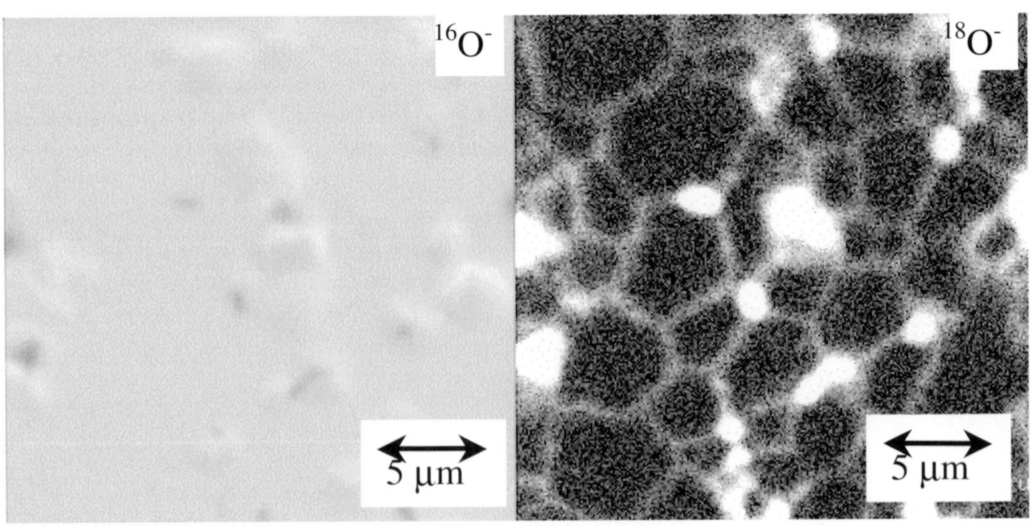

Fig. 12 Secondary ion images of the etched surface (depth = ca. 0.1 µm) of $La_{0.90}Ca_{0.11}CrO_{3-\delta}$ annealed in T = 1073 K in $p(^{18}O_2) = 2.13 \times 10^4$ Pa for 300 s.

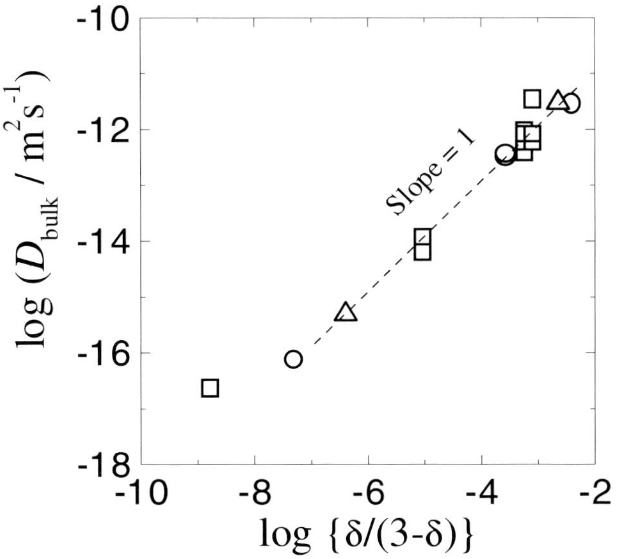

Fig. 13 Tracer diffusion coefficietns of oxide ion in $La_{1-x}Ca_xCrO_{3-\delta}$ bulk as a function of normalized oxygen vacancy content, δ/(3-δ). □ $x = 0.1$, ○ $x = 0.2$, Δ $x = 0.3$, $T = 1273$ K.

If the vacancy diffusion coefficient (D_V) is independent of oxygen vacancy content (δ), log D_{bulk} is linear to log{δ/(3-δ)}, and the slope of the line should be always unity. The intercept at log{δ/(3-δ)} = 0 represents the *log* fD_V. The deviation from linear dependence may indicates that some other effect must be included, such as clustering of vacancy, or existence of extrinsic defects.

The present data are re -plotted as a function of log{δ /(3-δ)} in Figure 13. The oxygen vacancy content (δ) was quoted from the latest literature data for $La_{1-x}Ca_xCrO_{3-\delta}$.[19] It is noteworthy that all data can be fitted on one line regardless of calcium content. The intercept of the line corresponds to log fD_V and the obtained value was ca. 1.5×10^{-9} m^2s^{-1}, which is in good agreement with the D_V obtained from other experimental results such as relaxation method,[26] and oxygen permeability measurement.

The slopes of logD_{bulk} vs. log{δ/(3-δ)} plots were very close to unity in the concentrated region at log{δ/(3-δ)} >-7, which indicates that the vacancy diffusion coefficient is constant in this region. However, it tends to be lower in the diluted region of log{δ/(3-δ)} <-7. This indicates that the correlation factor (*f*) or vacancy diffusion coefficient (D_V) may have some dependence on vacancy content, or the oxygen vacancy content has some uncertainty. In oxidizing atmospheres at $p(O_2) > 10^2$ Pa, the obtained bulk diffusion coefficients do not exhibit significant $p(O_2)$ dependence, and they rather increased with oxygen partial pressure as shown in Figure 11. Assuming that the D_V is constant, it seems that the oxygen vacancy content in $La_{1-x}Ca_xCrO_{3-\delta}$ has lower limit in high $p(O_2)$ region, for example, 10^{-8} for

$La_{0.9}Ca_{0.1}CrO_{3-\delta}$ and 10^{-6} for $La_{0.7}Ca_{0.3}CrO_{3-\delta}$. It indicates that the oxygen vacancy formation in oxidizing atmosphere does not follow the simple point defect model, and it also affects the slopes in Figure 13.

It should be noted that TG data was obtained only in the region of $\delta > 10^{-6}$, thereby the validity of estimated δ lower than $<10^{-6}$ could not be checked. In an oxidizing atmosphere, the oxide vacancy content is very difficult to be estimated, because the possible effects of extrinsic defect sources such as hexavalent chromium ion (Cr^{6+}) are not negligible. Unfortunately at present, there are no experimental methods available for determining the oxygen vacancy content smaller than 10^{-6}. Note also that the estimated concentration of A-site vacancies is in the order of 10^{-5} as described in the previous chapter, which is larger than that of oxygen vacancy. The interaction of A-site and oxygen vacancies should be clarified.

3. 4 Comparison with Oxygen Permeation Flux

The presently obtained diffusion coefficients are very important to understand the oxygen transport in $La_{1-x}Ca_xCrO_{3-\delta}$ under a gradient of oxygen chemical potential, for example, in an SOFC operation. Alkaline -earth substituted lanthanum chromites are generally regarded as electronic conductors over a wide range of oxygen potential, and the oxide ion conduction hardly becomes dominant. However, this minor oxide ion conduction may result in oxygen permeation even in a completely dense polycrystalline: In SOFC, a dense plate of lanthanum chromite is placed under a large gradient of oxygen partial pressure at high temperature ($p(O_2)$ = 10^5 - 10^{-13} Pa at T = 1273 K). Under such a condition, a gradient of oxygen concentration is generated, so that the oxygen in the lattice diffuses from the high $p(O_2)$ side to the low $p(O_2)$ side. Since the oxygen in $La_{1-x}Ca_xCrO_{3-\delta}$ crystal is charged as O^{2-}, electronic hole ($h^\bullet$) simultaneously migrates to an opposite direction to maintain the charge neutrality. As a result, oxygen permeates from the high $p(O_2)$ side to the low $p(O_2)$ side.

Since the hole diffusion is much faster than the oxide ion diffusion in $La_{1-x}Ca_xCrO_{3-\delta}$, the oxygen permeation flux (J) is determined as integration of oxide ion conductivity between two oxygen partial pressures.

$$J_{O^{2-}} = -\frac{1}{4FL}\int_{p_{O_2}(y=0)}^{p_{O_2}(y=L)} \sigma_{O^{2-}}\, d\mu_{O_2} \tag{9}$$

Here F is the Faraday constant, L is thickness of a dense interconnect, σ is oxide ion conductivity, and μ is chemical potential. According to the Nernst-Einstein equation, the oxide ion conductivity can be expressed as:

$$\sigma_{O^{2-}} = \frac{4F^2(3-\delta)D_{\mathrm{O}}}{RTv_m} = \frac{4F^2\delta D_{\mathrm{V}}}{RTv_m} \tag{10}$$

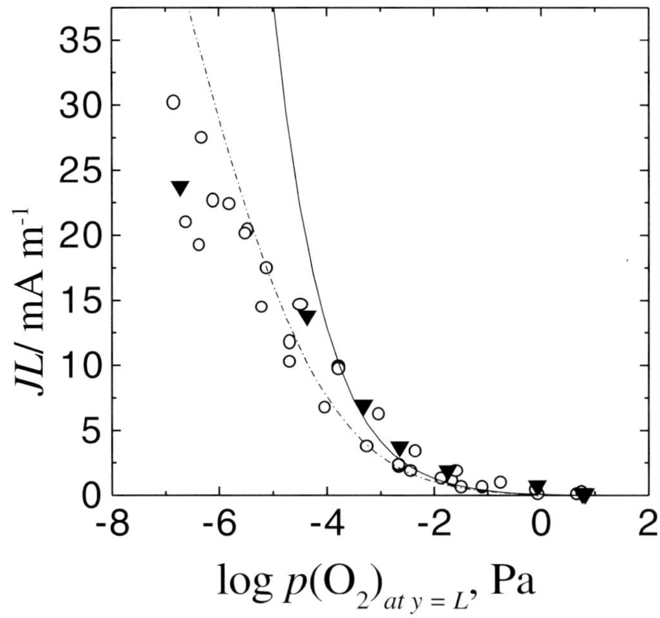

Fig. 14 Oxygen permeation flux through $La_{0.75}Ca_{0.25}CrO_{3-\delta}$. $p(O_2)$ = 4.5 Pa at y = 0. T = 1273 K, Thickness = 0.54 mm. ○, ▲ experimental data, Calculated data from eqns. 9 and 10. —— $D_V = 1.5 \times 10^{-9}$ m^{2}s^{-1}, -·-·-· $D_V = 1.5 \times 10^{-9}$ m^{2}s^{-1} surface reaction limited.

Here δ is oxygen vacancy content, D_O is oxygen diffusion coefficient, R is the gas constant, T is temperature and ν_m is molar volume of $La_{1-x}Ca_xCrO_{3-\delta}$.

The oxygen permeation flux in a steady state under a gradient of oxygen partial pressure has been measured by an electrochemical method,[5] and results are plotted as JL against oxygen partial pressure at $y = L$ in Figure 14. The oxygen partial pressure at $y = 0$ was fixed at 4.5 Pa. The solid and dashed lines are calculated oxygen permeation by using eqns. 9 and 10. Solid lines are calculated with the constant oxygen vacancy diffusion coefficient, $D_V = 1.5 \times 10^{-9}$ m^{-2}s^{-1}. In the $p(O_2)$ region from 1 to 10^{-4} Pa, the solid line is in good agreement with the measured oxygen permeation fluxes shown as symbols in Figure 14. However, the surface reaction becomes the rate determining step when $p(O_2) < 10^{-4}$ Pa, in which region the permeation flux is lower than the theoretically calculated value.

The absolute values of observed oxygen permeation fluxes was higher than theoretical value at high $p(O_2)$ region from 0 to 10^5 Pa as shown in Figure 15. It represents the effect of grain boundary diffusion. As shown in Figure 11, diffusion coefficient via grain boundary is much higher than that of bulk, but its effect is dominant only in oxidizing atmospheres, because bulk diffusion coefficient drastically increases with decreasing oxygen partial pressure. The flux of oxygen permeation via grain boundary was calculated and shown as lines in Figure 15. The length of

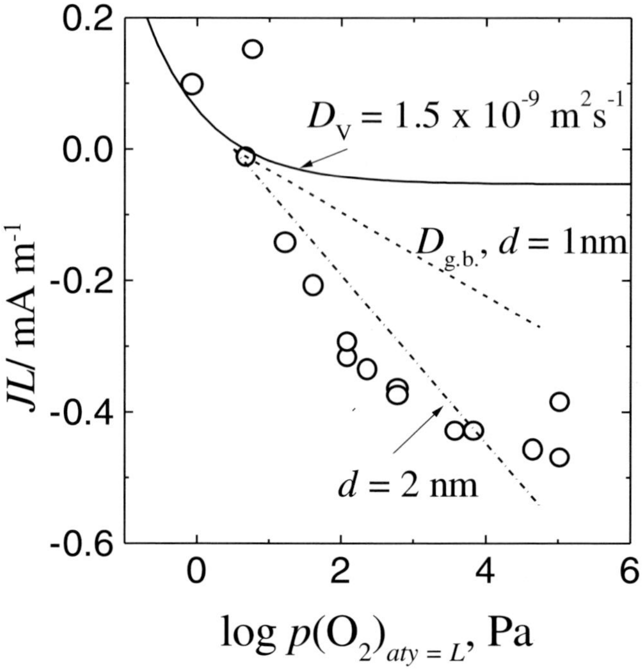

Fig. 15 Effect of grain boundary diffusion on oxygen permeability. Sample: $La_{0.75}Ca_{0.25}CrO_{3-\delta}$. T = 1273 K. O, Experimental data, —— D_V = 1.5 × 10^{-9} m^{2}s^{-1}, ... $D^{\bullet}_{g.b.}$ = 9.3 × 10^{-12} m^{2}s^{-1}, width of grain boundary = 1 nm, -·-·-·- width of grain boundary = 2 nm.

grain boundary per unit area was estimated as 3.51 × 10^5 m^{-1}, and the fluxes were calculated when the width of grain boundary(d) is 1 nm (solid line) and 2 nm (dotted line). The experimental oxygen permeation flux can be well fitted in case of d = 2 nm, which is in good agreement with TEM or XRD data.

4.0 SUMMARY

The present results show that SIMS depth profiling is a very powerful technique to investigate the diffusion behaviour in solid materials. Especially for ceramics which are used under a large gradient of chemical potential, the effect of grain boundaries as fast diffusion paths in materials transport is very important to be investigated. SIMS shallow depth profiling made it possible to separate the bulk and grain boundary effects. It will help the precise understanding of the diffusion mechanism in ionic crystals and along grain boundaries with combining the experimental data of defect concentrations and valence states of components.

5.0 REFERENCES

1. D. B. MEADOWCROFT: Some Properties of Strontium-Doped Lanthanum Chromite, *Brit. Journal of Applied Physics, Journal of Physics D*, **2**(2), 1969, pp.1225-1233.
2. N. Q. MINH: Ceramic Fuel Cells, *Journal of American Ceramic Society*, **76**(3), 1993, pp.563-588.
3. I. YASUDA and T. HIKITA: Precise Determination of the Chemical Diffusion Coefficient of Calcium-Doped Lanthanum Chromites by Means of Electrical Conductivity Relaxation, *Journal of Electrochemical Society*, **141**(5), 1994, pp.1268-1273.
4. T. KAWADA, T. HORITA, N. SAKAI, H. YOKOKAWA and M. DOKIYA: Experimental Determination of Oxygen Permeation Flux Through Bulk and Grain Boundary of $La_{0.7}Ca_{0.3}CrO_3$, *Solid State Ionics*, **79**, 1995, pp.201-207.
5. N. SAKAI, T. HORITA, H. YOKOKAWA, M. DOKIYA and T. KAWADA: Oxygen Permeability Measurement of $La_{1-x}Ca_xCrO_{3-\delta}$ by Using an Electrochemical Method, *Solid State Ionics*, **86-88**, 1996, pp.1273-1278.
6. T. HORITA, N. SAKAI, T. KAWADA, H. YOKOKAWA and M. DOKIYA: Grain Boundary Diffusion of Strontium in (La, Ca)CrO_3 Perovskite -Type Oxide by SIMS, *Journal of American Ceramic Society*, **81**(2), 1998, pp.315-320.
7. T. HORITA, M. ISHIKAWA, K. YAMAJI, N. SAKAI, M. YOKOKAWA and M. DOKIYA, Cation Diffuision in (La, Ca)CrO_3 Perovskite by SIMS, *Solid State Ionics*, **108**, 1998, pp.383-390.
8. N. SAKAI, K. YAMAJI, T. HORITA, M. ISHIKAWA, H. YOKOKAWA, M. DOKIYA and T. KAWADA: Influence of Oxygen Partial Pressure on Oxygen Diffusivity in $La_{1-x}Ca_xCrO_{3-\delta}$ *High Temperature Mixed Conducting Ceramics III*, T. A. Ramanarayanan, W. L. Worrell, H. L. Tuller, A. C. Khandkar, M. Mogensen and W. Gopel, The Electrochemical Society Proceedings, The Electrochemical Society Inc., Pennington, NJ. U.S.A., **PV97-24**, 1997, pp.819-826.
9. N. SAKAI, K. YAMAJI, T. HORITA, M. ISHIKAWA and H. YOKOKAWA: Determination of Tracer Diffusion Coefficient of Chromium in Lanthanum Chromites, *The Proceedings of the 3rd European SOFC Forum*, P. Stevens, ed., The European Solid Oxide Fuel Cell Forum, 1998, pp.333-342.
10. N. SAKAI, T. KAWADA, H. YOKOKAWA, M. DOKIYA and T. IWADA: Sinterability and Electrical Conductivity of Calcium Doped Lanthanum Chromites, *Journal of Materials Science*, **25**, 1990, pp. 4531-4534.
11. N. SAKAI, T. KAWADA, H. YOKOKAWA, M. DOKIYA and I. KOJIMA: Liquid-Phase - Assisted Sintering of Calcium-Doped Lanthanum Chromites, *Journal of American Ceramic Society*, **76**(3), 1993, pp.609-616.
12. N. SAKAI, T. TSUNODA, N. FUKUMOTO, I. KOJIMA, K. YAMAJI, T. HORITA, M. ISHIKAWA, H. YOKOKAWA and M. DOKIYA: TEM, XPS and SIMS Analyses on Grain Boundary of Lanthanum Chromites, *Journal of Electroceramics*, **4**(1), 1999, pp.119-126.
13. I. KOJIMA and M. KURAHASHI: Application of Asymmetrical Gaussian / Lorentzian Mixed Function for X-ray Photoelectron Curve Synthesis, *J. Elec. Spectr. Relat.*

Phenom., **42**, 1987, pp.177-181.

14. D. A. SHIRLEY: High Resolution X-ray Photoemission Spectrum of the Valence Bands of Gold, *Phys. Rev.*, **B5**, 1972, pp. 4709-4714.
15. W. Y. HOWNG and R. J. THORN: Investigation of the Electronic Struture of $La_{1-x}(M^{2+})_xCrO_3$, CrO_3, Cr_2O_3 and La_2O_3 by X-ray Photoelectron Spectroscopy, *Journal of Phys. Chem. Solids*, **41**, 1980, pp. 75-81.
16. J. CRANK: *The Mathematics of Diffusion*, 2nd edition, Oxford Science Publications, 1975, pp.1-36.
17. A. D. LE CLAIRE: The Analyses of Grain Boundary Diffusion Measurements, *Brit. J. Appl. Phys.*, **14**, 1968, pp.351-356.
18. T. AKASHI, M. NANKO, T. MARUYAMA, Y. SHIRAISHI and J. TANABE: Solid-State Reaction Kinetics of $LaCrO_3$ from the Oxides and Determination of La^{3+} Diffusion Coefficient, *Journal of Electrochemical Society*, **145**(6), 1998, pp.2090-2094.
19. T. AKASHI, M. NANKO and T. MARUYAMA: The Concentration and the Enthalpy of the Formation of the La Vacancy in $LaCrO_3$, *65th Annual Meeting of the Electrochemical Society of Japan*, Tokyo, 1998, p.325.
20. J. MIZUSAKI, S. YAMAUCHI, K. FUEKI and A. ISHIKAWA: Non-Stoichiometry of the Perovskite -Type Oxide $La_{1-x}Sr_xCrO_{3-\delta}$, *Solid State Ionics*, **12**, 1984, pp.119-124.
21. B. F. FLANDERMAYER, M. M. NASRALLAH, D. M. SPARLIN and H. U. ANDERSON: High Temperature Stability of Magnesium-Doped Lanthanum Chromite, *High Temp. Sci.*, **20**, 1985, pp.259-269.
22. S. ONUMA, S. YASHIRO, A. KAIMAI, K. KAWAMURA, Y. NIGARA, T. KAWADA, J. MIZUSAKI, N. SAKAI and H. YOKOKAWA: Nonstoichiometry of $La_{1-x}Ca_xCrO_{3-\delta}$, *5th SOFC Symposium in Japan*, 1997, pp.121-124.
23. I. YASUDA, K. OGASAWARA and M. HISHINUMA: Oxygen Tracer Diffusion in Polycrystalline Calcium Doped Lanthanum Chromites, *Journal of American Ceramic Society*, **80**(12), 1997, pp.3009-3012.
24. H. YOKOKAWA, S. YAMAUCHI and T. MATSUMOTO: Thermodynamic Database MALT2 and its Applications to High Temperature Materials Chemistry, *Thermochimica Acta*, **245**, 1994, pp. 45-55.
25. T. MATSUMOTO and H. YOKOKAWA: Solving the Chemical Equilibrium Problem Using the Gradient Projection Method, *Netsu Sokutei (Calorimetry and Thermal Analysis)*, **18**(4), 1992, pp.170-173.
26. I. YASUDA and M. HISHINUMA: Chemical Diffusion in Polycrystalline Calicum-Doped Lanthanum Chromites, *Journal Solid State Chem.*, **115**, 1995, pp.152-157.

High-Temperature Creep of Polycrystalline CeO_2 Doped with Yttrium

YASURO IKUMA, HIROYUKI HIRANAKA and ERIKO SHIMADA
Kanagawa Institute of Technology
Atsugi, Kanagawa 243-0292, Japan

ABSTRACT

High-temperature dead-load four-point creep was studied in Y-doped CeO_2 ($Ce_{0.98}Y_{0.02}O_{1.99}$ and $Ce_{0.92}Y_{0.08}O_{1.96}$) at Po_2 = 20.3 kPa and at temperatures between 1100 and 1300°C. At stress levels of 4.9-19.6 MPa, the stress exponent of the creep rate was $m \fallingdotseq 1$ where $\dot{\varepsilon} \propto \sigma^m$ ($\dot{\varepsilon}$ is the creep rate and σ is the stress), indicating diffusional creep. These creep rates and those reported in the literature were normalized at 4.9 MPa and 1 μm and were compared with each other. As CeO_2 was doped substitutionally with *Y*, the creep rate decreased. The observed creep rate was several orders of magnitude smaller than the creep rate calculated from the oxygen diffusion coefficient using the Nabarro-Herring creep equation and was several orders of magnitude larger than the creep rate calculated from the expected cation diffusion coefficient. Cation grain boundary diffusion may control the creep of CeO_2 and Y-doped CeO_2 in the experimental range studied. A summary of the diffusion studies of CeO_2 is also given.

1.0 INTRODUCTION

Metal oxides with the fluorite structure are useful due to many interesting properties that are related to the ionic conductivity of oxides of this structure. For example, UO_2 is an important material in the field of nuclear science[1] and ZrO_2 is used as a solid electrolyte in solid oxide fuel cells.[2] ZrO_2 is also used as an oxygen sensor in the steel-making industry. Consequently, there are many studies of materials with the fluorite structure, most of which are concentrated on UO_2 and ZrO_2. A major creep study of the fluorite structure also concentrated on UO_2 and ZrO_2.[3] After Wakai et al.[4] demonstrated that ZrO_2 can be deformed more than 100%, the superplastic behaviour of ZrO_2 has also attracted attention. Fewer studies have been performed on CeO_2. Of these studies on CeO_2, most have concentrated on the investigation of nonstoichiometry[5-8] and electrical conductivity.[9-12] Combining these results, some researchers[6, 9, 13] have discussed the defect structure of the oxide.

A material based on CeO_2 is now being considered[14, 15] as a possible material for electrolyte in solid oxide fuel cells. As the temperature is raised in these cells, the difference in the thermal expansion coefficients may give rise to stress in the cell components. Consequently, investigation of the high-temperature creep behaviour of this oxide is required. Since CeO_2 is not considered to be a structural material, the creep at low stress might be interesting. There has been only a limited number of studies on the creep of cerium oxide. One of them is by Poluboyarinov et al.,[16] who investigated the four-point bending creep of polycrystalline CeO_2 between 1350 and 1450°C. They found that the creep rate depended on the stoichiometry of the oxide. For CeO_{2-x}, where $x = 0.10$, the creep rate ($\dot{\varepsilon}$) was higher than that of stoichiometric CeO_2 ($x = 0.00$). Depending on the stoichiometry, the stress exponent (m in $\dot{\varepsilon} \propto \sigma^m$, σ is the stress) varied from 1.2 at $x = 0.10$ to 2.2 at $x = 0.00$ and the temperature dependence of the creep varied from 163 kJ · mol^{-1} at $x = 0.10$ to 387 kJ · mol^{-1} at $x = 0.00$. Ikuma and Uchida[17] studied the creep of stoichiometric CeO_2. They concentrated on the creep behaviour of CeO_2 at low temperatures and stresses. They treated the specimens in such a way that CeO_2 was of stoichiometric composition. At temperatures between 1075 and 1150°C, the value of m was 0.83 to 0.94 with an activation energy of 314 kJ · mol^{-1}.

CeO_2 may be nonstoichiometric depending on oxygen partial pressure (Po_2)[5, 7] and the impurity content.[8] At a constant Po_2, the material deviates from its stoichiometric composition to form oxygen-deficient oxide as the temperature increases. At constant temperature, however, the degree of non-stoichiometry increases with decreasing Po_2. Iwasaki and Katsura[7] reported that, at $Po_2 = 20.3$ kPa, the x in CeO_{2-x} is 0.000 between room temperature and 1300°C. The results of Iwasaki and Katsura[7] suggest that, in order to perform creep tests using stoichiometric CeO_2 at $Po_2 = 20.3$ kPa, temperatures have to be lower than 1300°C. The work of Kamiya et al.[18] indicated that, if CeO_2 was sintered at 1400°C and $Po_2 = 20.3$ kPa, the lattice constant of CeO_2 was 0.54105-0.54113 nm. From the work of Brauer and Gingerich[19] these lattice constants correspond to $x = 0.00$ in CeO_{2-x}.

CeO_2 is known to exhibit a wide range[20] of solubility with Y_2O_3. Electrical conductivity[21] and oxygen diffusion[22] in Y-doped CeO_2 have already been studied. However, the creep behaviour of Y-doped CeO_2 (= $Ce_{1-z}Y_zO_{2-z/2-x}$) is not known. The objective of the present investigation was to study the effect of Y on the creep of CeO_2 in the temperature region where the stoichiometry of CeO_2 remains constant. The intention was to investigate the effect of yttrium addition (i.e., the effect of z in $Ce_{1-z}Y_zO_{2-z/2-x}$) but not the effect of nonstoichiometry (i.e., the effect of x in $Ce_{1-z}Y_zO_{2-z/2-x}$) created by the change in Po_2 and/or temperature. Thus, an intimate solution of CeO_2 and Y_2O_3 was prepared by sintering at 1400°C and the creep tests were performed at $Po_2 = 20.3$ kPa (in air) and at temperatures lower than 1300°C.

2.0 EXPERIMENTAL PROCEDURE

To make 2% and 8% Y-doped CeO_2 ($Ce_{0.98}Y_{0.02}O_{1.99}$ and $Ce_{0.92}Y_{0.08}O_{1.96}$), the coprecipitation method was used. An appropriate amount of $Ce(NO_3)_3{\cdot}6H_2O$ (Rare Metallic, 99.9%) and $Y(NO_3)_3{\cdot}6H_2O$ (Rare Metallic, 99.99%) was dissolved into deionized water. An excess amount of hexamethylenetetramine was added to the solution. The temperature of the solution was increased to 75°C where the hexamethylenetetramine decomposed[23] to form NH_4OH, which precipitated both the Ce and Y compounds. The precipitate was filtered and dried. Finally it was calcined at 850°C for 1.5 h to form a mixture of CeO_2 and Y_2O_3 and part of them might form solid solution. The hot pressing technique was not used to obtain a higher density because the associated low oxygen partial pressure would have reduced the oxide to make x larger than 0.00. Reconstitution of the stoichiometric composition of the sample would have required annealing of several weeks.[16] Instead, these powders were uniaxially cold pressed into a bar of 4 × 5 × 30 mm at 100 MPa and sintered at 1400°C for 1 h in air. The sintering was also performed at 1500 and 1600°C in an effort to obtain a higher density. However, the density did not improve. Furthermore, the higher temperature will increase the value of x.[5,7] In order to remain in the region where $x = 0.00$, the Y-doped CeO_2 was sintered at 1400°C. The density of the rod was 6776 kg·m^{-3} (96.0% theoretical density) for $Ce_{0.98}Y_{0.02}O_{1.99}$ and 6763 kg·m^{-3} (95.8 % theoretical density) for $Ce_{0.92}Y_{0.08}O_{1.96}$. The rod was ground and polished using #1200 diamond paste to yield a creep specimen of approximately 3 × 2 × 23 mm.

The dead-load four-point bending creep tests were conducted on the bar in air (Po_2 = 20.3 kPa) at stresses of 4.9~19.6 MPa and at temperatures between 1100 and 1300°C. At this Po_2 and these temperatures, the CeO_2 was stoichiometric[7] (x in CeO_{2-x} is 0.000). Schneider et al.[24] reported that the nonstoichiometry (x) of CeO_2 doped with Ca ($Ce_{1-z}Ca_zO_{2-z-x}$) and CeO_2 doped with Sm ($Ce_{1-z}Sm_zO_{2-z/2-x}$) was $x = 0.0016$ at $Po_2 = 1 \times 10^{-7}$ Pa. The nonstoichiometry (x) due to intrinsic defects (thermal nonstoichiometry) in $Ce_{1-z}Y_zO_{2-z/2-x}$ at Po_2 = 20.3 kPa is, therefore, expected to be smaller than 0.0016 owing to the higher Po_2. For the creep test, the bar specimen was placed in the creep furnace and heated (13°C/min) to a high temperature with a very small load to keep the loading devices in place. The loading points (two outer points) and supporting points (two inner points) were made of single-crystal sapphire rods (diameter = 3 mm). After the specimen was equilibrated at the required temperature for about two hours, a constant load (4.9~19.6 MPa) was applied to the specimen. The deflection at the loading point was continuously measured. Immediately after the loading, there was elastic deformation followed by transition creep. Steady-state creep was then observed. During bending creep, maximum stress is observed in the outer (bending) surface of the bar, where the maximum strain exists. In the present study, the strain in the outer surface at the end of the creep test was always less than 1%.

The stress (σ) was calculated using the following equation[25]:

$$\sigma = \frac{3(L-a)}{2wh^2} p \tag{1}$$

where L is the distance between the support points, a is the distance between the load points, w is the width of the beam specimen, h is the thickness of the beam, and P is the applied load. The strain rate ($\dot{\varepsilon}$) was calculated using[25]:

$$\dot{\varepsilon} = \frac{6h\dot{y}}{(L-a)(L+2a)} \tag{2}$$

where $\dot{y}$ is the deflection rate in the steady-state region. The equation is valid only for the total strain of less than 3%. This condition was always satisfied in the present study. Following the creep test, the surface of maximum stress was examined using scanning electron microscopy (SEM, Jeol JXA-8600).

3.0 RESULTS

The maximum strain was observed on the outer surface of the test specimen, as expected from the four-point bending creep test that was used. The surface of the sample with the maximum strain was examined by SEM after the creep test. Some typical results are shown in Figure 1. In Y-doped CeO_2, grains after the creep test were still equiaxed, supporting the indication that the strain in the samples was less than 1%. This justifies the equation used for the calculation of stress and strain[25] later in the present study. The average grain sizes of the specimens were 1.2 μm for $Ce_{0.98}Y_{0.02}O_{1.99}$, and 0.97 μm for $Ce_{0.92}Y_{0.08}O_{1.96}$.

The strain rates were calculated from the steady-state curves using equation (2) and were plotted in Figure 2 as a function of applied stress at various temperatures. At all compositions, the creep rate increased with stress and a good linear relationship existed. The lines in the figure are the results of a least squares analysis of the data points. The slope (m) of the lines is the stress exponent for $\dot{\varepsilon} \propto \sigma^m$. The creep results[16, 17] for stoichiometric CeO_2 are also shown for comparison. As seen in Figure 2, the stress exponents (m) of the creep of CeO_2 and, $Ce_{0.98}Y_{0.02}O_{1.99}$, and $Ce_{0.92}Y_{0.08}O_{1.96}$ at temperatures 1200°C and less are all ~0.9. For the creep of Al_2O_3 doped with Cr, Fe, and Ti, the stress exponents were 0.8~1.48.[26, 27] Materials that are subjected to stress at high temperatures can deform plastically in a variety of ways. In polycrystalline ceramics, diffusional creep is the main mode of deformation at high temperatures. Depending on the transport path, diffusional creep is divided into two groups[28]: Nabarro-Herring creep

$$\dot{\varepsilon} = \frac{14\Omega D_l \sigma}{kTd^2} \tag{3}$$

and Coble creep

$$\dot{\varepsilon} = \frac{47\Omega \delta D_{gb} \sigma}{kTd^3} \tag{4}$$

where Ω is the atomic volume, D_l is the lattice diffusion coefficient, d is the grain diameter, δ is the grain boundary width and D_{gb} is the grain boundary diffusion coefficient. This creep is characterized by viscous behaviour, i.e. $m = 1$. Dislocation plasticity is another mode of creep in materials. It is characterized by $\dot{\varepsilon} \propto \sigma^m/d^n$ with m greater than 3 (non-viscous) and n equal to zero. Dislocation plasticity is operative in materials with large grain size and at high stress, whereas the diffusional creep is favorable in materials with small grain size and at low stress. Since $m = 1$ is predicted only for diffusional creep, the results for Al_2O_3 were interpreted by the investigators to be diffusional. Thus, the creep of CeO_2 by Ikuma and Uchida[17] and $Ce_{0.98}Y_{0.02}O_{1.99}$

(a)

(b)

Fig. 1 Scanning electron micrographs of Y-doped CeO_2 after creep test.(a) $Ce_{0.98}Y_{0.02}O_{1.99}$ and (b) $Ce_{0.92}Y_{0.08}O_{1.96}$.

and $Ce_{0.92}Y_{0.08}O_{1.96}$ in the present study may reasonably be considered to be viscous, i.e., diffusional in nature. As shown in Figure 2, Poluboyarinov et al.[16] observed slightly non-viscous behaviour (m = 2.2) for the creep of CeO_{2-x} (x = 0.00, bulk density = 96%, grain size = 15 μm) at a higher temperature (1350°C) and under lower stresses (0.62~2.49 MPa). Because the differences in bulk density are not significant enough to account for the change in the creep mechanism, the difference in the stress exponent for the results of the present study and Ikuma and Uchida

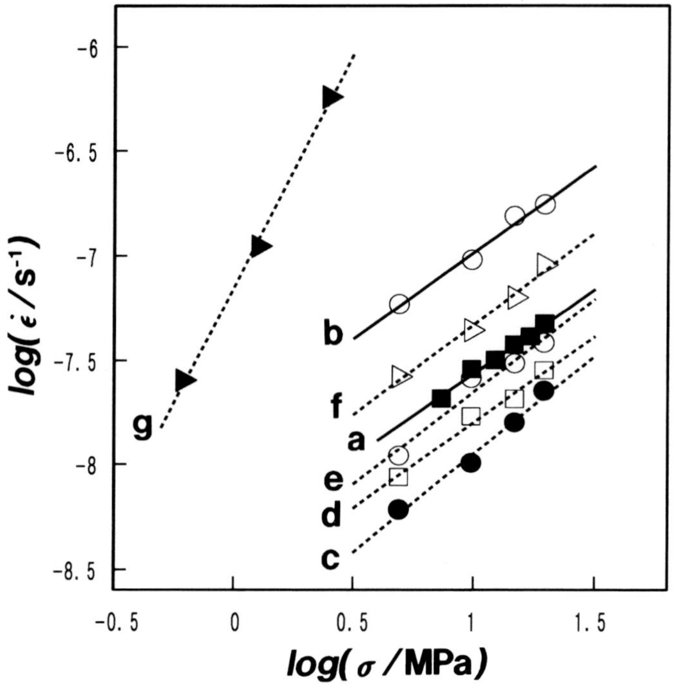

Fig. 2 Strain rate - stress relationship. (a) $Ce_{0.98}Y_{0.02}O_{1.99}$ at 1150°C, m = 0.81, (b) $Ce_{0.92}Y_{0.08}O_{1.96}$ at 1200°C, m = 0.82, (c) CeO_2 at 1075°C, m = 0.94, (d) CeO_2 at 1100°C, m = 0.83, (e) CeO_2 at 1125°C, m = 0.89, (f) CeO_2 at 1150°C, m = 0.87 and (g) CeO_2 at 1350°C, m = 2.2. Lines a and b are the results of present study, lines c, d, e and f are the results of Ikuma and Uchida[17] and line g is the result of Poluboyarinov et al.[16]

compared to those of Poluboyarinov et al. is probably due to either the stoichiometry (changed by purity of the specimen) or grain size (large grain size favors dislocation creep).[26, 29]

Arrhenius plots of the strain rate at various stresses are shown in Figure 3. Most of the data are represented by straight lines although slight deviations exist. The apparent activation energy was calculated using a least squares analysis. For both $Ce_{0.98}Y_{0.02}O_{1.99}$ and $Ce_{0.92}Y_{0.08}O_{1.96}$, the apparent activation energy is about 380 kJ · mol.$^{-1}$ The creep results of Poluboyarinov et al.[16] and Ikuma and Uchida[17] are also shown in Figure 3. Ikuma and Uchida obtained 314 kJ·mol^{-1} for the creep of stoichiometric CeO_2 at 9.8 MPa and 300 kJ · mol^{-1} for the average activation energy of the creep of CeO_2 at 4.9~19.6 MPa. Poluboyarinov et al. obtained 387 kJ · mol^{-1} for the creep of CeO_2 at 1.34 MPa (x = 0.00 in CeO_{2-x}, 15 μm, 96 % theoretical density). Although the difference between the activation energies of the present study and those of Poluboyarinov et al. is small, they may have different meanings with respect to each other because the stress exponents observed in the two studies also differ.

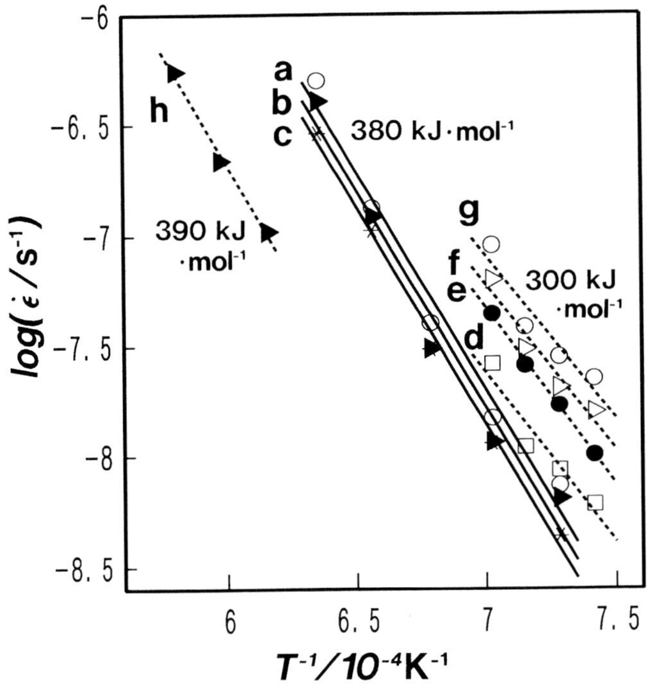

Fig. 3 Temperature dependence of strain rate. (a) $Ce_{0.98}Y_{0.02}O_{1.99}$ (grain size = 1.4 μm) at 4.9 MPa, (b) $Ce_{0.92}Y_{0.08}O_{1.96}$ (grain size = 0.91 μm) at 4.9 MPa, (c) $Ce_{0.92}Y_{0.08}O_{1.96}$ (grain size = 1.1 μm) at 4.9 MPa, (d) CeO_2 at 4.9 MPa, (e) CeO_2 at 9.8 MPa, (f) CeO_2 at 14.7 MPa, (g) CeO_2 at 19.6 MPa, (h) CeO_2 (grain size = 15 μm) at 1.3 MPa. Lines a, b, and c are the results of present study, lines d, e, f and g are the results of Ikuma and Uchida[17] for the grain size = 3.7 μm, and line h is the result of Poluboyarinov et al.[16] The apparent activation energies are shown for the three sets of data.

4.0 DISCUSSION

Comparing the creep data with those reported in the literature is difficult because the creep rate depends on many factors such as stress, temperature, microstructure (grain size), and oxygen partial pressure. If the creep were diffusional and the mechanism were known, the diffusion coefficient could be calculated to compare the results. To do this, the grain size exponents (n in $\dot{\varepsilon} \propto \sigma^m/d^n$) for the creep of the sample must be known. However, the preparation of $Ce_{0.98}Y_{0.02}O_{1.99}$ and $Ce_{0.92}Y_{0.08}O_{1.96}$ with different grain sizes was difficult and was not accomplished in the present study. An alternative way to perform such a comparison is to normalize the data with respect to variables such as the stress and grain size and then make a comparison with the predicted creep rate.

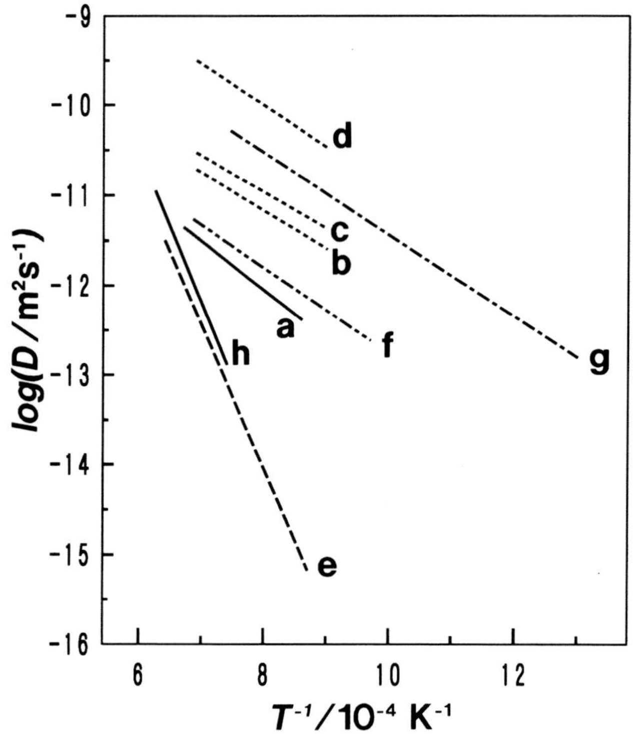

Fig. 4 Diffusion coefficients of CeO_2. (a) D_{oxy} in CeO_2 (Floyd[22]), (b) D_{oxy} in $Ce_{0.9}Y_{0.1}O_{1.95}$ (Floyd[22]), (c) D_{oxy} in $Ce_{0.8}Y_{0.2}O_{1.9}$ (Floyd[22]), (d) D_{oxy} in $Ce_{0.6}Y_{0.4}O_{1.8}$ (Floyd[22]), (e) D_{chem} in CeO_2 (Vinokurov[30]), (f) D_{chem} in $Ce_{0.997}Gd_{0.003}O_{1.9985}$ (Vinokurov[30]), (g) D_{oxy} in $Ce_{0.9}Gd_{0.1}O_{1.95}$ (Manning et al.[31]) and (h) D_{oxy} in CeO_2 (Kamiya et al.[32]).

To predict the diffusional creep rate using the available diffusion data, the diffusion data were reviewed. The diffusion coefficients of CeO_2 reported in the literature are summarized in Figure 4. Floyd[9, 22] has determined the oxygen diffusion coefficient (D_{oxy}) in CeO_2 and $Ce_{1-z}Y_zO_{1-z/2}$. The oxygen diffusion coefficient in undoped CeO_2 is shown by line (a) in Figure 4 with the activation energy of 89 kJ·mol.$^{-1}$ As CeO_2 is doped with Y (lines (b), (c) and (d)), the oxygen diffusion coefficient increased but the activation energy did not change much. Vinokurov[30] determined the chemical diffusion coefficient (D_{chem}) in single-crystal CeO_2 (line (e)) and Gd-doped CeO_2 (line (f)). He utilized the fact that, as CeO_2 changes from the nonstoichiometric state to the stoichiometric state, it undergoes a change in color. He simply followed the speed at which the color front moved. For CeO_2, the activation energy of the chemical diffusion observed by Vinokurov was high. As CeO_2 was doped with Gd, the chemical diffusion increased but the activation energy changed to a smaller value. The activation energy for the chemical diffusion in Gd-doped CeO_2 was close to that

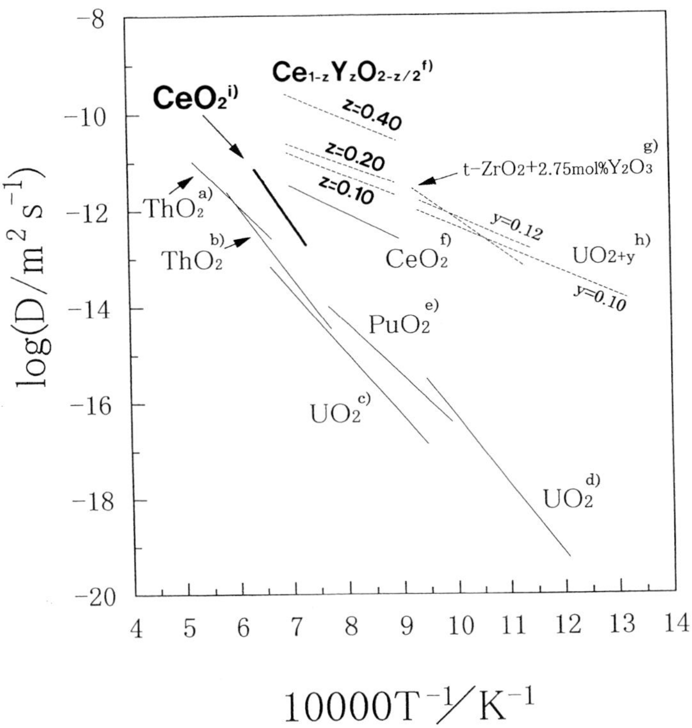

Fig. 5 Oxygen diffusion coefficients in oxides with fluorite structure.[32] D_{oxy} in ThO_2 (a) Ando et al.[33], (b) Edwards et al.[34], D_{oxy} in UO_2 (c) Marin and Contamin,[35] (d) Auskern and Belle[36], D_{oxy} in PuO_2 (e) Deaton and Wiedenheft[37], D_{oxy} in CeO_2 and $Ce_{1-z}Y_zO_{2-z/2}$ (f) Floyd,[22] (i) Kamiya et al.[32] D_{oxy} in t-ZrO_2 + 2.75% Y_2O_3, (g) Ikuma et al.[38], and D_{oxy} in UO_{2+y} (h) Contamin et al.[39] (Reproduced with permission from Ref.32).

reported by Floyd[22] for Y-doped CeO_2. Manning et al.[31] determined the oxygen diffusion coefficient in $Ce_{0.9}Gd_{0.1}O_{1.95}$ over a wide range of temperatures (line (g)). Although it was higher than the oxygen diffusion coefficient for $Ce_{0.9}Y_{0.1}O_{1.95}$ (line (b)), the activation energies were almost the same as those of most of the doped CeO_2 samples. Recently, Kamiya et al.[32] determined the oxygen diffusion coefficient in undoped CeO_2 (line (h)). Their result is different from that reported by Floyd for undoped CeO_2 (line (a)). From the figure, it is clear that the oxygen diffusion increased as CeO_2 was doped with Y or Gd. However, an accurate oxygen diffusion coefficient in undoped CeO_2 remains unknown. Furthermore, there are no data for cation diffusion in CeO_2. These are issues to be addressed.

The first issue was dealt with by Kamiya et al.,[32] who summarized most of the oxygen diffusion coefficients[22, 32-39] in oxides with the fluorite structure, as shown in Figure 5. These data can be separated into two groups: one with high activation energies (smaller diffusion coefficients) and the other with lower activation energies

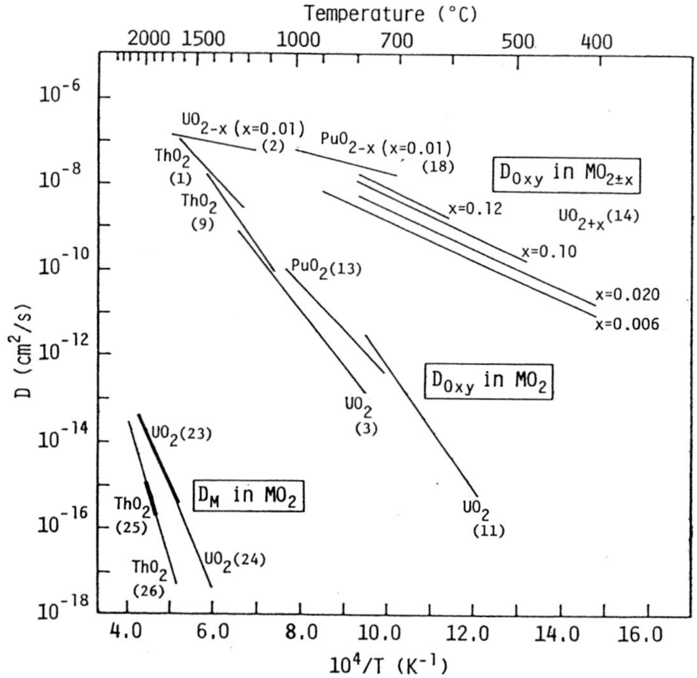

Fig. 6 Diffusion coefficients in oxides of the fluorite structure.[40] D_{oxy} in ThO_2 (1) Ando et al.,[33] (9) Edwards et al.,[34] D_{oxy} in UO_2 (3) Marin and Contamin,[35] (11) Auskern and Belle,[36] D_{oxy} in PuO_2 (13) Deaton and Wiedenheft,[37] D_{oxy} in UO_{2-x} (2) Kim,[41] D_{oxy} in UO_{2+x} (14) Contamin et al.,[39] D_{oxy} in PuO_{2-x} (18) Bayoglu and Lorenzelli,[42] D_M in UO_2 (23) Reimann and Lundy,[43] (24) Matzke,[44] and D_M in ThO_2 (25) King,[45] (26) Matzke.[46] (Reproduced with permission from Ref. 40).

(larger diffusion coefficients). As seen in the figure, the diffusion coefficients with the higher activation energies are associated with oxides of stoichiometric composition (the activation energy includes the defect formation energy and defect migration energy). On the other hand, the diffusion coefficients with lower activation energies are associated with oxides of nonstoichiometric composition (the activation energy is related only to the defect migration energy). The data in the figure support the hypothesis proposed by Ando and Oishi,[40] which states that activation energy of oxygen diffusion in stoichiometric oxides with the fluorite structure, if they have similar melting temperature, is 250 kJ · mol^{-1} but for nonstoichiometric oxides it is 87 kJ · mol^{-1}. Obviously, the activation energy of oxygen diffusion in undoped CeO_2 reported by Floyd[22] is in the regime of the non-stoichiometric composition. Kamiya et al.[32] have concluded that the CeO_2 used by Floyd must have been influenced by the impurities. The result of Kamiya et al. is for oxygen diffusion in stoichiometric CeO_2.

Ando and Oishi[40] have summarized the cation diffusion as well as oxygen diffusion[33-37, 39, 41-46] in UO_2, PuO_2 and ThO_2 (Figure 6). All the diffusion data can be

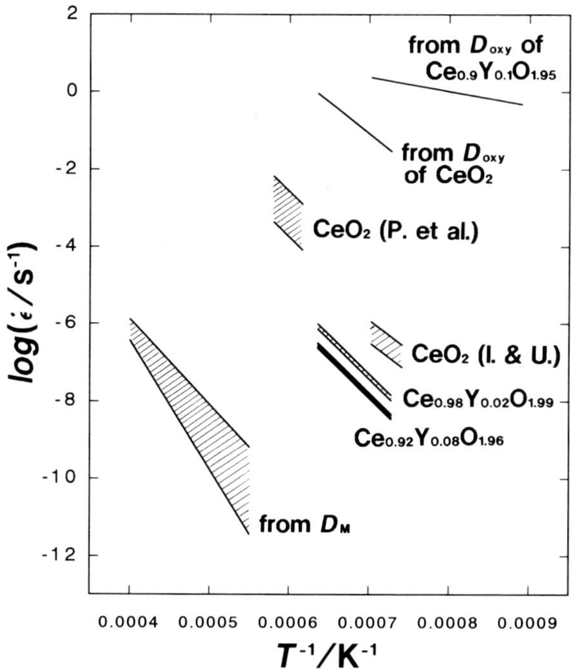

Fig. 7 Temperature dependence of strain rate for CeO_2. The data in Figure 3 were normalized in this figure at 4.9 MPa and 1 μm. Some of the lines were calculated from lattice diffusion data[22, 32, 40] (D_{oxy} and D_M) using equation (3).

grouped into three groups: the oxygen diffusion in $MO_{2\pm x}$, oxygen diffusion in MO_2, and cation diffusion (D_M) in MO_2. In particular, the cation diffusion in MO_2 does not depend on the cation species. They further compared the diffusion coefficient of species on the face-centered cubic sublattice in both fluorite and anti-fluorite crystals as a function of T_m/T where T_m is the melting temperature of each crystal. All the species on the face-centered cubic sublattice exhibited similar diffusion coefficient. This is the characteristic of diffusion in fluorite and anti-fluorite structures. Since the melting temperature of CeO_2 is not much different from that of UO_2 and ThO_2, the cation diffusion coefficients in CeO_2, UO_2 and ThO_2 are expected to be similar. Therefore, we may estimate with some confidence the cation diffusion in CeO_2 to be in the range of these D_M data.

The creep data normalized at 4.9 MPa and 1 μm are shown in Figure 7. To normalize the data, it was assumed that the creep data follow the diffusional creep equations (equations 3 or 4). The data included in the figure are the creep rate of polycrystalline CeO_2 reported by Poluboyarinov et al.[16] (original data were shown in Figure 3 at σ = 1.34 MPa, d = 15 μm), Ikuma and Uchida[17] (original data in Figure 3 at σ = 4.9 MPa, d = 3.7 μm), and the present study (original data in Figure 3 at σ = 4.9 MPa, d = 0.91~1.4 μm). Since we don't know whether the present creep data follow the Nabarro-Herring creep or Coble creep, both equations were used. Consequently the normalized data are given by bands in the figure. Substituting the diffusion coefficients for oxygen (D_{oxy})[22, 32] and the cation (D_M) into equation (3), the creep rates were calculated. These results are also plotted in Figure 7. Extrapolation of the Poluboyarinov et al.[16] data to the temperature examined by Ikuma and Uchida[17] revealed that the result of Ikuma and Uchida was slightly smaller than that of Poluboyarinov et al. This may be due to the difference in stoichiometry and/or difference in creep mechanisms. It should be noted in the figure that, as CeO_2 was doped with *Y*, the creep rate decreased. The effect of *Y* on the oxygen diffusion was the opposite, where oxygen diffusion increased[22] as CeO_2 was doped with *Y* (see Figure 4). This implies that oxygen is not rate-limiting to the creep of CeO_2. The observed creep rate was several orders of magnitude smaller than the creep rate calculated from the oxygen diffusion coefficient using equation (3) and was several orders of magnitude larger than the creep rate calculated from the expected cation diffusion coefficient. Thus, the creep of CeO_2 was limited neither by the cerium lattice diffusion nor the oxygen lattice diffusion. Since grain boundary diffusion is generally faster than lattice diffusion, the creep of CeO_2 was not limited by the grain boundary diffusion of oxygen. Instead, the cation grain boundary diffusion was most likely the rate-determining one.

5.0 CONCLUSIONS

The creep of Y-doped CeO_2 ($Ce_{0.98}Y_{0.02}O_{1.99}$ and $Ce_{0.92}Y_{0.08}O_{1.96}$) was studied at 1100-1300°C in the stress range of 4.9-19.6 MPa using the four-point dead-load creep test. The creep of *Y*-doped CeO_2 was characterized by viscous creep ($m \fallingdotseq 1$) as opposed to non-viscous behaviour for the creep of CeO_2 at higher temperatures. The creep data normalized at 4.9 MPa and 1 μm have indicated that, as CeO_2 was doped with *Y*, the creep rate decreased. Comparing the observed creep rate with the creep rate calculated from the oxygen and cation diffusion coefficients, it was inferred that the creep of CeO_2 was controlled by cation diffusion through the grain boundary.

6.0 ACKNOWLEDGEMENTS

This work was supported in part by the Grant-in-Aid for Scientific Research (C), the Japanese Ministry of Education, Science, Sports and Culture (Grant No. 10650673).

7.0 REFERENCES

1. K. Ando: *Japanese Journal of Applied Physics*, Series 2, 1989, pp.113-123.
2. S. C. Singhal: "*Science and Technology of Zirconia V*", S. P. S. Badwal,

M. J. Bannister, and R. H. J. Hannink, eds., Technomic Pub., Basel, Switzerland, 1993, pp.631-651.

3. W. R. Cannon and T. G. Langdon: *Journal of Materials Science,* **18**, 1983, pp.1-50.
4. F. Wakai, S. Sakaguchi and Y. Matsuno: *Advanced Ceramics Materials,* **1**, 1986, pp.259-263.
5. D. J. M. Bevan and J. Kordis: *Journal of Inorganic Nuclear Chemistry,* **26**, 1964, pp.1509-1523.
6. P. Kofstad and A.Z. Hed: *Journal of American Ceramic Society,* **50**, 1967, pp.681-682.
7. B. Iwasaki and T. Katsura: *Bulletin of Chemical Society of Japan,* **44**, 1971, pp.1297-1301.
8. J. W. Dawicke and R. N. Blumenthal: *Journal of Electrochemical Society,* **133**, 1986, pp.904-909.
9. B. C. H. Steele and J.M. Floyd: *Proceedings of British Ceramic Society,* **19**, 1971, pp.55-76.
10. R. N. Blumenthal and R. K. Sharma: *Journal of Solid State Chemistry,* **13**, 1975, pp.360-364.
11. E. H. Baker, M. Iqbal, and B. E. Knox: *Journal of Materials Science,* **12**, 1977, pp.305-310.
12. H. L. Tuller and A. S. Nowick: *Journal of Electrochemical Society,* **126**, 1979, pp.209-217.
13. H. L. Tuller and T. S. Stratton: *Nato ASI Ser. B,* **129**,1985, pp.87-99.
14. K. Eguchi, T. Setoguchi, T. Inoue and H. Arai: *Solid State Ionics,* **52**, 1992, pp.165-172.
15. D. Schneider, M. Godickemeier and L.J. Gauckler: *Journal of Electroceramics,* **1**, 1997, pp.165-172.
16. D. N. Poluboyarinov, E.Ya. Shapiro, V. S. Bakunov and F. A. Akopov: *Neorganic Materials,* **2**, 1966, pp.336-342.
17. Y. Ikuma and T. Uchida: *Journal of Material Science Letters,* **16**, 1997, pp1637 - 1639.
18. M. Kamiya, E. Shimada and Y. Ikuma: *Journal of Materials Synthesis Processing,* **6**, 1998, pp.283-286.
19. G. Brauer and K. A. Gingerich: *Journal of Inorganic Nuclear Chemistry,* **16**, 1960, pp.87-99.
20. D. J. M. Bevan, W. W. Barker, R. L. Martin and T. C. Parks: *Proceedings of 4th Conference on Rare Earth Research,* L. Eyring, ed., 1965, pp.441-468.
21. R. T. Dirstine, R. N. Blumenthal and T. F. Kuech: *Journal of Electrochemical Society,* **126**, 1979, pp.264-269.
22. J. M. Floyd: *Indian Journal of Technology,* **11**, 1973, pp.589-594.
23. P. L. Chen and I. W. Chen: *Journal of American Ceramic Society,* **76**, 1993, pp.1577-1583.
24. D. Schneider, M. Godickemeier and L. J. Gauckler: *Br. Ceramics Proceedings,* **56**, 1996, pp.103-112.
25. G. W. Hollenberg, G. R. Terwilliger and R. S. Gordon: *Journal of American Ceramic Society,* **54**, 1971, pp.196-199.

26. P. A. Lessing and R. S. Gordon: *Journal of Materials Science,* **12**, 1977, pp.2291-2302.
27. Y. Ikuma and R. S. Gordon: *Journal of American Ceramic Society,* **66**, 1983, pp.139-147.
28. Y. M. Chiang, D. P. Birnie III and W. D. Kingery, *"Physical Ceramics"*, John Wiley & Sons, New York, 1997, pp.245-251.
29. J. D. Hodge, P. A. Lessing and R. S. Gordon: *Journal of Materials Science,* **12**, 1977, pp.1598-1604.
30. I. V. Vinokurov: *Neorganic Material*, **6**, 1970, pp.31-36.
31. P. S. Manning, J. D. Sirman and J. A. Kilner: *Solid State Ionics,* **93**, 1997, pp.125-132.
32. M. Kamiya, E. Shimada and Y. Ikuma: *Journal of Ceramics Society Japan,* **106**, 1998, pp.1023-1026.
33. K. Ando, Y. Oishi and Y. Hidaka: *Journal of Chemical Physics,* **65**, 1976, pp.2751-2755.
34. H. S. Edwards, A. F. Rosenberg and J. T. Bittel: Report No. ASD-TDR-63-635, Aeronautical System Division, Wright-Patterson Air Force Base, Ohio, 1963.
35. J. F. Marin and P. Contamin: *Journal of Nuclear Materials,* **30**, 1969, pp.16-35.
36. A. B. Auskern and J. Belle: *Journal of Nuclear Materials,* **3**, 1961, pp.267-276.
37. R. L. Deaton and C. J. Wiedenheft: *Journal of Inorganic Nuclear Chemistry,* **35**, 1973, pp.649-650.
38. Y. Ikuma, Y. Tsubaki and T. Masaki: **99**, 1991, pp.101-103.
39. P. Contamin, J. J. Bacmann and J. F. Marin: *Journal of Nuclear Materials,* **42**, 1972, pp.54-64.
40. K. Ando and Y. Oishi: *Journal of Nuclear Science and Technology,* **20**, 1983, pp.973-982.
41. K. C. Kim: Lawrence Berkeley Lab., California Univ. Report, 1980, LBL-11905.
42. A. S. Bayoglu and R. Lorenzelli: *Journal of Nuclear Materials* **82**, 1979, pp.403-410.
43. D. K. Reimann and T. S. Lundy: *Journal of American Ceramic Society,* **52**, 1969, pp.511-512.
44. H. Matzke: *Journal of Nuclear Materials,* **114**, 1983, pp.121-135.
45. A. D. King: *Journal of Nuclear Materials,* **38**, 1971, pp.347-349.
46. H. Matzke: *Journal of Physics,* Paris, Colloq., *7,* 1976, pp.452-457.

Measurement of Chemical Diffusivity of $BaTiO_{3-\delta}$ by a Conductivity Relaxation Technique

HAN-ILL YOO and CHANG-ROCK SONG
Solid State Ionics Research Laboratory
School of Materials Science and Engineering
Seoul National University, Seoul, 151-742, Korea

ABSTRACT

The chemical diffusivity of "undoped" polycrystalline $BaTiO_{3-\delta}$ has been determined via a conductivity relaxation technique, at elevated temperatures (800°C $\leq T \leq$ 1100°C) as a function of the ambient oxygen partial pressure in the range of 10^{-16} atm $\leq P_{O_2} \leq$ 1 atm including an n-to-p type transition regime. Mathematical formulation has been developed to convert a conductivity relaxation to the corresponding non-stoichiometry (δ) relaxation in the transition regime. It has been found that the chemical diffusivity appears to exhibit a maximum at the n-to-p transition point where the electronic, minimum conductivity falls, and that surface reaction becomes more rate-determining than diffusion as the transition point is approached from both n-type and p-type branches. Experimental details are given and the results are exhaustively compared with the reported data till date on the "undoped" $BaTiO_3$.

1.0 INTRODUCTION

$BaTiO_3$ enjoys wide applications as an important material in modern electronics technology: positive temperature coefficient resistors (PTCR), capacitor dielectrics and ferroelectric memories are only a few to name. Due to the technological importance of this perovskite oxide, its defect structure and electrical transport properties have been extensively studied.[1-11] Notwithstanding its practical importance, particularly in processing, the chemical diffusivity of $BaTiO_3$ has seldom been studied and hence, still not known in the entire stability range of the phase even at a single temperature. $BaTiO_{3-\delta}$ is a nonstoichiometric compound which turns from n-type to p-type semiconducting as ambient oxygen partial pressure (P_{O_2}) increases from 10^{-16} atm at an elevated temperature.[1] The chemical diffusivity here refers to a measure of diffusion kinetics governing the process of (oxygen) nonstoichiometry (δ) re-equilibration after a change of the ambient oxygen potential.

We have measured the chemical diffusivity of "undoped" $BaTiO_3$ via a conductivity relaxation technique, as a function of oxygen partial pressure in the range of

10^{-16} atm $\leq P_{o2} \leq 1$ atm, including an n-to-p type transition regime, at 800, 900, 1000 and 1100°C, respectively. $BaTiO_3$ is one of the rare oxides that undertake a distinctive n-to-p transition within its stability range. Mathematical formulation is, thus, developed and applied to convert a conductivity relaxation into the corresponding nonstoichiometry relaxation in the transition regime. Experimental details are reported and the results exhaustively compared with the reported data. Defect chemical analysis of the present results will be given elesewhere.[12]

2.0 THEORETICAL BACKGROUND

2.1 Solution to Diffusion Equation

Let us consider a parallelpiped specimen of $BaTiO_{3-\delta}$ measuring $2a \times 2a \times 2c$, where $c >> a$ and hence, diffusion along the length (2c) may be reasonably neglected. For an abrupt change of surrounding oxygen partial pressures, $\Delta logP_{O_2}$, which corresponds to a non-stoichiometry change from an initial (equilibrium) value $\delta(0)$ (at $t = 0$) to the final one $\delta(\infty)$ (as $t \to \infty$), the average value of non-stoichiometry $\bar{\delta}(t)$ varies with time t as[13]

$$\frac{\bar{\delta}(t) - \delta(0)}{\delta(\infty) - \delta(0)} = 1 - \left[\sum_{n=1}^{\infty} \frac{2L^2 \exp\left(-\frac{\beta_n^2 \tilde{D} t}{a^2} \right)}{\beta_n^2 (\beta_n^2 + L^2 + L)} \right]^2 \quad (1)$$

where β_n's are the positive roots of

$$\beta \tan\beta = L; \; L \equiv \frac{ak}{\tilde{D}} \quad (2)$$

It is here assumed that the surface reaction be of the first order with the rate constant k [such that $\partial\delta(x, y; t)/\partial x|_{xy=\pm a} = k(\delta(\pm a, \pm a; t) - \delta(\infty))$] and the chemical diffusion coefficient $\tilde{D}$ constant for a $\Delta logP_{O_2}$ chosen. Note that $1/k$ and $a/\tilde{D}$ correspond respectively to the kinetic resistance of the surface reaction and solid state diffusion that are the consecutive kinetic steps in the non-stoichiometry re-equilibration process.

As L tends to zero or $k << \tilde{D}/a$, the surface reaction becomes rate-determining and Eqn. 1 takes the form

$$\frac{\bar{\delta}(t) - \delta(0)}{\delta(\infty) - \delta(0)} = 1 - \exp\left(-\frac{2kt}{a} \right) \quad (3)$$

As L tends to infinity or $k >> \tilde{D}/a$, solid state diffusion becomes rate-determining and Eqn. 1 takes the form

$$\frac{\bar{\delta}(t)-\delta(0)}{\delta(\infty)-\delta(0)} = 1-\frac{64}{\pi^4}\left[\sum_{n=0}^{\infty}\frac{1}{(2n+1)^2}\exp\left(-\frac{(2n+1)^2\pi^2\tilde{D}t}{4a^2}\right)\right]^2. \tag{4}$$

Equations (3) and (4), therefore, serve both extreme cases of Eqn. (1) depending on L or the relative resistance of the two consecutive reaction steps.

2.2 Correlation Between Nonstoichiometry and Conductivity

In determining $\tilde{D}$ from a conductivity relaxation by using Eqn. 1, it is crucial to establish a correct correlation between a mean conductivity and the corresponding, mean nonstoichiometry $\bar{\delta}(t)$. When a system is an electron/hole mixed conductor, the correlation is by no means trivial, and $BaTiO_3$ is such a conductor in the n-to-p transition regime.[1]

The possible defects in "undoped" $BaTiO_3$ may be taken as electrons (e'), holes ($h^{\bullet}$), oxygen vacancies ($V_o^{\bullet\bullet}$), cation vacancies (V_{Ba}'' and V_{Ti}'''') and acceptor-type background impurities (A_C').[1,14] Due to overall charge neutrality, one has

$$n+2[V_{Ba}'']+4[V_{Ti}'''']+[A_C'] = p+2[V_O^{\bullet\bullet}] \tag{5}$$

where n, p and $[j]$ are the concentration of electrons, holes and ionic defects of type j, respectively. Defining the effective density of acceptors as $[A'] \equiv 2[V_{Ba}''] + 4[V_{Ti}''''] + [A_C']$ for the sake of simplicity,[15] one may represent the lattice molecule as $(Ba_{1-x}A_x)(Ti_{1-y}A_y)O_{3-(x+y)/2-\delta}$. The oxygen nonstoichiometry, δ, is then given as

$$\begin{aligned}\frac{N_A\delta}{V_m} &= [V_O^{\bullet\bullet}]-\frac{1}{2}[A'] \\ &= \frac{1}{2}(n-p)\end{aligned} \tag{6}$$

where N_A and V_m denote Avogadro's number and the lattice-molecular volume, respectively. It is noted that the second equality of Eqn. 6 is always valid whatever types of ionic defects are presumed.

Let

$$\alpha \equiv \frac{\sigma_p}{\sigma_n} = \frac{p}{nb}; \quad b = \frac{\mu_n}{\mu_p}; \quad K_i = np \tag{7}$$

where σ_k and μ_k are the partial conductivity and mobility of electrons ($k = n$) and holes ($k = p$). It then follows[16] that

$$p = \sqrt{K_i\alpha b}; \; n = \sqrt{\frac{K_i}{\alpha b}} \tag{8}$$

The oxygen nonstoichiometry, Eqn. 6, is, thus, rewritten in terms of the conductivity ratio α as

$$\frac{N_A\delta}{V_m} = -\sqrt{K_i}\,sinh\left(\frac{1}{2}ln\,\alpha b\right) \tag{9}$$

The electronic total conductivity, $\sigma_{el}(=\sigma_n+\sigma_p)$, is also written as [1,17]

$$\frac{\sigma_{el}}{\sigma_{el,m}} = cosh\left(\frac{1}{2}ln\,\alpha\right) \tag{10}$$

where $\sigma_{el,m}$ is the minimum electronic conductivity falling at the n-to-p transition point (where $\sigma_n = \sigma_p$). Obviously, the local value for the electronic conductivity at a time, $\sigma_{el}(x, y; t)$, can be correlated to the corresponding nonstoichiometry, $\delta(x, y; t)$, by means of $\alpha(x, y; t)$. It is noted, however, that what one is monitoring is not the local value $\sigma_{el}(x,y;t)$, but its spatial average $\bar{\sigma}_{el}(t)$ or $\overline{cosh(1/2\,ln\,\alpha)}$, and that $\overline{cosh(1/2\,ln\,\alpha)} \neq cosh(1/2\,\overline{ln\,\alpha})$ and $\overline{sinh(1/2\,ln\,\alpha)} \neq sinh(1/2\,\overline{ln\,\alpha})$, in general.

It has been shown via computer simulation[18] that if the variation of the conductivity or the conductivity ratio for a given $\Delta logP_{O_2}$ is limited to the region of either $\alpha > 1$ or $\alpha < 1$, one may approximate as

$$\overline{cosh\left(\frac{1}{2}ln\,\alpha\right)} \approx cosh\left(\frac{1}{2}\overline{ln\,\alpha}\right);\ \overline{sinh\left(\frac{1}{2}ln\,\alpha b\right)} \approx sinh\left(\frac{1}{2}\overline{ln\alpha b}\right) \tag{11}$$

within an error of less than a few %. Accordingly, the mean conductivity variation, $\bar{\sigma}_{el}(t)$, can be correlated to the mean nonstoichiometry variation, $\bar{\delta}(t)$, via Eqns. 9 and 10 as

$$\frac{\bar{\delta}\ (t) - \delta\ (0)}{\delta\ (\infty) - \delta\ (0)} = \frac{sinh\left(\frac{1}{2}\overline{\ln\alpha(t)b}\right) - sinh\left(\frac{1}{2}\ln\alpha(0)b\right)}{sinh\left(\frac{1}{2}\ln\,\alpha(\infty)b\right) - sinh\left(\frac{1}{2}\ln\alpha(0)b\right)} \tag{12}$$

where $\overline{ln\alpha(t)}$, $ln\alpha(0)$ and $ln\alpha(\infty)$ are the conductivity ratios corresponding to $\bar{\delta}(t)$, $\delta(0)$ and $\delta(\infty)$, respectively. Even when $\ln\alpha$ is varied crossing 0 for a given $\Delta logP_{O_2}$, the validity of Eqn. 12 is guaranteed within the same error bound in the time after $\bar{\sigma}_{el}(t)$ passes its minimum. In the entire time range of relaxation across $ln\alpha = 0$, however, the approximation of Eqn. 11 normally causes an error of some tens of % to the final results. A detailed analysis will be given elsewhere.[18]

When $\alpha b >> 1$ or in the regime of exclusively p-type ($\sigma_p \approx \sigma_{el}$), Eqn. 12 takes the form

$$\frac{N_A\delta}{V_m} = -\frac{p}{2} = -\frac{\sigma_p}{2e\mu_p} \tag{13a}$$

and, when $\alpha b << 1$ or in the regime of exclusively n-type ($\sigma_n \approx \sigma_{el}$),

$$\frac{N_A \delta}{V_m} = \frac{n}{2} = \frac{\sigma_n}{2e\mu_n} . \tag{13b}$$

Assuming that the mobilities, μ_p and μ_n are independent of the nonstoichiometry or oxygen partial pressure at a given temperature, one can, thus, establish the correlation in the exclusively n- or p-type region as

$$\frac{\bar{\delta}(t) - \delta(0)}{\delta(\infty) - \delta(0)} = \frac{\overline{\sigma_{el}}(t) - \sigma_{el}(0)}{\sigma_{el}(\infty) - \sigma_{el}(0)} \tag{14}$$

where $\bar{\sigma}_{el}(t)$, $\sigma_{el}(0)$ and $\sigma_{el}(\infty)$ are total electronic conductivities corresponding to $\bar{\delta}(t)$, $\delta(0)$ and $\delta(\infty)$, respectively. Eqn. 14 is what has been routinely employed in the conductivity relaxation technique.[19,20]

3.0 EXPERIMENTAL

3.1 Conductivity Relaxation Measurement

"Undoped" $BaTiO_3$ powder of 99.995% purity (Lot 05717 BW, Aldrich) was used for the preparation of specimens. According to the manufacturer's product information, concentrations of Fe, Si and Al as the nonvolatile impurities are 5, 5 and 1 ppm, respectively. The total concentration of the background acceptors, say, Fe and Al, can, thus, be estimated to be 6 ppm at best. The powder, with 3 w/o PVA added, was moulded into plates of ca. 3 mm thickness × 17 mm width × 24 mm length and subsequently pressed uni-axially under a pressure of 250 MPa. Sintering was followed at 1400°C in air for 5 hours. The relative density and average grain size of the as-sintered samples are 94 ± 1% and 43 ± 8 μm, respectively. The as-sintered sample was cut with a low speed saw into parallelpiped specimens, measuring 2.2 × 2.2 × 14.7 mm, for conductivity and/or relaxation measurements. The four electrodes, platinum wires of 200 μm diameter, were attached to the specimen as described elsewhere.[1]

The entire measurement setup is as illustrated in Figure 1. The specimen (#9 in Figure 1) was hung onto the four Pt-lead wires (200 μm dia.) which were previously passed through a 4-bore alumina tubing (#8), ca. 4 mm in diameter. This assembly, together with an S-type thermocouple (#10) passing through another alumina tubing, was inserted into a one-end closed mullite tubing (working chamber, #7), ca. 13 mm inner-diameter × 250 mm length. The dead volume inside the working chamber was no greater than 33 cm^3. An atmosphere gas was made to flow directly onto the specimen through another 4-bore alumina tubing (#11) of ca. 4 mm outer diameter. The flow rate was fixed at approximately 300 cc/min. While an atmosphere gas was being passed through the working chamber for initial equilibration, another one with a different P_{O_2}-value was made standing by, flowing through the auxiliary bypass line. Atmospheres within the working chamber were then exchanged with the aid of

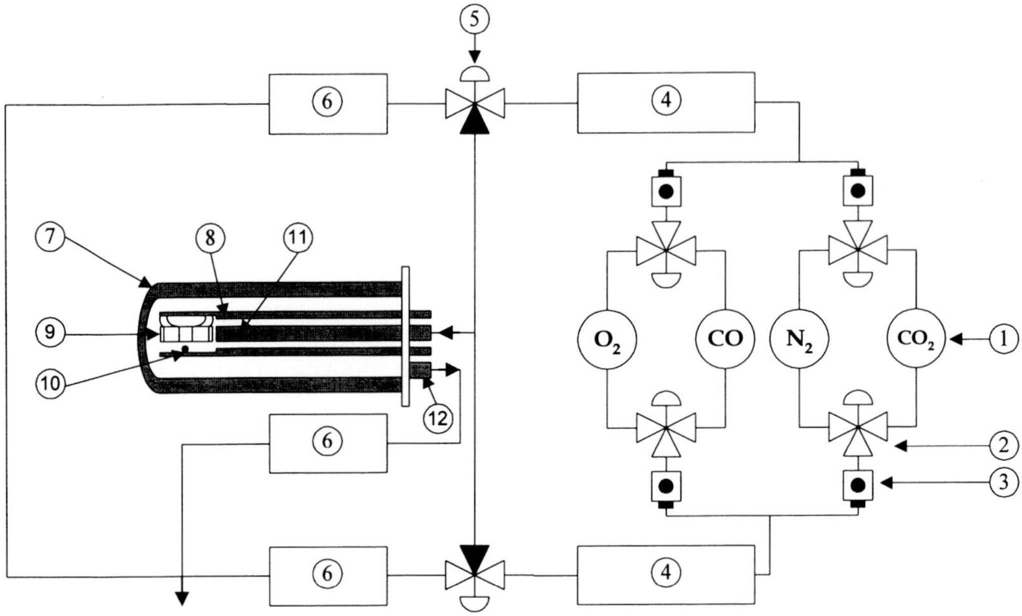

Fig. 1 Schematic of the experimental setup for conductivity relaxation measurement. 1. gas cylinders, 2. 3-way valves for gas mixing, 3. flowmeters, 4. ZrO_2 cells, 5. 3-way valves for atmosphere exchange, 6. gas bubblers, 7. one-end-closed tubing, 8. 4-bore alumina tubing for sample hanging, 9. specimen, 10. thermocouple, 11. 4-bore alumina tubing for gas inlet and 12. gas outlet.

the two 3 way valves (#5) each located on each gas line prior to the chamber. In this way, a change in oxygen partial pressures ($\Delta log P_{O_2}$) could be achieved within less than 1 second. The atmosphere gases were N_2/O_2 or Co/CO_2 mixtures, the oxygen partial pressures of which were monitored by zirconia-based electrochemical cells (#6). The change of oxygen partial pressure, $\Delta log(P_{O_2})$, was controlled to be no greater than 1 except in the interval between nominally pure N_2 [$log(P_{O_2}/atm) \approx -4.4$] and CO_2 [$log(P_{O_2}/atm)=-7.4$] e.g., at 1000°C. The temperatures of both zirconia cells and the working chamber were maintained the same within ±3°C, and the temperature of the sample itself was controlled within ± 1°C.

Equilibrium conductivity and its relaxation were measured by a conventional 4-probe technique. A specimen was first equilibrated in a predetermined P_{O_2}-atmosphere and its equilibrium conductivity determined from the slope of current vs. voltage variation. A constant current was then passed by using a current source (Keithley 220) and the potential drop between the inner probes was monitored against time after an abrupt change of the oxygen partial pressure, using an electrometer (Keithley 617) that is interfaced with a personal computer. The current was chosen in the range of 10 μA to 1 mA so that potential drop was not greater than 200 mV. After the specimen was completely re-equilibrated in the new atmosphere, the equilibrium conductivity [$\sigma(\infty)$] was again determined in the same manner as the initial equilibrium conductivity [$\sigma(0)$]. That equilibrium conductivity was in agreement, within 1% at worst, with the conductivity evaluated from the single potential drop (due to the fixed current). The measurements were carried out in the range of $-16 \le log(P_{O_2}/atm) \le 0$ at 1100, 1000, 900 and 800°C.

3.2 Analysis of Relaxation Profiles

Eqn. 1 per se is not appropriate for the nonlinear-least-squares (NLLS) fitting of a relaxation profile $\bar{\delta}(t)$ as determined via Eqn. 12 from a conductivity relaxation profile, $\bar{\sigma}_{el}(t)$. It is because Eqn. 2 has no analytical solution and one never knows the exact starting moment (t_0) of the conductivity relaxation except for its upper bound, i.e. $t_0 < 1$ sec in the present experiment. We thus modify Eqn. 1 by taking the actual time zero (t_0) as the third fitting parameter in addition to β_1 and $\tilde{D}$ as

$$\frac{\bar{\delta}(t) - \delta(0)}{\delta(\infty) - \delta(0)} = 1 - \left[\sum_{n=1}^{\infty} \frac{2(\beta_n \tan\beta_n)^2 \exp\left(-\frac{\beta_n^2 \tilde{D}(t-t_0)}{a^2}\right)}{\beta_n^2(\beta_n^2 + (\beta_n \tan\beta_n)^2 + \beta_n \tan\beta_n)} \right]^2 \tag{15}$$

Note that $(n-1)\pi \le \beta_n (n = 1,2,3...) \le (n-1/2)\pi$ and, once a value for β_1 is chosen for a given value for L, the remaining β_n's ($n \ge 2$) are subsequently fixed by Eqn. 2. As L decreases (i.e., the surface reaction becomes more rate-determining), β_n approaches $(n-1)\pi$ faster as n becomes larger, and hence, the first term of Eqn. 15 becomes more and more predominating. As $\beta_1 \to 0$ as an extreme case, Eqn. 15 takes the form corresponding to Eqn. 3

$$\frac{\bar{\delta}(t) - \delta(0)}{\delta(\infty) - \delta(0)} = 1 - \exp\left(-\frac{2k(t-t_0)}{a}\right) \tag{16}$$

with the two fitting parameters of k and t_0. As $\beta_1 \to \pi/2$, as the other extreme, Eqn. 15 takes the form corresponding to Eqn. 4

$$\frac{\bar{\delta}(t) - \delta(0)}{\delta(\infty) - \delta(0)} = 1 - \frac{64}{\pi^4}\left[\sum_{n=0}^{\infty} \frac{1}{(2n+1)^2} \exp\left(-\frac{(2n+1)^2 \pi^2 \tilde{D}(t-t_0)}{4a^2}\right)\right]^2 \tag{17}$$

with the two fitting parameters of $\tilde{D}$ and t_0.

In order to facilitate the NLLS fitting of a relaxation profile to Eqn. 15, we first fitted the profile to both Eqns. 16 and 17 in order to determine whether $L<1$ (more surface-reaction-controlling) or $L>1$ (more diffusion-controlling). Because these two equations have the same number of fitting parameters [k and t_0 in Eqn. (16), and $\tilde{D}$ and t_0 in Eqn. 17], the size of the fitting errors evaluated was employed as a criterion to determine whether $L > 1$ or $L < 1$: If k turned out to be more precise than $\tilde{D}$, then $L < 1$ and *vice versa*. The best-fitted values of k and t_0, for $L<1$, and those of $\tilde{D}$ and t_0, for $L > 1$, were subsequently chosen as the initial values for $\tilde{D}$ and t_0 of Eqn. 15 while fully fitting the profile, $\bar{\delta}(t)$, to Eqn. 15. The initial value for β_1 was chosen such that $L = 1$ in Eqn. 2. This is only for the sake of fitting efficiency: By doing this way, the fitting parameters ($\tilde{D}$, β_1, t_0) could be made converge faster. It turned out that fitting to the fifth term ($n = 5$) in the summation of Eqns. 15 or 17 was sufficient enough for any conductivity relaxation.

The surface reaction rate constant, k, is finally calculated from the best-fitted values for $\tilde{D}$ and β_1 by using Eqn. 2. It should be pointed out that if $L<0.1$ or >10, it is practically impossible to determine both $\tilde{D}$ and k by fitting to Eqn. 15 because of a relatively large fitting error (~10%) being induced. In such a case, either Eqns. 16 or 17 obviously work better.

In order to confirm the reproducibility, the conductivity relaxation was repeated four times at each $\Delta logP_{O_2}$ chosen. We thus had four sets of data of the chemical diffusivity and/or the surface reaction rate constant. The weighted averages of those were taken as the best estimate of $\tilde{D}$ and k.

4.0 RESULTS AND DISCUSSION

4.1 Equilibrium Electrical Conductivity

Figure 2 shows the equilibrium electrical conductivity as measured against oxygen partial pressure at different temperatures. The oxygen-partial-pressure exponent of the conductivity, m_σ (such that $\sigma \sim P_{O_2}^{\ m_\sigma}$), changes apparently from -1/4 to +1/4 across the conductivity minimum (or the n-to-p transition point) as P_{O_2} increases. This trend is in perfect agreement with the results reported earlier.[1,3,5,7,9,15] According to the defect structure proposed,[1,15] the majority type of disorder in this n-to-p transition regime is $2[V_o^{\bullet\bullet}] \approx [A_C']$. The total conductivity, σ, can, thus, be fitted in the Po_2 region around the minimum conductivity to the equation,[21]

$$\sigma = \frac{\sigma_{el,m}}{2}\left\{\left(\frac{P_{O_2}}{P^*_{O_2}}\right)^{-1/4} + \left(\frac{P_{O_2}}{P^*_{O_2}}\right)^{+1/4}\right\} + \sigma_{ion} \qquad (18)$$

where σ_{ion} denotes the ionic partial conductivity that is constant in this regime, and $P^*_{O_2}$ the oxygen partial pressure corresponding to the n-to-p transition point. The solid lines in Figure 2 are the best fitted, and the parameters as evaluated, $\sigma_{el,m}$, σ_{ion} and $P^*_{O_2}$, are listed in Table 1. The values for σ_{ion} are in a reasonable agreement with the literature values[14] for "undoped" polycrystalline $BaTiO_3$ which have been determined in the same manner. One can now calculate the conductivity ratio, α, from Eqn. 10 with the use of these fitted results. The result, $log\alpha$ vs. $logP_{O_2}$, is shown in Figure 3. The slope, $dlog\alpha/dlogP_{O_2}$, that corresponds to $2|m_\sigma|$, is 0.5 in the region around the conductivity minimum and deviates to a smaller value as P_{O_2} further

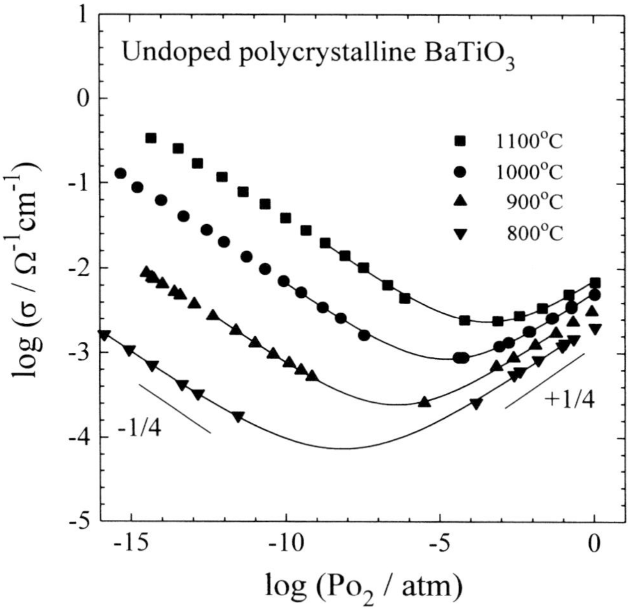

Fig. 2 Equilibrium, total conductivity vs. oxygen partial pressure at different temperatures. The solid lines are the best-fitted to Eqn. (18) in the text.

Table 1 Parameters, $\sigma_{el,m}$, σ_{ion} and $P^*_{O_2}$ evaluated from the conductivity isotherms in Fig. 2.

T/°C	Log ($\sigma_{el,m}$ / $\Omega^{-1}cm^{-1}$)	Log (σ_{ion} / $\Omega^{-1}cm^{-1}$)	Log ($P^*_{O_2}$ / atm)
800	- (4.401 ± 0.008)	- (4.463 ± 0.089)	- (8.174 ± 0.020)
900	- (3.756 ± 0.008)	- (4.155 ± 0.065)	- (6.400 ± 0.020)
1000	- (3.191 ± 0.010)	- (3.681 ± 0.083)	- (4.772 ± 0.025)
1100	- (2.740 ± 0.015)	- (3.272 ± 0.100)	- (3.447 ± 0.041)

decreases. It indicates a change of the majority type of disorder from ($V_o^{\bullet\bullet}$, A') to (e', $V_o^{\bullet\bullet}$) for the present system.[1]

4.2 Conductivity Relaxation Kinetics

Figures 4 and 5 show the typical relaxation of the mean electronic conductivity, $\bar{\sigma}_{el}(t)$ during both oxidation and reduction, in an exclusively n-type regime ($1.33 \times 10^{-5} \leq \alpha \leq 2.66 \times 10^{-5}$ for the P_{O_2} interval, $-14.77 \leq logP_{O_2} \leq -14.01$), and in an exclusively p-type regime ($55.4 \leq \alpha \leq 112.4$ for $-1.32 \leq logP_{O_2} \leq -0.72$), respectively, at, e.g. 1000°C. In that n-type regime, the mean total conductivity as measured

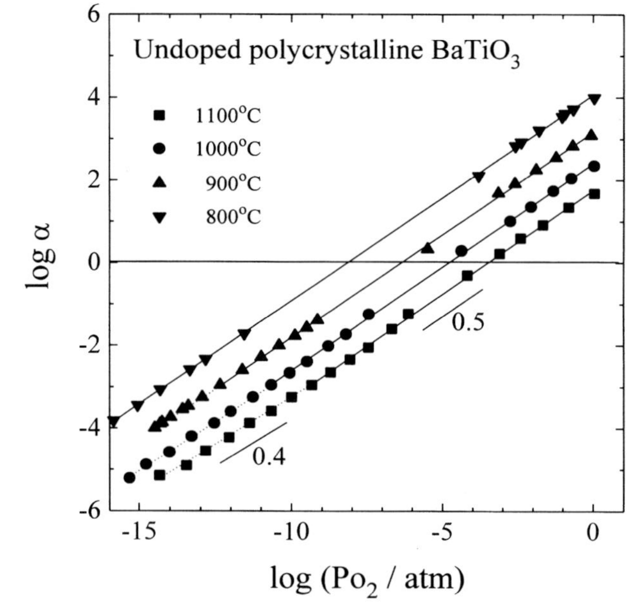

Fig. 3 The conductivity ratio, $\log\alpha$ vs. $\log P_{O_2}$. The solid lines with an ideal slope of $2|m_\sigma|=0.5$ represent the regions of the majority type disorder (A', $V_O^{\bullet\bullet}$) at different temperatures, and the dotted lines with a smaller slope indicate the shift of the majority disorder to (e', $V_O^{\bullet\bullet}$).

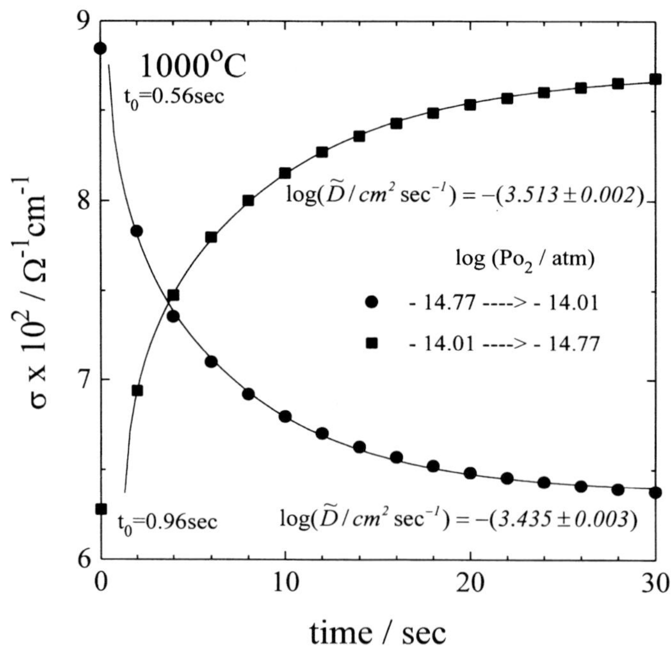

Fig. 4 Typical conductivity relaxation profiles in exclusively n-type regime of undoped $BaTiO_3$ during reduction and oxidation at 1000°C. The solid lines are the best-fitted to Eqn. 17.

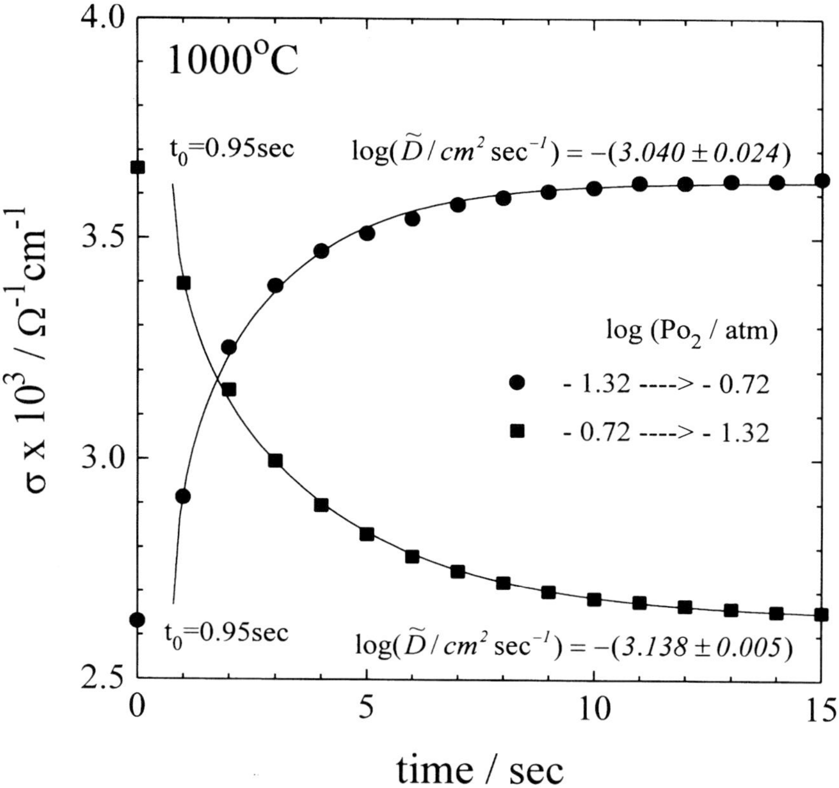

Fig. 5 Typical conductivity relaxation profiles in exclusively p-type regime of undoped $BaTiO_3$ during reduction and oxidation at 1000°C. The solid lines are best-fitted to Eqn. 17.

could be taken as the mean electronic conductivity (i.e., $\bar{\sigma}_{el}(t) \approx \bar{\sigma}(t)$) as the ionic transference number $t_{ion} < 0.003$. In the p-type regime, however, the constant ionic conductivity ($2.08 \times 10^{-4}\ \Omega^{-1}cm^{-1}$) that was determined by Eqn. 18 was previously subtracted from the mean total conductivity to get the mean electronic conductivity (i.e., $\bar{\sigma}_{el}(t) = \bar{\sigma}(t)\text{-}\sigma_{ion}$) as $t_{ion} \leq 0.08$.

As $\alpha << 1$ or $>> 1$ in these P_{O_2} ranges chosen, the mean electronic conductivity relaxation, $\bar{\sigma}_{el}(t)$, can be directly transformed to the non-stoichiometry relaxation via Eqn. 14. All the relaxation profiles are much better fitted to Eqn. 17 rather than to Eqn. 16 as depicted by the solid lines. It has indeed turned out that when fitted to Eqn. 15, the parameter β_1 converges to $\pi/2$ with $t_o = 0.56\sim0.96$ sec. One can thus conclude that the diffusion is practically rate-determining in these specific cases.

In contrast with the regimes of exclusively n- or p-type where $\bar{\sigma}_{el}(t)$ is monotonically decaying with time, the relaxation across the n-to-p transition point appears rather complicated. A typical example is as shown in Figure 6a, which illustrates the

Table 2 $\widetilde{D}$ and k evaluated from conductivity relaxation across the n-to-p transition point at 1000°C assuming b = 0.1, 1.0, 10, respectively

b	Reduction			Oxidation		
	$\tilde{D}$ / cm^2 sec^{-1}	k / cm-sec^{-1}	Eq.	$\tilde{D}$ / cm^{2}sec^{-1}	k /cm-sec^{-1}	Eq.
0.1	—	—		(2.39 ± 0.15) × 10^{-4}	(1.15 ± 0.23) × 10^{-3}	(15)
1.0	—	(1.23 ± 0.01) × 10^{-3}	(16)	—	(2.29 ± 0.12) × 10^{-3}	(16)
10	(5.30 ± 0.04) × 10^{-5}	(3.41 ± 0.09) × 10^{-3}	(15)	—	(1.29 ± 0.08) × 10^{-3}	(16)

__ : it cannot be evaluated.

relaxations on both reduction and oxidation legs between α = 0.06 (corresponding to $logP_{O_2}$ = -7.44) and 1.97 (corresponding to $logP_{O_2}$ = -4.37) at 1000°C (see Figures 2 and 3). Upon each leg, the mean electronic conductivity first decreases and then increases leaving a minimum in between. It is simply attributed to the fact that the local conductivity, $\sigma(x, y; t)$, varies across the conductivity minimum along the 1000°C isotherm in Figure 2.

Obviously, the conductivity relaxation in Figure 6a cannot be transformed to the corresponding nonstoichiometry relaxation via Eqn. 14 which is monotonic (see Eqn. 15); the transformation may be accomplished through Eqn. 12, despite a rather big uncertainty stemming from the assumptions of Eqn. 11, only provided that the mobility ratio b is known. For the system of $BaTiO_3$, however, the value for b is not correctly known yet. We have thus tried three different values b = 0.1, 1.0, 10 in order to see the effect of the magnitude of b on the transformation. Figures 6b and c show the transformed nonstoichiometry relaxation profiles depending on the b-values during reduction and oxidation, respectively. The dashed, solid and dotted lines are the best fitted to Eqn. 15 (mixed-controlled), Eqn. 16 (surface-reaction controlled) and Eqn. 17 (diffusion-controlled), respectively. The fitting results, $\widetilde{D}$ and k, are as summarised in Table 2.

One can recognise from Table 2 that in any case, the relaxation behaviour cannot be consistently described by diffusion kinetics only (Eqn. 17). When b = 10, diffusion is more rate-controlling during reduction (Eqn. 15 with $L \approx 7$), while the surface reaction totally rate-determining during oxidation (Eqn. 16). Given that the values for k are of the same order of magnitude ($k \approx 10^{-3}$cm sec^{-1}) for both directions of relaxation, it is implied that the chemical diffusivity during the oxidation be larger at least by an order of magnitude than that during the exactly reverse process, reduction. This is hard to believe to be the case in reality. When b = 0.1, on the other hand, the relaxation during oxidation can be well described by Eqn. 15 with $L \approx 5$, meaning that diffusion is more rate-determining. However, the reduction relaxation (Figure 6b) turns out to be non-monotonic and hence, can be described by none of Eqns. 15-17. These inconsistencies associated with both b = 0.1 and 10 may suggest that $0.1<b<10$ at the temperature of 1000°C.

When b = 1.0, both the reduction and oxidation relaxations can be consistently, well described by Eqn. 16, that is, both are practically controlled by the surface reaction as depicted by the solid lines in Figures 6b and c. It has indeed been found that β_1 converges to 0 (or $L \rightarrow 0$) with $t_0 = 0.12$ sec (reduction) and 0.25 sec (oxidation) when fitted to Eqn. 15. This consistency may indicate that b is not far from 1.

We have, thus, taken the results for b = 1.0. These results have turned out to be in good agreement with the results extrapolated from those in the exclusively n- or p-type regimes as explained below. It has actually been believed on the basis of electrical properties of the present system[22,23] that the mobility ratio is not much different from b = 1.0 at 1000 and 1100°C. We have later found that b = 1.55 at 1000°C through the defect chemical analysis of the isotherms of conductivity and chemical diffusivity.[12]

Nevertheless, one may still argue that the fitting consistency solely with b = 1.0 may be an artifact due to relatively large error associated with the transformation of $\bar{\sigma}_{el}(t)$ to $\bar{\delta}(t)$ via Eqn. 11, but it is by no means the case now. According to a detailed analysis,[18] the minimum value of the mean conductivity equals the minimum value of the equilibrium conductivity i.e. $\bar{\sigma}_m = \sigma_m$ if the surface-reaction alone is governing the overall kinetics during conductivity relaxation along a conductivity isotherm across a conductivity minimum, otherwise always $\bar{\sigma}_m > \sigma_m$. These facts can be used even as a criterion to decide whether a nonstoichiometry re-equilibration kinetics are totally governed by surface-reaction or not. It should be noted for the relaxations in Figure 6a, that the minimum value of the conductivity on each relaxation leg (8.47×10^{-4} and $8.36 \times 10^{-4}\ \Omega^{-1}cm^{-1}$ during reduction and oxidation, respectively) is in agreement within the experimental error with the true minimum in equilibrium state ($8.52 \times 10^{-4}\ \Omega^{-1}cm^{-1}$) in Figure 2. When the surface reaction is totally rate-determining, obviously $\delta(x, y, t) = \bar{\delta}(t)$ and hence, $\sigma(x, y, t) = \bar{\sigma}(t)$. Therefore Eqn. 11 is exact, no longer an approximation.

4.3 Chemical Diffusivity and Surface Reaction Rate Constant

Most of the earlier authors[24-28] have assumed that the surface reaction is much faster than chemical diffusion for the systems of $BaTiO_3$ and $SrTiO_3$ at elevated temperatures. If the same had also been assumed, then one would have extracted only the chemical diffusivity of $BaTiO_3$ through the NLLS-fitting of the conductivity relaxation profiles to Eqn. 17 ignoring the fitting accuracy. The result at, e.g., 1000°C would be as illustrated in Figure 7. Small arrows therein indicate the data obtained from the transformed (via Eqn. 12) relaxation profiles with the assumption, b = 1, in the n-p mixed regime. The value of $logP_{O_2}$ on the abscissa represents the central value of $\Delta logP_{O_2}$, chosen during reduction and oxidation and an error bar represents the width of the $logP_{O_2}$ window or $\Delta logP_{O_2}$. One would have a bizarre result that the chemical diffusivity appears to be of an "M"-shape against $logP_{O_2}$. We have already recognised that the surface reaction is more rate-determining in the n-to-p transition regime. The value for "*ka*" would thus have been mistaken for chemical diffusivity under the assumption that the surface reaction is fast enough.

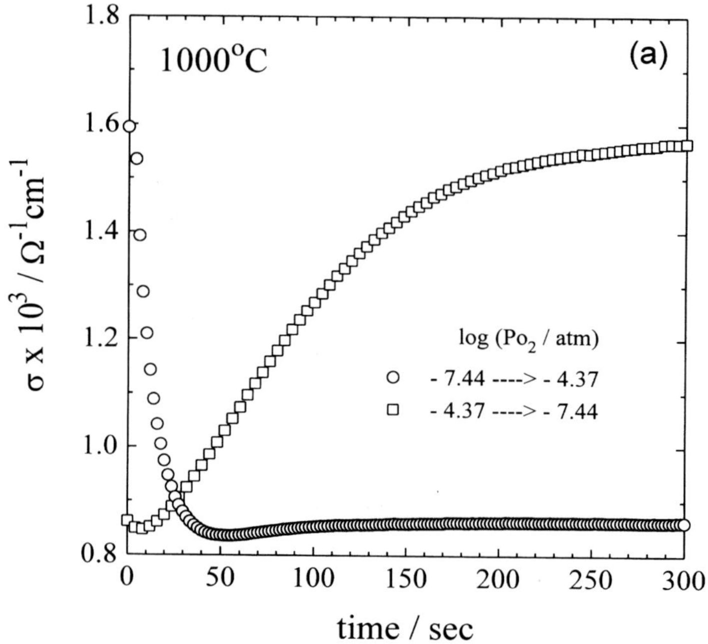

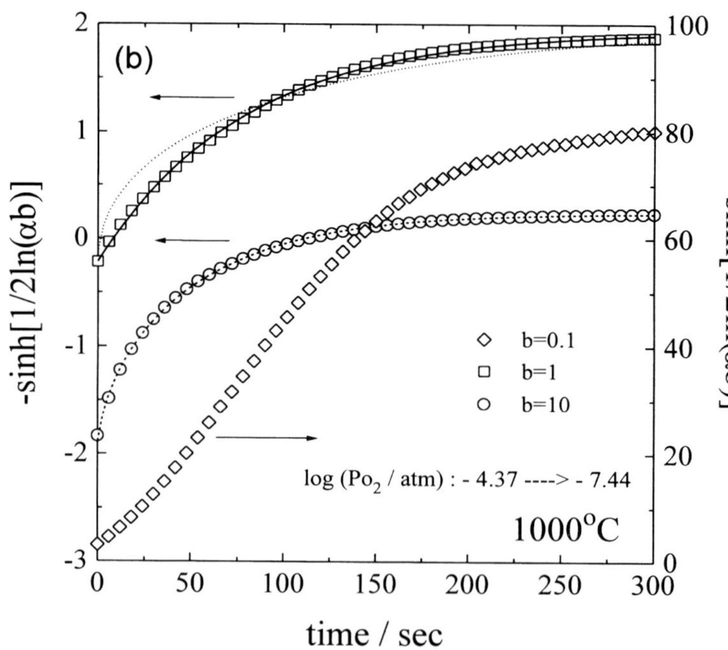

Fig. 6 (a) Typical conductivity relaxation profiles during both reduction and oxidation across the n-to-p transition point of undoped $BaTiO_3$ at 1000°C. They are transformed to nonstoichiometry relaxation profiles with different mobility-ratio values, b =0.1, 1.0, 10, during reduction (b) and oxidation. (See (c) in next page).

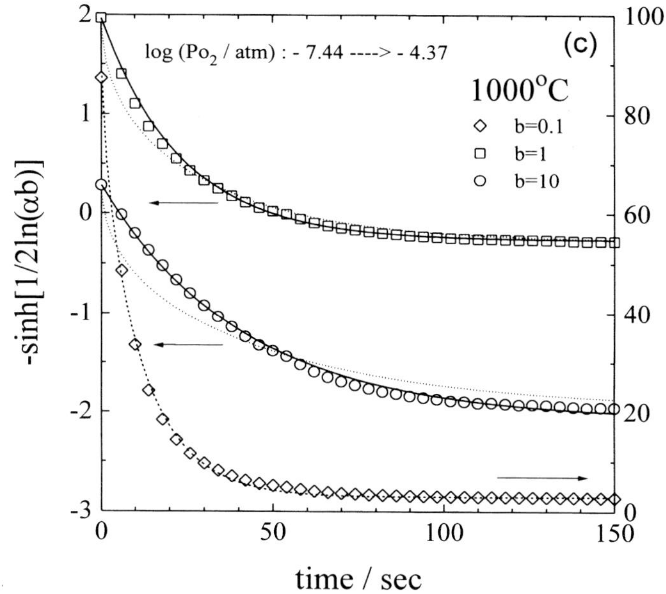

Fig. 6c The dashed, solid and dotted lines are best-fitted to Eqns. 15, 16 and 17, respectively.

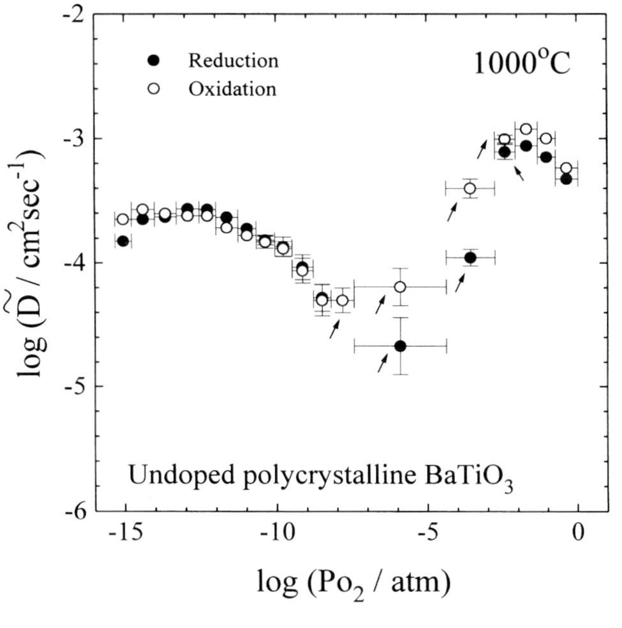

Fig. 7 Chemical diffusivity vs. oxygen partial pressure at 1000°C when evaluated via Eqn. 17 assuming the surface reaction is fast enough. A horizontal error bar represents the $\Delta log P_{O_2}$ window of each datum. Small arrows indicate the data being obtained from non-stoichiometry relaxation profiles transformed with the value of b=1.

Figures 8a-d shows the chemical diffusivity ($\widetilde{D}$) and surface reaction rate constant (k) evaluated on the basis of Eqn. 15 at 1100, 1000, 900 and 800°C, respectively. The relative error of each datum is no greater than 10%. The value of $logP_{O_2}$ on the x-axis represents again the median value of $\Delta logP_{O_2}$ for a datum of $\widetilde{D}$ (a horizontal bar encloses the width $\Delta logP_{O_2}$), and the value of $logP_{O_2}$ imposed in the surrounding for a datum of k. Small arrows indicate the data that were obtained from the transformed nonstoichiometry relaxation profiles (via Eqn. 12) in the electrons/ holes mixed regime. It is emphasised that these are quite consistent, in trend as well as in magnitude, with those obtained in either exclusively p-type or n-type regimes.

Upon comparison with Figure 2, it is clearly seen that chemical diffusion is rate-determining in the regimes of exclusively n-type and p-type, and that the surface reaction becomes more and more rate-determining as the n-to-p transition point is approached at a given temperature. Furthermore, it is strongly suggested that the isothermal variation of $\widetilde{D}$ be convex upward in the n-to-p transition regime where $2|m_\sigma|=0.5$ and become more or less flat to the left of this regime. One may be tempted to expect the maximum to occur at the stoichiometric composition $\delta = 0$, where the conductivity minimum occurs if $b = 1$. There have, thus far, been known only two compounds where the chemical diffusivity displays a maximum. They are α-Ag_2S[29] and α'-Ag_2Te,[30] and the maximum falls at the stoichiometric composition of each. Becker et al.[29] attributes the latter to the thermodynamic factor becoming maximum at the stoichiometric composition. It seems that a similar phenomenon is now observed, for the first time, among oxide systems. A detailed analysis will be given elsewhere.[12]

4.4 Comparison with the Reported Data

Very limited amount of chemical diffusion data has been reported till date, on "undoped" $BaTiO_3$. Wernicke [27] measured the chemical diffusivity of polycrystalline $BaTiO_3$ in an oxygen partial pressure range, 0.1 atm $\le P_{O_2} \le$ 1 atm, at temperatures, 750°C $\le T \le$ 1000°C, by a conductivity relaxation technique. Maier et al.[31] reported on the chemical diffusivities of both polycrystalline and single crystal $BaTiO_3$ in air atmosphere at temperatures, 600°C $\le T \le$ 660°C , as determined by a polarisation technique in electron-blocking cells. Recently, Müller and Härdtl [28] have determined the chemical diffusivity of both polycrystalline and single crystal $BaTiO_3$ in an oxygen partial pressure range, 0.1 atm $\le P_{O_2} \le$ 0.2 atm, at temperatures in the range of 700°C $\le T \le$ 1200°C, via the conductivity relaxation technique. In addition, Shirasaki et al.[32] reported on the oxygen tracer diffusivity of polycrystalline $BaTiO_3$ that was determined in an oxygen partial pressure, P_{O_2} = 0.05 atm, as a function of temperature in the range of 770°C $\le T \le$ 1430°C by using an O^{18}/O^{16} gas-exchange technique.

As is seen, the chemical diffusivity of $BaTiO_3$ has so far been measured only in a limited range of oxygen partial pressure, $P_{O_2} \ge$ 0.1 atm, where the "undoped" $BaTiO_3$ is exclusively of p-type,[1,11] see Figure 2. The present results corresponding only to air atmosphere are thus compared with those reported in Figure 9, where the oxygen tracer diffusivity data are also compiled. They appear to be in reasonable agreement

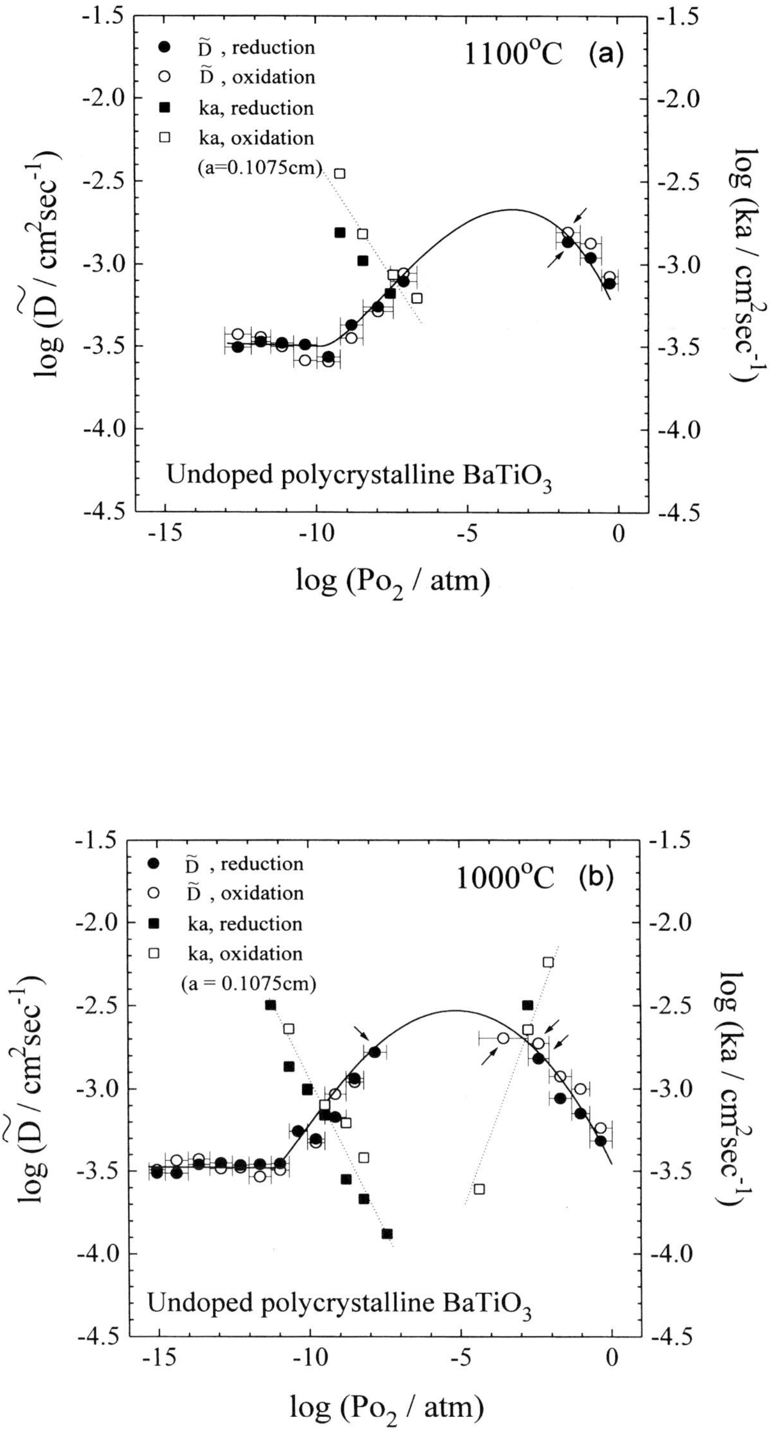

Fig. 8a-b Chemical diffusivity ($\tilde{D}$) and surface-reaction-rate constant (k) vs. oxygen partial pressure for undoped $BaTiO_3$ at (a) 1100 and (b) 1000. (Figures (c, d) contd. in next page).

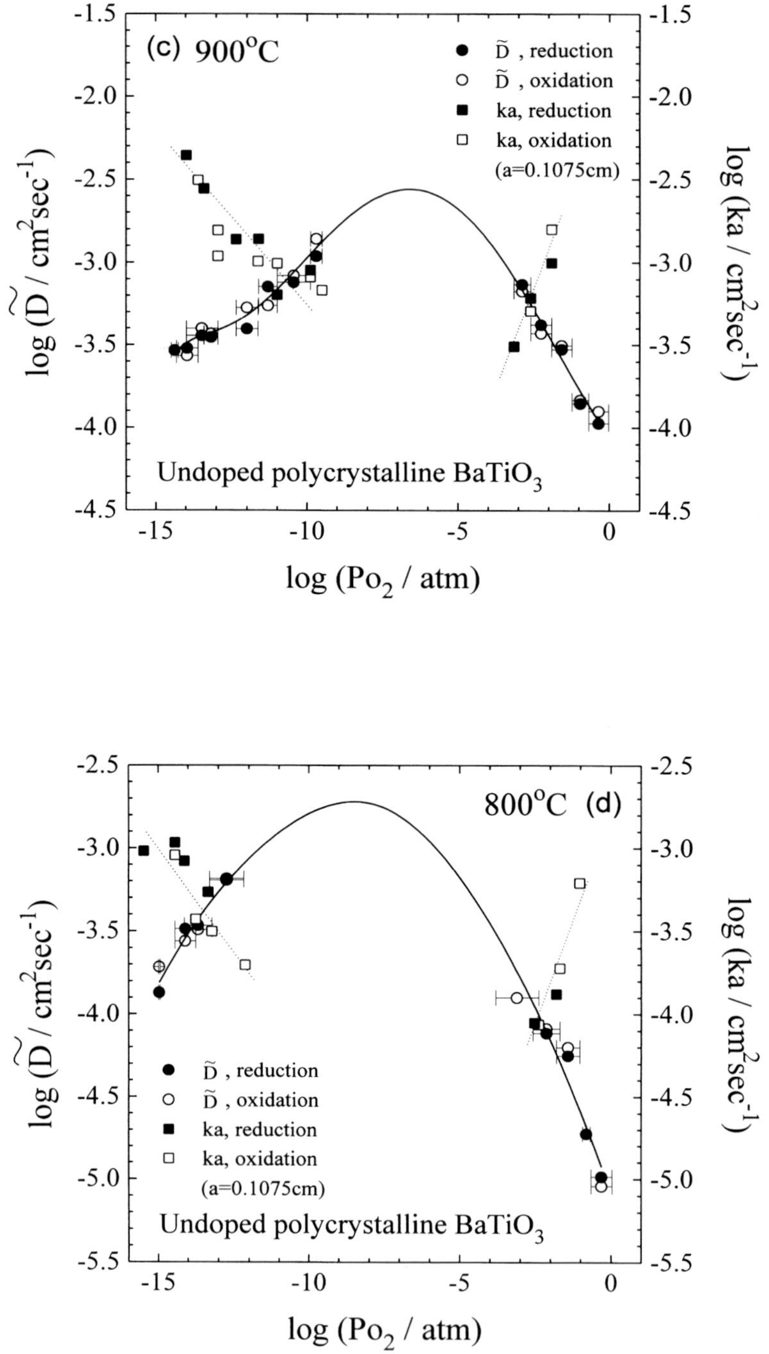

Fig. 8c-d Chemical diffusivity ($\tilde{D}$) and surface-reaction-rate constant (k) vs. oxygen partial pressure for undoped $BaTiO_3$ at (c) 900 and (d) 800°C, respectively. The dotted and solid lines are for visual guidance only. A horizontal error bar denotes the $\Delta logP_{O_2}$ window of each datum.

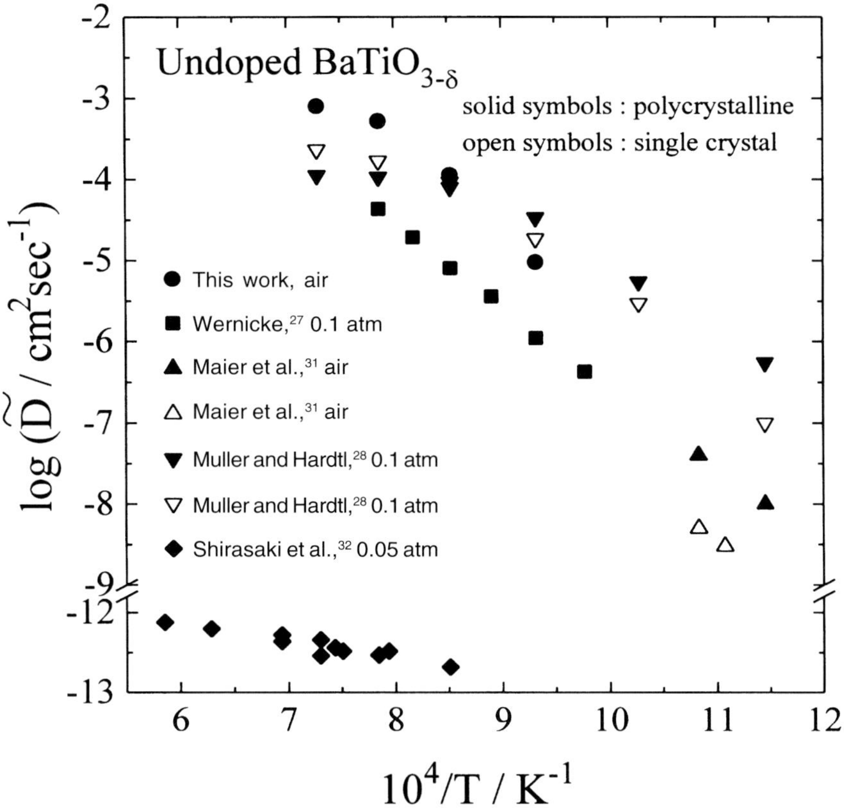

Fig. 9 Comparison of the present results for the air atmosphere with all the reported values on the chemical diffusivity of undoped $BaTiO_3$. For further comparison, the data on oxygen tracer diffusivity (◆) are also compiled.

with the reported chemical diffusivities. However they reach the maximum at 1000 and 1100°C. It seems to be consistent that the trend of variation of $\widetilde{D}$ vs. $1/T$ deviates (elbow-upwardly) from the linearity around 1000°C and higher. This deviation may be of thermodynamic origin, which will be fully explored elsewhere.[12] The present results in the temperature range of $T \leq 1000$°C may be best estimated, for the purpose of comparison, as

$$\widetilde{D}(\text{cm}^2\,\text{sec}^{-1}) = (1.38 \pm 0.34) \times 10^6 \exp\left(-\frac{2.37 \pm 0.21\,\text{eV}}{kT}\right) \tag{19}$$

The activation energies reported are 2.09 eV[27] and 2.70 eV[28], which are comparable with the present value, 2.4 eV.

5.0 CONCLUSION

As P_{O_2} decreases from 1 atm at elevated temperatures, the chemical diffusivity of $BaTiO_3$ varies convex-upwardly across the p-to-n transition regime and then becomes more or less flat. The maximum diffusivity appears to fall at the oxygen partial pressure where the conductivity minimum falls. The diffusivity in air atmosphere is best estimated in the range of $T \leq 1000°C$ as

$$\widetilde{D}(\mathrm{cm}^2\,\mathrm{sec}^{-1}) = (1.38 \pm 0.34) \times 10^6 \exp\left(-\frac{2.37 \pm 0.21\mathrm{eV}}{kT}\right)$$

which is in reasonable agreement with the reported data thus far.

6.0 ACKNOWLEDGEMENTS

This work has been financially supported by the Korea Science and Engineering Foundation through the Centre for Interface Science and Engineering of Materials.

7.0 REFERENCES

1. J. Y. Kim, C. R. Song and H. I. Yoo: *Journal of Electrochemical Society,* **1**, 1997, p.27.
2. J. Novak and H. Arend: *Journal of American Ceramics Society,* **47,** 1964, p.530.
3. S. A. Long and R. N. Blumenthal: *Journal of American Ceramics Society*, **54,** 1971, p.515.
4. G. J. Conger and H. U. Anderson: *Journal of American Ceramics Society*, **55,** 1972, p.539.
5. A. M. J. H. Seuter, *Philips Research Reports, Supplement*, **3,** 1974, p.1.
6. G. V. Bois, N. A. Mikhailova, E. I. Prodavtsova and V. A. Yusova: *News. Acad. Science USSR, Inorg. Mater.*, **12,** 1976, p.1302.
7. J. Daniels and K. H. Härdtl: *Philips Research Reports,* **31,** 1976, p.489.
8. H. Ihrig and D. Hennings: *Physics Review B*, **17,** 1978, p.4593.
9. N. G. Eror and D. M. Smyth: *Journal of Solid State Chemistry*, **24,** 1978, p.235.
10. J. F. Baumard and P. Abelard: *Solid State Ionics,* **12,** 1984, p.47.
11. J. Nowotny and M. Rekas: *Ceramics International,* **20,** 1994, p.225.
12. C. R. Song and H. I. Yoo: *Physical Review B*, **61**, 2000, p.3975.
13. J. Cran: *The Mathematics of Diffusion,* 2nd Ed., Clarendon Press, Oxford, U.K., 1975, p.60.
14. J. Nowotny and M. Rekas: *Solid State Ionics,* **49,** 1991, p.135.
15. N. H. Chan, R. K. Sharma and D. M. Smyth: *Journal of American Ceramics Society,* **65,** 1982, p.167.
16. J. H. Becker and H. P. R. Frederikse: *Journal of Applied Physics*, **33,** 1962, p.447.

17. H. I. YOO and C. S. KIM: *Solid State Ionics,* **53-56,** 1992, p.583.
17a. C. S. KIM and H. I. YOO: *Journal of Electrochemical Society,* **143,** 1996, p.2863.
18. C. R. SONG and H. I. YOO: Submitted to *Solid State Ionics.*
19. P. E. CHILDS, L. W. LAUB and J. B. WAGNER: *Proceedings of British Ceramics Society,* **19,** 1971, p.29.
20. F. MORIN and R. DIECKMANN: *Zeitschrift für Physikalische Chemie N.F.,* **129,** 1982, p.219.
21. E. K. CHANG, A. METHA and D. M. SMYTH: *Proceedings of the Symposium on Electro-Ceramics and Solid State Ionics,* H. L. Tuller and D. M. Smyth eds., The Electrochemical Society, Princeton, NJ, 1988, p.35.
22. N. H. CHAN, R. K. SHARMA and D. M. SMYTH: *Journal of American Ceramics Society,* **64,** 1981, p.556.
23. E. DUVERGER, B. JANNOT, M. MAGLIONE and M. JANNIN: *Solid State Ionics,* **73,** 1994, p.139.
24. A. E. PALADINO: *Journal of American Ceramics Society,* **48,** 1965, p.476.
25. L. C. WALTER and R. E. GRACE: *Journal of Physics and Chemistry of Solids,* **28,** 1967, p.245.
26. D. B. SCHWARZ and H. U. ANDERSON: *Journal of Electrochemical Society,* **122,** 1975, p.707.
27. R. WERNICKE: *Philips Research Reports,* **31,** 1976, p.526.
28. A. MÜLLER and K. H. HÄRDTL: *Applied Physics A,* **49,** 1989, p.75.
29. K. D. BECKER, H. SCHMALZRIED and V. VON WURMB: *Solid State Ionics,* **11,** 1983, p.213.
30. W. SITTE: *Solid State Ionics,* **94**, 1997, p.85.
31. J. MAIER, G. SCHWITZGEBEL and H. J. HAGEMANN: *Journal of Solid State Chemistry,* **58**, 1985, p.1.
32. S. SHIRASAKI, H. YAMAMURA, H. HANEDA, K. KAGEKAWA and J. MOORI: *Journal of Chemical Physics,* **73**, 1980, p.4640.

Impact of Segregation-Induced Effects on the Transfer of Charge and Matter Across the Gas/Electrolyte Interface: The Oxygen/Zirconia System

JANUSZ NOWOTNY
Centre for Materials Research in Energy Conversion
School of Materials Science and Engineering
The University of New South Wales
Sydney, NSW 2052, Australia

NOBUYASU MIZUTANI
Tokyo Institute of Technology
Department of Inorganic Materials
2-12-1, O-oakayama, Meguro-ku
Tokyo 152, Japan

CHARLES C SORRELL
School of Materials Science and Engineering
The University of New South Wales
Sydney, NSW 2052, Australia

ABSTRACT

This paper considers a model which describes the effect of segregation on defect chemistry of the boundary layer of ionic conductors and the impact of this layer on performance of electrochemical devices, such as solid oxide fuel cells. It is shown that segregation may have a dual effect on functionality of solid electrolytes involving:

(i) generation of an electrical potential barrier which acts as a diffusive resistance for the transport of charged defects across the interface and
(ii) the formation of a mixed-type conductivity within the boundary layer.

Specifically, this paper considers the effect of segregation on defect chemistry of the boundary layer of yttria-stabilised zirconia and its effect on the formation of concentration gradient in the interface layer.

1.0 INTRODUCTION

The transfer of charge (electrons) and matter, such as oxygen, is determining the performance of electrochemical devices based on ionic conductors, such as solid oxide fuel cells (SOFCs), electrochemical gas sensors, electrochemical gas separators and electro-catalysts. Their basic components include:

i. solid electrolyte,

ii. electrode material and

iii. gas phase/involving fuel on one side and air on the other side.

These three phases are forming a system involving all the functions required for electrochemical reaction to take place. Assuming that zirconia is a pure ionic conductor, the reaction site (where the transfer of charge and matter occurs) is limited to a three phase boundary (TPB) formed of the phases (i), (ii) and (iii). Accordingly, the performance of the electrochemical devices is determined by the area of the TPB which, in consequence, is determined by the properties of the materials used as electrodes and electrolytes (the electrode, as a separate phase, is not required when the surface layer of the solid electrolyte exhibits mixed conductivity). Properties of the materials used for SOFCs were recently reviewed by Badwal and Foger[1] and Kawada and Yokokawa.[2]

The focus of the present paper is on interface properties of the solid electrolyte and, specifically, on the transfer of charge and matter across the gas/electrolyte interface and its impact on the performance of electrochemical devices based on this electrolyte. The purpose of the present paper is to indicate the importance of segregation-induced effects at the gas/solid interface on processing of solid electrolytes for electrochemical devices.

This paper does not intend to address the details of interface engineering of solid electrolytes aiming at

i. the formation of mixed conductivity within the interface layer and

ii. decrease of the segregation-induced diffusion resistance for the transport of matter across the gas/solid interface.

So far, too little is known in this matter. Solution of this problem requires accumulation of more experimental data on interface properties.

The main objective of this paper is to consider the segregation-induced effects and their impact on the gas/solid kinetics, specifically for the oxygen/oxygen conductor system. Awareness is growing that an increase of the present state of understanding on the effect of segregation on interface chemistry is required in order to develop solid electrolytes with enhanced properties via interface engineering.

The present considerations are performed for zirconia as a model electrolyte and the oxygen/zirconia interface as a model interface.

2.0 POSTULATION OF THE PROBLEM

Zirconia involving acceptor-type additions, such as yttria, calcia and magnesia, is a good oxygen conductor with transference number of oxygen ions close to unity within a wide range of oxygen partial pressures.[1-4] Other materials, which are candidates for application in SOFC's, include ceria, thoria and their solid solutions and composites.[1] Also mullite and mullite/zirconia composites exhibit good ionic conductivity, especially at high temperatures.[5] So far, however, the most commonly applied solid electrolyte in electrochemical devices is stabilised zirconia (tetragonal or cubic). Its bulk defect chemistry and related properties are relatively well known. In contrast to the extensive knowledge on bulk properties of zirconia, little is known about its interface properties and the impact of interfaces on the transport of charge and matter. As shown below, this impact may be significant and, therefore, cannot be ignored.

It has been documented that performance of electrochemical devices are strongly influenced, or even determined, by the transfer of electrons at the TPB.[3,4] In the case of metal electrodes, such as Pt, the TPB is limited to the perimeter of the electrode. Such electrode is satisfactory for the determination of the Nernst signal of solid state galvanic cells when the transport of matter through the cell is very low as it is the case for electrochemical gas sensors. However, for SOFCs, when a high current is to be passed through the cell, its performance is determined by the TPB.

In the case of electrodes made of oxide materials, such as (La, Sr)MnO_3 (LSM), the TPB area is all the more critical because of their low ionic conduction.[6,7] The TPB of the LSM electrodes may be increased via controlled microstructure and related porosity. Then oxygen can be transported through the pores to the TPB located at the electrode/electrolyte interface.

By replacement of Mn in LSM with Co we obtain (La, Sr)CoO_3 (LSC). The LSC exhibits substantially higher ionic conductivity than that of LSM by 4-6 orders of magnitude.[1,2] Accordingly, oxygen reduction and related charge transfer for the LSC electrodes occurs on the entire area of the electrode/gas interface with lower polarization losses.

Application of oxide electrodes, such as LSM and LSC, requires a good adhesion between the electrode material and the solid electrolyte in order to enable an efficient transfer of charge and matter through the contact between the two. Another problem related with the application of perovskite-type electrodes concerns their stability. Both LSM and LSC may react with the electrolyte material thus resulting in deterioration of the properties of both electrode and electrolyte.

Another approach, aiming at elimination of metal electrodes, consists of functionalizing the boundary layer of solid electrolyte, such as zirconia, in order to combine the functions of both oxygen conductor and of metallic conductor.[6, 8-12] The electrode serves as a catalyst which provides (or removes) electrons to (or from) the reaction site whenever such supply is required. It is well known that zirconia exhibits mixed conductivity at very low $p(O_2)$ corresponding to about 10^{-23} Pa at 1273 K and 10^{-32} Pa at 1073 K.[2] Then metallic electrodes are not required. However, the intention of efforts of many researchers is to process zirconia in order to increase its electronic conductivity within the boundary layer under the conditions, such as $p(O_2)$ and temperature, comparable to the operational conditions of SOFC, i.e. at 10^5 Pa and 10^{-15} Pa for the air and the fuel electrode, respectively. The search for zirconia of enhanced electronic conduction involves introduction of different dopants which result either in an increase in the concentration of electronic charge carriers at the oxygen/zirconia interface or in an increase of their mobility. A clear advantage of this approach, over those which require application of electrode as a separate phase, is that an electrolyte within a single phase combines two functions:

i. of an ionic conductor (electrolyte) and
ii. of an electronic conductor (electrode).

It is argued that such functionalization may be achieved via interface engineering using segregation-induced effects to impose required interface composition rather than through bulk chemistry. It is also argued that knowledge of segregation is required to functionalize the electrolyte interface in desired way.

The purpose of this paper is to bring together the concepts of interface segregation in ionic solids with the concepts of electrochemical devices, such as SOFC. Specifically, this paper considers the effects of segregation-induced concentration gradients and related electric fields on the local transfer of charge and matter across the oxygen/electrolyte interface. Detailed considerations are performed for the oxygen/zirconia interface.

3.0 REACTIONS AT THE OXYGEN/ZIRCONIA INTERFACE

The processes which take place at the oxygen/zirconia interface, corresponding to cathode (the air electrode in SOFC), may be considered within several steps, such as

i. Physical adsorption,
ii. Ionization of adsorbed oxygen leading to the formation of chemisorbed species,
iii. Incorporation of oxygen into zirconia involving a reaction between oxygen vacancies and adsorbed oxygen ions and
iv. Transport of oxygen vacancies within zirconia.

These processes will be considered below.

3.1 Adsorption and Dissociation

Oxygen molecules are adsorbed on active adsorption sites of the electrode material or electrolyte, such as surface cations, and then decompose into atoms. Oxygen adsorption may be described by the following reactions:

$$O_{2(gas)} \Leftrightarrow O_{2(ads)} \tag{1}$$

$$O_{2(ads)} \Leftrightarrow 2O_{(ads)} \tag{2}$$

It is assumed that the shift of equilibria (1) and (2) into the right is very fast and, therefore, is not rate-controlling the process of oxygen incorporation into zirconia.

Oxygen dissociation seems to be easier on Ag than on Pt electrodes.[13, 14] According to Moyhadam and Stevenson[13] oxygen atoms are soluble in silver while in the case of Pt the adsorbed atoms/molecules have to diffuse along the surface to TPB in order to be ionized.

The formation of other neutral forms of adsorbed oxygen, such as molecular species $O_{3(ads)}$ and $O_{4(ads)}$, were postulated at lower temperatures.[15]

3.2 Chemisorption

Oxygen chemisorption may be represented by the following reactions.[15, 16]

$$O_{2(gas)} + e' \Leftrightarrow O^{-}_{2(ads)} \tag{3}$$

$$O^{-}_{2(ads)} + e' \Leftrightarrow O^{2-}_{2(ads)} \tag{4}$$

$$O^{2-}_{2(ads)} \Leftrightarrow 2O^{-}_{(ads)} \tag{5}$$

$$O_{ads} + e' \Leftrightarrow O^{-}_{ads} \tag{6}$$

$$O^{-}_{ads} + e' \Leftrightarrow O^{2-}_{ads} \tag{7}$$

$$O^{2-}_{ads} \Leftrightarrow O^{2-}_{lattice} \tag{8}$$

As seen, a shift of equilibria (3),(4),(6) and (7) into the right requires consumption of electrons which must be provided to the site of the electrochemical reaction. The chemisorbed species are formed at the surface of the electrode or electrolyte and then are transported to the electrode/electrolyte interface involving

i. Surface diffusion in the case of metallic electrodes or

ii. Lattice diffusion in the case of mixed conductors.

The mechanisms of oxygen transfer, involving processes of adsorption, dissociation, ionization, surface diffusion and incorporation into zirconia have been considered by Riess and Schoomman.[17]

The energy of different oxygen ions, (i) in the gas phase, (ii) in the crystal lattice and (iii) at the surface and is shown in Figure 1.[15] According to this diagram the doubly ionized oxygen ions, O^{2-}, are not stable at the surface as chemisorbed species and may reach stability in the crystal field. Due to their high stabilisation energy the lattice oxygen ions are the most stable form of oxygen.

With regard to the stabilisation energy one should distinguish the O^{2-} lattice ions in the bulk phase and in the boundary layer. The transfer between the two may be strongly affected by the segregation-induced electric fields. According to Wiemhofer[13] the O_2^{2-} and O_2^{-} species can be stabilised at the surface by the protons of adsorbed water. He also argues that below 600°C surface dissociation becomes the rate determining step.

Other forms of chemisorbed oxygen include molecular species, such as $O^{-}_{2(ads)}$ and $O^{+}_{2(ads)}$.[15] These forms are stable at lower temperatures. Increase of temperature results in a shift of chemisorption equilibria into the right leading to dissociation of molecular species and their ionization.

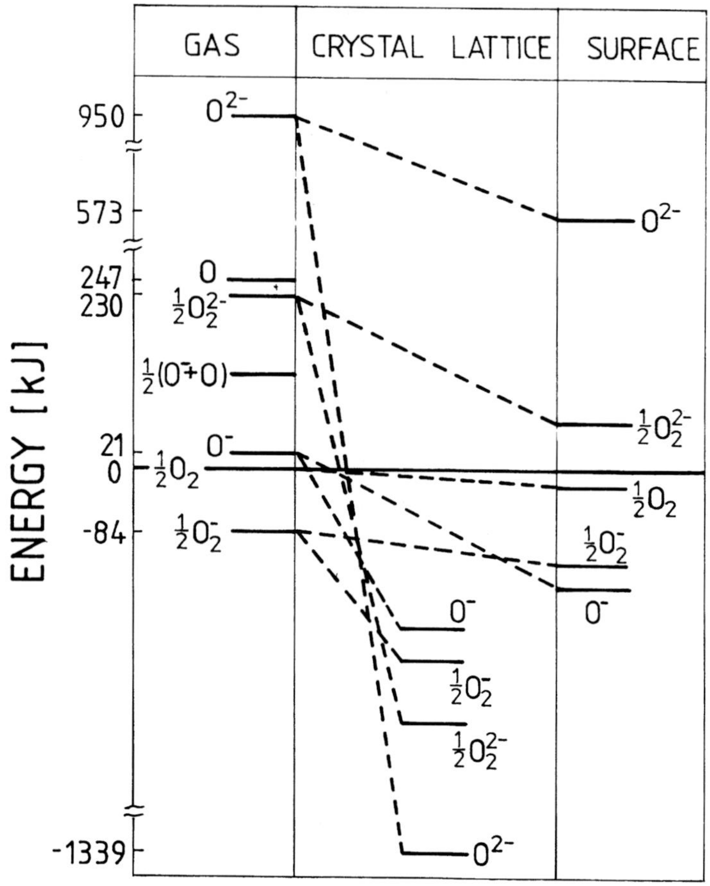

Fig. 1 The energy of different species of oxygen and their stabilisation in the crystal field and at the surface of binary oxides.[15]

3.3 Oxygen Incorporation

The reaction (7) may also be considered as a reaction between oxygen ions, O^-, and oxygen vacancies. According to the Kroeger-Vink notation[18] Eqn. 7 assumes the form:

$$O_i' + (V_O^{\cdot\cdot})_s + e' \Leftrightarrow O_O^x \qquad (9)$$

where $(V_O^{\cdot\cdot})_s$ denotes an oxygen vacancy in the surface layer. Reaction (9) results in consumption of oxygen vacancies at the gas/solid interface. In analogy to equilibrium (7), the shift of the equilibrium (9) into the right requires consumption of electrons.

Concordantly, a shift of equilibria (3)-(7) and (9) into the right is determined by supply of both oxygen vacancies and electrons to TPB.

The effect of segregation on properties of zirconia, specifically on those properties which may have a marked effect on the supply of both electrons and oxygen vacancies to the gas/solid interface, will be analysed below.

3.4 Lattice Transport

Oxygen vacancies, which are formed at anode, have to be transported to the oxygen/zirconia interface corresponding to cathode, where oxygen is reduced according to the reaction (9). This transport is very fast as far as the bulk phase is concerned.[19-25]

Chemical composition and structure of the bulk phase differs substantially from that in the boundary layer as a result of segregation. Due to different compositional and structural environment one should distinguish oxygen vacancies in the bulk and in the boundary layer:

$$(V_O^{\cdot\cdot})_s \Leftrightarrow V_O^{\cdot\cdot} \tag{10}$$

where $V_O^{\cdot\cdot}$ denote oxygen vacancies in the bulk phase. Concordantly, in considering the lattice transport of oxygen vacancies one should distinguish two different kinetic regimes:

1. Diffusion in the bulk phase and
2. Diffusion in the boundary layer across the segregation-induced concentration gradients and related electric fields.

The transport of the vacancies across the interface layer may be rate determining the transfer of matter rather than the bulk mobility of oxygen vacancies, if segregation results in the formation of a diffusive resistance. This effect will be analysed below.

3.5 Three-Phase System

A shift of equilibria (3),(4),(6),(7) and (9) into the right requires availability of the following three phases:

1. Gas phase, such as oxygen,
2. Oxygen conductor, such as zirconia, and
3. Metallic conductor, such as Pt, which provides electrons at cathode and removes electrons at anode (the metallic conductor is not required when the surface layer of the solid electrolyte exhibits mixed conductivity).

A microsystem involving these TBP is shown in Figure 2.[3, 4] Accordingly, for the case of metallic electrodes, the TPB is limited to the line along the electrode. It is assumed that the performance of electrochemical devices, such as SOFC, is determined by the charge transfer within the TPB which, in consequence, is determined by the limited area of the system. This limitation has generated efforts in functionalising the boundary layer of zirconia in order to make it electronically

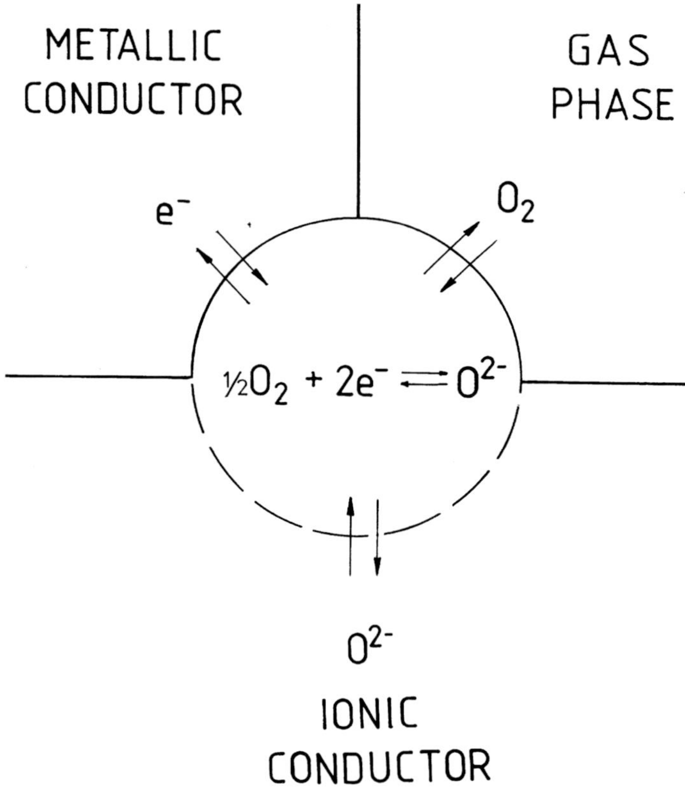

Fig. 2 Three-phase microsystem involving solid electrolyte, metallic conductor and gaseous oxygen.[3,4]

conducting. This may result in a substantial increase of the charge transfer kinetics and, in consequence, lead to efficient oxygen incorporation within the entire oxygen/zirconia interface.

Figure 3 illustrates a two-phase boundary between the gas phase and a mixed ionic-electronic conductor exhibiting the functions of both electrolyte and the electrode. Concordantly, this two phase boundary involves all the functions of the TPB.

The objective of this paper was to consider the effect of segregation-induced concentration gradients of defects, such as dopants, impurities and host lattice ions, on

i. The charge transfer at the oxygen/electrolyte interface and
ii. The transport of oxygen vacancies across the boundary layer.

The considerations which are performed for zirconia, are valid for other solid electrolytes as well.

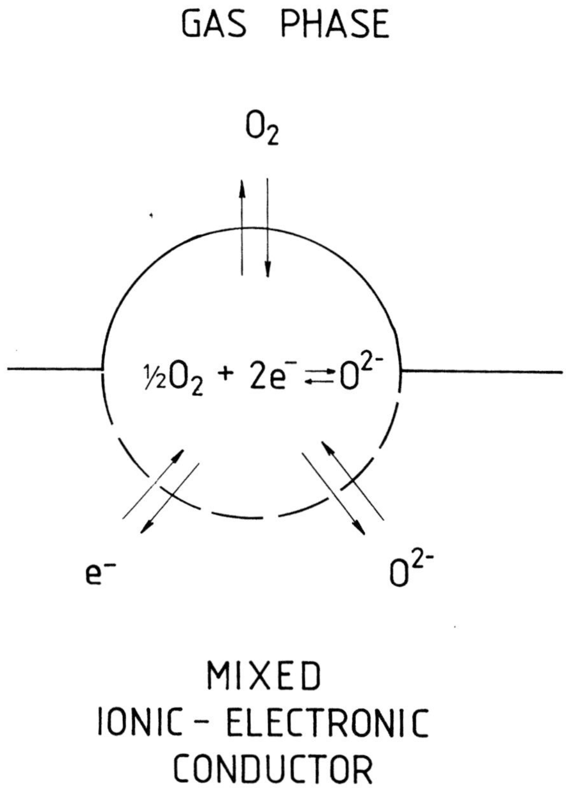

Fig. 3 Two-phase microsystem involving a mixed ionic-electronic conductor exhibiting two functions of both electrolyte and electrode as well as gaseous oxygen.

4.0 BULK DEFECT CHEMISTRY

Bulk defect chemistry for calcia- and yttria-stabilised zirconia was developed by Smith et al.[19] Kitazawa and Coble,[20] Burke et al.[21] Worrell et al.,[6, 8, 11] Weppner et al.[22-24] and Park and Blumenthal.[25] Figure 4 illustrates both ionic and electronic conductivity in yttria-stabilised (8 mol.%) zirconia in the range 1073-1323 K according to Park and Blumenthal.[25] As seen from this diagram the electronic conductivity at the $p(O_2)$ corresponding to the air and fuel electrodes (2.1×10^4 and 10^{-15} Pa, respectively) is very low. Concordantly, the electronic transference number within the boundary layer of zirconia to be very low. This does not allow the charge transfer, required to reaction (9), to occur. Therefore, there are efforts aiming at increasing the local electronic conductivity of the boundary layer and, consequently, at increasing the charge transfer kinetics at the oxygen/zirconia interface without presence of a separate phase of an electronic conductor.

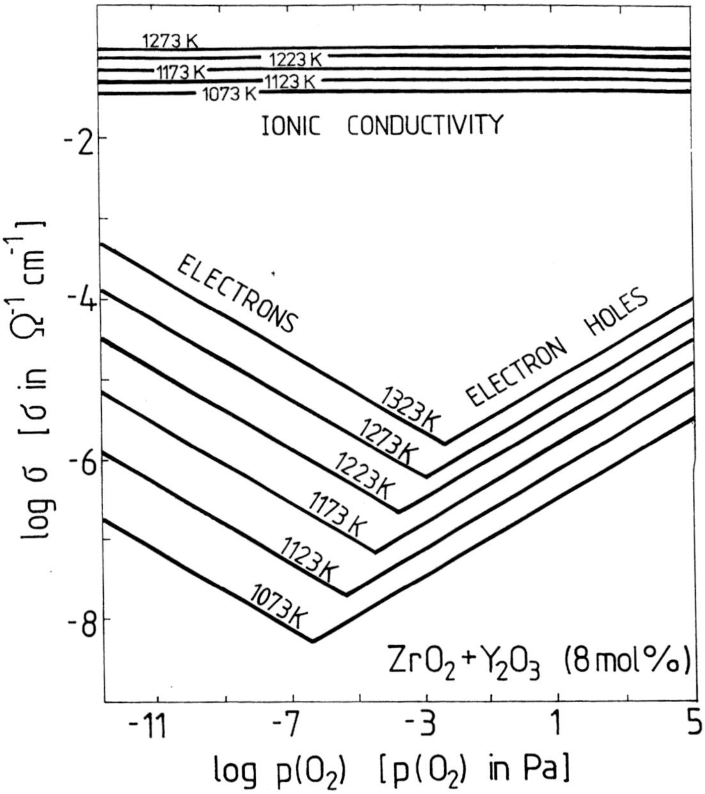

Fig. 4 Ionic and electronic conductivity of yttria-doped zirconia (8 mol %) according to Park and Blumenthal.[25]

Defects concentration and related properties of zirconia may be modified by bulk doping. Figure 5 illustrates schematically the conductivity both before and after the modification (solid and broken lines, respectively). The modification of zirconia aims at the incorporation of such dopants that result in increasing the concentration of electronic defects at the oxygen/zirconia interface to a level that their transference number is comparable to that of ionic defects. In other words the doping procedure aims at forming the boundary layer, which, under the oxygen potential barrier adequate to operational condition of SOFC, assumes a p-type conductivity at cathode, n-type conductivity at anode and pure ionic conductor in the bulk (Figure 6).

An enhancement of electronic conduction requires that the cathode site is doped with acceptors and the cathode site is doped with donors. Accordingly, bulk doping allows modification in the desired way only one electrode at the time. On the other hand simultaneous modification of both electrodes in the desired way requires that a local doping procedure is applied to respective zirconia interfaces.

Defects in Y-doped zirconia may be described by the following general charge neutrality:

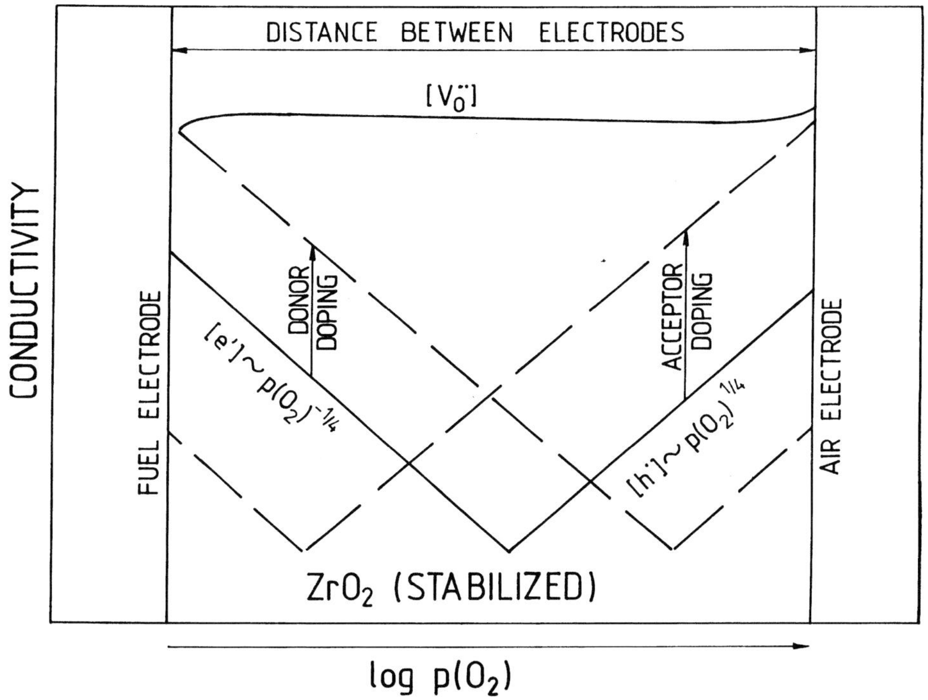

Fig. 5 Schematic defect diagram of zirconia illustrating the effect of bulk doping with donors and acceptors on defects concentration under the $p(O_2)$ gradient corresponding to operation conditions of SOFC.

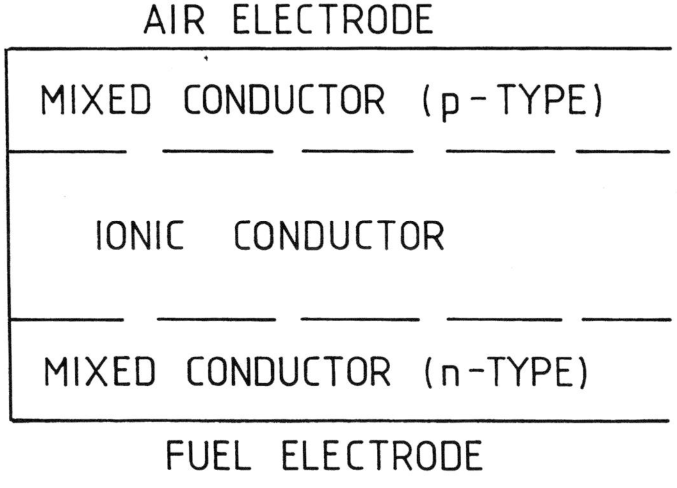

Fig. 6 Illustration of the intentional effect of doping of zirconia on its conductivity under $p(O_2)$ gradient involving n-type conductivity at the interface acting as anode, pure ionic conductor in the bulk and p-type conductivity at the interface acting as cathode.

$$2[V_O^{\cdot\cdot}] + [h^{\cdot}] + [D^{\bullet}] = [e'] + [A'] + [Y'_{Zr}] \quad (11)$$

where $[D^{\bullet}]$ and $[A']$ denote the concentration of donors and acceptors (different than yttrium). The condition (11) requires that there is a full charge compensation at any point in the bulk phase. In a wide range of $p(O_2)$ the condition (11) assumes a simple form:

$$2[V_O^{\cdot\cdot}] = [Y'_{Zr}] \quad (12)$$

This charge compensation is not valid for very reduced zirconia when the transference number of both electrons and ions assumes comparable values. There have been efforts in doping zirconia in order to assume comparable transference numbers for charge carriers at the $p(O_2)$ which correspond to operation conditions of SOFCs, such as $p(O_2)=10^{-15}$ Pa (anode) and around $p(O_2)=10^5$ Pa (cathode).

The charge neutrality conditions (11) and (12) are valid for the bulk phase. Different rules, however, apply to defect chemistry of the interface layer. The effect of segregation on defect chemistry in the interface layer of oxide materials and, specifically, of zirconia and its solid solutions, is considered below.

5.0 SEGREGATION

5.1 General

Most of the reports on segregation in solids have been accumulated for metals and alloys.[26,27] According to a simple model, based on regular solution approximation, there is the following relation between the concentration of a solute (1) and solvent (2) at the surface (s) and in the bulk phase (b), (expressed in mole fraction, x):

$$\frac{x_s^1}{x_s^2} = \frac{x_b^1}{x_b^2}\, exp\left(-\frac{\Delta H_{seg}}{RT}\right) \quad (13)$$

where ΔH_{seg} is the heat of segregation.

In metals, the segregation-induced concentration gradient is limited to one lattice layer. In ionic solids, however, this distance over which the space charge develops, may be substantially larger. In the case of insulating materials, the space charge may extend over tens of nanometers.

Segregation results in a stable modification of surface composition and related properties, corresponding to equilibrium. However, practical use of the phenomenon of segregation in engineering interface properties requires better knowledge of thermodynamics of segregation, specifically with regard to oxide materials.

In the case of ionic solids the segregation-induced enrichment of the interface layer in defects results in substantial interactions between the defects leading in consequence in the formation of bidimensional interface structures and, in extreme cases, tridimensional phases which may be extremely stable.[28] These structures have

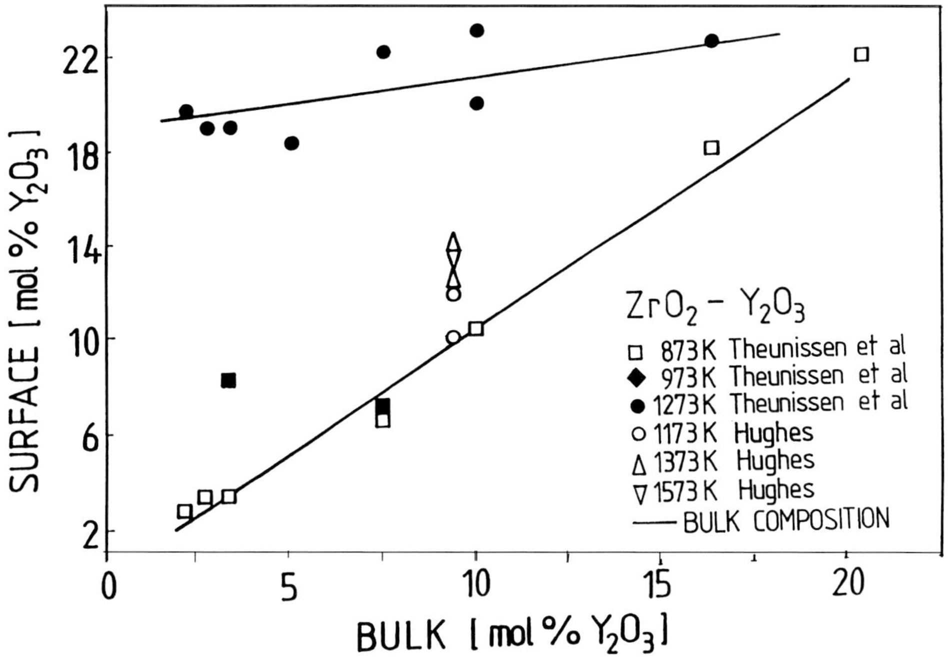

Fig. 7 Surface vs bulk concentration of yttrium in yttria-stabilised zirconia according to AES analysis of Theunissen et al.[44] and XPS data of Hughes.[46]

different properties to the bulk. Identification and characterisation of these structures is of importance for the preparation of materials with controlled properties.

So far, the most extensive segregation studies were reported for binary oxides, such as NiO,[28-30] CoO,[31-34] Al_2O_3,[35,36] MgO[37] and their solid solutions. An extensive overview of segregation in oxide materials was reported by Nowotny,[28] Wynblatt and McCune[37] and Butggraaf and Winnubst.[38]

5.1 Segregation in Zirconia

5.1.1 Effect on Composition and Structure

There is an accumulation of experimental data on segregation in zirconia and its solid solutions.[39-52] There is also a general agreement that yttrium segregates to the surface of zirconia. However, there is a lack of agreement concerning the extent of this segregation.

The results of the most systematic studies of Theunissen et al.[44] and Hughes[46] are shown in Figure 7. As seen the results of AES surface analysis reported by

Theunissen et al. indicate that

i. Segregation equilibrium cannot be reached below 1273 K and
ii. Yttria-stabilised zirconia has a tendency to assume constant yttrium concentration at the surface (ca 30 mol.%) which is independent of bulk composition.

The results of XPS analysis reported by Hughes[46-48] confirm that yttrium segregates to the surface of zirconia. However, his enrichment data are lower than those of Theunissen et al. The results of XPS studies performed by Zhang[51] are in agreement with those of Hughes rather than those of Theunissen.

Axelsson et al.[45] have shown that the extent of segregation of ions, such as Si, depends on the gas phase composition. They observed that heating of undoped zirconia in H_2O followed by oxidation in oxygen results in the formation of a silicate, $ZrSiO_4$, overlayer.

Badwal, Drennan and Hughes[39, 40, 46-50] have reported extensive studies on the effect of segregation on the formation of interface structures and grain boundary phases during processing of zirconia. These studies, reported for both single crystals and ceramic specimens, involve the determination of temperature coefficients of segregation-induced enrichment for both solutes and impurities. It has been postulated that segregation results in the formation of surface crystalline structures which are different than those in the bulk phase.[46, 47] It was observed that Si segregation has a deteriorating effect on oxygen conduction of zirconia.[39, 40, 48, 49] Hughes[46-48] argues that attractive interactions between segregation Si and Y results in the formation of yttria-rich silicate overlayer structure which is very stable.

Badwal, Drennan and Hughes[39, 40] have observed that segregation of Si results in the formation of silicate glass which is responsible for the formation of grain boundary weak links for oxygen conduction.

So far, little is known on the effect of the surface, specifically surface segregation, on the mechanism and kinetics of gas/solid processes. It was shown that the segregation-induced electric field may have a strong effect on the rate of transport of charged defects across the gas/solid interface, such as oxygen/metal oxide interface.[53] Concordantly, there is a need to evaluate this effect on performance of electrochemical devices based on zirconia.

It is known that segregation of undesired elements, such as Si, may block ionic transport. However, more studies are needed on evaluation of the effect of segregation on defect chemistry of the interface layer and related local electrical properties.

5.1.2 Charge Compensation

The conditions (11) and (12) do not apply for the interface layer. In this case the charge neutrality requires that the charge located at the interface is compensated by the charge in the space charge layer. It seems that for the interface layer of yttria-doped zirconia the following charge neutrality applies:

$$\iint [Y'_{Zr}]_s \, dS = \iiint 2[V_O^{\cdot\cdot}]_{sc} \, dV \tag{14}$$

where S denotes the area of the surface layer, V denotes the respective volume of the space charge layer, and subscripts '*s*' and '*sc*' correspond to the surface layer and the space charge layer, respectively.

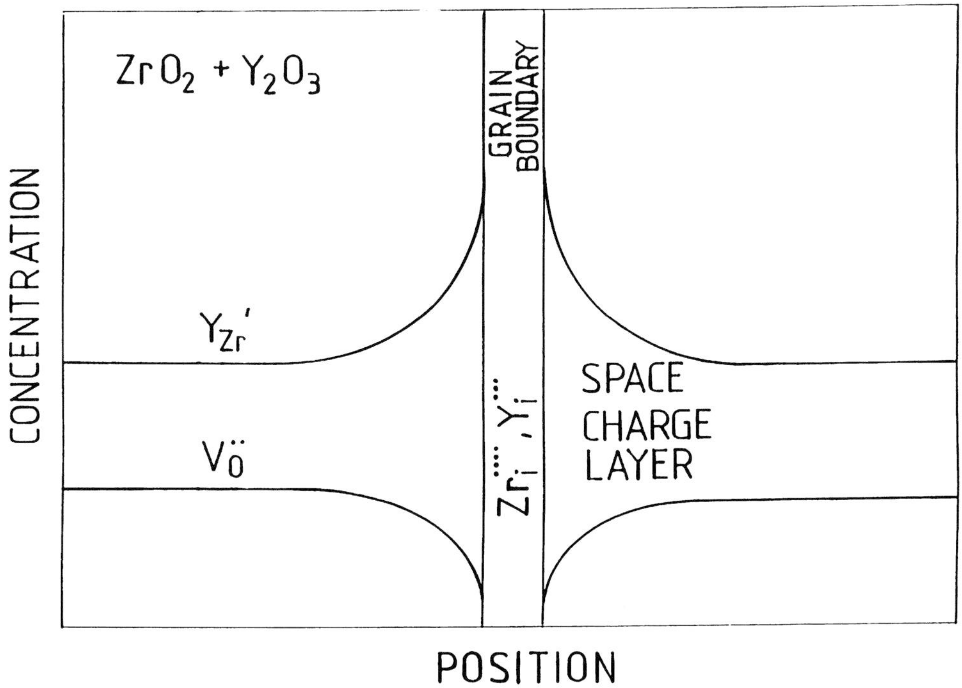

Fig. 8 Interface vs bulk defect model of yttria-stabilised zirconia illustrating the grain boundary layer, enriched in positively charged cations, such as $Zr_i^{\bullet\bullet\bullet\bullet}$ and $Y_i^{\bullet\bullet\bullet}$, and the space charge layer enriched in yttrium and depleted in oxygen vacancies.[52]

An intention of the research involving doping zirconia with different ions, in order to functionalize its surface, is to modify the surface charge compensation condition in such a way that the transference number of electrons or electron holes are elevated to the level comparable to that of oxygen vacancies. In this case the desired functionalization should lead to the following charge compensation conditions for anode and cathode, respectively:

$$\iint([Y'_{Zr}]_s + [A']_s)dS]_{sc} = \iiint[h^\bullet]_{sc} + 2[V_O^{\bullet\bullet}]_{sc})dV \qquad (15)$$

$$\iint([D']_s + 2[V_O^{\bullet\bullet}]_s)dS = \iiint([e']_{sc} + [Y'_{Zr}]_{sc})dV \qquad (16)$$

According to Worrell et al.[6, 8] doping with titania and terbia results in an increased electronic n- and p-type conductivity of the boundary layer, respectively.

Verification of interface charge compensation conditions requires knowledge of the picture of segregation with respect to all lattice defects involving dopants, impurities and intrinsic (Schottky- and Frenkel-type) defects. It has also been shown that segregation in zirconia results in an enrichment of interfaces in interstitials of Zr and Y (Figure 8).[52]

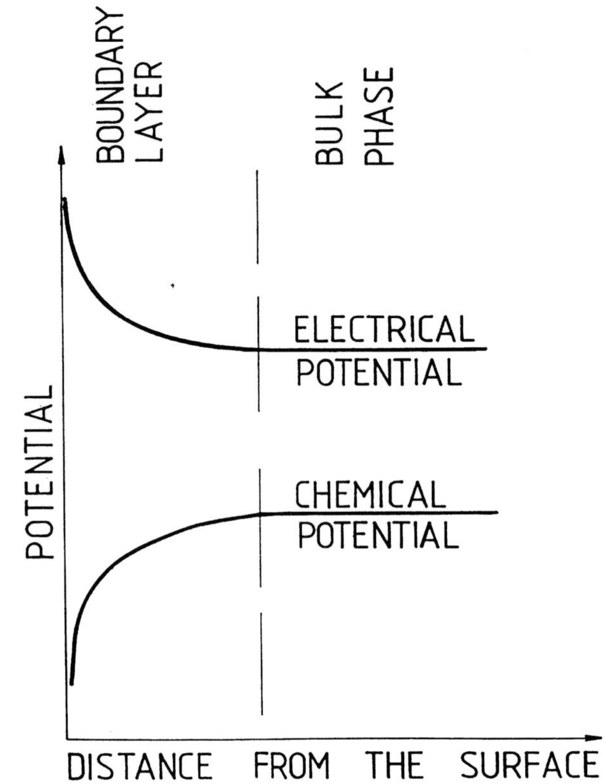

Fig. 9 Schematic illustration of the segregation-induced electrical and chemical potential gradients of defects within the interface layer.

6.0 SEGREGATION-INDUCED EFFECTS

6.1 Effect on Concentration Gradients Within Boundary Layer

Segregation of defects to interfaces results in the formation of concentration gradients within the boundary layer and related electrical potential gradients (Figure 9). This segregation leads to charging the surface and a change of surface potential. Figure 10 illustrates the effect of segregation on the formation of a negative and positive surface charge component of work function, Φ_s.

Segregation-induced enrichment of the interface layer in yttria (Figure 8) results in an increase of silicon segregation both involved in the formation of insulating interface glassy phases which has deteriorating effect on the transfer of charge and matter.

It has been argued that mixed conductivity of the interface layer could be achieved after doping with an ion which results in electronic compensation, such as terbium

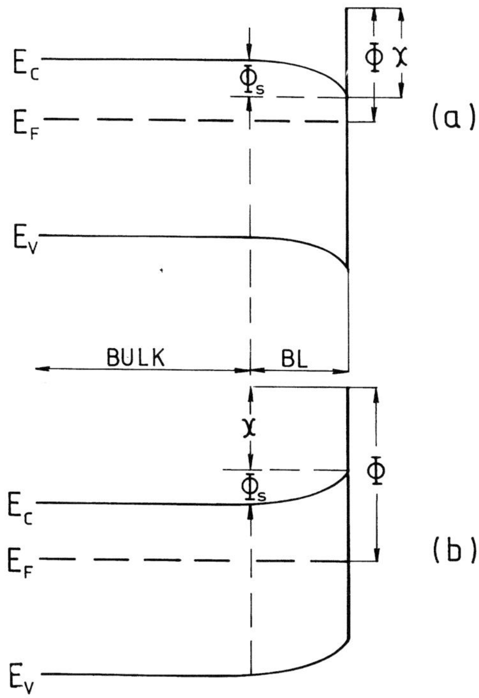

Fig. 10 The band model of an n-type semiconductor illustrating a positive (a) and negative (b) surface charge-related work function components.

and titanium,[6, 8, 9] however, correct understanding of their effect on interface properties requires that the effect of segregation of both ions on the local defect chemistry be examined.

Evaluation of defect chemistry in the boundary layer requires knowledge of the segregation-induced concentration gradients of both electronic and ionic defects under specific $p(O_2)$. Concordantly, there is an urgent need to determine these gradients within the zirconia boundary layer corresponding to both zirconia/gas and zirconia/electrode interfaces. These studies are of strategic importance for the development of zirconia with enhanced properties with respect to its performance in electrochemical devices.

6.2 Effect on Transport Across Interface

The transport across interfaces involves two basic components:

i. Thermal diffusion term and
ii. Migration term which corresponds to transport across the segregation-induced electric fields.[53, 54]

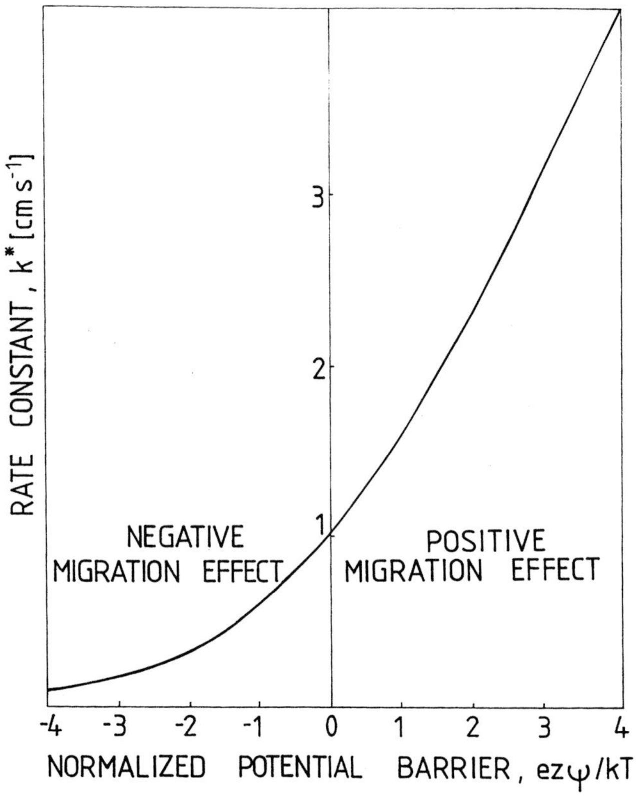

Fig. 11 The relationship between the rate constant of the migration of charged defects across interface, k^*, and electric potential, Ψ, for negatively charged defects, such as metal vacancies.[54]

This transport plays an important role in heterogeneous gas/solid processes, such as equilibration processes.

It has been generally assumed that the rate of gas/solid re-equilibration at elevated temperatures is rate controlled by the bulk transport kinetics while the processes that take place at the interface or within the interface layer are relatively fast. This assumption, however, is not valid when segregation results in the formation of a retarding potential barrier. The retarding effect may assume substantial values even at elevated temperatures.[53, 54]

The effect of the segregation-induced electric fields on the transport kinetics of negatively charged defects is illustrated in Figure 12 in terms of rate constant, k^*, as a function of normalized electric potential, Ψ^*.[54] The relationship between the two may be expressed as:

$$k^* = (D/\delta)\left|\Psi^*\right|\exp(\Psi^*) \tag{17}$$

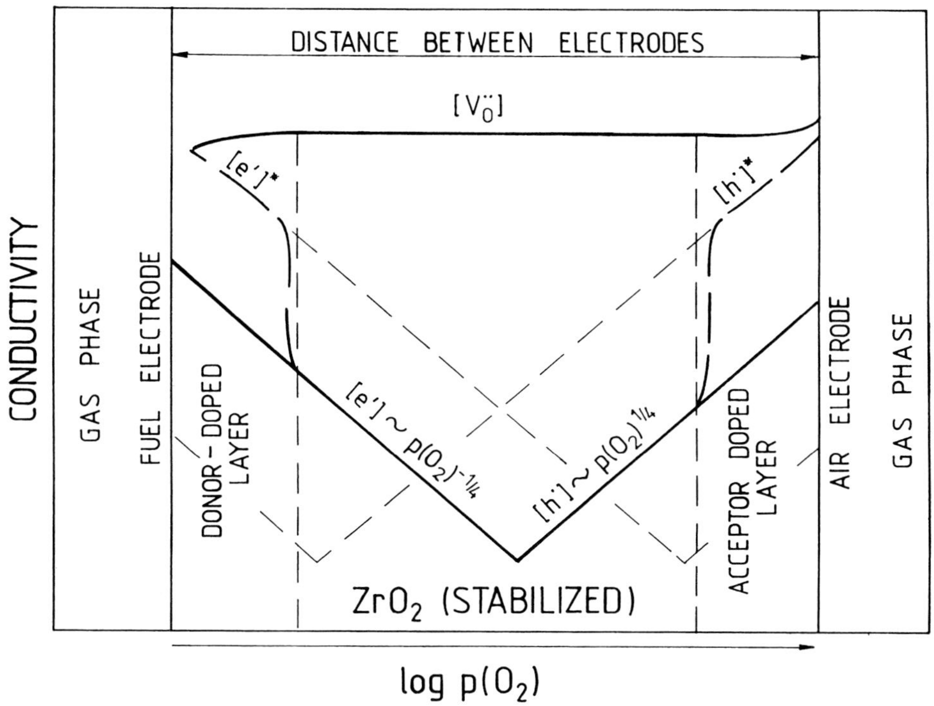

Fig. 12 Schematic defect diagram of zirconia illustrating the segregation-induced enhancement effect involving an increase of electronic conductivity in the boundary layers acting as cathode and anode under the p(O_2) gradient corresponding to operation conditions of SOFC ([e']* and [h']* denote, respectively, the concentrations of electrons and electron holes in the boundary layer.

where D is the diffusion coefficient of defects and d is the thickness of the boundary layer involving the electrical potential gradient, and the normalized electrical potential is defined as:

$$\Psi^* = \frac{ez\Psi}{kT} \tag{18}$$

where e is elementary charge, z is valency of defects and k is Boltzmann constant.

When the migration component assumes positive values (accelerating effect) then the gas/solid reaction is controlled by bulk diffusion. However, when the migration term assumes negative values, then the transport within the electric field in the boundary layer may control the entire gas/solid kinetics. The migration effect, expressed in terms of the rate constant k^* is shown in Figure 11. When:

$$k^* >> \frac{D}{\delta} \tag{19}$$

then the gas/solid kinetics are rate controlled by bulk diffusion. However, if:

$$k^* << \frac{D}{\delta} \tag{20}$$

then the surface potential barrier results in the formation of a near-surface diffusive resistance. This resistance may assume substantial values even at high temperatures. When the condition (20) applies, then the transport in solids must be described using general diffusion equation, involving the thermal diffusion term and the migration term.[54]

The effect of segregation-induced potential barrier, ψ, may result in a substantial decrease of the local concentration of defects and, consequently, may lead to the formation of a diffusive resistance.

6.3 Effect on Nernst Potential

Electromotive force (EMF) of solid state galvanic cells is determined by the electrochemical reactions taking place at the electronic conductor/electrolyte interface. Specifically, the EMF is determined by the activities of all the reaction products and substrates including the electrons involved in the reaction:

$$E = E^0 + \frac{RT}{nF} \sum ln \nu_i a_i \tag{21}$$

where n is the number of electrons involved in the reaction of the cell, E^0 is a constant EMF related to standard activities and a_i and ν_i denote activities and stoichiometric coefficients, respectively. The electrochemical reactions taking place at the TPB are determined by the local interface properties rather than by the properties of the bulk phase. Concordantly, correct understanding of the physical meaning of EMF requires knowledge of the local properties at the gas/solid interface where the electrochemical reactions and related charge transfer take place and where the Nernst signal is being generated.

7.0 NON-EQUILIBRIUM DOPING PROCESSES

There have been efforts to modify surface composition using the procedures, such as implantation[10, 12] and impregnation.[55] Both procedures consist of imposition of local composition which does not correspond to equilibrium. Consequently, thus imposed composition has a tendency to change in order to assume ultimately the level which corresponds to equilibrium.

Both Scholten[10] and van Hassel[12] have performed pioneering work aiming at understanding the effect of the local interface composition on the charge transfer kinetics at the oxygen/zirconia interface.

Ion implantation allows introduction of a dopant into the crystal at a controlled level (dose). The implantation procedure results in destruction of the original structure which, however, may be restored during subsequent heating. Implantation allows doping of the surface layer resulting in the concentration profile which exhibits its maximum below the surface. At higher doses, the implantation process results in the formation of amorphous layers and complex microstructure. Accordingly, subsequent thermal treatment is required to restore the initial structure of the material. However, in the case of zirconia even annealing at 1100°C does not result in re-crystallization.[10]

Impregnation is a simple procedure which allows imposition of desired element at the surface and then to propagate it into the surface layer during heat treatment resulting in the formation of a diffusion-controlled gradient.

Both these procedures result in imposition of concentration gradients within the boundary layer. Annealing results in a decrease of these gradients in time as the dopant is propagated into the bulk. Concordantly, this kind of doping may be effective in the modification of zirconia in the case when one of the following two scenarios apply:

1. The dopant has a tendency to segregate. Then the boundary limit of the decrease of the concentration gradient during annealing corresponds to the segregation-induced enrichment.
2. The diffusion coefficient of the dopant is so low that the rate of the decrease of the imposed concentration gradient in the operation condition of the electrolyte is negligible.

Scholten[10] and van Hassel[12] have implanted both Fe and Ti into yttria-stabilised zirconia. Thus formed Fe concentration profile remains stable for several hours at 800°C. It has been documented that the Fe^{3+} ions are incorporated into Zr^{4+} sites up to 4.5×10^{21} Fe/cm^3 which appears to be solubility level at 1500°C. It was observed that Fe at this concentration results in mixed conductivity for both solid solutions obtained in a chemical route and implanted specimens. Implantation with Ti results in conductivity increase by 2-3 orders of magnitude, if the specimens are annealed in a reducing atmosphere. It is not clear, however, to what extent the effect of Fe and Ti concerns the local concentration of electronic charge carriers in the boundary layer. Experimental verification of this effect is required.

The advantage of these doping procedures, over bulk doping, is that they may be applied to individual interfaces involving doping with donors or acceptors, as required.

8.0 EFFECT OF SEGREGATION ON DEFECT CHEMISTRY OF THE BOUNDARY LAYER AND ITS ELECTRONIC CONDUCTIVITY

The effect of bulk doping on electronic conductivity of zirconia, under the specific $p(O_2)$ corresponding to operational conditions of SOFC, has been derived in Figure 5 assuming that zirconia is free of segregation-induced concentration gradients in the boundary layer. The purpose of this section is to consider the effect of segregation on defect chemistry of this layer and related electronic transport kinetics.

Both enhancement and detrimental effects with regard to the charge transfer, may be considered in terms of the effect of segregation on the transport mechanism and kinetics of electronic charge carriers. Segregation of mixed valency cations may result in an increase of electronic conductivity via hopping mechanism. Alternatively, the effect of segregation may also be considered in terms of the formation of a bi-dimensional surface structure which is able to transport electronic charge carriers rapidly according to a band mechanism within this structure.

8.1 Enhancement Effect

Segregation of defects results in the formation of a chemical potential gradient within the boundary layer and related gradients of electrical potential (Figure 10). Accordingly, the segregation-induced effects may be used to impose desired concentration of electronic charge carriers in the boundary layer as an alternative to bulk doping.

The procedure of functionalising the zirconia surface may be considered within two possible scenarios:

1. Zirconia is bulk doped with traces of elements whose segregation strongly depends on $p(O_2)$. The elements are selected in such a way that donors segregate at lower $p(O_2)$ (fuel electrode) and acceptors segregate at higher $p(O_2)$ (air electrode).
2. Doping is limited to boundary layers serving as electrodes. Then the cathode and anode sides are doped with donors and acceptors, respectively. This kind of doping may be successful when the transport kinetics of the dopants in the operation conditions of the device is quenched and, therefore, does not result in any substantial changes in composition.

The effect of the segregation-induced concentration gradients, resulting in enhancement of electronic conductivity at both air and fuel electrodes, is schematically shown in Figure 12. The enhancement effect may be considered within two possible transport mechanisms,

i. The hopping mechanism and
ii. The band transport mechanism.

The hopping mechanism was proposed by Han and Worrell[6] as a result of incorporation of mixed valency cations, such as *Ti*. One may also expect that a structural deformation of the interface layer may result in fast transport within a bidimensional layer as a result of a modification of the band model within this layer.

It appears that segregation of *Y*, although it results in imposition of surface charge, results in deterioration of electronic conduction within the boundary layer rather than in its enhancement. This is due to lateral interaction of *Y* with Si resulting in the formation of an insulating layer. Therefore, the effect of *Y* segregation on surface electrical properties of zirconia should be evaluated for zirconia which is free of Si.

8.2 Detrimental Effect

The segregation-induced concentration gradients may have a detrimental effect on electronic conductivity within the boundary layer even if bulk doping results in an increase of the electronic conductivity. In this case segregation results in the formation

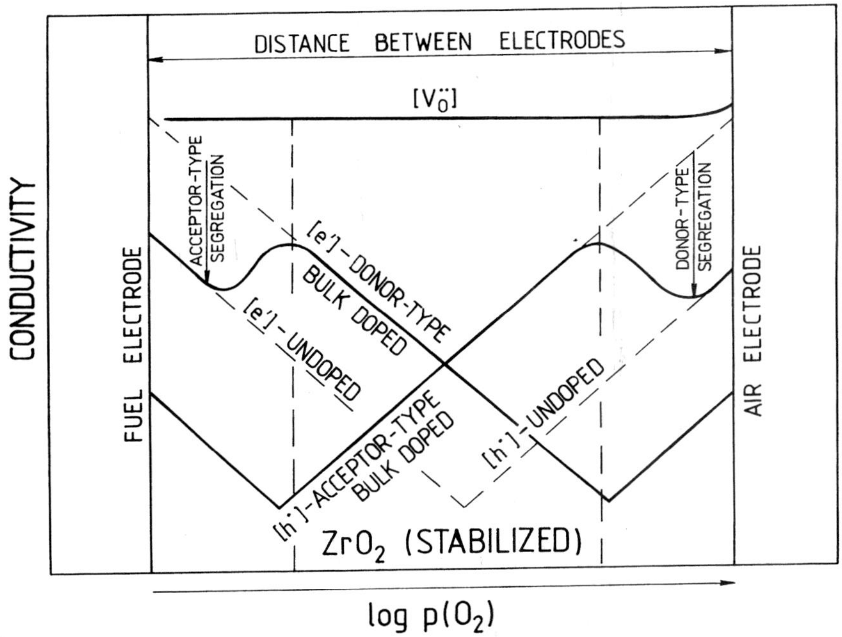

Fig. 13 Schematic defect diagram of zirconia illustrating the segregation-induced deteriorating effect involving a decrease of electronic conductivity in the boundary layers acting as cathode and anode under the p(O_2) gradient corresponding to operation conditions of SOFC.

of an insulating surface layer as shown schematically in Figure 13. This layer exhibits a retarding effect on charge transfer. Therefore the boundary layer should be kept free of these defects which lower the concentration of electronic charge carriers. This may be achieved by doping the electrolyte with the elements which

i. Have the segregation driving force much stronger than that of undesired elements, and
ii. Exhibit repulsive interactions between the two.

9.0 PROBLEMS TO BE ADDRESSED

Development of zirconia with enhanced properties via interface engineering requires the following problems to be addressed:

1. There is a need to increase the present state of understanding between the local interface composition and structure on one hand and properties of functional materials on the other hand. Construction of a more sophisticated models describing the relationship between the two requires accumulation of larger empirical material on

i. Segregation in oxide materials and
ii. The impact of segregation on materials functions.

2. Development of new functional materials with enhanced properties for specific application may be achieved via control of chemical composition of the interface region. In this respect there is a need to
   i. Develop the analytical techniques that allow determination of the concentration of impurities at the ppm or even below this level,
   ii. Block the segregation of undesired elements from interfaces, such as gas/solid interface and grain boundaries and
   iii. Impose segregation-induced concentration gradients of the elements which lead to the modification of interface properties in the desired way.

   In this respect it becomes increasingly important to evaluate the effect of segregation on properties.
3. There is a need to characterise the low dimensional interface structures which are formed as a result of defect segregation.
4. With regard to zirconia it becomes urgent to develop the processing procedures that allow functionalisation of their interfaces in order to enhance the transfer of charge and matter through the gas/solid interface. This may be achieved via imposition of desired chemical concentration gradients across the interfaces. In this respect there is a need to develop defect chemistry of zirconia for the interface layer. Specifically, the present state of understanding on the local transport mechanisms in this layer should be enhanced.

10 SUMMARY AND CONCLUSIONS

10.1 General

A general model of the phenomena at the gas/electrolyte interface was proposed. According to this model, a bi-functional solid electrolyte, combining the functions of an ionic conductor and an electrode, may be developed either via bulk chemistry or interface chemistry. In both cases, however, knowledge of segregation of defects, such as solutes, impurities and intrinsic defects, is required. In this respect, there is a need to enhance the present state of understanding on thermodynamics and kinetics of segregation in oxide materials. Specifically, the following items should be addressed:

1. Principles of defect chemistry for the interface layer should be understood.
2. The segregation-induced low dimensional interface structures should be characterised.
3. Processing procedures that are able to modify interface composition and related properties in a desired way should be developed.
4. Lateral interactions between defects within the interface layer should be established and their impact on multi-segregation should be evaluated.

10.2 Zirconia

The model of the phenomena at the gas/electrolyte interface was considered for the oxygen/zirconia system. It was shown that mechanism and kinetics of these phenomena are strongly influenced by segregation-induced concentration gradients in the interface layer.

Segregation results in the formation of concentration gradients within the boundary layer and related electric fields which may have either detrimental or enhancement effect on

i. The charge transfer kinetics at the zirconia/oxygen interface and
ii. Transport of oxygen vacancies through this layer.

Therefore the effects produced by segregation should be taken into account in the attempt to modify the electronic transport within the boundary layer.

The attempts aiming at increasing the local electronic conductivity within the boundary layer through bulk doping may be successful only, if zirconia is free of segregation-induced detrimental effects. Evaluation of these effects requires enhancement of our understanding of the effect of segregation on the local defect chemistry of the interface layer of zirconia.

11 ACKNOWLEDGEMENTS

Thanks are due to Dr. S.P.S. Badwal and Dr. M. Rekas for reviewing this paper and for remarks on several aspects of SOFCs. The paper was supported by the NEDO international grant. This support is gratefully acknowledged. Thanks are also due to Mrs. Janina Nowotny for preparation of the illustration material.

12 REFERENCES

1. S. P. S. Badwal and F. Foger: *Materials Forum,* in print.
2. T. Kawada and H. Yokokawa: *Key Engineering Materials,* **187,** 1997, pp.125-126.
3. J. Foultier, P. Fabry and M. Kleitz: *Journal of Electrochemical Society,* **123**, 1976, p.204.
4. E. J. L. Schouler and M. Kleitz: *Journal of Electrochemical Society,* **134**, 1987, p.1445.
5. K. Yamana, S. Nakamura, J. Nowotny, Z. Zhang and F. Terasaki: *Journal of Ceramic Society*, Japan, **103**, 1995, p.189.
6. P. Han and W. L. Worrell: *Journal of Electrochemical Society,* **142**, 1995, p.4235.
7. J. Nowotny and M. Rekas: *Journal of American Ceramics Society,* in print.
8. K. E. Swider and W. L. Worrell: *Journal of Electrochemical Society,* **142**, 1996, p.3706.
9. A. Kopp, H. Naefe, W. Weppner, P. Kountouros and H. Schubert: *Science and*

Technology of Zirconia, S. P. S. Badwal, M. J. Bannister and R. H. J. Hannink, eds., Technomic Publ. Co., Lancaster, 1993, p.567.
10. D. Scholten: Ph.D. Thesis, Twente University, Enschede, 1987.
11. W. L. Worrell: *Solid State Ionics*, **28-30**, 1988, p.1215.
12. B. A. Van Hassel: Ph.D. Thesis, Twente University, Enschede, 1990.
13. H. D. Wiemhofer: *Solid State Ionics*, **75**, 1995, p.167.
14. F. K. Moyhadam and D. A. Stevenson: *Journal of Electrochemical Society*, **133**, 1986, p.1329.
15. A. Bielanski and J. Haber: *Catalysis Review - Science Engineering*, **19**(1), 1979, pp.1-41.
16. J. Nowotny: *Journal of Materials Science*, **12**, 1979, p.1143.
17. I. Riess and J. Schoonman: *Electrochemistry of Solids*, P. Gellings and H. J. M. Bouwmeester, eds., CRC Press, Boca Raton, 1997, p.269.
18. F. A. Kroeger and H. J. Vink: *Solid State Physics*, F. Seitz and D. Turnbull, eds., Academic Press, New York, **3**, 1956, p.307.
19. A. W. Smith, F. W. Meszaros and C. D. Kamata: *Journal of American Ceramic Society*, **49**, 1966, p.240.
20. K. Kitazawa and R. L. Coble: *Journal of American Ceramic Society*, **57**, 1974, p.360.
21. L. D. Burke, H. Rickert and R. Steiner: *Z. Physics Chem.*, NF, **74**, 1971, p.146.
22. W. Weppner: *Zeitschrift für Naturforschung*, **31**a, 1971, p.1336.
23. W. Weppner: *Goldschmidt Inform.*, **2**, 1983, p.16.
24. B. Y. Liaw and W. Weppner: *Journal of Electrochemical Society*, **138**, 1991, p.2478.
25. J. H. Park and R. N. Blumenthal: *Journal of Electrochemical Society*, **136**, 1989, p.2867.
26. F. Cabane and J. Cabane: '*Science of Ceramic Interfaces*', J. Nowotny, ed., Elsevier, Amsterdam, 1991, p.1.
27. D. McLean: *Grain Boundaries in Metals*, Oxford University Press, London, U.K., 1957, p.116.
28. J. Nowotny: *Science of Ceramic Interfaces*, J. Nowotny, ed., Elsevier, Amsterdam, 1991, p.79.
29. D. M. Duffy and P. W. Tasker: *Philosophical Magazine A*, **50**, 1984, p.155.
30. W. Hirschwald, I. Sikora and F. Stolze: *Surface Interface Analysis*, **7**, 1985, p.155.
31. J. Haber, J. Nowotny, I. Sikora and J. Stoch: *Applied Surface Science*, **17**, 1984, p.324.
32. I. Sikora, F. Stolze and W. Hirschwald: *Surface Interface Analysis*, **10**, 1987, p.424.
33. A. Bernasik, J. Nowotny, S. Scherrer and S. Weber: *Journal of American Ceramic Society*, **80**(2), 1997, pp.343-348.
34. A. Bernasik, J. Nowotny, S. Scherrer and S. Weber: *Journal of American Ceramics Society*, **80**(2), 1997, pp.349-356.
35. R. I. Taylor, J. P. Coad and R. J. Brook: *Journal of American Ceramics Society*, **57**, 1974, p.539.
36. S. M. Mukhopadhyay, A. P. Jardine, J. M. Blakely and S. Baik: *Journal of American Ceramics Society*, **71**, 1988, p.358.
37. P. Wynblatt and R. C. McCune: *Surface and Near-Surface Chemistry of Oxide*

Materials, J. Nowotny and L. C. Dufour, eds., Elsevier, Amsterdam, 1988, p.247.

38. A. J. BURGGRAAF and A. J. A. WINNUBST: Ref. 37., p.449.
39. S. P. S. BADWAL, J. DRENNAN and A. E. HUGHES: Ref. 28., p.227.
40. S. P. S. BADWAL and J. DRENNAN: *Science of Ceramic Interfaces II*, J. Nowotny, ed., Elsevier, Amsterdam, 1994, p.71.
41. J. NOWOTNY, M. REKAS and T. BAK: *Zirconia Engineering Ceramics*, E. Kisi, ed., Trans Tech Publications, 1997, (in print).
42. A. J. A. WINNUBST, P. J. M. KROOT and A. J. BURGGRAAF: *Journal of Physics Chemical Solids*, **44**, 1983, p.99.
43. A. J. BURGGRAAF, M. VAN HEMERT, D. SCHOLTEN and A. J. A. WINNUBST: *Materials Science Monographs*, 28, P. Barret and L. C. Dufour, eds., Elsevier, Amsterdam, 1985, p.797.
44. G. S. A. M. THEUNISSEN, A. J. A. WINNUBST and A. J. BURGGRAAF: *Journal of Material Science Letters*, **8**, 1989, p.55.
45. K. O. AXELSSON, P. J. M. KECK and B. KASEMO: *Applied Surface Science*, **25**, 1986, p.217.
46. A. E. HUGHES: Ph.D. Thesis, Royal Melbourne Institute of Technology, 1991.
47. A. E. HUGHES: *Journal of American Ceramics Society*, **78**, 1995, p.369.
48. A. E. HUGHES: *Science of Ceramic Interfaces II*, J. Nowotny, ed., Elsevier, Amsterdam, 1994, p.183.
49. A. E. HUGHES and S. P. S. BADWAL: *Solid State Ionics*, **46**, 1991, p.265.
50. A. E. HUGHES and S. P. S. BADWAL: *Journal of Europian Ceramics Society*, **10**, 1992, p.115.
51. Z. ZHANG: Unpublished Results.
52. X. GUO: *Journal of Material Science Letters*, **15**, 1996, p.2017.
53. J. NOWOTNY and J. B. WAGNER, Jr.: *Oxidation of Metals*, **15**, 1981, p.169.
54. Z. ADAMCZYK and J. NOWOTNY: *Journal of Physics Chemical Solids*, **47**, 1986, p.11.
55. T. BAK, X. LI, J. NOWOTNY and Z. ZHANG: *Proceedings of PACRIM2 on Ceramics*, Cairns, 1996.

SECTION III

Interfacial Phenomena and Kinetics

Chemical Kinetics of Phase Boundaries in Solids

H. SCHMALZRIED
Institut für Physikalische Chemie
Universität Hannover, Germany

ABSTRACT

The kinetics of solid/solid-interfaces is of major importance in the context of solid state reactions. Beside being sites of chemical interface reactions, solid/solid-interfaces offer high diffusivity path for the crystal compounds and thus represent two-dimensional reaction media. In addition, they are normally sources of elastic stress and thus may interact with dislocations, other interfaces and outer surfaces. We summarise the essential kinetic situations. Corresponding experimental examples are presented, dealing with static and moving phase boundaries. This includes: Relaxation phenomena of structural elements, exchange fluxes across interfaces, boundary controlled solid state reactions, interface morphology, nonlinear phenomena and reactions in boundaries.

1.0 INTRODUCTION AND STATEMENT OF THE PROBLEM

The theme of this paper is the role which solid heterophase boundaries (*b*) play in the kinetics of solid state reactions. Two main questions arise:

1. How to handle the phenomenological kinetics?
2. How to explain the kinetic coefficients in terms of the atomistic structure and dynamics? Recent reports are found in refs. 1 and 2.

Let us state a number of facts which are relevant here: The interface energy, in equilibrium, depends on the interface crystallography. Since it contains normally elastic contributions, the boundary is not necessarily flat. Since furthermore both, structure and morphology of a moving *b*, depend on the velocity *v(b)*, the equilibrium properties of *b* may even not be really relevant in solid state reactions.

In view of the multitude of structurally conceivable *b*, let us generalise by treating only three types of *b*: coherent, semicoherent and incoherent ones. Semicoherent *b* contain dislocations moving along with the *b* during reactions. Two modes of dislocation motions exist: Glide (conservative) and climb (nonconservative). Climb involves the diffusional transport of whole lattice molecules,[1] and is correspondingly slow. Thus, depending on the orientation and degree of coherency, the *b* - mobility

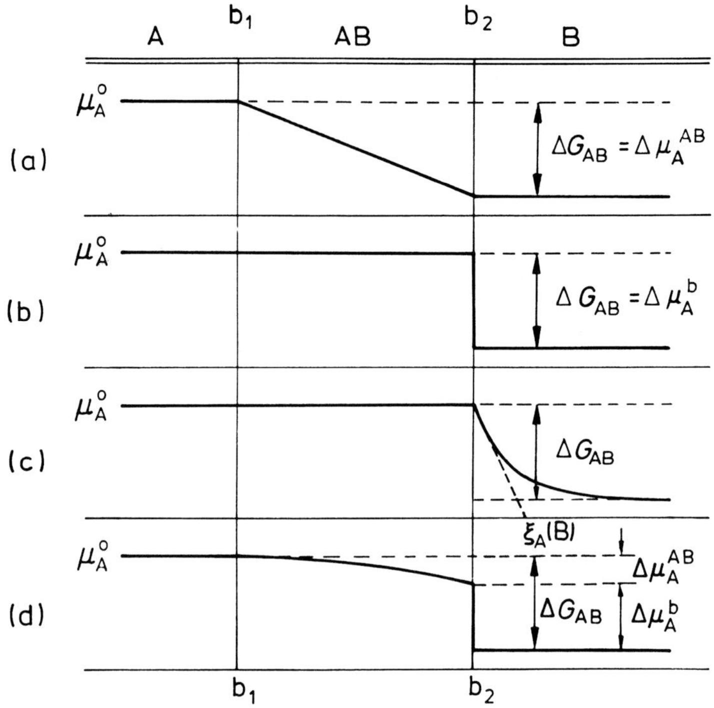

Fig. 1 The chemical potential of component A in phase AB and across the boundary AB/B during the solid state reaction A + B = AB, (a) diffusion control, (b) only interface control, (c) relaxation in zone ξ_A(B) and (d) mixed control of kinetics.

can change drastically.[3] In the kinetics of the solid state, the *b* plays a twofold role. It is a series resistance for matter transport across *b* and will thus control the kinetics of the *b* if it is larger than the bulk resistance. Also it is a medium for two-dimensional lateral transport and reaction in the *b* itself.[2] Atomistically, *b* has to achieve essentially two purposes:

1. To perform the structural transformation.
2. To equilibrate the structure elements (SE) which have crossed the *b* during reaction.

Figure 1 explains the course of a component chemical potential during the solid state reaction $A + B = AB$, assuming (partial) interface control of the kinetics. One notes that the gradients are larger if interface control prevails. Large driving forces, however, can easily lead to nonlinear kinetic behaviour. In what follows we will exemplify the basic problems of interface kinetics which are by and large understood. The aim is to rationalize and explain the measured kinetics and the evolution of the boundary morphology. This is done in terms of transport and reaction of SE across *b* in the bulk, including relaxation and stability phenomena.

2.0 THE RESTING BOUNDARY

Let us consider the boundary between AX and AY crystals if $D_A >> D_X, D_Y$ (A is the mobile cation). Under the action of an electrochemical potential gradient

$$\nabla\eta_{A^+} = \nabla\mu_{A^+} + F\nabla\varphi,$$

A^+ cations will cross the resting b. In this case we are only concerned with the relaxation of SE after crossing b in order to attain the steady state and its local equilibrium concentrations.

Since it is unlikely that the one-step crossing of b (which is a region of increased energy and therefore less tightly bound than the bulk) determines the kinetics, it is most probably the relaxation of point defects (SE) after the crossing proper which is rate determining.[4] The basic kinetic quantity of a resting interface is the exchange current for which we have

$$j_A^o(b) = |\vec{j}_A(b)| = |\overleftarrow{j}_A(b)| \quad or \quad |\vec{j}_A(b)| - |\overleftarrow{j}_A(b)| = 0$$

$j_A^o(b)$ is kinetically as basic for b as the selfdiffusion coefficient is for the bulk.

The mobile irregular SE (= point defects) are either interstitials or vacancies (in this example: $A_i^+, V_A^{'}$). Their concentrations are coupled by the Frenkel equilibrium. As long as the lattice remains undisturbed, we have

$$j_A(b) = (j_i - j_v) \tag{1}$$

Mass and site conservation requires that $\nabla j_i = r_i$, where r_i is the recombination rate; see Figure 2. In a linearised form we may set,

$$\gamma_i = -k_R \cdot \frac{C_i - C_i^o}{C_i^o} \tag{2}$$

Integration of $\nabla j_i = r_i$ then yields (4)

$$j_A(b) = -C_i^o \frac{D_i + D_V}{RT} \cdot \frac{\Delta\varphi . F}{\xi_R} \tag{3}$$

D denotes the defect diffusion coefficient; ξ_R is the relation length,

$$\xi_R = \sqrt{2\tau_R \overline{D}}\,;\ \overline{D} = D_i \cdot D_V/(D_i + D_V),$$

and the relaxation time $\tau_R = c_i^o/k_R$. One notes that $\Delta\varphi \cdot F/\xi_R$ is the average thermodynamic potential gradient in the recombination (relaxation) zone. Comparing Eqn. (3) and the defining flux equation,

$$j_A(b) = j_A^o(b) \cdot \frac{\Delta\eta_{A^+}}{RT} \tag{4}$$

one obtains the elementary kinetic parameter $j_A^o(b)$ of the resting interface b as ($D_i >> D_V$):

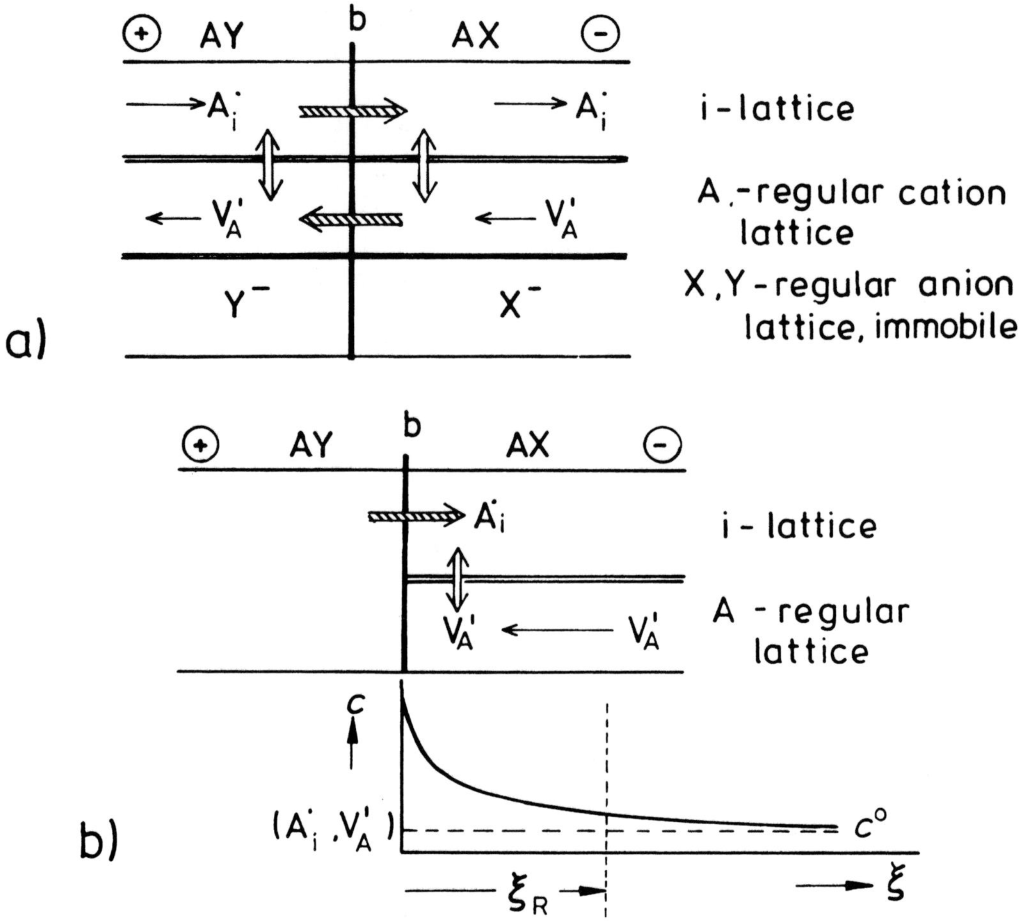

Fig. 2 Reaction scheme of the resting boundary AX/AY, (a) processes on the structure element level and (b) relaxation zone ξ_R where the Frenkel equilibrium is established for a steady state load, injecting A-interstitials from AY as the rate determining step.

$$j^o_A(b) = \frac{c^o_i \cdot (D_i + D_v)}{\sqrt{2\tau_R . \overline{D}}} \cong \frac{c^o_i . D_i}{\sqrt{2\tau_R . \overline{D}_v}} \tag{5}$$

Figure 3 illustrates the experimental setup for the determination of $j^o(b)$, Figure 4 shows experimental results.[4, 5] For the AgI/Ag_2S interface at 260°C one obtains $j^o_{Ag}(b) = 2 \times 10^{-5}$ mol./cm^{2}S. In other words, during one single lattice vibration, a fraction of 10^{-10} of a lattice plane in Ag^+ ions will cross the equilibrium interface to and fro. More data are given in refs. 4 and 5.

We have briefly considered the kinetic behaviour of a resting *AX/AY* interface, without discussing its thermodynamic state under load. This has been done elsewhere.[6]

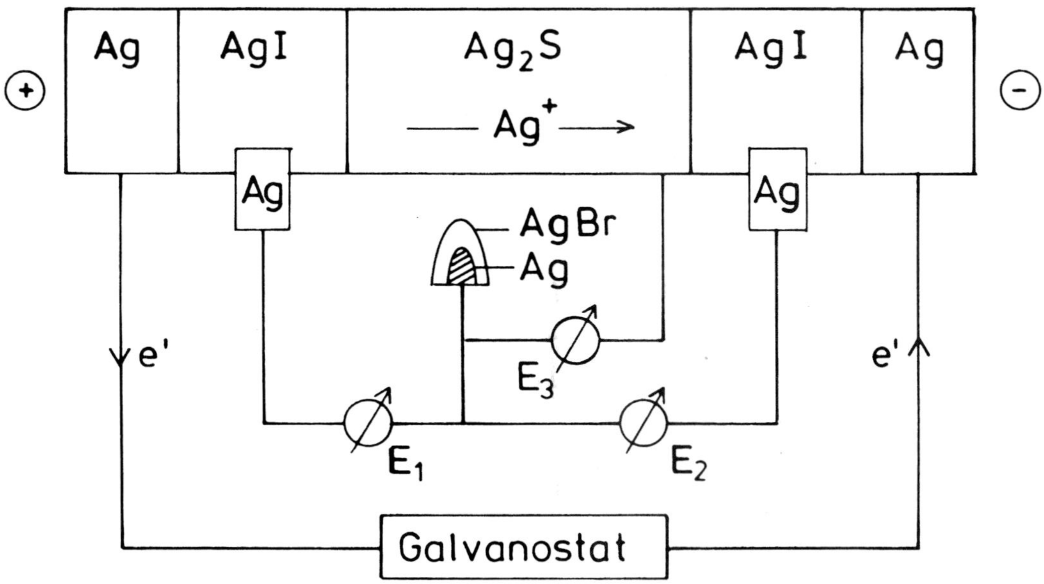

Fig. 3 Electric circuit for the determination of $j_A(b)$ as a function of $\Delta\eta^b_{Ag^{\cdot}}$. The driving force across b is obtained from an extrapolation of the probe *(Ag/AgBr)*-emf to the phase boundary.

Let us, however, briefly state that as long as the number of lattice sites remain conserved, and no decomposition of *AX* or *AY* occurs, the interface is the site of a maximum or a minimum of the chemical potential of the common component *A*, despite the fact that the electrodes are kept on a constant A-potential, e.g. by contacting *AX* and *AY* with A-electrodes. The reason for this behaviour, which eventually leads to the decomposition of the ionic crystals by reduction or oxidation at this interface, is the necessary change in the (small) transference number of the electronic charge carriers across the boundary.

3.0 NONLINEAR KINETICS

Nonlinear interface kinetics is well known for liquid/solid and gas/solid interfaces (passivation; heterogeneous catalysis). Much less is known empirically about solid/solid-interface kinetics. If one visualizes the moving b under nonequilibrium conditions as a relaxation box for SE which is coupled to the two adjacent phases by matter transport, (see Figure 5), one can expect situations which result in nonlinear behaviour including positive feedback.

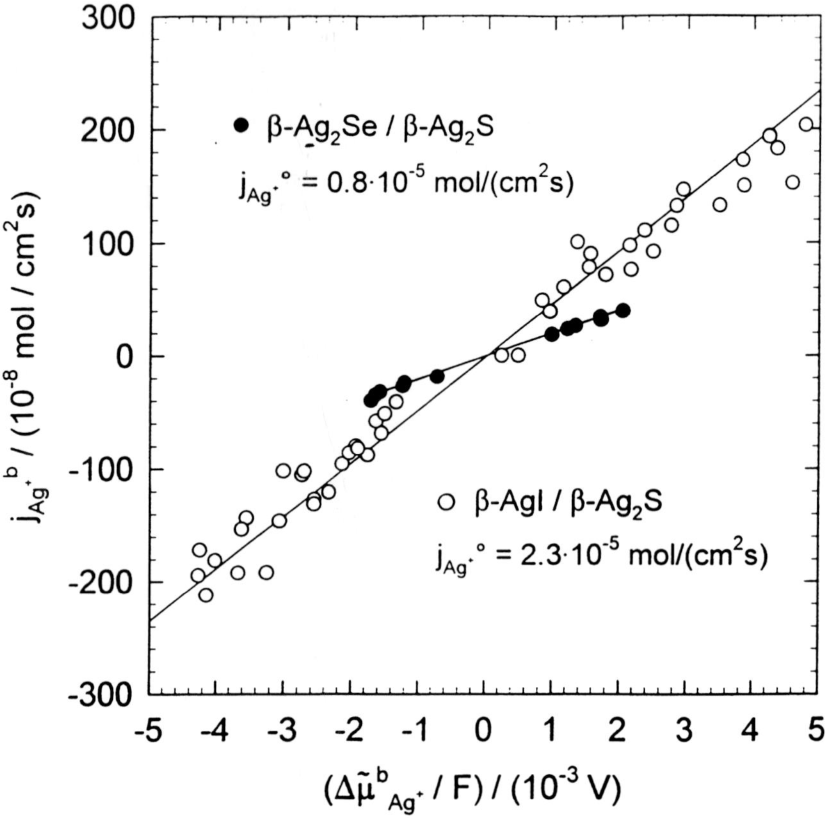

Fig. 4 Silver ion flux density vs. the change of the electrochemical potential of silver ions across the AgI/Ag_2S-interface at 266°C and the Ag_2Se/Ag_2S-interface at 188°C, as studied experimentally.[4, 5]

An experimental example is provided by the Ag_2S-transformation from its high temperature form (β) to its low temperature form (α). Chemical potential probes placed along the transforming crystal register a nonmonotonic behaviour of the Ag-potential as a function of time (see Figure 6), while the α/β-interface is moving from one end of the crystal to the other. The origin of these oscillations is still not fully clear. Several suggestions have been put forward for the nonlinear coupling of SE-transfer and relaxation; a nonmonotonic motion of the dislocation network associated with the semicoherent *b*; a discontinuous drag of impurities and excess silver along with the *b*.[7]

Another experimental example is provided by the work of Janek and coworkers on the electrochemical polarisation of the Ag/AgI interface.[8] This work is in continuation of a study by Fischbach,[9] who studied the interface Ag/β-Ag-Al_2O_3. Figure 7 illustrates, for a galvanostatic experiment, the voltage drop across the Ag/AgI interface. One observes periodic oscillations of the overvoltage. The

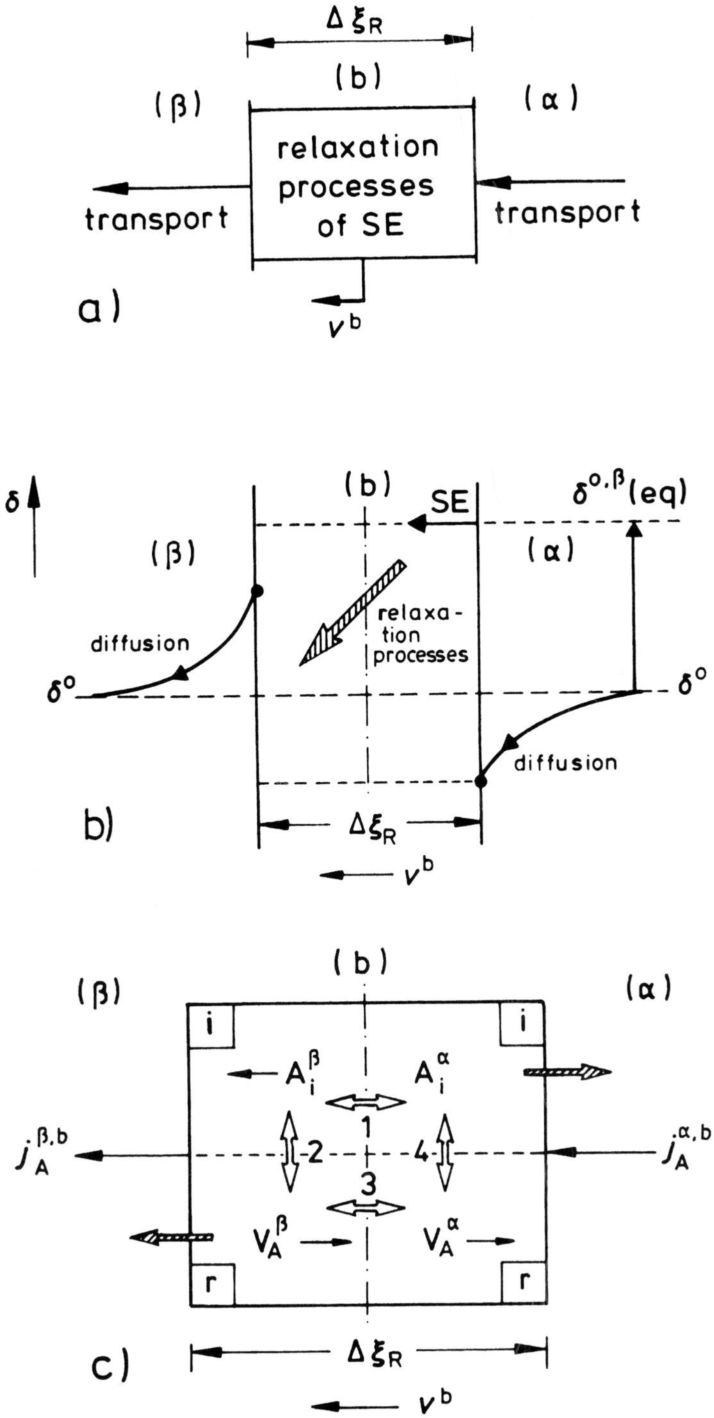

Fig. 5 Scheme of a phase boundary which moves during reaction. (a) general, (b) on the component scale and (c) on the structural element scale.

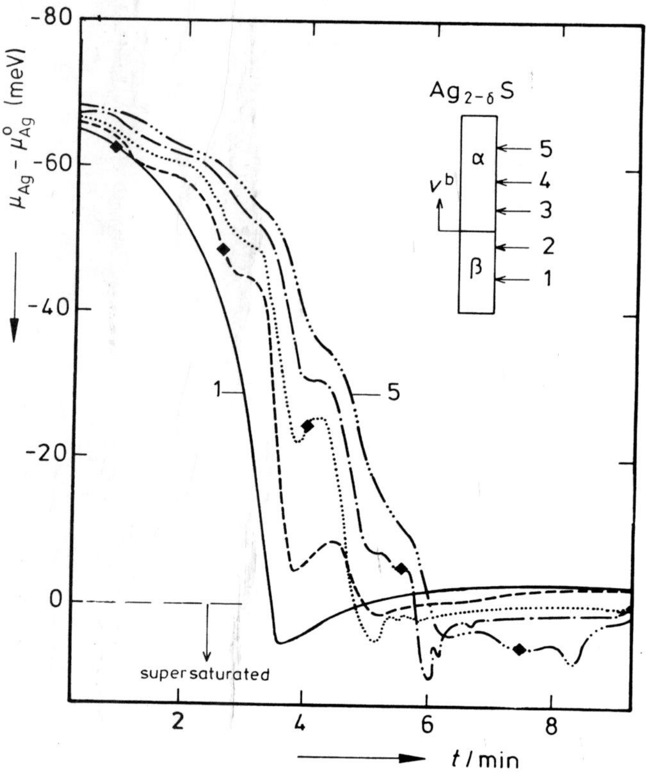

Fig. 6 Low frequency oscillations of the Ag-component chemical potential during a continuous transformation of Ag_2S from its high temperature (β) form to its low temperature (α) form. The various curves represent Ag-chemical potentials registered as emf from local miniaturised potential probes (Ag/AgBr) as a function of time. The cooling rate in a small temperature gradient was 7°C/min. ◆ indicates the time when the β/α-boundary passes the corresponding Ag/AgBr-probe.

simultaneously measured length change which is indicative of the Ag-dissolution under anodic load, which is periodic as well with the same frequency, reveals the nonmonotonic behaviour of the dissolution process. The data analysis shows that at least three different microscopic processes are active during polarisation. Since the interface is incoherent, as SEM-pictures document, the three different reaction steps are associated with:

a. Ad-atom dissolution coupled with surface diffusion,
b. Direct Ag-transfer at contact points and
c. Charge transfer by anodic electrolysis of AgI.

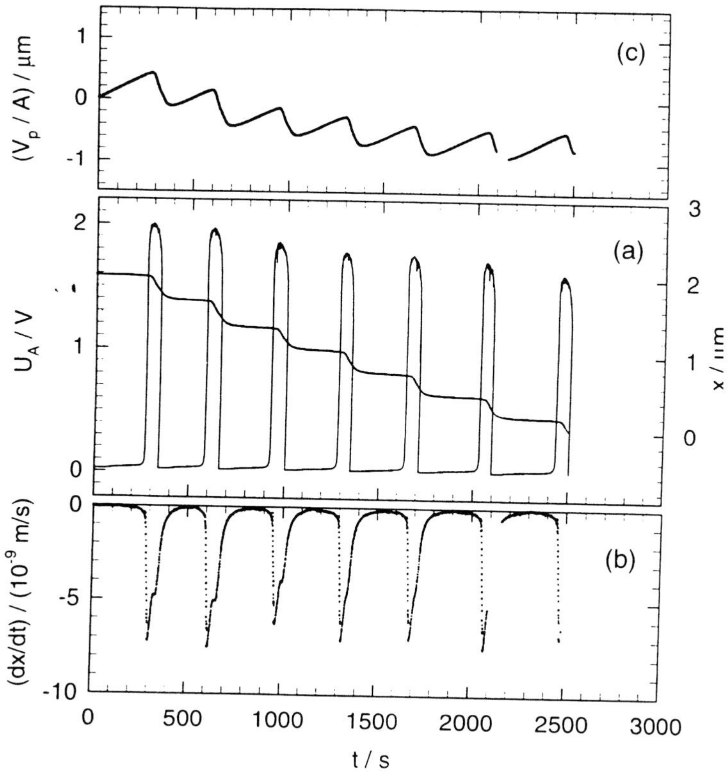

Fig. 7 Periodic reactions at the Ag/AgI-interface under galvanostatic electric load. (a) Voltage across the boundary and length change of the Ag-electrode as a function of time, (b) the rate of Ag-consumption as a function of time in terms of the length change of Ag and (c) the pore volume as derived from the ionic current and the length change as a function of time. $T = 260°C$, $i = 1.2$ mA/cm^2.[8]

Processes (a) and (b) lead, if the silver vacancy relaxation in the interface is not fast enough, to porous material. The breakdown of the pores under some external pressure is thus the main reason for the overall periodic behaviour of the interface resistance. Similar behaviour has been observed during anodic polarisation of Pt/CoO-interfaces.[10] It can be predicted that the galvanic cell A/AX/AY/A under load, neglecting the electrode behaviour, will also yield periodic behaviour during its internal decomposition at the interface AX/AY, if the electric current density is chosen to be high enough.[6] All these phenomena are quite sensitive to the applied pressure, in particular the frequency of the oscillations.

4.0 MORPHOLOGIES

Heterogeneous reaction rates, matter fluxes and boundary motions are coupled. What sort of changes does the moving boundary experience under nonequilibrium conditions? We have changes both on the macroscopic and the microscopic scale. The microscopic *b*-structure concerns the dislocations, ledges, atomistic pores[11] and the interface crystallography. It has been shown, for example, that the misfit angle between the basal planes on the two sides of a moving boundary depends on the *b*-velocity. In other words: we expect the atomic structure of moving *b* to depend on its rate of advancement. Here, however, we will be concerned mainly with the macroscopic *b* morphology when a heterogenous reaction takes place. By setting up the (differential) transport equations and the boundary conditions it is seen that we are dealing with a stability problem in the first place. The transport equations for the bulk phases are well known (e.g. Fick's laws). The coupling conditions at the *b* are simply the continuity equations, i.e.

$$v^b = \frac{\Delta j_i}{\Delta c_i} = \frac{j_i^\beta - j_i^\alpha}{c_i^\beta - c_i^\alpha} \tag{6}$$

In a first order treatment, one assumes local equilibrium to be established and neglects effects due to the elastic nature of the contiguous solids. Straightforward qualitative rules for the discussion of morphological stability have been formulated in,[1] and are illustrated in Figure 8. They stem essentially from a Laplacian solution of the concentration field or the chemical potential field. The simple question of stability or nonstability of interfaces which move during solid state reactions can be answered straightforward by calculating the reaction path under the assumption of stable boundaries. If the reaction path enters into the miscibility gap as shown in Figure 9, the moving *b* is morphologically unstable. Simulations result in the illustrations as produced in ref. 12. One may enhance stability or nonstability by the application of an electric field.

Several experimental examples of morphological instabilities of moving phase boundaries are available. The first example has been thoroughly discussed in ref. 13. It is the motion of a slab of single crystalline CoO placed between two different oxygen potentials. In line with the stability rules, the gas/solid-interface is morphologically unstable at the reducing side (low P_{O_2})· Figure 10 illustrates this interface after motion. At the start it was a planar, cleaved (100)-surface of CoO. The interface reactions read:

$$\frac{1}{2} O_2 + Co_{Co}^x = CoO + V_{Co}^{'} + h$$

It is worthwhile to note that by changing the reducing gas atmosphere, for example from N_2O_2 to CO/CO_2, the structure of the unstable surface changes as well. This indicates that the surface energy (which is altered by adsorption even at high temperatures) plays an important role in the specific morphology of unstable interfaces.

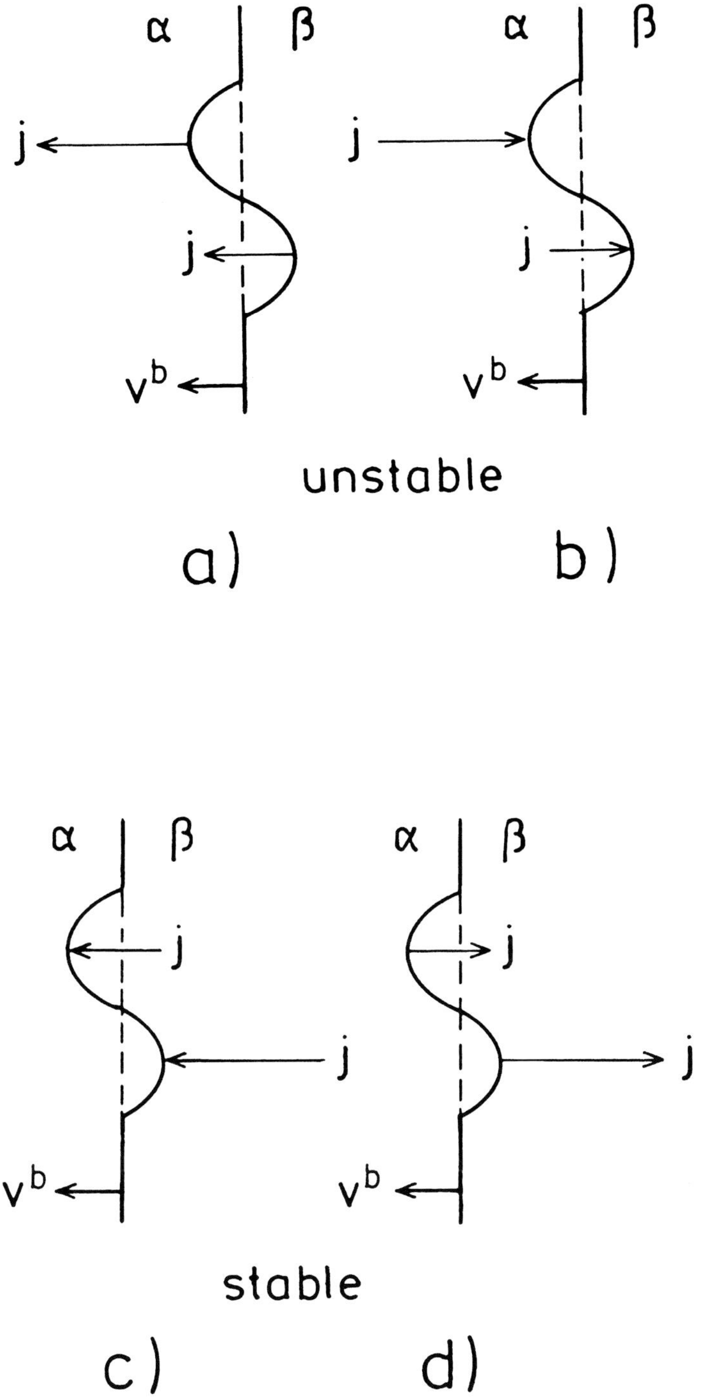

Fig. 8 The morphological stability of a phase boundary moving with velocity v^b. j_i are the reactive fluxes which are responsible for the boundary motion in the contiguous phases α and β.

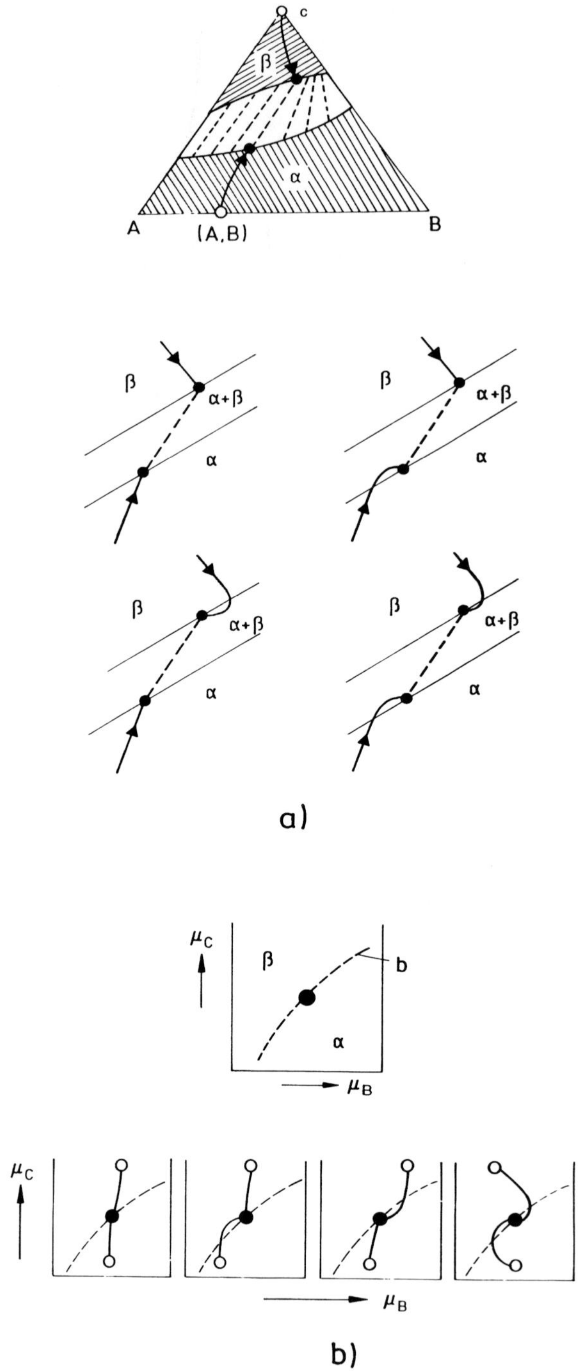

Fig. 9 (a) Ternary (Gibbs) phase diagram with starting compositions and reaction path for a linear reaction couple. The tie lines in the miscibility gap are indicated. Except for the first situation, the interfaces are morphologically unstable. (b) Same situation in a phase diagram of third kind (μ_c vs. μ_B).

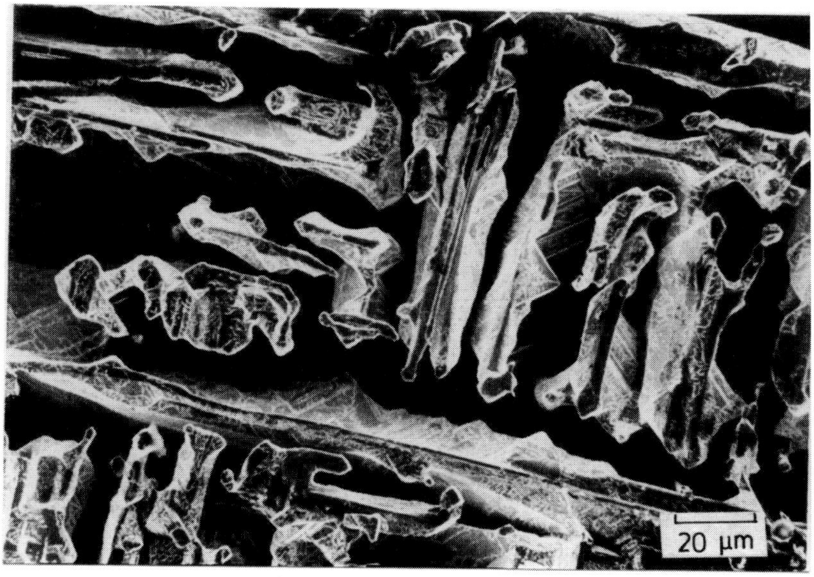

Fig. 10 CoO-single crystal slab with surface b' and b'' exposed to an oxygen chemical potential gradient. The SEM picture shows the unstable reducing boundary (1000°C).[13]

Other examples in this context concern the internal reduction front driven by a cathodic reduction of YSZ,[14] the motion of the KCl/AgCl interface driven by the transport of cations across the interface by the application of an external electric field in a galvanic cell,[15] displacement reactions of the type Cu_2O + Co = CoO + 2Cu,[16,17] or ternary interdiffusion experiments in a two-phase system, e.g. $(Fe_y\,Cr_{1-y})_2O_3/(Fe_xMn_{1-x})_3O_4$.[18]

5.0 PHASE BOUNDARY CONTROLLED SOLID STATE REACTIONS

In textbooks one can read: Interface controlled solid state reactions obey a linear rate law. Or the reverse: Linear rate laws mean interface control. This, however, implies that the state of the *b* is stationary during reaction, which is not at all true by necessity, as we have seen before. In order to tackle the basic problem, namely the thermodynamic state of *b* in higher than binary systems, let us first inspect a simple boundary *AX*/*AY*, which is crossed by cations A^+ driven by an externally applied electric field. Both phases, in addition, exhibit electronic conduction. As a limiting case – to make the point clear - let us assume that *AY* is essentially a semiconductor ($t_h \cong 1$), while *AX* is essentially an ionic conductor ($t_h << 1$)·

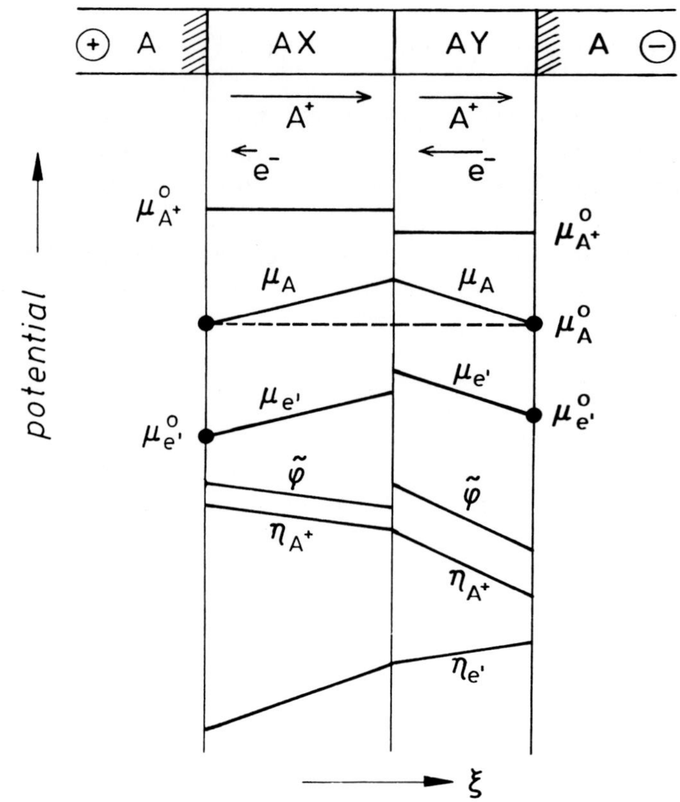

Fig. 11 The course of various thermodynamic potentials in the phase sequence A/AX/AY/A under electric load, if the transference numbers for electronic defects differ in AX and AY.

Without much formalism we can immediately see from the definition of the electrochemical potential $\eta_i = \mu_i + z_i\widetilde{\varphi}$, that if $i = A^+$ and μ_{A^+} = constant which follows approximately from Raoults law, $\nabla\varphi$ = constant in AX as well as in AY, because $j_i = L_i\nabla\eta_i$ = Constant for A^+ in the bulk phases, as long as the crystals do not decompose. However, since $j_h \cong 0$ because of the filtering action of AY, $\nabla\widetilde{\varphi} = \nabla\mu_h$.

Finally, in view of $A + h = A^+$, $\nabla\mu_A = -\nabla\mu_h = \nabla\widetilde{\varphi}$. In other words, the A-metal chemical potential drops with the electric potential in the semiconductor AY, and the chemical potential of Y increases accordingly. This is shown in Figure 11. At the AY/AX-interface, $\mu_{1/2}$ is at a minimum or maximum, depending on the sign of the polarisation of the galvanic cell. Metallic A or bubbles of $Y_2(g)$ may even be formed if

$$p^b_{Y_2} >> 1\,\text{bar, since}(\mu^b_{Y_2} - \mu_{Y_2}(A/AY)) = RT \cdot \ln \frac{p^b_{Y_2}}{p_{Y_2}(A/AY)}$$

But if no decomposition takes place, the *b* will nonetheless respond structurally if the chemical potential of a component at the *b* changes due to the electrical load (current density).

It has been verified that rate laws of solid state reactions become linear for sufficiently thin reaction layers.[3,19] The application of the concept of the SE relaxation and the relaxation length ξ_R as discussed in section 3 and illustrated in Figure 1 allows to calculate the linear rate of the boundary advancement v^b if the relaxation time for the redistribution of SE in the solids near *b* are known. Relaxation times have been determined by X-ray or optical spectroscopy.[20,21] The simplest model yields

$$v^b = \frac{\Delta G_R}{RT} \cdot \sqrt{D_A(B)\tau_R}$$

for the boundary rate of advancement.[1]

Carter and coworkers have recently illustrated that in forming $NiAl_2O_4$ from NiO and Al_2O_3 (single crystal), the $Al_2O_3/NiAl_2O_4$ interface determines the reaction rate in the regime of linear kinetics which is found as long as the product layer is thin (< 100 nm). It was found that the growth rate is smallest if this interface is bounded by (000.1)-Al_2O_3, whereas it is up to 200 times larger if it is bounded by $(1\bar{1}0.2)$-Al_2O_3.[3,22]

In view of the fact that the *b* is a surface of increased energy and is correspondingly looser than the equivalent bulk plane, one has to assume that again relaxation processes, either of structural transformation (hcp → fcc) or local SE equilibration are rate determining and not the crossing proper of the ion fluxes over a single boundary plane.

There are two possibilities here of a rate determining boundary crossing: that of Ni^{2+} or that of Al^{3+}.

If the Al^{3+} cations crossing *b* and injected into the spinel product relax onto their proper sites in a cubic-isotropic system it is difficult to understand why the rates would depend on the orientation of the Al_2O_3/spinel interface. However, if the crossing of *b* and the injection into unisotropic Al_2O_3 of the rate determining Ni^{2+} ions is the prevailing reaction mechanism, one would indeed expect the orientation dependence of the linear rate law to be observed.

On the other hand, structural transformation may also become the rate determining step in the thin product regime. This has been suggested quite early, and a later more and systematic study[11] has stressed this point again. Recently, Hesse and coworkers[23] have illustrated the validity of such a concept by invoking interface dislocation glide and climb through a systematic variation of the reactant-product misfit in spinel forming reactions. Whereas the reaction rates can be fast if the structural transformation can occur with dislocation glide only (on appropriate combinations of glide planes in suitable orientations, see Figure 12), dislocation climb slows down the boundary motion if it is the only mechanism of structure transformation since whole lattice molecules including the slow diffusing SE (for example oxygen ions in oxide spinels) have to be transported.

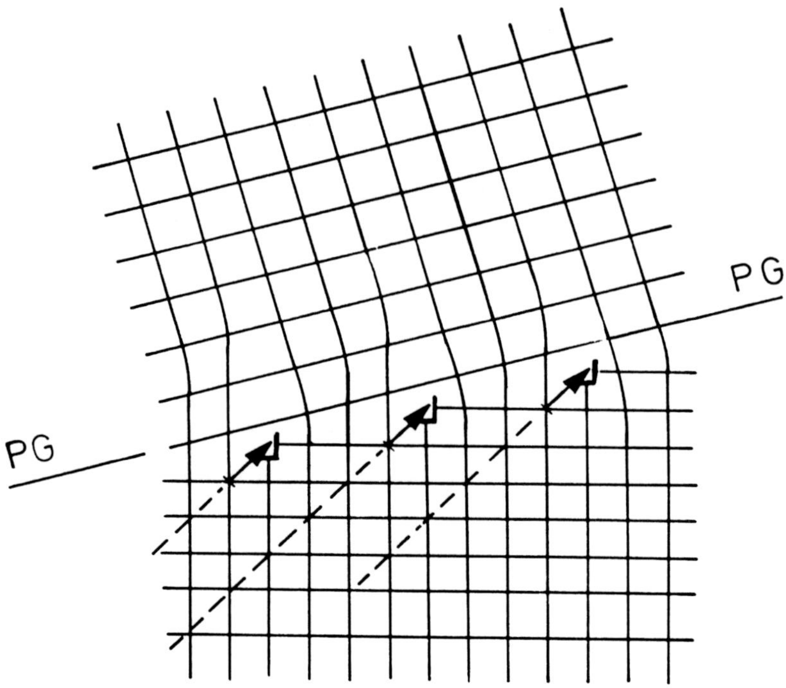

Fig. 12 Schematic illustration of a structural transformation during a phase boundary motion induced by dislocation glide only.

6.0 REFERENCES

1. H. Schmalzried: *Chemical Kinetics of Solids*, VCH-Weinheim, 1995.
2. H. Schmalzried and J. Janek: Chemical Kinetics of Phase Boundaries in Solids, *Ber. Bunsenges. Physics Chem.*, **102**, 1998, pp.127-143.
3. P. G. Kotula: Ceramic Thin Films Reactions, Ph.D Thesis, University of Minnesota, Department of Chemical Engineering Materials Science, 1995.
4. H. Schmalzried, et al.: Investigations of Metal Transport Across Ag/Ag_2S-Interfaces, *Solid State Ionics*, **51**, 1992, pp.91-96.
5. G. Beck: Messung Von Durchtrittswiderständen an Inneren Phasengrenzen, *Diplamarbeit*, Institut für Physikalische Chemie, Universität Hannover.
6. H. Schmalzried and S. Smolin: On Electric Field Induced Processes in Ionic Compounds, *Ber. Bunsenges. Physics Chem.*, **102**, 1998, In Print.
7. H. Schmalzried and H. Reye: The Role of Phase Boundaries in Heterogeneous Solid State Reactions III, *Ber. Bunsenges. Physics Chem.*, **83**, 1979, pp.53-59.

8. S. MAJONI and J. JANEK: Investigation of Charge Transport Across the Ag/α-AgI-Interface, *Ber. Bunsenges. Physics Chem.*, **99**, 1995, pp.14-20.
9. H. FISCHBACH: Leerstelleninjektion in Silber Während Elektrolytischen Abtrags Bei Hohen Temperaturen (Injection of Vacancies in Silver by Electrochemical Etching at High Temperatures), *Z. Metallk.*, **71**, 1980, pp.115-118.
10. J. JANEK: Oscillating Reactions at Metal Electrodes in Solid Electrolytes, *Solid State Ionics*, **101-103**, 1997, pp.721-727.
11. 'B' C. CARTER and H. SCHMALZRIED: The Growth of Spinel into Al_2O_3, *Phil. Mag.*, **52**, 1985, pp.207-229.
12. M. MARTIN: Transport and Degradation in Transition Metal Oxides in Chemical Potential Gradients, *Materials Science Engineering Report*, **7**, 1991, pp.1-86.
13. M. MARTIN and H. SCHMALZRIED: Cobaltous Oxide in an Oxygen Potential Gradient: Morphological Stability of the Phase Boundaries, *Ber. Bunsenges. Physics Chemical*, **89**, 1985, pp.124-130.
14. J. JANEK and C. KORTE: Personal Communication.
15. S. SCHIMSCHAL-THÖLKE, et al.: Instability of Moving Interfaces Between Ionic Crystals KC1/Agcl, *Ber. Bunsenges. Physics Chem.*, **99**, 1995, pp.1-7.
16. G. YUREK, et al.: Kinetics of the Displacement Reaction Between Iron and Cu_2O, *Metallurgical Transactions*, **4**, 1973, pp.1293-1299.
17. Y. UESHIMA, et al.: Demixing in Oxygen Potential Gradients, *Ber. Bunsenges. Physics Chem.*, **93**, 1989, pp.24-29.
18. M. BACKHAUS and H. SCHMALZRIED: Morphological Stability in the Course of Solid State Reactions, *Ber. Bunsenres. Physics Chem.*, **89**, 1985, pp.1323-1330.
19. C. A. DUCKWITZ and H. SCHMALZRIED: Reaktionen von Festen Oxiden Under Einschluss von Gastransport, *Z. Physics Chem.*, N.F.**, 76**, 1971, pp.173-193.
20. R. ROETTGER, et al.: An X-Ray Diffraction Study on the Kinetics of the Cation Redistribution in the Spinel $NiAl_2O_4$, *Ber. Bunsenges. Physics Chem.*, **101**, 1997, pp.1083-1089.
21. K. D. BECKER and F. RAU, Cation Disorder and Cation Kinetics in $NiAl_2O_4$, *Ber. Bunsenges. Physics Chem.*, **91**, 1987, pp.1279-1282.
22. P. G. KOTULA, et al.: Thin-Film Reactions, *Z. Physics Chem.*, **206**, 1998, pp.73-99.
23. H. SIEBER, et al.: Misfit Accomodation Mechanism at Moving Reaction Fronts During Topotaxial Spinel-Forming Thin-Film Solid-State Reactions: A High Resolution Transmission Electron Microscopy Study of Five Spinels of Different Misfits, *Phil. Mag.*, **A75**, 1997, pp.889-908.
24. H. SIEBER, et al.: The Atomic Structure of the Reaction Front as a Function of the Kinetic Regime of a Spinel-Forming Solid-State Reaction, *Phil. Mag.*, **A75**(4), 1997, pp.909-924.

ZrO_2 As Catalyst and Catalyst Support

K. Dyrek*, A. Adamski and Z. Sojka
Faculty of Chemistry
Jagiellonian University
Cracow, Poland

ABSTRACT

A brief overview of zirconia catalytic chemistry with the emphasis on the V_2O_5/ZrO_2 system is given. It is shown that both the chemical state and physical properties like crystallographic form and specific surface area of ZrO_2-based catalysts can be varied by appropriate doping and pretreatment. The influence of various factors on appearance and stabilisation of the specific polymorphs, morphology and surface area of ZrO_2 is also discussed. As the result, these materials can acquire several new characteristics such as superacidity (SO_4^{2-}/ZrO_2), oxygen storage and transport capacity (CaO-ZrO_2) and catalytic redox properties (V_2O_5/ZrO_2). It was found that the molecular structure and catalytic function of zirconia-supported surface oxovanadium species is very sensitive to the V_2O_5 loading and thermal treatment of V_2O_5/ZrO_2 samples.

1.0 INTRODUCTION

Zirconium dioxide (ZrO_2) chemistry has developed rapidly and much of the progress achieved can be attributed to advances in catalysis and surface science. The unique chemical, thermal and mechanical properties render zirconium dioxide and its modifications attractive for fundamental and applied investigations. It has high melting point (~3000 K), low thermal conductivity and high resistance to corrosion even in the presence of sulphur.[1,2,3] Another interesting property of zirconium dioxide is its high conductivity of oxygen, controlled by the concentration of very mobile oxygen vacancies.[4,5]

Diverse applications of ZrO_2 include production of catalysts, refractory materials, piezoelectric devices, pigments, ceramics and oxygen sensors.[1,3] Very important for the reactivity of ZrO_2 are peculiar acid-base properties of its surface that can be modified by introduction of various additives and appropriate thermal treatment.

In this paper the structure and properties of bare and modified ZrO_2 will be discussed using a catalytic chemistry perspective.

*Regional Laboratory of Physicochemical Analyses and Structural Research, Cracow, Poland

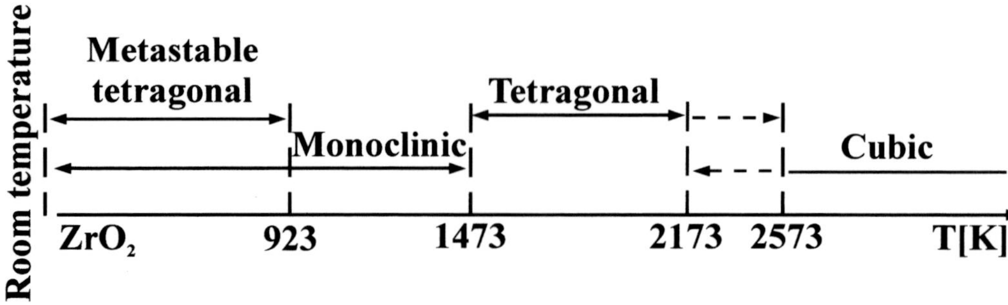

Fig. 1 Temperature ranges of the stability of ZrO_2 polymorphs and corresponding phase transitions.

2.0 PROPERTIES OF ZIRCONIA AND ITS POLYMORPHS

2.1 Phase Transitions of ZrO_2

Depending on the temperature, pure ZrO_2 can exist in various polymorphic structures (Figure 1).[1,3,6,7,8] A monoclinic phase (*M*) is stable up to 1473 K, while tetragonal form is present in the temperature range from 1473 K to 2173-2573 K.[1,3,6] Above 2600 K only a cubic zirconia is observed. The tetragonal phase cannot be preserved at ambient temperature by quenching, because below 1473 K it transforms into the monoclinic one spontaneously.[6] However, a metastable tetragonal zirconia (T), described first by Ruff and Ebert,[9] can be prepared by precipitation from aqueous solutions followed by calcination below 923 K.[1,6] Four ZrO_2 polymorphs in total can thus be distinguished: the monoclinic, the metastable low temperature tetragonal, the stable high temperature tetragonal and the cubic one. In the case of powders same phase transitions are accompanied by corresponding decrease in the specific surface area. This effect is related to facile sintering of high-surface undoped zirconia, which results in an increase of the average size of the crystallites.

Following Wells,[6] the cubic ZrO_2 has a fluorite structure while the tetragonal polymorph represents distorted fluorite lattice with two sets of four oxygen ions at distance of 2.065 and 2.455 Å, respectively. In the monoclinic polymorph Zr^{4+} cations exhibit sevenfold coordination, and O^{2-} anions are tri- ($d_{Zr\text{-}O}$ = 2.07 Å) and tetra-coordinated ($d_{Zr\text{-}O}$ = 2.21 Å).

2.2 Stabilisation of the Low-temperature Tetragonal Phase

Stabilisation of the low-temperature tetragonal phase still remains not fully understood. Some authors claim that this effect is related to the presence of additives or anionic impurities.[10-14] Garvie et al., however, based on thermodynamic considerations attribute this phenomenon to the influence of the ZrO_2 surface energy.[15,16] When compared with the monoclinic phase, the tetragonal form of comparable specific surface area exhibits lower surface energy. Moreover, important role is attributed to strains which may appear in polydomain structure of the tetragonal phase.[16-18] The T → M phase transition occurring at constant temperature and pressure is determined by the following thermodynamic relation:

$$(G_T - G_M) + (S_T\gamma_T - S_M\gamma_M) + (V_T - V_M) \leq 0 \quad (1)$$

where G, S, γ and V denote, Gibbs energy of the corresponding phase, specific surface area, surface tension and lattice strain energy, respectively. The first term corresponds to the change in Gibbs energy during T → M phase transition, the next one to the changes in surface energy while the last term to the effect of different lattice strains possibly present within both phases. Because at room temperature the difference of the Gibbs energies is positive, stabilisation of the tetragonal phase may result from the negative value of the other components only.[15,18] In the case when no considerable lattice strain occurs, the last term of the eqn. (1) may be neglected. Thus the stabilisation of the low temperature tetragonal phase is possible owing to the favourable surface energy difference between the monoclinic and tetragonal ZrO_2. This occurs above a threshold value of the surface area $S > (G_T - G_M)/(\gamma_M - \gamma_T)$ at the temperature when both phases are in equilibrium (e.g., at 537 K for crystallites of 300 Å).[6,15,18] However, despite these predictions it was found that in certain cases crystallites of the tetragonal ZrO_2 may be bigger (> 1000 Å) while those of the monoclinic phase smaller than the critical value,[18,19] indicating that the thermodynamic transformation may be hindered by kinetic constrains.

The size of ZrO_2 crystallites is controlled by the precipitation conditions such as pH of the solution and the kind of the precursor and its concentration.[3] It was found that ageing of the precipitate favours formation of the monoclinic phase.[20]

The role of preparation conditions may be accounted for by the ability of zirconium compounds to undergo hydrolysis and condensation reactions in aqueous solutions leading to the formation of amorphous $Zr_nO_p(OH)_m$ of various structures and degrees of polymerisation.[21] The actual nature of these species determine the type of ZrO_2 polymorph obtained after thermal treatment of the amorphous phase. For instance, the so-called α-form is a precursor of the tetragonal while a γ-form is a precursor of the monoclinic ZrO_2 (Figure 2).[22]

Another explanation of the effects related to the influence of the preparation conditions on the nature of the zirconia obtained, uses the structural analogy between amorphous and tetragonal ZrO_2.[23,24] The mechanism involves crystallisation of the tetragonal ZrO_2 on the germinal amorphous phase. Further growth of the metastable tetragonal ZrO_2 occurs according to a topotactic mechanism.[19]

2.3 Influence of Doping with Metal Oxides on the Properties of ZrO_2

Specific surface areas of pure ZrO_2 are typically in the range of 10-100 m^2g^{-1} and strongly depend on the calcination temperature.[1,3] It was shown that doping with several mol.% of bi and tri-valent metal oxides (MgO, CaO, Y_2O_3, Sc_2O_3, La_2O_3)

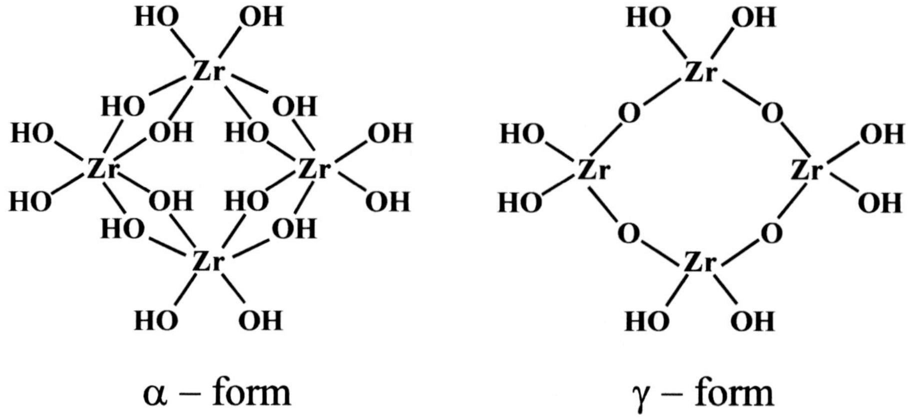

Fig. 2 Zirconium polycationic precursors of zirconia according to Zaitsev.[22]

stabilises the metastable tetragonal phase and the specific surface area.[1-3,25,26] Such partially stabilised zirconia containing two phases, a dominant tetragonal and a minor cubic is usually denoted PSZ (partially stabilised zirconia), whereas monophasic, cubic ZrO_2 of higher content of stabilising cations is called FSZ (fully stabilised zirconia).[1,26] Pure tetragonal ZrO_2 can be obtained, for instance, from precipitate containing 3 mol.% Y_2O_3. According to Hirata et al., by increasing yttria content to 8 mol.%, the FSZ is formed.[25, 26] The presence of dopants influences phase transitions through modification of the surface energy of the samples.[27]

An example of the beneficial action imparted by doping on the value of the specific surface area is provided by ZrO_2 doped with 6 % SiO_2, where an increase from 20-30 m^2g^{-1} for pure ZrO_2 to 140-150 m^2g^{-1} for SiO_2/ZrO_2 was observed.[3]

Aliovalent cationic additives in the zirconia solid solutions with other oxides give rise to high ionic conductivity of such materials, which above 900 K is comparable with that of fused salts. The major reason for such properties is the formation of oxygen vacancies $V_O^{\cdot\cdot}$ at high concentrations through which O^{2-} ions can diffuse. CaO-stabilised zirconia $Ca_xZr_{1-x}O_{2-x}(V_O^{\cdot\cdot})_x$, where $0.1 \leq x \leq 0.2$, may serve here as an example. Typical conductivity for 15% CaO containing ZrO_2 is $5 \times 10^{-2}\ \Omega^{-1}cm^{-1}$ at 1300 K with the activation energy equal to 1.3 eV.[28]

2.4 Influence of Anions on the Value of ZrO_2 Surface Area

A method frequently used for preparation of ZrO_2 catalysts is the precipitation of hydrous oxide from aqueous solutions of zirconium salts. Specific surface areas of zirconia obtained in this way depend on the presence of anions associated in the precipitate. According to Norman et al. sulphates (VI) and carbonates influence formation of the tetragonal phase and the value of specific surface area of zirconia.[27] On the basis of the series of thermal stability of Zr^{4+} complexes with various anions

(CO_3^{2-} < OH^- << SO_4^{2-}) proposed by Aiken,[29] it may be assumed that the formation of Zr-O-Zr bridges occurs in carbonate-containing compounds at lower temperatures than in the case of sulphates. Indeed, carbonates favour nucleation, sintering and crystal growth, while sulphates considerably inhibit all these processes. The Zr^{4+} -Zr^{4+} separation in the presence of SO_4^{2-} is larger (3.5-4.3 Å) than in the case of OH^- groups (3.3–3.7 Å), which may hinder the formation of tetragonal phase by favouring development of α-type structures.

Although Garvie[15] contests the influence of OH^- groups present in the bulk of zirconia on the phase composition, Cyprés et al.[10] attribute the effect of surface stabilisation in the metastabile tetragonal ZrO_2 to the presence of small amounts (1 wt. %) of OH^- or other anionic groups like SO_4^{2-} within the solid. In fact, Mitsuhashi et al.[18] show that in the tetragonal phase a broad IR band at 3200-3600 cm^{-1} attributed to stretching vibrations of OH^- groups located at the boundary of domains and interacting through hydrogen bonds, and a line at 3770 cm^{-1} due to structural hydroxyls appear.[18]

3.0 ZrO_2 AS A SUPPORT

Among principal parameters characterising the properties of ZrO_2 as a support the point of zero charge (~ 6.4) and the surface tension which is higher than that for SiO_2 and TiO_2 and comparable with γ-Al_2O_3 belong to the most important.[30] This provides the basis for versatile methods of the active phase deposition on ZrO_2 surfaces, which include anionic and cationic adsorption and thermal spreading. Another advantage of ZrO_2 as a support consists in its remarkable oxygen conduction and storage properties. The earlier drawback related to a low specific surface area[1,3] has been recently overcome by application of the sol-gel method,[31] especially when using an organic micelle templation. In the latter case mesoporous ZrO_2 catalysts of high surface area (up to ca. 600 m^2g^{-1}) may be obtained by crosslinking colloidal $Zr(OH)_4$ deposited onto the surface of the template and subsequent removing the organic component by calcination.[32]

ZrO_2 is a good support for monolayer vanadium catalysts for reduction of NO_x by ammonia or hydrocarbons.[1,3] The VO_x/ZrO_2 catalysts are used for selective oxidation of methanol to methylformate occurring in the presence of stabilising addition of Y_2O_3 at 473 K with the selectivity of 75%.[33] Low temperature adsorption of oxygen on this system shows that the number of active centres is much higher for ZrO_2 than for other supports indicating better dispersion of VO_x on ZrO_2.[7]

A spectacular example of the advantage of using ZrO_2 as a carrier is provided by the catalytic oxidative dehydrogenation of propane over zirconia supported perovskites (La-Sr-Co-O).[34] Very high dispersion of the active phase achieved onto ZrO_2, resulted in a 10 times increase of the activity as compared to that observed for unsupported perovskite (Figure 3).[35]

By the same token, in the reaction of alcoxylane with the ZrO_2 surface, a 1 nm thin layer of strongly dispersed SiO_2 is formed while onto TiO_2 the same reaction leads just to the formation of a mixed oxide layer.[36]

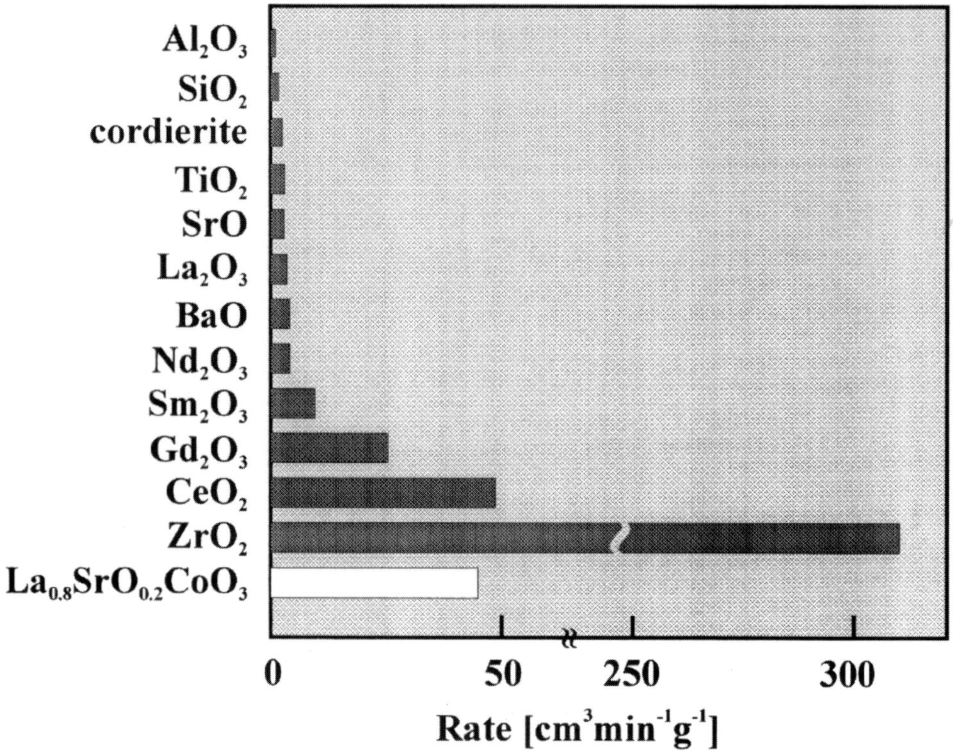

Fig. 3 Catalytic activities of $La_{0.8}Sr_{0.2}CoO_3$ perovskites, massive and supported on various metal oxides in ODP reaction, adapted from Iizuka et al.[40]

Interesting results were obtained by Yamaguchi et al.,[37] who has shown that on ZrO_2 support formation of bidimensional octa- and pentacordinated CrO_x surface species occurs in contrast to the 3D structures present on the surface of SiO_2, even at low Cr_2O_3 loading. As the result catalytic performance of zirconia supported CrO_3 in the oxidation of CO was much higher than in the case of SiO_2 or Al_2O_3 supports.

ZrO_2 is also a competitive support of nobel metals (Pt, Pd, Rh) in three way catalyst (TWC).[3] In the Ford laboratories it was found that ZrO_2 significantly improves thermal stability of the catalysts working at 1173 K preventing migration and agglomeration of the active metallic phase that leads to catalysts deactivation.[3,38] Fornasiero et al.[39] investigating the system Rh-CeO_2-ZrO_2 have revealed an important role of the zirconia support as an oxygen buffer due to its high oxygen storage capacity. This property can enhance the air/fuel ratio of the TWC catalyst in oxygen lean conditions.

Pure and stabilised ZrO_2 are also used as a support for catalysts containing rhenium, rhodium and platinum ions, which are far more active in hydrogenation of CO_x or oxidation of methane than those deposited onto Al_2O_3, TiO_2 or SiO_2.[40-42]

4.0 ZrO_2 AS A CATALYST

4.1 Acid-Base and Redox Properties of ZrO_2

Characteristic feature of ZrO_2 catalysts is the coexistence of moderate Lewis acid, coordinatively unsaturated Zr^{4+} ions and base, O^{2-}, surface centres as well as OH groups of different chemical character, which can be involved in catalytic processes either separately or jointly. [1,3] Their presence may be revealed by adsorption of CO_2 and NH_3.[43]

Besides the chemical modification, two mutually related parameters control the strength of surface acid-base centres on ZrO_2, namely the calcination temperature and the degree of sintering. Primarily they influence the number and the nature of defects on the edges of zirconia microcrystals and the type of the crystallographic planes exposed.[44] Mazzocchia et al.,[45] stated that the base character of the ZrO_2 surface is stronger after calcination at 773 K than at 973 K. For instance, on the surface of ZrO_2 calcined at 873 K about 0.6 μmol m^{-2} of NH_3 was irreversibly adsorbed on the acid and 4 μmol m^{-2} of CO_2 on the basic sites. Acid-base properties were also investigated by IR using pyridine and CO_2 probes and by ESR technique combined with diphenylamine adsorption. In the latter case the amount of diphenylnitroso radicals formed in the presence of oxygen was used to quantify basic sites. The number of those centres was found to be higher than the acid ones.[46]

Acid-base properties of ZrO_2 catalysts undergo substantial modification depending on the method of their synthesis and the presence of dopants.[44,47] Thus, acid properties may be reduced by doping with alkali metals and enhanced by addition of chloride ions.[1] Remarkable improvement of acidic properties of ZrO_2 can be obtained by sulphation. In this way super acid SO_4^{2-}/ZrO_2 catalysts of the strength H_0 = -14.52 (as gauged by the Hammett method) can be obtained.[40] Yamaguchi and Parera et al. attribute super acidity to increase of the number and the strength of Lewis acid centres,[48-51] while Arata and Hino[52] relate their formation to Brønsted centres by reaction with sulphuric acid. The nature of sulphating agent, H_2SO_4 or $(NH_4)_2SO_4$, may also be important since it gives rise to the formation of Brønsted type HSO_4^{2-} surface centres or super acidic hydroxyl groups, respectively.[53]

ZrO_2 surface exhibits reducing and oxidising properties simultaneously. [1] Nakano et al. used ESR spectroscopy combined with adsorption of triphenylamine and nitrobenzene to determine the number of oxidising sites and reducing centres, respectively.[46] The concentration of oxidising centres was larger (150 nmole m^{-2}) than the reducing ones (43 nmole m^{-2}) for ZrO_2 calcined in air at temperatures below 573 K.

4.2 Applications of ZrO_2 in Catalysis

Specific acid-base and redox properties of ZrO_2 determine its wide applications in various catalytic processes. Typical reactions catalysed by zirconia-based systems are shown in Figure 4.

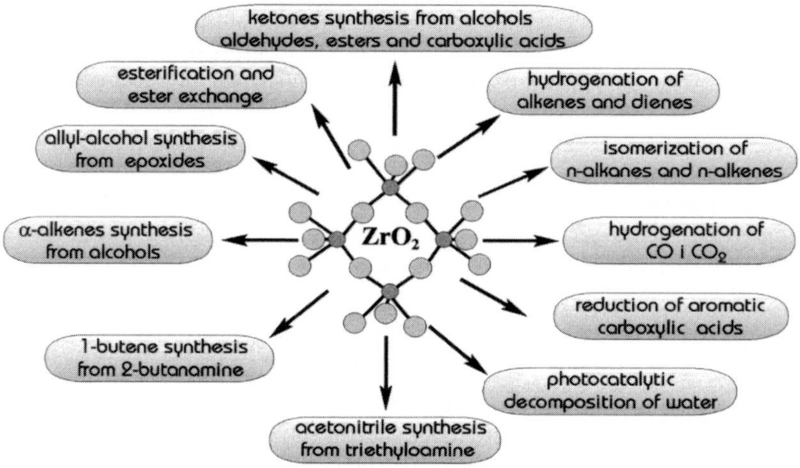

Fig. 4 Selected reactions catalysed by ZrO_2-based systems.

4.2a Bare Zirconia

Undoped ZrO_2 is often used in selective dehydration of alcohols.[1] For instance, in the dehydration of 1-cyclohexylethyl alcohol, vinylcyclohexane is produced with 42% conversion and 66% selectivity.[3] The latter increases upon introduction of small amounts of halogens[1] or alkali metals[3] which modify the strength of acid-base centres on the ZrO_2 surface. Catalytic activity of pure ZrO_2 in the isomerisation of 1-butene correlates well with the number of surface basic centres.[46]

A good example of the application of ZrO_2 in catalysis is provided by the industrial production of allylamine from 1-amino-2-propanol (Koei Chemicals) occurring at 673 K with 98% conversion and 64% selectivity.[1] Moreover, ZrO_2 is also active in CO_x hydrogenation to isoalkanes (isobutane), methanol and dimethylether.[54]

The cooperation of acid and basic centres on the ZrO_2 surface can be shown in acetonitrile synthesis from triethyloamine. This process is inhibited when centres of the one type only are present (e.g., acidic, like on SiO_2-Al_2O_3 or basic, as on MgO).[43]

Photocatalytic decomposition of water on ZrO_2 by UV-irradiation was studied by Sayama et al.[55-57] This reaction may serve as a basis for modelling of photosynthesis or as a potential source of cheap hydrogen. Using ZrO_2 as a photocatalyst changes the mechanism of the process as compared with that operating on previously studied systems like $A_4Nb_6O_{17}$ (A = K, Rb) or layer perovskites doped with Pt or Ni.[58-60]

4.2b Sulphated ZrO_2

Superacidic properties of SO_4^{2-}/ZrO_2 are important for many catalytic acid-base processes which then may occur at relatively low temperatures (below 373 K). The

Table 1. Typical Reactions Catalysed by Sulphated Zirconia.[40]

Reaction	Examples
Cracking	Propane → ethane + methane Neopentane → methane + propane
Isomerisation	Butane → isobutane Pentane → isopentane
Alkylation	Methane → ethylene → C_3–C_7 Isobutane + butenes → C_5–C_{11}
Acylation	Toluene + PhCOCl → methylbenzophenone Toluene + PhCOOH → methylbenzophenone Chlorobenzene + *o*-chlorobenzoyl chloride → bis(chlorophenyl ketone)
Estrification Oligomerisation	Ethanol + acrylic acid → ethyl acrylate

presence of strong acids in the zirconia-based systems favours high concentrations of carbenium cations, the key intermediates in these reactions.[61] Skeletal isomerisation of alkanes (e.g., n-butane to isobutane) acylation, Friedel-Crafts alkylation, dehydration and many others may serve here as good examples.[3, 20, 61-64] Some important examples of the reactions catalysed by sulphated zirconia are given in Table 1.

The main problem, however, is low stability of the superacid centres that rapidly deactivate during catalytic processes. Most probably these effects are due to the removal of sulphate groups or formation of carbonaceous deposits. Doping the catalyst with platinum or carrying out the reactions in the presence of hydrogen protects the catalyst from deactivation.[1]

4.2c ZrO_2 in Binary Oxide Systems

ZrO_2 is used as a catalyst in combination with other oxides, like SiO_2 or TiO_2. The acidity of the ZrO_2-SiO_2 catalyst is one of the highest among binary oxides, whatever the method of preparation, sol-gel or coprecipitation.[63]

ZrO_2-TiO_2 catalyst was used in alcohol dehydration, isomerisation of n-alkenes and dealkylation of cumene.[63] The ZrO_2-TiO_2 system was tested in non-oxidising dehydrogenation of ethylbenzene to styrene.[65] In all these cases the catalytic activity strongly depends on the concentration ratio of both oxides used. The results of TPR and diffuse reflectance IR studies performed by Miller and Ko for isomerisation of 1-butene[66] indicate that increasing ZrO_2 content augments the number of Lewis acid centres, while the increase in the SiO_2 concentration enhances the number of

Brønsted acid centres. These observations are in good agreement with expectations derived from Tanabe model.[67] For ZrO_2-rich systems the excess charge is equal to +2 and is balanced by the associated Lewis surface sites, whereas for silica-rich systems the excess charge of -4 is balanced by protons giving rise to appearance of Brønsted centres.[66]

ZrO_2 is also known as an activator for other catalysts. Small amounts of ZrO_2 added to K_2CO_3 protect this catalyst from the formation of carbonaceous deposits during the transformation of biomass by water vapour.[3,68]

4.2d ZrO_2 Pillars in Intercalated Materials

Promising results were obtained by introducing ZrO_2 as pillars into the interlayer space of clay minerals. Such systems reveal features similar to those found in zeolites and are potential catalysts of hydrocracking and dehydration of alcohols.[3,69,70] Upon pillaring, the interlayer gallery is enlarged and even after the calcination at 673 K a basal spacing of 18 Å may be achieved. The dimensions of channels created after pillaring depend on the presence of transition metal ions like Sm^{3+}, Nb^{3+} and La^{3+} in the clay gallery. These additives simultaneously stabilise the specific surface area of the catalysts.[71]

5.0 V_2O_5/ZrO_2 DOPED WITH ALKALI METALS AS MODEL CATALYST FOR OXIDATIVE DEHYDROGENATION OF ALKANES

It has been shown that numerous factors can influence surface and bulk properties of ZrO_2 and significantly change its catalytic behaviour. In this section as a case study we will discuss modifications of ZrO_2 induced by the presence of V_2O_5 and various alkali cations. The V_2O_5/ZrO_2 system was investigated by XRD and spectroscopic methods (ESR, UV-VIS and RS)[72,73] from the viewpoint of its applications as a catalyst of oxidative dehydrogenation of propane.[74]

A series of undoped and Li-, Na- and K-doped zirconia supported V_2O_5 catalysts were prepared by wet impregnation. The details of the preparation method are given elsewhere.[75] The ZrO_2 matrix was obtained by precipitation of $ZrO(NO_3)_2 \cdot xH_2O$ with 25% aqueous solution of ammonia (at pH = 8.4). Catalysts with various vanadia loading (1, 3, 5 and 10 mol.% V_2O_5) were calcined in air at 973 K for 6 hours (undoped samples) and at 873 K for 6 hours (alkali doped samples).

5.1 Structure of the Surface Oxovanadium Phase

Spectroscopic results supported by XRD data confirmed that the composition and thermal treatment strongly influences the heterogeneity of surface oxovanadium species.[76] With the increasing V_2O_5 loading an agglomeration of isolated vanadyl entities occurs, leading to the formation of larger decavanadate-like clusters. At vanadia concentrations between 3 and 5 mol.% tetra and octahedral vanadium species coexist on the surface. Above the monolayer coverage, V_2O_5 microcrystals with penta- and octa-coordinated vanadium appear. High V_2O_5 loading (above 5 mol.%) and calcination temperatures above 773 K favour reaction between V_2O_5 and the support, resulting in the formation of ZrV_2O_7 with tetrahedrally coordinated vanadium.

In all investigated samples small amount (ca. 10 mol. %) of paramagnetic $V^{4+}(3d^1)$ ions are present. ESR spectra of the catalysts with various vanadium loadings are rather complex and can be interpreted as a superposition of several signals. Simulations of the spectra provided detailed information about the electronic structure and the symmetry of paramagnetic vanadium species involved.[73] For isolated V^{4+} ions, the signals with well resolved hyperfine structure due to ^{51}V (I = 7/2, 99.97%) are observed. It is found that V^{4+} ions can be dispersed on the surface in the form of monomeric vanadyl complexes or spread out within zirconia matrix as $V_xZr_{1-x}O_2$ solid solution.[73,76] The ratio of the surface to bulk vanadium depends on the calcination temperature and the total vanadium content.

5.2 Phase Composition of the V_2O_5/ZrO_2 Catalysts

Analysis of the x-ray diffraction patterns allows the phase composition of the ZrO_2 matrix and its changes upon thermal treatment and doping to be determined. The parent zirconia and the matrix of the catalyst containing 1 mol.% of V_2O_5 calcined at 873 K consists of two phases, a metastable tretragonal and a monoclinic one (Figures 5a and b). The ratio of the intensity of two diagnostic peaks $(11\bar{1})_M/(111)_T$ changes when V_2O_5 is deposited onto ZrO_2 surface. At high calcination temperatures (ca. 1350 K) the $(111)_T$ line disappears (Figures 6a and b) indicating that the phase transition T → M has occurred. These changes can be better revealed in a variable temperature XRD study, when the samples were heated stepwise up to 1477 K *in situ* (Figure 7). The ratio of the $(11\bar{1})_M/(111)_T$ peak intensities remains constant up to about 850 K for both parent zirconia and the 1 mol.% V_2O_5/ZrO_2 catalysts, while above this temperature the tetragonal polymorph starts to disappear. The threshold temperature of this phase transition determined as shown in Figure 8, is distinctly higher (T = 1025 K) for the sample containing 1 mol.% of V_2O_5 than for bare ZrO_2 (T = 883 K). In the case of sample containing 3 mol.% V_2O_5 doped with potassium (K/V = 0.3) this temperature is lower (T = 974 K) than that for 1 mol.% V_2O_5/ZrO_2 (T = 1025 K). Thus the T → M phase transition of zirconia is strongly related to the presence of potassium which modifies the diffusivity of vanadium species into the ZrO_2 lattice.[72]

5.3 Temperature Induced Migration of V^{4+} Ions into the Zirconia Support

Depending on the calcination temperature of the V_2O_5/ZrO_2 the specific interactions between V_2O_5 and ZrO_2 give rise to partial reduction of the system, leading to the formation of V^{4+} ions. [74]

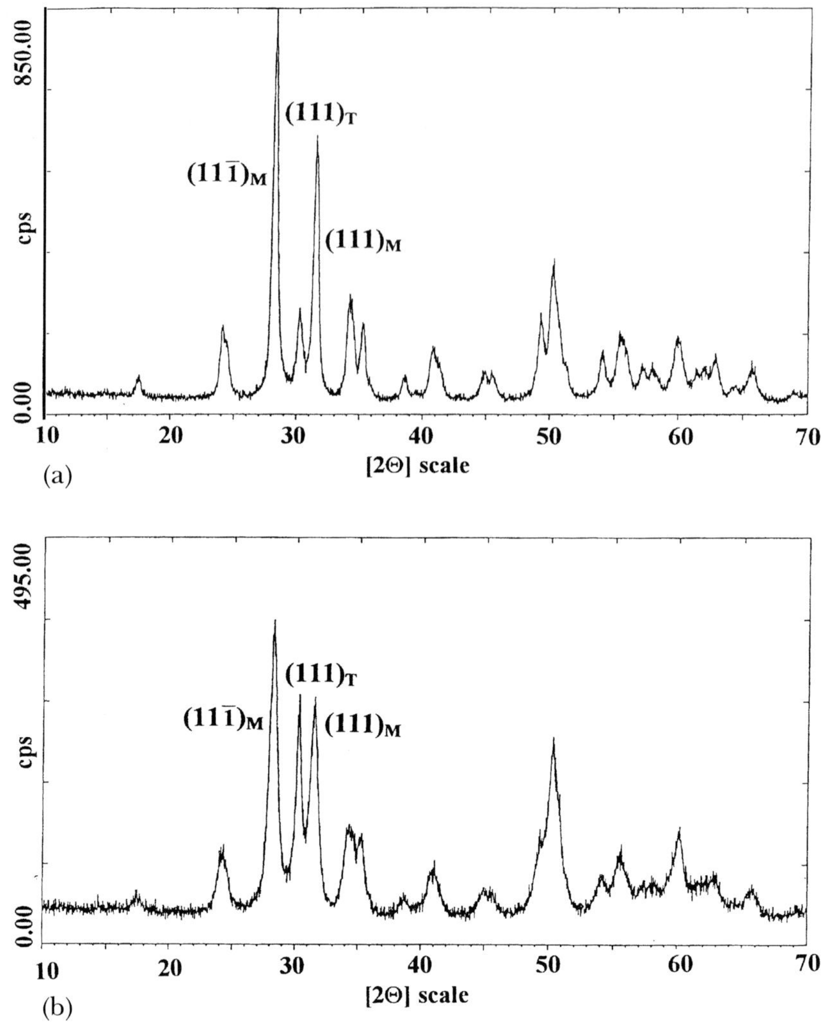

Fig. 5 XRD patterns of the samples calcined at 873 K. (a) Bare ZrO_2 and (b) V_2O_5/ZrO_2 catalyst containing 1 mol. % of V_2O_5.

Combined variable temperature XRD and ESR experiments performed for the catalyst containing 1 mol.% V_2O_5 show that the reduction of surface vanadium and simultaneous diffusion of the V^{4+} ions into the ZrO_2 lattice occur at the same threshold temperature (1015 K) at which the phase transition of the metastable tetragonal to monoclinic ZrO_2 is observed. This process leads to depletion of the active vanadium phase from the surface and is responsible for the loss of the catalytic activity. It was found that alkali metal additives influence the depletion by lowering the phase transition temperature.

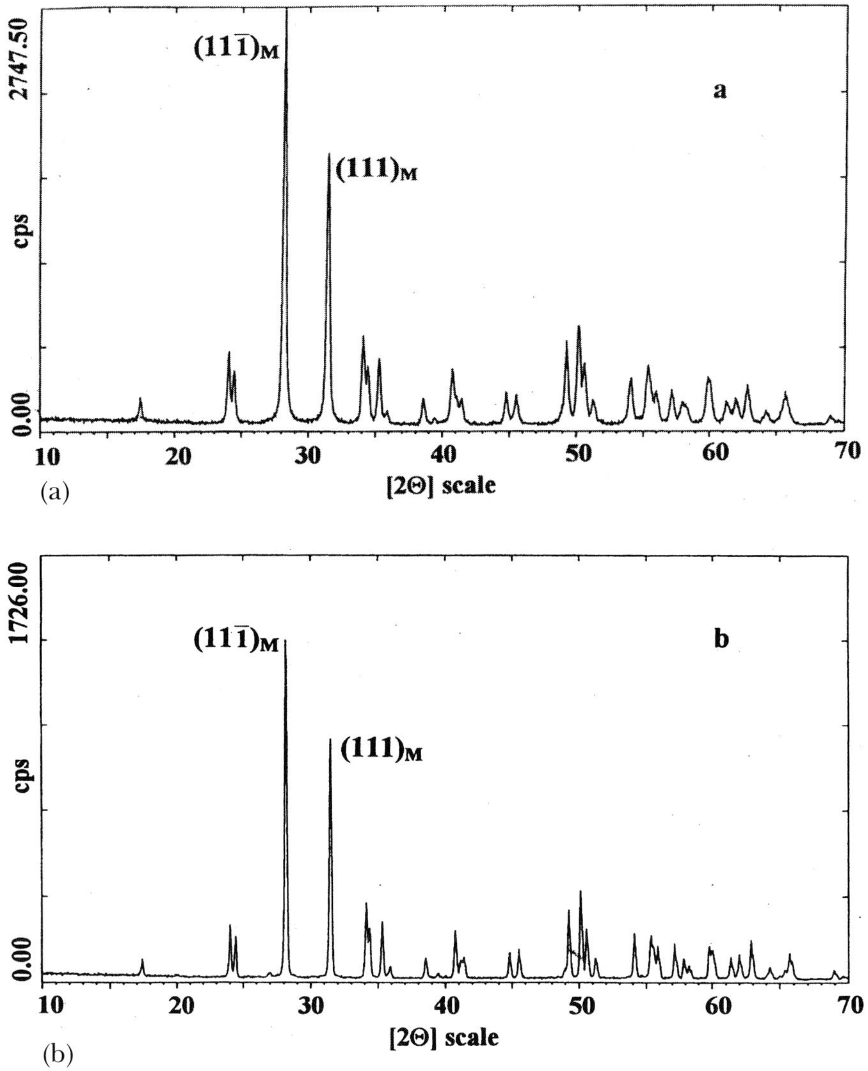

Fig. 6 XRD patterns of the samples calcined at 1350 K. (a) Bare ZrO_2 and (b) V_2O_5/ZrO_2 catalyst containing 1 mol. % of V_2O_5

6.0 CONCLUDING REMARKS

The field of zirconia and zirconia-based materials and catalysts, although mature, continues to attract significant attention. This paper summarises the developments in their recent catalytic applications focusing particularly on the V_2O_5/ZrO_2 system. The properties and the chemical state of ZrO_2 catalysts can be readily modified to

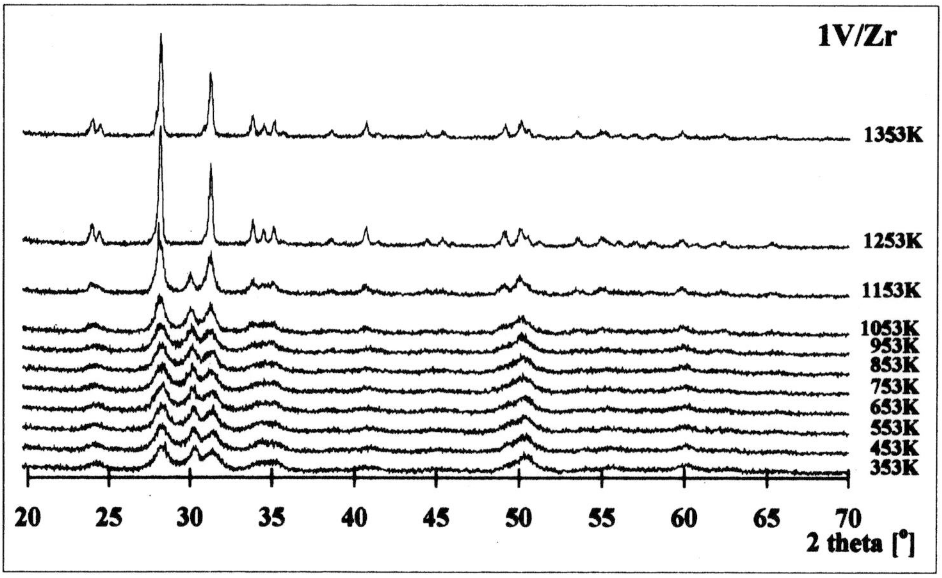

Fig. 7 Thermal evolution of XRD patterns of the V_2O_5/ZrO_2 catalyst containing 1 mol. % of V_2O_5 calcined in the temperature range 353–1353 K

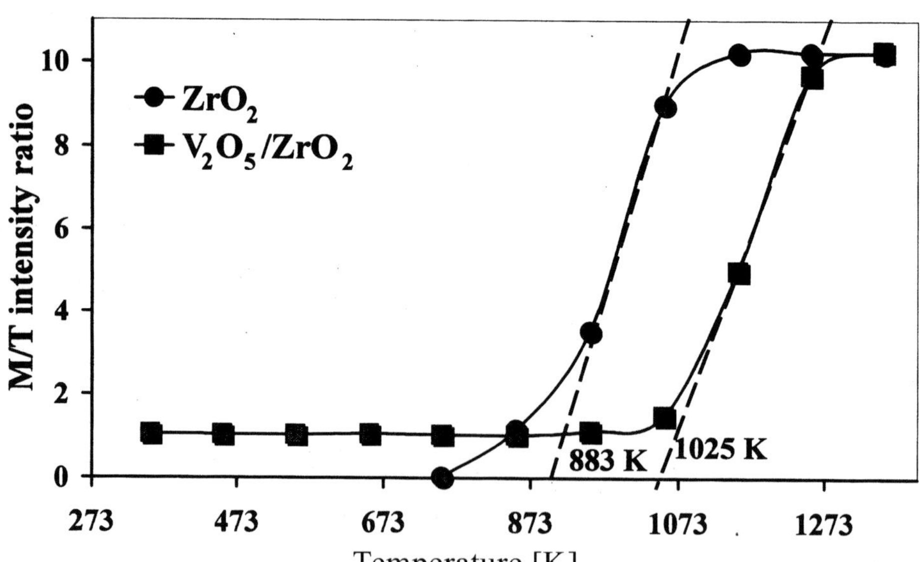

Fig. 8 The rate of disappearing of the metastable zirconia as a function of the calcination temperature: ZrO_2 and V_2O_5/ZrO_2 catalyst containing 1 mol. % of V_2O_5 (adapted from[72])

large extent by appropriate doping and thermal treatment. Advances will require continued developments in characterisation and extension of synthetic methods to further broaden the class of these materials.

7.0 ACKNOWLEDGEMENTS

This work was financially supported by the Polish Committee for Scientific Research (KBN) (Grant No. T09A 104 15).

The authors sincerely thank Dr. J. Sokolowski (Faculty of Chemistry of the Jagiellonian University) for his help in the XRD experiments.

8.0 REFERENCES

1. T. Yamaguchi: 'Application of ZrO_2 as a Catalyst and a Catalyst Support', *Catal. Today,* **20**, 1994, pp.199-218.
2. G. Monrós, J. Carda, M. A. Tena, P. Escribano, F. Ribot and C. Sánchez: 'Vanadium Clusters in Doped ZrO_2-SiO_2 Toughened Ceramic Composites Obtained from Alkoxides', *Solid State Ionics,* **63-65**, 1993, pp.218-225.
3. A. Lewicki, T. Paryjczak and J. Rynkowski: 'Zirconium Dioxide in Catalysis. Properties and Applications', *Wiadomosci Chemiczne (Polish Chemical News),* **50**(11-12), 1996, pp.879-902.
4. T. Bak, X. Li, J. Nowotny, C. C. Sorrell and Z. Zhang: 'Surface Dynamics of Zirconia at Room Temperature', (in print).
5. T. Bak, X. Li, J. Nowotny, C. C. Sorrell and Z. Zhang: 'Work Function of Yttria-Stabilized Zirconia at 900°C', (in print).
6. A. F. Wells: *Structural Inorganic Chemistry*, (Polish version) WNiT, Warsaw, 1993.
7. K. V. R. Chary, B. Rama Rao and V. S. Subrahmanyam: 'Characterization of Supported Vanadium Oxide Catalysts by a Low Temperature Oxygen Chemisorption Technique III, The V_2O_5-ZrO_2 System', *Applied Catal.,* **74**(1), 1991, pp.1-13.
8. M. J. Torralvo, M. A. Alario and J. Soria: 'Crystallization Behaviour of Zirconium Oxide Gels', *Journal of Catal.,* **86**, 1984, pp.473-476.
9. O. Ruff and F. Ebert: 'Die Formen des Zirkondioxyds', *Z. Anorg Allg Chem.,* **180**, 1929, pp.19-41.
10. R. Cyprés, R. Wollast and J. Raucq: 'Contribution on the Polymorphic Conversion of Pure Zirconia', *Berichte Deutscher Keramischen Gesselschaft,* **40**, 1963, pp.527-532.
11. G. L. Clark and D. H. Reynolds: 'Chemistry of Zirconium Dioxide, X-Ray Diffraction Studies', *Ind. Eng. Chem.,* **29**, 1937, pp.711-715.
12. R. Srinivasan and B. H. Davis: 'Influence of Zirconium Salt Precursor on the Crystal Structures of Zirconia', *Catal. Letters,* **14**, 1992, pp.165-170.
13. E. D. Whitney: 'Kinetics and Mechanism of the Transition of Metastable

Tetragonal to Monoclinic Zirconia', *Trans. Farad. Society,* **61**, 1965, pp.1991-2000.

14. A. Clearfield: 'Crystalline Hydrous Zirconia', *Inorg. Chem.,* **3**(1), 1964, pp.146-148.
15. R. C. Garvie: 'Occurrence of Metastable Tetragonal Zirconia as a Crystalline Size Effect', *Journal of Physical Chemistry,* **69**(4), 1965, pp.1238-1243.
16. R. C. Garvie: 'Stabilization of the Tetragonal Structure in Zirconia Microcrystals', *Journal of Physical Chemistry,* **82**(2), 1978, pp.218-224.
17. P. D. L. Mercera, J. G. Van Ommen, E. B. M. Doesburg, A. J. Burggraaf and J. R. H. Ross: 'Zirconia as a Support for Catalysts. Evolution of the Texture and Structure on Calcination in Air', *Applied Catal.* **57**, 1990, pp.127-148.
18. T. Mitsuhashi, M. Ichihara and U. Tatsuke: Characterization and Stabilization of Metastable Tetragonal ZrO_2', *Journal of American Ceramic Society,* **57**(2), 1974, pp.97-101.
19. E. Tani, M. Yoshimura and S. Sómiya: 'Formation of Ultrafine Tetragonal ZrO_2 Powder Under Hydrothermal Conditions', *Journal of American Ceramics Society,* **66**(1), 1983, pp.11-14.
20. W. Stichert and F. Schüth: 'Influence of the Crystal Phase on the Catalytic Properties of Zirconia', *International Congress on Catalysis EuropaCat III* (poster), Kraków, 1997.
21. J. A. Jutson, R. M. Richardson, S. L. Jones and C. J. Norman: *Better Ceramics Through Chemistry IV,* MRS, 1990.
22. L. M. Zaitsev: 'Zirconium Hydroxides', *Zh. Nieorg. Khim., Russian Journal of Inorganic Chemistry,* **11**, 1966, pp.1684-1692.
23. J. Livage, K. Doi and C. Maziéres: 'Nature and Thermal Evolution of Amorphous Hydrated Zirconium Oxide', *Journal of American Ceramic Soc.,* **51**(6), 1968, pp.349-353.
24. Y. Murase, E. Kato and H. Matsumoto: 'Crystallization and Phase Transformation of Low Temperature Form Zirconia Prepared from Different Starting Materials', *Nippon Kagaku Kaishi,* **12**, 1972, pp.2329-2336.
25. T. Hirata, E. Asari and M. Kitajima: 'Infrared and Raman Spectroscopic Studies of ZrO_2 Polymorphs Doped with Y_2O_3 or CeO_2', *Journal of Solid State Chem.,* **110**, 1994, pp.201-207.
26. C. J. Howard, R. J. Hill and B. E. Reichert: 'Structures of the ZrO_2 Polymorphs at Room Temperatures by High Resolution Neutron Powder Diffraction', *Acta Cryst.,* **B44**, 1988, pp.116-119.
27. C. J. Norman, P. A. Goulding and I. Mcalpine: 'Role of Anions in the Surface Area Stabilization of Zirconia', *Catal. Today,* **20**, 1994, pp.313-322.
28. A. R. West: *Solid State Chemistry,* John Wiley & Sons, New York, 1992.
29. B. Aiken, W. P. Hsu and E. Matijevic: 'Preparation and Properties of Uniform Mixed and Coated Colloidal Particles. Part V. Zirconium Compounds', *Journal of Material Science,* **25**, 1990, pp.1886-1894.
30. S. H. Overbury, P. A. Bertrand and G. A. Somorjai: 'Surface Composition of Binary Systems, Prediction of Surface Phase Diagrams of Solid Solutions', *Chem. Review,* **75**, 1975, pp.547-560.
31. M. J. Hudson and J. A. Knowles: 'Preparation and Characterization of

Mesoporous, High-Surface-Area Zirconium (IV) Oxide', *Journal of Material Chem.*, **6**(1), 1996, pp.89-95.
32. R. K. Iler and H. J. Mcqueston: U.S. Patent No.4010 242, 1972.
33. R. Ukropec, J. G. Vanommen, P. D. L. Mercera, K. Seshan and J. R. H. Ross: 'Influence of Doped ZrO_2 on the Selective Oxidation of Methanol to Methylformate on Vanadium Oxide Catalysts', *Stud. Surface Science Catal.*, 1994, p.383.
34. H. Fujii, N. Mizuno and M. Misono: 'Pronounced Catalytic Activity of Lanthanum Strontium Cobalt Oxide ($La_{1-x}Sr_xCoO_3$) Highly Dispersed on Zirconium Dioxide for Complete Oxidation of Propane', *Chem. Letters*, 1987, pp.2147-2150.
35. T. Okuhara and M. Misono: *Inorganic Chemistry Encyclopedia,* R. B. King, ed., *Journal of Wiley*, Chichester, **6**, 1994.
36. T. Okuhara, T. Jin and J. M. White: *Applied Surface Science,* **29**, 1987, p.223.
37. T. Yamaguchi, M. Tan-no and K. Tanabe: *Preparation of Catalysts V,* Elsevier, Amsterdam, 1991.
38. S. Semainides and M. Stoukides: 'Catalytic Oxidation of Methane on Polycrystalline Palladium Supported Over Rhodium Catalysts', *Journal of Catal.*, **98**, 1986, pp.540-549.
39. P. Fornasiero, R. Dimonte, G. R. Rao, J. Kaspar, S. Meriani, A. Trovarelli and M. Graziani: 'Rh-Loaded CeO_2-ZrO_2 Solid Solutions as Highly Efficient Oxygen Exchangers: Dependence of the Reduction Behaviour and the Oxygen Storage Capacity on the Structural Properties', *Journal of Catal.,* **151**, 1995, pp.168-177.
40. T. Iizuka, Y. Tanaka and K. Tanabe: 'Hydrogenation of CO and CO_2 over Rhodium Catalysts Supported on Various Metal Oxides', *Journal of Catal.,* **76**, 1982, pp.1-8.
41. T. Iizuka, Y. Tanaka and K. Tanabe: 'Hydrogenation of Carbon Monoxide and Carbon Dioxide Over Supported Rhodium Catalysts', *Journal of Mol. Catal.*, **17**, 1982, pp.381-389.
42. T. Iizuka, M. Kojima and K. Tanabe: 'Hydrogenation of Carbon Monoxide and Carbon Dioxide over Supported Rhodium Catalysts', *Journal of Chemical Society, Chemical Communication,* 1983, pp.638-639.
43. B. Q. Xu, T. Yamaguchi and K. Tanabe: 'Formation of Acetonitrile from Triethyloamine on Oxide Catalysts', *Chemical Letters,* 1988, pp.1053-1056.
44. C. Morterra, G. Cerrato, V. Bolis, C. Lamberti, L. Ferroni and L. Montanaro: 'Surface Characterization of Yttria-Stabilized Tetragonal ZrO_2, Part 2. - Adsorption of CO', *Journal of Chemical Society Faraday Trans.,* **91**(1), 1995, pp.113-123.
45. C. Mazzocchia, E. Tempesti, P. Gronchi, L. Giuffré and L. Zanderighi: 'Hydrogenation of CO over ZrO_2-Supported Rh Catalysts: Role of Experimental Parameters in Modifying the C_2H_5OH/CH_4 Product Ratio', *Journal of Catal.,* **111**, 1988, pp.345-352.
46. Y. Nakano, T. Iizuka, H. Hattori and K. Tanabe: 'Surface Properties of Zirconium Oxide and Its Catalytic Activity for Isomerization of 1-Butene', *Journal of Catal.*, **57**, 1979, pp.1-10.

47. C. Morterra, G. Cerrato and L. Ferroni: 'Surface Characterization of Yttria-Stabilized Tetragonal ZrO_2, Part 3-CO_2 Adsorption and the CO_2-CO Interaction', *Journal of Chemical Society, Faraday Trans.*, **91**(1), 1995, pp.125-132.
48. T. Yamaguchi, K. Tanabe and Y. C. Kung: *Material Chemical Physics,* **16**, 1990, p.67.
49. J. M. Parera: 'Promotion of Zirconia Acidity by Addition of Sulfate Ion', *Catal. Today,* **15**, 1992, pp.481-490.
50. J. C. Yori, M. A. d'Amato, G. Costa and J. M. Parera: 'Isomerization of n-Butane on Pt/SO_4^{2-}-ZrO_2 and Mechanical Mixtures of Pt/Al_2O_3 + SO_4^{2-}-ZrO_2', *Journal of Catal.,* **153**, 1995, pp.218-223.
51. J. C. Yori and J. M. Parera: 'Isomerization of n-Butane over Ni/SO_4^{2-}-ZrO_2', *Applied Catal. A,* **129**, 1995, pp.83-91.
52. K. Arata and M. Hino: 'Preparation of Superacids by Metal Oxides and Their Catalytic Action', *Material Chemical Physics,* **26**, 1990, pp.213-237.
53. A. Clearfield, G. P. D. Serrette and A. H. Khazi-syed: 'Nature of Hydrous Zirconia and Sulfated Hydrous Zirconia', *Catal. Today,* **20**, 1994, pp.295-312.
53. A. Clearfield, G. P. D. Serrette and A. H. Khazi-syed: 'Nature of Hydrous Zirconia and Sulfated Hydrous Zirconia', *Catal. Today,* **20**, 1994, pp.295-312.
54. M. Y. He and J. G. Ekerdt: 'Methanol Formation on Zirconium Dioxide', *Journal of Catal.*, **90**, 1984, pp.17-23.
55. K. Sayama, A. Tanaka, K. Domen, K. Maruya and T. Onishi: 'Improvement of Nickel–Loaded Potassium Niobium Oxide ($K_4Nb_6O_{17}$) Photocatalyst for the Decomposition of Water', *Catal. Lett.,* **4**, 1990, pp.217-222.
56. K. Sayama and H. Arakawa: *Prepr. 70th Meeting of Catalysis,* Catalysis Society of Japan, 1992, 4F519.
57. K. Sayama and H. Arakawa: 'Photocatalytic Decomposition of Water and Photocatalytic Reduction of Carbon Dioxide Over Zirconia Catalyst', *Journal of Phys. Chem.,* **97**, 1993, pp.531-533.
58. K. Domen. A. Kudo, A. Shinozaki, A. Tanaka, K. Maruya and T. Onishi: 'Photodecomposition of Water and Hydrogen Evolution from Aqueous Methanol Solution over Niobate Photocatalysts', *Journal of Chemical Society, Chem. Communication,* 1986, pp.356-357.
59. K. Domen, A. Kudo, A. Shibata, A. Tanaka, K. Maruya and T. Onishi: 'Novel Photocatalysts: Ion-Exchanged $K_4Nb_6O_{17}$ with a Layer Structure', *Journal of Chemical Society, Chemical Communication,* 1986, pp.1706-1707.
60. A. Kudo, A. Tanaka, K. Domen, K. Maruya, M. Aika and T. Onishi: 'Photocatalytic Decomposition of Water Over NiO- $K_4Nb_6O_{17}$ Catalyst', *Journal of Catal.,* **111**, 1988, pp.67-76.
61. V. Adeeva, J. W. De Haan, J. Janchen, G. D. Lei, V. Schunemann, L. J. M. van De Ven, W. M. H. Sachtler and R. A. Van Santen: 'Acid Sites in Sulfated and Metal-Promoted Zirconium Dioxide Catalysts', *Journal of Catal.,* **151**, 1995, pp.364-372.
62. T. Yamaguchi: 'Recent Progress in Solid Superacids', *Applied Catal.,* **61**, 1990, pp.1-25.
63. K. Tanabe, H. Hattori and T. Yamaguchi: *Crit. Rev. Surface Chem.,* **1**, 1990, p.1.

64. K. Arata: 'Solid Superacids', *Advances Catal.*, **37**, 1990, pp.165-212.
65. I. Wang, W. F. Chang, R. J. Shiau, J. Ch. Wu and Ch. S. Chung: 'Nonoxidative Dehydrogenation of Ethylobenzene over TiO_2-ZrO_2 Catalysts. I. Effect of Composition', *Journal of Catal.*, **83**, 1983, pp.428-436.
66. J. B. Miller and E. I. Ko: 'Acidic Properties of Silica-Containing Mixed Oxide Aerogels: Preparation and Characterization of Zirconia-Silica and Comparison to Titania-Silica', *Journal of Catal.*, **159**, 1996, pp.58-68.
67. K. Tanabe, T. Sumiyoshi, K. Shibata, T. Kiyoura and J. Kitagawa: 'New Hypothesis Regarding the Surface Acidity of Binary Metal Oxides', *Bulletin Society Chemical Japan.*, **47**(5), 1974, pp.1064-1066.
68. K. Andersen: 'Gasification Process of Organic Wastewater', PCT International Applied WO 89 08, 1989, p.138.
69. D. E . W. Vaughan, R. J. Lussier and J. S. Magee: 'Smectic Clay Cracking Catalysts', Ger. Offen. 2 825 769, 1979, p.42.
70. R. Burch and C. I. Warburton: 'Zr-Containing Pillared Interlayer Clays. II. Catalytic Activity for the Conversion of Methanol into Hydrocarbons', *Journal of Catal.*, **97**, 1986, pp.511-515.
71. F. Figueras, A. Mattrod-bashi, G. Fetter, A. Thrierr and J. V. Zanchetta: 'Preparation and Thermal Properties of Zr-Intercalated Clays', *Journal of Catal.*, **119**, 1989, pp.91-96.
72. A. Adamski, Z. Sojka, K. Dyrek and M. Che: 'An XRD and ESR Study of V_2O_5/ZrO_2 Catalysts: Influence of the Phase Transitions of ZrO_2 on the Migration of V^{4+} Ions into Zirconia', *Solid State Ionics*, **117**, 1999, pp.113-122.
73. A. Adamski, Z. Sojka, K. Dyrek, M. Che, G. Wendt and S. Albrecht: 'Surface Heterogeneity of Zirconia-Supported V_2O_5 - The Link Between Structure and Catalytic Properties', *Langmuir*, **15**, 1999, pp.5733-5741.
74. S. Albrecht, G. Wendt, G. Lippold, A. Adamski and K. Dyrek: 'Characterization of Modified V_2O_5-ZrO_2 Catalysts for the Oxidative Dehydrogenation of Propane', *Solid State Ionics*, **101-103**, 1997, pp.909-914.
75. S. Albrecht: *Doktorarbeit (Ph.D. Thesis)*, Universität Leipzig, Leipzig 1998.
76. A. Adamski: *Praca Doktorska (Ph.D. Thesis)*, Jagiellonian University, Cracow 1999.

Predictive and Explanatory Phase Equilibria in Ceramic–Metal Bonding

C. C. SORRELL
School of Materials Science and Engineering
University of New South Wales, Sydney, NSW 2052
Australia

1.0 INTRODUCTION

In an earlier publication by the author,[1] the usefulness of phase equilibria data to ceramic-metal joining was described. This instructional paper reviewed the range of different types of phase equilibria data that are available or that can be derived readily. For each example, a reported ceramic-metal joining study was compared to a reported or derived phase diagram, with the aim of explaining the experimental results in terms of the phase equilibria. These comparisons were used to demonstrate the usefulness of phase equilibria data in explaining or predicting ceramic-metal joining combinations.

The present instructional work represents a continuation of these concepts. In this case, three systems have been selected for discussion. Each of these systems involved the bonding of nonoxide ceramics (BN and Si_3N_4) to metals (Ag, Sn and Cu) alloyed with a reactive metal (Ti). The systems studied were:

Ti-Ag-B-N

Ti-Sn-Si-N

Ti-Cu-Si-N

None of the relevant quaternary phase diagrams for these systems have been reported, although the majority of the binary and ternary subsystems have been determined at specific temperatures. For all three systems, quaternary diagrams at relevant temperatures have been derived using existing phase equilibria data and geometric considerations; no thermodynamic calculations were done.

The purpose of the present work is to describe how these complex phase equilibria data can be derived relatively simply using only existing phase equilibria data, geometric considerations and several assumptions. These tentative phase diagrams can be used to explain interfacial reactions observed during experimental studies and, perhaps more usefully, to allow prediction of potential alloys and alloy compositions appropriate for bonding to ceramics.

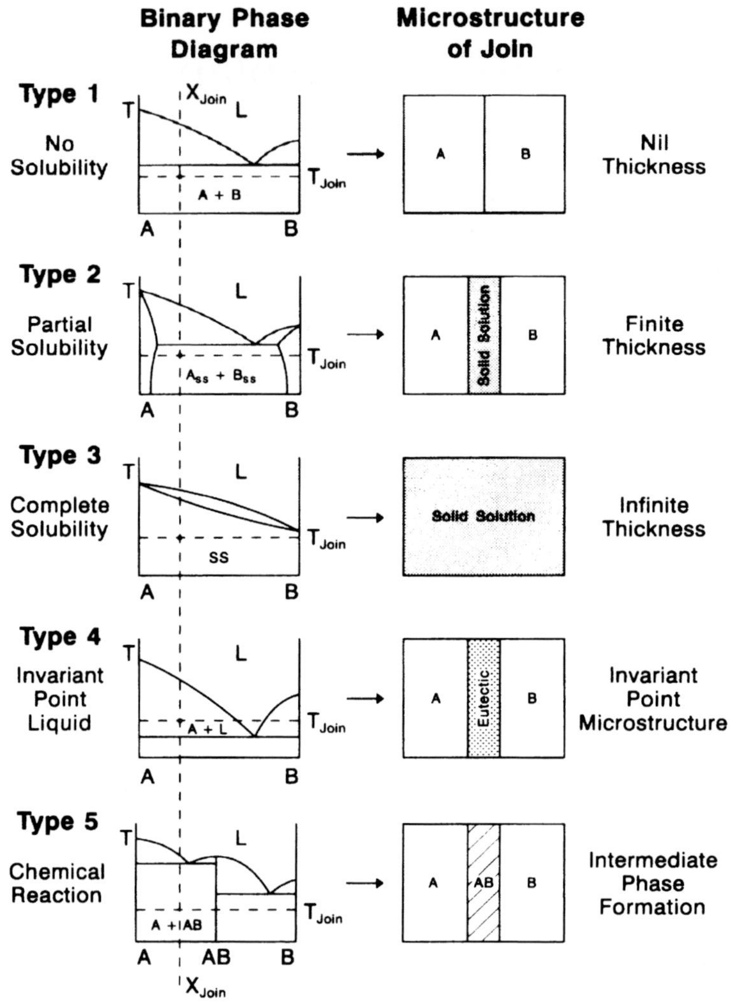

Fig. 1 Five basic interface types in terms of corresponding phase diagrams and indicative microstructures (join is of nominal composition).

2.0 PRINCIPLES

In terms of the phase equilibria and microstructure, there are five basic types of ceramic-metal interfaces, which are indicated in Table 1. Examples corresponding to these five types are shown in Figure 1.

Each of these five types of interfaces can be predicted beforehand or explained afterwards by using phase equilibria data because phase diagrams indicate compatibilities (solid-solid and solid-liquid), reactions, solubilities (solid and liquid),

Table 1 Five basic types of ceramic-metal interfaces.

Type	Chemical Behaviour of Phases	Chemical Nature of Interface	Physical Nature of Interface
1	Compatible	No Solubility	Nil Thickness
2	Compatible	Partial Solubility	Finite Thickness
3	Compatible	Compete Solubility	Potential Infinite Thickness
4	Compatible	Invariant Point Liquid	Invariant Point Microstructure
5	Not Compatible	Chemical Reaction	Intermediate Phase(s) Formation

liquid formation (melting and invariant point), and decomposition, both qualitatively and quantitatively.

Since most ceramics of interest in joining are binary compounds, combination with a metal generates a ternary system; the addition of a reactive metal expands it to a quaternary system. In general, most reported phase equilibria data are for binary and ternary systems owing to the complicated nature of higher order systems and the time commitment necessary to determine them. While isothermal quaternary systems can be illustrated with the use of tetrahedra, higher order systems are problematic since there are no suitable techniques for fully illustrating the equilibria in these systems.

In order to derive an isothermal, quaternary, condensed phase equilibrium diagram, ideally, it is necessary to know the following at the temperatures of interest (viz., at the temperature of joining, after cooling and during use). This information is summarised in Table 2.

There are few instances where all of these data are available. The binary and ternary phase equilibria data can be obtained from some of the many standard compilations of phase diagrams of metal[2,3] and ceramic[4] systems. The existence of reported quaternary compounds can be determined using the Landolt-Bornstein tables.[5] The existence of quaternary amorphous, liquid, or metastable phases is more difficult to determine since there are few data reporting them. The existence of sizable regions of solid solubility may be indicated on the ternary phase diagrams but it generally is unknown if these extend into the quaternary diagram. Therefore, the quaternary phase equilibria can be derived if certain assumptions are made:

1. The reported binary and ternary phase equilibria are correct.
2. Condensed phase equilibrium conditions exist, so no metastable phases are present and no mass is lost through volatilisation.
3. There are no unknown binary, ternary, or quaternary crystalline or amorphous phases.
4. The compatibilities reported at one temperature are identical to those at another (similar) temperature; for subsolidus equilibria, this generally is correct.

Table 2 Data necessary to determine an isothermal, quaternary, condensed phase equilibrium diagram.

Type	Necessary Data	Position on/in Tetrahedron
1	All Binary Phase Equilibria	Edges
2	All Ternary Phase Equilibria	Faces
3	Existence of Any Quaternary Compounds	Interior
4	Existence of Any Quaternary Amorphous or Liquid Phases	Interior
5	Existence of Any Metastable Phases	All Regions
6	Existence of Any Sizable Solid Solubility Regions	Faces & Interior

5. The formation of liquid and solid solution homogeneity regions does not alter the reported compatibilities; i.e., most liquid and solid solution regions can be reduced to point compounds that show unaltered compatibilities.

Fortunately, it frequently is the case that all of the data listed in Table 2 are not necessary owing to geometric requirements. Many quaternary systems demonstrate ternary compatibilities (tetrahedron faces) such that simple geometric considerations force the quaternary tie line configurations (tetrahedron interior) to an unambiguous array. The same situation often exists for ternary systems. In this case, where a ternary isothermal section is unknown, the binary edges may force the ternary tie line configuration to a fixed array.

The principles of derivation of phase diagrams are based upon the geometric requirements that an open angle formed by tie lines must be closed to create a triangle (ternary system or subsystem), a triangle (ternary system or subsystem) can contain only three phases in equilibrium and a tetrahedron (quaternary system or subsystem) can contain only four phases in equilibrium. These derivations are illustrated in Figure 2 and they are clarified in Table 3.

In Figure 2a, it can be seen that, if only two of the four ternary faces are known, the angle formed by the tie lines B-AC and AC-CD creates a plane, which thus requires compatibility between B and CD to close the triangle. In this way, it is possible to generate complete isothermal ternary diagrams in the absence of published data.

Figure 2b demonstrates that the existence of the compound ABC creates four phases (B + AC + CD + ABC) in a ternary subsystem (B-AC-CD), which is not allowed. Thus, CD-ABC compatibility is required to reduce the system to two ternary subsystems. In this way, it is possible to derive ternary diagrams and generate tie lines that are located in the interior of a quaternary tetrahedron, again in the absence of published data.

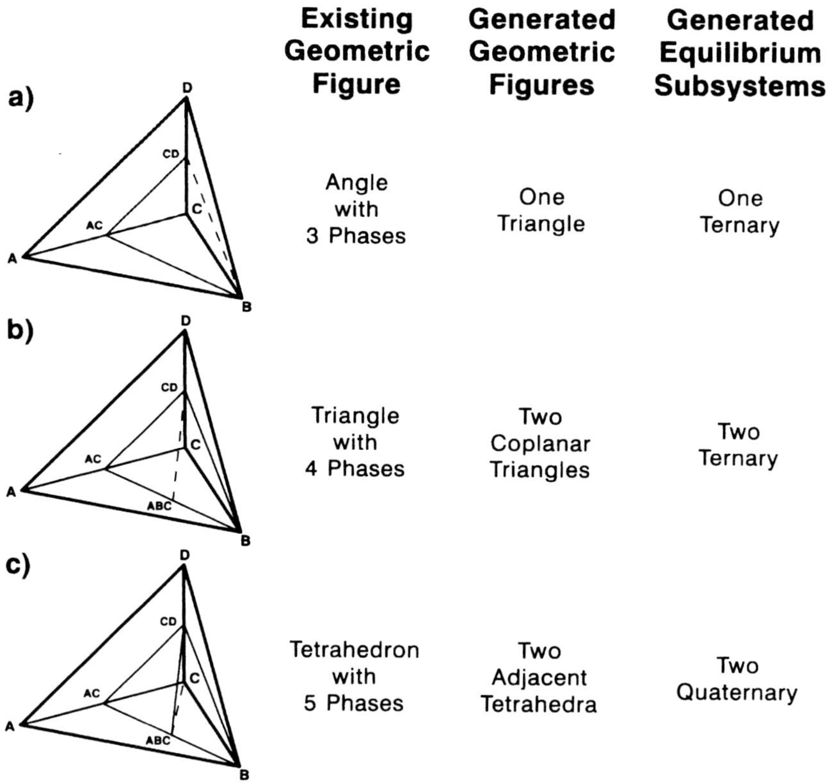

Fig. 2 Illustration of the requirements of geometric derivation of phase diagrams.

Finally, Figure 2c depicts five phases (B + C + AC + CD + ABC) in a quaternary subsystem (B-C-AC-CD). With CD-ABC compatibility being indicated, C-ABC compatibility is required to reduce the system to two quaternary subsystems. This also can generate unknown tie lines in ternary and quaternary systems.

These three steps show the simplicity of generating undetermined tie lines using the geometric requirements of the phase rule used in conjunction with known experimental data.[6] Complete ternary and quaternary diagrams can be derived by working across the diagram using only the three procedures shown in Figure 2. In many cases, these diagrams can be constructed in only one possible way. This can be confirmed by ensuring that all ternary compatibilities on the tetrahedron faces and within the tetrahedron are matched correctly like a puzzle as the complete diagram is resolved step by step. When one or more ternary· phases are unknown, this procedure can allow the generation of unknown ternary equilibria data with no ambiguity.

The specific procedures used to generate quaternary diagrams are as follows:

1. Draw the tetrahedron with all known intermediate phases present at the temperature of interest on the diagram edges.

Table 3 Clarification of geometric derivation of phase diagrams according to Figure 2.

Figure No.	Tie Lines		Geometric Figures		Existing Phases	Generated Equilibrium Subsystems	
	Existing	Generated	Existing	Generated		Types	Phases
2a	B - AC AC - CD	B - CD	Angle with 3 Phases	One Triangle	B AC CD	One Ternary	B - AC - CD
2b	B - CD B - ABC AC - CD AC - ABC	CD - ABC	Triangle with 4 Phases	Two Coplanar Triangles	B AC CD ABC	Two Ternary	B - CD - ABC AC - CD - ABC
2c	B - C B - CD B - ABC C - AC C - CD AC - CD AC - ABC CD - ABC	C - ABC	Tetrahedron with 5 Phases	Two Adjacent Tetrahedra	B C AC CD ABC	Two Quaternary	B - C - CD - ABC C - AC - CD - ABC

2. Concerning solid and liquid solutions, some care must be taken owing to the possibility of alteration of compatibilities resulting from the occlusion of tie lines. With limited homogeneity regions of only a few percent, there is little likelihood of alteration of the compatibilities. However, with more extensive homogeneity regions, the solubility region may extend significantly into the ternary and quaternary diagrams, thus making possible the occlusion of nearby tie lines. The guidelines are.
   a. *Limited Binary Solid or Liquid Solution*
      Reduce to a point compound
   b. *Limited Ternary Solid or Liquid Solution*
      Reduce to a point compound
   c. *Extensive Ternary Solid or Liquid Solution in a ternary corner of the diagram*
      Reduce to a point compound: ensure that the envelopment of other phases by the homogeneity region, if present, is incorporated
   d. *Extensive Ternary Solid or Liquid Solution on a binary edge of the diagram*
      Convert to a compatibility join consisting of the end members being the pure compound (on the edge of the diagram) and a saturated solution (in the interior of the diagram); ensure that the envelopment of other phases by the homogeneity region, if present, is incorporated.
4. Fill in the tetrahedron with all known ternary compatibilities on the diagram faces.
5. Locate angles formed by three phases; close these angles to generate the ternary subsystems. The new tie lines will create more angles, thus allowing the generation of more ternary subsystems.
6. If the phase diagram has not been completely resolved after all angles have been closed, then locate any triangles with four phases; since there is only one possibility for the joining of the necessary tie line, this resolves each four-phase subsystem into two ternary subsystems.
7. If the phase diagram has not been completely resolved after all four-phase triangles have been resolved into two ternary subsystems, then locate any tetrahedra with five phases; since there is only one possibility for the joining of the necessary tie line, this resolves each five-phase subsystem into two quaternary subsystems.
8. Confirm the correctness of the phase diagram in two steps:
   a. List all ternary subsystems on the four faces of the tetrahedral diagram and identify the apex (fourth phase) for the base of each triangular subsystem.
   b. Note all coplanar angles formed by two tie lines generated in the interior of the diagram. If the associated tetrahedra do not have a plane on one of the faces of the diagram, then these quaternary subsystems must be added to the list.

Ti-Ag-B-N SYSTEM AT 1000°C

Benko[7] studied the joining of Ti-1.5Ag to cubic BN at 1000°C under a vacuum of 3×10^{-3} Pa. The contact angle decreased from 100 to 0° in 15 min. Energy dispersive

Table 4 Binary and ternary subsystems in the quanternary system Ti-Ag-B-N.

Binary Subsystems (Edges)		Ternary Subsystems (Faces)	
Ti - Ag	Ag - B	Ti – Ag - B	Ti – B - N
Ti - B	Ag - N	Ti – Ag - N	Ag – B - N
Ti - N	B - N		

Table 5 Phases that are present at 1000°C according to reported binary phase equilibria data[2].

System	End Member	Intermediate Phases	End Member
Ti - Ag	βTi(Ag)	TiAg	Ag(liquid)
Ti - B	βTi	TiB TiB_2	βB
Ti - N	αTi(N) βTi(N)	Ti_2N TiN	N

spectroscopy indicated that the alloy composition at the interface had altered to Ti-2.0Ag. A chemical reaction at the interface was reported but no X-ray diffraction data were presented. The author stated that, according to thermodynamic calculations, the reaction products should be TiB_2 and TiN. Energy dispersive spectroscopy data indicated that the interface had become enriched in Ti, B and N.

There is no reported Ti-Ag-B-N phase diagram at any temperature. However, owing to the well known stabilities of TiB_2 and TiN, these phases would be expected to form, as suggested by the thermodynamic analysis. In the present case, if this quaternary phase diagram had been reported, it would have been possible for the author to have predicted the reaction products at the interface. Similarly, the author could have presented a more convincing assessment of the reaction products in the absence of confirmation by x-ray diffraction. Less than 10 min. were required to derive this quaternary phase diagram.

The Ti-Ag-B-N system consists of the subsystems as given in Table 4. Of the six binary phase diagrams, Massalski et al.[2] give reliable data only for the systems Ti-Ag, Ti-B and Ti-N. According to these data, the phases that should exist at 1000°C are as given in Table 5.

Massalski et al.[2] give a speculative pressure-temperature diagram for BN and brief mention is made of the reported compounds $B_{25}N$, B_3N and BN_6 which are unlikely to exist at 1000°C, if at all. In the Ag-B system, the compound AgB_2 was mentioned but appears unlikely to exist. In the Ag-N system, the compounds AgN_3 and Ag_3N were reported not to be stable above 300 and 25°C, respectively. From these assessments, it can be concluded that the only intermediate phase in these systems at 1000°C is BN.

Of the ternary subsystems, the minimal two systems (for angle generation) Ag-B-N at 800°C[8] and Ti-B-N at 1090[8] and 1500°C[9] have been reported. The two Ti-B-N diagrams are essentially identical, indicating that phase relations at widely differing temperatures can be identical.[1] No ternary compounds were reported to exist. The former system is of note because BN is the only intermediate phase, thus giving four phases in a ternary system and so requiring Ag-BN compatibility at all temperatures up to the decomposition temperature of BN. As will be seen, knowledge of the Ti-Ag-B and Ti-Ag-N system is not necessary (assuming no unknown ternary compounds) owing to the unambiguous nature of the geometrically derived quaternary phase diagram.

Figure 3 shows the relatively simple Ti-Ag-B-N phase diagram as derived solely by geometry based on experimental data. For an experimental join between Ti-1.5Ag and BN, it can be seen that the corresponding join on the phase diagram crosses eight tetrahedral subsystems:

Ag - BN - TiB_2 - TiN	Ag - TiB - Ti_4N_3 - Ti_3N_2
Ag - TiB_2 - TiN - Ti_3B_4	Ag - TiB - Ti_3N_2 - αTi
Ag - TiN - Ti_3B_4 - TiB	TiAg - TiB - Ag - αTi
Ag - TiB - TiN - Ti_4N_3	TiAg - TiB - αTi - βTi

For each tetrahedral subsystem that the join crosses, a reaction will occur at equilibrium. Consequently, a join between Ti-1.5Ag and BN at equilibrium could generate all of the following phases in approximate sequence:

BN - TiB_2 - TiN - Ag - Ti_3B_4 - TiB - Ti_4N_3 - Ti_3N_2 - αTi - TiAg - βTi

In principle, all of the above phases could form as the interface composition shifts from Ti-1.5Ag toward BN or *vice versa*, with a new phase or phases forming as the overall composition enters each compatibility tetrahedron. Such a join might be expected to be problematic owing to the potential formation of such a large number of intermediate phases and the resultant microstructural disruption and subsequent stresses arising from differential thermal expansion.

However, the phases that actually are observed depend principally on the thermodynamic stabilities, kinetics of formation and resultant reaction path. In more practical terms, the final phase assemblage is determined mainly by the thermodynamic stabilities of the intermediate phases. As stated, the most stable of these phases are TiB_2 and TiN, and so these would he expected to form relatively quickly and be retained. Since Figure 3 shows Ag-BN-TiB_2-TiN compatibility, inspection of this compatibility tetrahedron suggests that the interface composition would remain closer to BN than to the Ti alloy, so requiring Ti and/or Ag diffusion to the interface rather than B and/or N diffusion from BN. Since TiB_2 and TiN would be expected to form, an interface rich in Ti, B, and N would be expected, and this is what was observed.[7]

Table 6 lists the known compatibilities that were used to generate the unknown compatibilities. For each plane created by an angle of compatibilities dictated by the

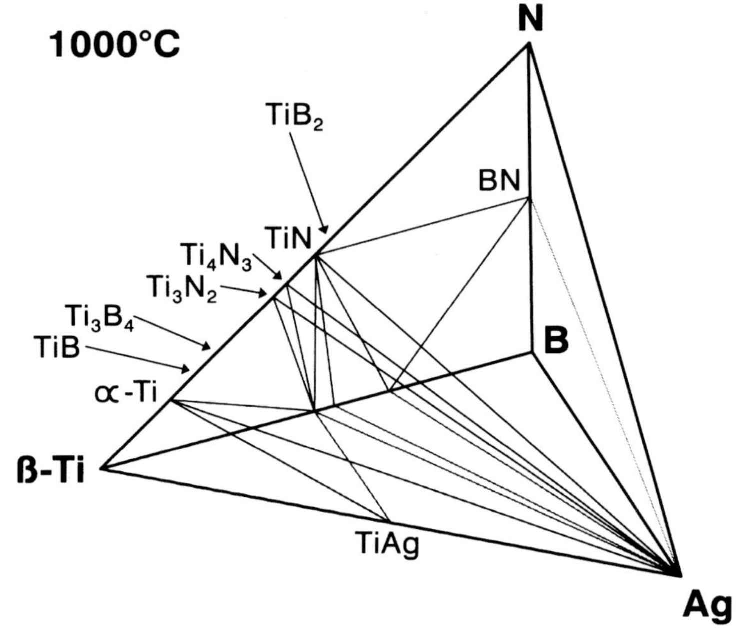

Fig. 3 Derived phase equilibria in the system Ti-Ag-B-N at 1000°C.

two known ternary diagrams (Ag-B-N and Ti-B-N), an unknown tie line was generated. The exception is Plane No. 8, which could not be derived this way because the triangle containing the unknown TiB-TiAg tie line contains only one known tie line (αTi-TiB). Hence, the Ag-TiB-βTi triangle, which contains a fourth phase (TiAg), dictates TiB-TiAg compatibility. The resultant compatibility tetrahedra are given in Table 7.

Another benefit of this type of analysis is the generation of unknown phase equilibria. In this case, Figure 3 shows that unknown isothermal sections have been generated for the systems Ti-Ag-B and Ti-Ag-N at 1000°C.

4.0 Ti-Sn-Si-N SYSTEM AT 950°C

Kapoor and Eagar[10] studied the reactive metal bonding of a range of Sn-based alloys (0.5, 2.0, 5.0 and 7.0 wt.% Ti) to Si_3N_4 at 950°C under a vacuum of 8×10^{-3} Pa. The nature of the Si_3N_4 was not specified. The contact angle decreased from 80-150° to ~10-40° in 60 min. for the alloys containing 2.0-7.0 wt.% Ti; the alloy with 0.5 wt.% Ti did not wet the Si_3N_4. Energy dispersive spectroscopy indicated that the interface matrix composition consisted of pure Sn with Ti_6Sn_5 interspersed within it. No X-ray diffraction data were presented. It was noted that, thermodynamically,

Table 6 Existing and generated tie lines in the Ti-Ag-B-N system at 1000°C.

Plane No.	Existing Tie Lines	Generated Tie Line
1	Ag - BN - TiB_2	Ag - TiB_2
2	Ag - TiB_2 - TiN	Ag - TiN
3	Ag - TiN - Ti_3B_4	Ag - Ti_3B_4
4	Ag - TiN - TiB	Ag - TiB
5	Ag - TiB - Ti_4N_3	Ag - Ti_4N_3
6	Ag - TiB - Ti_3N_2	Ag - Ti_3N_2
7	Ag - TiB - αTi	Ag - αTi
8	Ag - TiB - βTi	TiB - TiAg
9	αTi - TiB - TiAg	αti - TiAg

Table 7 Quaternary compatibilities in the system Ti-Ag-B-N system at 1000°C.

Tetrahedron No.	Compatibility Tetrahedron
1	Ag - BN - βB - TiB_2
2	Ag - BN - TiB_2 - TiN
3	Ag - BN - TiN - N
4	Ag - TiN - TiB_2 - Ti_3B_4
5	Ag - TiN - Ti_3B_4 - TiB
6	Ag - TiB - TiN - Ti_4N_3
7`	Ag - TiB - Ti_4N_3 - Ti_3N_2
8	Ag - TiB - Ti_3N_2 - αTi
9	TiAg - TiB - Ag - αTi
10	TiAg - TiB - αTi - βTi

TiN is more stable than Si_3N_4 at these temperatures and so should decompose the latter, although this was not observed.

There is no reported Ti-Sn-Si-N phase diagram at any temperature. However, the well known stabilities of Ti_5Si_3, $TiSi_2$ and TiN suggest that these phases could be expected to form. Again, the existence of this quaternary phase diagram would allow the prediction of the reaction products at the interface. Likewise, the description of the reaction product (Ti_6Sn_5) without confirmation by X-ray diffraction would

Table 8 Binary and ternary subsystems in the quaternary system Ti-Sn-Si-N.

Binary Susbsystems (Edges)		Ternary Subsystems (Faces)	
Ti - Sn	Sn - Si	Ti - Sn - Si	Ti - Si - N
Ti - Si	Sn - N	Ti - Sn - N	Sn - Si - N
Ti - N	Si - N		

Table 9 Phases that are present at 950°C according to reported binary phase equilibria data.[2]

System	End Member	Intermediate Phases	End Member
Ti - Sn	βTi(Sn)	Ti_3Sn Ti_2Sn Ti_5Sn_3 βTi_6Sn_5	Sn (Liquid)
Ti - Si	βTi(Si)	Ti_3Si Ti_5Si_3 Ti_5Si_4 TiSi $TiSi_2$	Si
Ti - N	αTi(N) βTi(N)	Ti_2N TiN	N
Sn - Si	Sn (Liquid)	None	Si
Si - N	Si	Si_3N_4	N

have been more persuasive. Approximately 30 min. were required to derive this quaternary phase diagram.

The system Ti-Sn-Si-N consists of the subsystems as given in Table 8. Massalski et al.[2] give reliable data for all of the binary systems except Sn-N. According to these data, the phases that should exist at 950°C are as given in Table 9.

For the Sn-N system, it was stated that the compound Sn_3N_4 decomposes at 800°C and liquid Sn does not dissolve N at 800°C.[2] Also, at 950°C, liquid Sn dissolves only ~2 mol.% Si.[2] These solubility data are useful to know because the expected large homogeneity region of liquid Sn, which melts at 232°C, could occlude any nearby tie lines. However, the Ti-Sn-N isothermal section at 900°C[11] confirms that this does not happen. Liquid Sn dissolves only Ti (~29 mol.%) at 900°C, so no homogeneity region projects into the quaternary diagram. That is, the homogeneity region for liquid Sn, which effectively dissolves Ti but not Si or N, can be considered a binary solid solution and therefore represented as a point compound.

Three of the ternary subsystems have been reported, these being Ti-Sn-N at 900°C,[11] Ti-Si-N at 1000°C[8] and Sn-Si-N at 200°C.[8] The former two systems are notable for the existence of significant solid solutions of N in Ti_3Sn (12 mol.%) and N in Ti_5Si_3 (4 mol.%), respectively. These solid solutions show narrow homogeneity regions that can be represented effectively as lines.

Many M_5Si_3 compounds, where M is a metal, are important in these derivations because these *Nowotny phases*[12-15] very commonly dissolve substantial amounts of B, C, N and O (X phases), thus projecting their homogeneity regions (effectively lines) into the ternary diagrams. Their potential to alter the compatibilities due to occlusion of nearby tie lines necessitates their retention rather than reduction to point compounds. In these cases, they can be simplified to two point compounds, the end members being the pure and saturated phases, such as M_5Si_3 and $M_5Si_3(X)$, respectively.

No stoichiometric ternary compounds were reported in the Ti-Sn-N and Ti-Si-N systems. As with the Ti-Ag-B-N system, knowledge of the ternary systems Ti-Sn-Si and Sn-Si-N (assuming no unknown ternary compounds) is not necessary owing to the unambiguous nature of the resultant quaternary phase diagram, which results from the geometric requirements of the derivation.

Figure 4 shows the Ti-Sn-Si-N phase diagram as derived solely by geometry based on experimental data. For experimental joins between the Ti-Sn alloys (0.5-7.0 wt.% Ti) and Si_3N_4, it is clear that the corresponding joins on the phase diagram cross only three quaternary subsystems:

Sn - Si_3N_4 - $TiSi_2$ - TiN

Sn - $TiSi_2$ - TiN - $Ti_5Si_3(N)$

Sn - TiN - $Ti_5Si_3(N)$ - βTi_6Sn_5

As for the Ti-Ag-B-N system, for each tetrahedral subsystem that the join crosses, a reaction will occur under equilibrium conditions. Thus, joins between the Ti-Sn alloys and Si_3N_4 at equilibrium could generate the following phases in approximate sequence:

Si_3N_4 - $TiSi_2$ - TiN - Sn - $Ti_5Si_3(N)$ - βTi_6Sn_5

Table 10 lists the known compatibilities that were used to generate the unknown compatibilities. For each plane created by an angle of compatibilities dictated by the two known ternary diagrams (Ti-Sn-N and Ti-Si-N), an unknown tie line was generated. This was done for the entire quaternary system by starting with the Si corner and working out toward the Ti-Sn edge. There was no necessity of using the existence of triangles with four phases or tetrahedra with five phases to resolve the system. Confirmation of these compatibilities is given in Tables 11a-f, which give the resultant compatibility tetrahedra.

This situation is interesting because the lines between the alloys and Si_3N_4 pass above the many compatibility tetrahedra that radiate from the Sn-$Ti_5Si_3(N)$ tie line and because none of the thermodynamically stable phases Ti_5Si_3, $TiSi_2$, or TiN were observed.[10] Since only Ti_6Sn_5 was observed, this suggests that this phase is thermodynamically stable and that the progress of the ceramic-metal reaction path was limited to the metal-rich end of the alloy-Si_3N_4 line, ending in the TiN-Sn-βTi_6Sn_5-$Ti_5Si_3(N)$ quaternary subsystem. Also, the presence of only one reaction product indicates that the joins formed were non-equilibrium or that the microstructural analysis was incomplete.

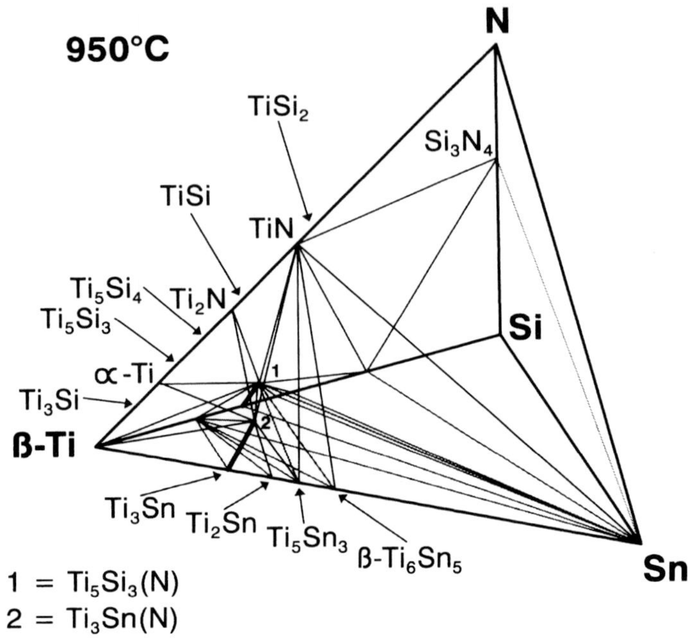

Fig. 4 Derived phase equilibria in the system Ti-Sn-Si-N at 1000°C.

Table 10 Existing and generated tie lines in the system Ti-Sn-Si-N at 950°C.

Plane No.	Existing Tie Lines	Generated Tie Line
1	Sn - Si_3N_4 - $TiSi_2$	Sn - $TiSi_2$
2	Sn - $TiSi_2$ - Ti_5Si_3(N)	Sn - Ti_5Si_3(N)
3	Sn - Ti_5Si_3 (N) - TiSi	Sn - TiSi
4	Sn - Ti_5Si_3 (N) - Ti_5Si_4	Sn - Ti_5Si_4
5	Sn - Ti_5Si_3 (N) - Ti_5Si_3	Sn - Ti_5Si_3
6	Sn - Ti_5Si_3 (N) - Ti_3Si	Sn - Ti_3Si
7	Ti_5Si_3(N) - TiN - βTi_6Sn_5	Ti_5Si_3(N) - βTi_6Sn_5
8	Ti_5Si_3(N) - TiN - Ti_5Sn_3	Ti_5Si_3(N) - Ti_5Sn_3
9	Ti_5Si_3(N) - TiN - Ti_3Sn(N)	Ti_5Si_3(N) - Ti_3Sn(N)
10	Ti_3Si - Ti_5Si_3(N) - βTi_6Sn_5	Ti_3Si - βTi_6Sn_5
11	Ti_3Si - Ti_5Si_3(N) - Ti_5Sn_3	Ti_3Si - Ti_5Sn_3
12	Ti_3Si - Ti_5Si_3(N) - Ti_3Sn(N)	Ti_3Si - Ti_3Sn(N)
13	Ti_3Si - Ti_3Sn(N) - Ti_2Sn	Ti_3Si - Ti_2Sn
14	Ti_3Si - Ti_3Sn(N) - Ti_3Sn	Ti_3Si - Ti_3Sn

Table 11a Derived compatibility tetrahedra in the system Ti-Sn-Si-N at 950°C.

Base System Ti-Si-N		
Tetrahedron No.	**Face Triangle**	**Apex**
1	Si_3N_4 - Si - $TiSi_2$	Sn
2	Si_3N_4 - $TiSi_2$ - TiN	Sn
3	Si_3N_4 - TiN - N	Sn
4	$Ti_5Si_3(N)$ - TiN - $TiSi_2$	Sn
5	$Ti_5Si_3(N)$ - $TiSi_2$ - TiSi	Sn
6	$Ti_5Si_3(N)$ - TiSi - Ti_5Si_4	Sn
7	$Ti_5Si_3(N)$ - Ti_5Si_4 - Ti_5Si_3	Sn
8	$Ti_5Si_3(N)$ - Ti_5Si_3 - Ti_3Si	Sn
9	$Ti_5Si_3(N)$ - TiN - Ti_2N	$Ti_3Sn(N)$
10	$Ti_5Si_3(N)$ - Ti_2N - αTi	$Ti_3Sn(N)$
11	$Ti_5Si_3(N)$ - αTi - βTi	$Ti_3Sn(N)$
12	$Ti_5Si_3(N)$ - βTi - Ti_3Si	$Ti_3Sn(N)$

Table 11b Derived compatibility tetrahedra in the system Ti-Sn-Si-N at 950°C.

Base System Ti-Si-N		
Tetrahedron No.	**Face Triangle**	**Apex**
1	Sn - Si - Si_3N_4	$TiSi_2$
3	Sn - Si_3N_4 - N	TiN

If Sn and βTi_6Sn_5 are the only phases present at the interface, this indicates that very little inward diffusion took place and that the alloy has remained a binary alloy and equilibrated in the Ti-Sn system. Since Ti was found to be absent from the main body of the alloy, being concentrated in an interfacial layer,[10] this would require partitioning of the phases into a non-equilibrium mixture of Ti + βTi_6Sn_5 + Sn. Diffusion of Ti to the interface should result in the formation of TiN, so its absence in the final phase assemblage is surprising in light of its considerable thermodynamic stability. Since data for the distribution of N were not given, it is possible that this phase was overlooked. However, this does not explain the failure to detect $Ti_5Si_3(N)$, which also is relatively stable, since an X-ray map for Si was given.

These observations suggest that the system was not at equilibrium and therefore the phase diagram would be of more limited usefulness in explaining the results. This conclusion can be considered to be appropriate in light of the limited experimental data presented.[10]

Table 11c Derived compactability tetrahedra in the system Ti-Sn-Si-N at 950°C.

Base System Sn-Ti-N		
Tetrahedron No.	**Face Triangle**	**Apex**
3	TiN - Sn - N	Si_3N_4
13	TiN - Sn - βTi_6Sn_5	$Ti_5Si_3(N)$
14	TiN - βTi_6Sn_5 - Ti_5Sn_3	$Ti_5Si_3(N)$
15	$Ti_3Sn(N)$ - TiN - Ti_5Sn_3	$Ti_5Si_3(N)$
16	$Ti_3Sn(N)$ - Ti_5Sn_3 - Ti_2Sn	Ti_3Si
17	$Ti_3Sn(N)$ - Ti_2Sn - Ti_3Sn	Ti_3Si
18	$Ti_3Sn(N)$ - Ti_3Sn - βTi	Ti_3Si
11	$Ti_3Sn(N)$ - βTi - αTi	$Ti_5Si_3(N)$
10	$Ti_3Sn(N)$ - αTi - Ti_2N	$Ti_5Si_3(N)$
9	$Ti_3Sn(N)$ - Ti_2N - TiN	$Ti_5Si_3(N)$

Table 11d Derived compactability tetrahedra in the system Ti-Sn-Si-N at 950°C.

Base System Sn-Si-Ti		
Tetrahedron No.	**Face Triangle**	**Apex**
1	Sn - Si - $TiSi_2$	Si_3N_4
5	Sn - $TiSi_2$ - TiSi	$Ti_5Si_3(N)$
6	Sn - TiSi - Ti_5Si_4	$Ti_5Si_3(N)$
7	Sn - Ti_5Si_4 - Ti_5Si_3	$Ti_5Si_3(N)$
8	Sn - Ti_5Si_3 - Ti_3Si	$Ti_5Si_3(N)$
19	Ti_3Si - Sn - βTi_6Sn_5	$Ti_5Si_3(N)$
20	Ti_3Si - βTi_6Sn_5 - Ti_5Sn_3	$Ti_5Si_3(N)$
16	Ti_3Si - Ti_5Sn_3 - Ti_2Sn	$Ti_3Sn(N)$
17	Ti_3Si - Ti_2Sn - Ti_3Sn	$Ti_3Sn(N)$
18	Ti_3Si - Ti_3Sn - βTi	$Ti_3Sn(N)$

Table 11e Derived compatibility tetrahedra in the system Ti-Sn-Si-N at 950°C.

Interior Tie Lines	
Line No.	**End Member**
1	$Ti_5Si_3(N)$ - Sn
2	$Ti_5Si_3(N)$ - Ti_6Sn_5
3	$Ti_5Si_3(N)$ - Ti_5Sn_3
4	$Ti_3Sn(N)$ - $Ti_5Si_3(N)$
5	$Ti_3Sn(N)$ - Ti_3Si

Table 11f Derived compatibility tetrahedra in the system Ti-Sn-Si-N at 950°C.

Base System Formed by Two Interior Tie Lines		
Tetrahedron No.	**Face Triangle**	**Apex**
13	Ti_5Si_3(N) - Sn - βTi_6Sn_5	TiN
19	Ti_5Si_3(N) - Sn - βTi_6Sn_5	Ti_3Si
14	Ti_5Si_3(N) - βTi_6Sn_5 - Ti_5Sn_3	TiN
20	Ti_5Si_3(N) - βTi_6Sn_5 - Ti_5Sn_3	Ti_3Si
15	Ti_5Si_3(N) - Ti_5Sn_3 - Ti_3Sn(N)	TiN
21	Ti_5Si_3(N) - Ti_5Sn_3 - Ti_3Sn(N)	Ti_3Si
12	Ti_5Si_3(N) - Ti_3Sn(N) - Ti_3Si	βTi
22	Ti_5Si_3(N) - Ti_3Sn(N) - Ti_3Si	Ti_5Sn_3

Although useful information still can be obtained since non-equilibrium and equilibrium systems often show similar phase equilibria,[1] fuller consideration of the phase diagram can explain the results and allow the conclusion that equilibrium in fact was reached. Examination of other data[10] shows that it is probable that this system achieved equilibrium. A study of the wetting angles during isothermal holds at 950°C for 60 min. was made, with the result that no changes were observed. Therefore, it can be assumed that equilibrium was reached, despite the microscopy results.

The joining study[10] indicated that the diffusion of Ti and/or Sn toward Si_3N_4 was somewhat limited and that there was no diffusion of Si and/or N from Si_3N_4 toward the alloys. These data can be interpreted straight forwardly by assuming the existence of limited alloy diffusion toward the ceramic, which would result in the presence of βTi_6Sn_5 and Sn. However, since the overall composition would be limited to regions close to the βTi_6Sn_5-Sn tie line but within the TiN-Sn-βTi_6Sn_5-Ti_5Si_3(N) quaternary subsystem, lever rule[6] considerations suggest that the levels of TiN and Ti_5Si_3(N) would be very low and probably below the limits of detection of the instrument used.

Figure 4 shows that an unknown isothermal section has been generated for the Ti-Sn-Si system at 950°C.

Ti-Cu-Si-N SYSTEM AT 800°C

Kim et al.[16] studied the reactivity, wettability and bonding of a range of Cu-based alloys (0.5, 1.0, 3.0, 5.0, 10, 15 and 50 wt.% Ti) to Si_3N_4 at 940-1160°C under a vacuum of $< 6.7 \times 10^{-3}$ Pa. The Si_3N_4 was pressureless sintered PLS-Si_3N_4 produced by Kyocera Corporation. Spreading was observed for Ti levels ≥ 3 wt.% (≥ 4 mol.%),

Table 12 Binary and ternary subsystems in the quaternary system Ti-Cu-Si-N.

Binary Subsystems (Edges)	Ternary Subsystems (Faces)
Ti - Cu	Ti - Cu - Si
Ti - Si	Ti - Cu - N
Ti - N	Ti - Si - N
Cu - Si	Cu - Si - N
Cu - N	
Si - N	

near-complete wetting for Ti levels ≥ 15 wt.% (≥ 19 mol.%), and complete wetting for the Ti level of 50 wt.% (57 mol.%). All samples were held at temperature for 60-120 min. A reaction layer was observed for all of the samples, with the thickness decreasing as the Ti level increased. The intermediate phases formed at the interface were determined by X-ray diffraction to consist of Ti_5Si_3 and TiN in the low-Ti alloys; CuSiTi was observed only in the Cu50Ti alloy.

The system Ti-Cu-Si-N is complicated owing to the large number of phases, particularly ternary, that exist. Thus, there is no reported phase diagram at any temperature. This diagram would be very useful because there is considerable interest in brazing with Cu-Ti alloys.[17-21] The derivation of this phase diagram was considerably more complicated than required for the previous two systems; over 8 hours were required to resolve this system.

The system Ti-Cu-Si-N consists of the subsystems as given in Table 12. Massalski et al.[2] give reliable data for all of the binary systems except Cu-N. A temperature lower than those used for joining (940-1160°C) was chosen since all of the intermediate phases melt or decompose at ≤ 869°C. A temperature below the solidus (802°C) was selected in order to show subsolidus phase relations that would be a reflection of the solidified join. The phases that should exist at 800°C are as given in Table 13.

For the Cu-N system, Massalaski et al.[2] mentioned work reporting the possible existence of the compounds Cu_3N, CuN_3 and CuN_6. It is unlikely that these phases exist at 800°C, if at all. There are several compounds with limited stability ranges in these systems. As will be seen, the geometry of the system Ti-Cu-Si-N at 800°C is such that the overall compatibilities do not change even when these compounds are present at other temperatures. The compounds of note are:

$TiCu_2$	Stable at 870 to ~890°C
$Ti_5N_3(\delta')$	Decomposes at 800°C
Cu_3Si	Melts congruently at 859°C
$Cu_{15}Si_4(\varepsilon)$	Decomposes at ~800°C
$Cu_{17}Si_4(\delta)$	Stable at 710 to 825°C
$Cu_5Si(\gamma)$	Decomposes at ~730°C
$Cu_6Si(\beta)$	Stable at ~785 to 852°C
$Cu_7Si(\kappa)$	Stable at 552 to ~840°C

Table 13 Phases that are present at 800°C according to reported binary phase equilibria data.[2]

System	End Member	Intermediate Phases	End Member
Ti - Cu	αTi βTi	Ti_2Cu TiCu Ti_3Cu_4 Ti_2Cu_3 $\beta TiCu_4$	Cu
Ti - Si	αTi(Si)	Ti_3Si Ti_5Si_3 Ti_5Si_4 TiSi $TiSi_2$	Si
Ti - N	αTi(N)	Ti_2N TiN	N
Cu - Si	Cu(Si)	Cu_3Si $Cu_{17}Si_4$ Cu_6Si Cu_7Si	Si
Si - N	Si	Si_3N_4	N

All four of the ternary subsystems have been reported, these being Ti-Cu-Si at 800°C,[3] Ti-Cu-N at 850°C,[22] Ti-Si-N at 1000°C,[8] and Cu-Si-N at 700°C.[8] Three of these systems are notable for important ternary compounds that influence the compatibilities:

1. Ti-Cu-Si There are two ternary compounds $CuSi_{19}Ti_{10}$ and CuSiTi.
   The Nowotny phase Ti_5Si_3(Cu) effectively acts as a ternary compound.
2. Ti-Cu-N There is one ternary compound Ti_3CuN.
3. Ti-Si-N The Nowotny phase Ti_5Si_3(N) effectively acts as a ternary compound.

Since Ti_5Si_3(Cu) and Ti_5Si_3(N) exist as Nowotny phases, it is likely that these two phases are mutually soluble. This creates the plane Ti_5Si_3-Ti_5Si_3(Cu)-Ti_5Si_3(N), which enters into the quaternary diagram, effectively creating a quaternary phase.

Since Cu dissolves substantial amounts of Ti and Si, it is possible that a homogeneity region exists in the Cu corner, although it is unlikely to dissolve N. Thus, this homogeneity region, limited to the Ti-Cu-Si plane, does not enter into the quaternary diagram and can be reduced to a point composition (pure Cu).

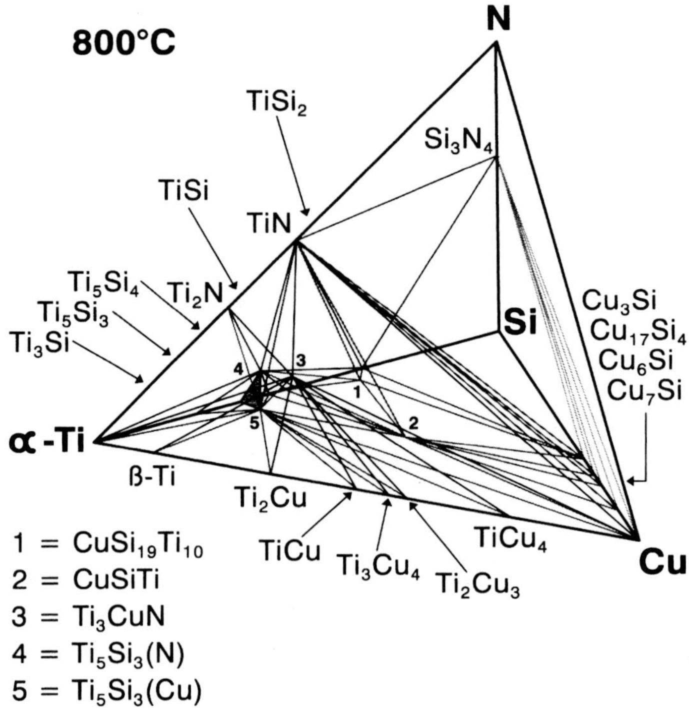

Fig. 5 Derived phase equilibria in the system Ti-Cu-Si-N at 800°C.

Figure 5 shows the Ti-Cu-Si-N phase diagram as derived solely by geometry based on experimental data. All of the alloy compositions, except Cu-50Ti (57 mol.% Ti), lie between $TiCu_4$ and Cu; the latter lies between Ti_2Cu and TiCu.

Although a relevant quaternary phase has been reported,[23, 24] its existence is rejected since the crystallographic indices for this phase match those of Ti_3CuN.[25]

For experimental joins between the Ti alloys and Si_3N_4, the situation is more complicated than for the previous two cases (Figures 3 and 4), so a more descriptive method of illustration is necessary. This is based on the five following principles:

1. The alloy-ceramic line crosses a number of tetrahedra.
2. The line enters and/or exits each tetrahedron through a ternary plane.
3. Each ternary plane is common to two adjacent tetrahedra.
4. The potential reaction path can be determined by working through the series of adjacent tetrahedral, using the identical plane pairs as guides.
5. When each pair of identical adjacent planes is located, the apices of the two adjacent tetrahedra can be identified.

Tables 14a and b illustrate the procedures used to determine the reaction paths for both the low-Ti and the Cu-50Ti alloys. These tables illustrate the potential reaction path sequence and products for an equilibrium join between the Ti-Cu alloys and Si_3N_4.

Table 14a Compatibility tetrahedra crossed by lines joining low-Ti alloys to Si_3N_4 at 800°C.

Tetra-hedron No.	Ceramic End	← Potential Reaction Path →											Metal End
	Si_3N_4		TiN		$CuSi_{19}Ti_{10}$		$Cu_{17}Si_4$		Cu_7Si		Ti_5Si_3 (Cu)		$\beta TiCu_4$
		$TiSi_2$		Cu_3Si		CuSiTi		Cu_6Si		Cu		Ti_3CuN	
2													
26		B	B	B	A								
27			B	B	B	A							
28			B	B		B	A						
29			B			B	B	A					
30			B			B		B	A				
31			B			B			B	A			
32			B			B				B	A		
16			B							B	B	A	
17										B	B	B	A

A = Phase that serves as the apex of the tetrahedron
B = Phases that form the base plane shared by adjacent tetrahedra

Table 14b Compatibility tetrahedra crossed by line joining Cu50Ti to Si_3N_4 at 800°C.

Tetra-hedron No.	Ceramic End		← Potential Reaction Path →												Metal End
	Si_3N_4		TiN		$CuSi_{19}Ti_{10}$		$Ti_5Si_3(N)$		Cu		$\beta TiCu_4$		Ti_3Cu_4		Ti_2Cu
		$TiSi_2$		Cu_3Si		CuSiTi		Ti_5Si_3 (Cu)		Ti_3CuN		Ti_2Cu_3		TiCu	
2															
26		B	B	B	A										
27			B	B	B	A									
42			B		B	B	A								
43			B			B	B	A							
32			B			B		B	A						
16			B					B	B	A					
17								B	B	B	A				
18								B		B	B	A			
19								B		B		B	A		
20								B		B			B	A	
21								B		B				B	A

A = Phase that serves as the apex of the tetrahedron
B = Phases that form the base plane shared by adjacent tetrahedra

Table 15a Derived compatibility tetrahedra in the system Ti-Cu-Si-N at 800°C.

Base System Ti-Si-N		
Tetrahedron No.	**Free Triangle**	**Apex**
1	Si_3N_4 - Si - $TiSi_2$	Cu_3Si
2	Si_3N_4 - $TiSi_2$ - TiN	Cu_3Si
3	Si_3N_4 - TiN - N	Cu
4	$Ti_5Si_3(N)$ - TiN - $TiSi_2$	$CuSi_{19}Ti_{10}$
5	$Ti_5Si_3(N)$ - $TiSi_2$ - TiSi	$CuSi_{19}Ti_{10}$
6	$Ti_5Si_3(N)$ - TiSi - Ti_5Si_4	CuSiTi
7	$Ti_5Si_3(N)$ - Ti_5Si_4 - Ti_5Si_3	$Ti_5Si_3(Cu)$
8	$Ti_5Si_3(N)$ - Ti_5Si_3 - Ti_3Si	$Ti_5Si_3(Cu)$
9	$Ti_5Si_3(N)$ - Ti_3Si - αTi	$Ti_5Si_3(Cu)$
10	$Ti_5Si_3(N)$ - αTi - Ti_2N	$Ti_5Si_3(Cu)$
11	$Ti_5Si_3(N)$ - Ti_2N - TiN	$Ti_5Si_3(Cu)$

Table 15b Derived compatibility tetrahedra in the system Ti-Cu-Si-N at 800°C.

Base System Cu-Si-N		
Tetrahedron No.	**Face Triangle**	**Apex**
1	Si_3N_4 - Si - Cu_3Si	$TiSi_2$
12	Si_3N_4 - Cu_3Si - $Cu_{17}Si_4$	TiN
13	Si_3N_4 - $Cu_{17}Si_4$ - Cu_6Si	TiN
14	Si_3N_4 - Cu_6Si - Cu_7Si	TiN
15	Si_3N_4 - Cu_7Si - Cu	TiN
3	Si_3N_4 - Cu - N	TiN

Tables 15a-f list the generated compatibility tetrahedra, which were derived using the principles involving angles to generate ternary subsystems, triangles containing four phases to generate two ternary subsystems, and tetrahedra containing five phases to generate two quaternary tetrahedra.

The experimental study of this system[10] indicated three phenomena that may be explained by the phase diagram:

1. The intermediate phases formed at the interfaces were Ti_5Si_3, TiN and CuSiTi.
2. There is a direct correlation between the degree of spreading and the Ti level.
3. There is a reverse correlation between the reaction layer thickness and the Ti level.

Table 15c Derived compatibility tetrahedra in the system Ti-Cu-Si-N at 800°C.

Base System Cu-Ti-N		
Tetrahedron No.	**Face Triangle**	**Apex**
3	TiN - Cu - N	Si_3N_4
16	Ti_3CuN - TiN - Cu	Ti_5Si_3(Cu)
17	Ti_3CuN - Cu - $\beta TiCu_4$	Ti_5Si_3(Cu)
18	Ti_3CuN - $\beta TiCu_4$ - Ti_2Cu_3	Ti_5Si_3(Cu)
19	Ti_3CuN - Ti_2Cu_3 - Ti_3Cu_4	Ti_5Si_3(Cu)
20	Ti_3CuN - Ti_3Cu_4 - TiCu	Ti_5Si_3(Cu)
21	Ti_3CuN - TiCu - Ti_2Cu	Ti_5Si_3(Cu)
22	Ti_3CuN - Ti_2Cu - βTi	Ti_5Si_3(Cu)
23	Ti_3CuN - βTi - αTi	Ti_5Si_3(Cu)
24	Ti_3CuN - αTi - Ti_2N	Ti_5Si_3(Cu)
25	Ti_3CuN - Ti_2N - TiN	Ti_5Si_3(Cu)

Although spreading usually is considered in terms of the thermodynamics of interfaces, it is useful to view spreading in terms of the physical factors that reflect the condition of the liquid, in particular the temperature and composition. In light of the dearth of phase equilibria data for quaternary systems, it is not surprising that the literature contains little consideration of the nature of liquid formation as a function of the movement of the composition along the diffusion or reaction path between metal and ceramic. In terms of the phase equilibria, there are two important considerations:

1. For each tetrahedron, there is an individual invariant point (eutectic or peritectic) that corresponds to the temperature and composition of liquid formation for that particular quaternary subsystem.
2. The location of the alloy-ceramic join through the quaternary subsystem relative to the position of the invariant point determines the amount of liquid present (according to the lever rule).[6]

Concerning the observed intermediate phases. Kim et al.[16] observed Ti_5Si_3 and TiN at the interfaces of the low-Ti alloys; they also observed CuSiTi only in the Cu50Ti alloy, but they considered that it was not part of the reaction layer. The phase diagram indicates that this assumption is incorrect because this phase occurs in many of the quaternary subsystems crossed by the ceramic-metal join line.

Intermediate Phases: The presence of Ti_5Si_3, TiN and CuSiTi at the interface of the Cu50Ti-Si_3N_4 join suggests that the equilibrium interfacial join was located in the CuSiTi-Cu-Ti_5Si_3(Cu)-TiN quaternary subsystem. This is not surprising since Ti_5Si_3 and TiN are thermodynamically stable; nothing is known of the stability of CuSiTi,

Table 15d Derived compatibility tetrahedra in the system Ti-Cu-Si-N at 800°C.

Base System Cu-Si-Ti		
Tetrahedron No.	**Face Triangle**	**Apex**
1	Cu_3Si - Si - $TiSi_2$	Si_3N_4
26	Cu_3Si - $TiSi_2$ - $CuSi_{19}Ti_{10}$	TiN
27	Cu_3Si - $CuSi_{19}Ti_{10}$ - CuSiTi	TiN
28	CuSiTi -Cu_3Si - $Cu_{17}Si_4$	TiN
29	CuSiTi - $Cu_{17}Si_4$ - Cu_6Si	TiN
30	CuSiTi - Cu_6Si - Cu_7Si	TiN
31	CuSiTi - Cu_7Si - Cu	TiN
32	CuSiTi - Cu - $Ti_5Si_3(Cu)$	TiN
33	$Ti_5Si_3(Cu)$ - Cu - $\beta TiCu_4$	Ti_3CuN
34	$Ti_5Si_3(Cu)$ - $\beta TiCu_4$ - Ti_2Cu_3	Ti_3CuN
35	$Ti_5Si_3(Cu)$ - Ti_2Cu_3 - Ti_3Cu_4	Ti_3CuN
36	$Ti_5Si_3(Cu)$ - Ti_3Cu_4 - TiCu	Ti_3CuN
37	$Ti_5Si_3(Cu)$ - TiCu - Ti_2Cu	Ti_3CuN
38	$Ti_5Si_3(Cu)$ - Ti_2Cu - βTi	Ti_3CuN
39	$Ti_5Si_3(Cu)$ - βTi - αTi	Ti_3CuN
9	$Ti_5Si_3(Cu)$ - αTi - Ti_3Si	$Ti_5Si_3(N)$
8	$Ti_5Si_3(Cu)$ - Ti_3Si - Ti_5Si_3	$Ti_5Si_3(N)$
7	$Ti_5Si_3(Cu)$ - Ti_5Si_3 - Ti_5Si_4	$Ti_5Si_3(N)$
40	CuSiTi - $Ti_5Si_3(Cu)$ - Ti_5Si_4	$Ti_5Si_3(N)$
6	CuSiTi - Ti_5Si_4 - TiSi	$Ti_5Si_3(N)$
41	CuSiTi - TiSi - $CuSi_{19}Ti_{10}$	$Ti_5Si_3(N)$
5	$CuSi_{19}Ti_{10}$ - TiSi - $TiSi_2$	$Ti_5Si_3(N)$

Table 15e Derived compatibility tetrahedra in the system Ti-Cu-Si-N at 800°C.

Interior Tie Lines			
Line No.	**End Members**	**Line No.**	**End Members**
1	TiN - Cu_3S_i	7	$Ti_5Si_3(Cu)$ - TiN
2	TiN - $Cu_{17}Si_4$	8	$Ti_5Si_3(Cu)$ - Ti_2N
3	TiN - Cu_6Si	9	$Ti_5Si_3(Cu)$ - $Ti_5Si_3(N)$
4	TiN - Cu_7Si	10	$Ti_5Si_3(Cu)$ - Ti_3CuN
5	TiN - $CuSi_{19}Ti_{10}$	11	$Ti_5Si_3(N)$ - $CuSi_{19}Ti_{10}$
6	TiN - CuSiTi	12	$Ti_5Si_3(N)$ - CuSiTi

Table 15f Derived compatibility tetrahedra in the system Ti-Cu-Si-N at 800°C.

Base System Formed by Two Interior Tie Lines		
Tetrahedron No.	**Face Triangle**	**Apex**
26	TiN - $CuSi_{19}Ti_{10}$ - Cu_3Si	$TiSi_2$
27	TiN - $CuSi_{19}Ti_{10}$ - Cu_3Si	CuSiTi
27	TiN - $CuSi_{19}Ti_{10}$ - CuSiTi	Cu_3Si
42	TiN - $CuSi_{19}Ti_{10}$ - CuSiTi	Ti_5Si_3(N)
4	TiN - $CuSi_{19}Ti_{10}$ - Ti_5Si_3(N)	$TiSi_2$
42	TiN - $CuSi_{19}Ti_{10}$ - Ti_5Si_3(N)	CuSiTi
27	TiN - CuSiTi - Cu_3Si	$CuSi_{19}Ti_{10}$
28	TiN - CuSiTi - Cu_3Si	$Cu_{17}Si_4$
28	TiN - CuSiTi - $Cu_{17}Si_4$	Cu_3Si
29	TiN - CuSiTi - $Cu_{17}Si_4$	Cu_6Si
29	TiN - CuSiTi - Cu_6Si	$Cu_{17}Si_4$
30	TiN - CuSiTi - Cu_6Si	Cu_7Si
30	TiN - CuSiTi - Cu_7Si	Cu_6Si
31	TiN - CuSiTi - Cu_7Si	Cu
42	TiN - CuSiTi - Ti_5Si_3(N)	$CuSi_{19}Ti_{10}$
43	TiN - CuSiTi - Ti_5Si_3(N)	Ti_5Si_3(Cu)
43	TiN - Ti_5Si_3(Cu) -Ti_5Si_3(N)	CuSiTi
11	TiN - Ti_5Si_3(Cu) -Ti_5Si_3(N)	Ti_2N
11	TiN - Ti_5Si_3(Cu) - Ti_2N	Ti_5Si_3(N)
25	TiN - Ti_5Si_3(Cu) - Ti_2N	Ti_3CuN
16	TiN - Ti_5Si_3(Cu) - Ti_3CuN	Cu
25	Ti_2N - Ti_5Si_3(Cu) - Ti_3CuN	Ti_2N
25	Ti_2N - Ti_5Si_3(Cu) -Ti_3CuN	TiN
24	TiN - Ti_5Si_3(Cu) - Ti_3CuN	αTi
43	Ti_5Si_3(N) - $CuSi_{19}Ti_{10}$ - CuSiTi	TiN
41	Ti_5Si_3(N) - $CuSi_{19}Ti_{10}$ - CuSiTi	TiSi

although it is likely to be relatively stable since its presence dominates the compatibilities in the centre of the diagram. By analogy, it is likely that all of the alloy-ceramic interfaces contained Ti_5Si_3,TiN and CuSiTi. Inspection of Figure 5 reveals that the experimental join between Cu-50Ti and Si_3N_4 passes through the CuSiTi-Cu-Ti_5Si_3(Cu)-TiN tetrahedron relatively closely to CuSiTi. However, the corresponding experimental joins for the low-Ti alloys pass through the same tetrahedron at a much greater distance from CuSiTi. Thus, according to lever rule[6] considerations, it would be expected that there would be more CuSiTi at the Cu50Ti-Si_3N_4 interface than would be at the interfaces between the low-Ti alloys and Si_3N_4. It is likely that the levels of CuSiTi in these latter samples were below the limits of detection of the instrument used.

Through this type of phase equilibria analysis, it can be seen that the phases expected to form at the interface can be identified and their relative amounts also can be estimated.

Degree of Spreading: Kim et al.[16] observed that the degree of spreading increased as the Ti level increased. The authors attempted to interpret this in terms of the thickness of the reaction layer. Again, these data can be interpreted by inspection of Figure 5. At equilibrium, the liquid must wet a solid interface whose components are compatible with the liquid. This wetting or spreading is influenced by the thermochemical nature of the solid phases. That is, if the solid (or the liquid) composition is altered, then the spreading is likely to be affected. The lever rule[6] demonstrates that the relative proportions of the four solid phases in the CuSiTi-Cu-Ti_5Si_3(Cu)-TiN quaternary tetrahedron are such that the levels of Ti_5Si_3(Cu), TiN, and CuSiTi all increase as the alloy composition increases in Ti level. However, as the starting alloys increase in Ti levels and end at Cu-50Ti, the experimental joins swing toward CuSiTi, so increasing its proportion more than those of Ti_5Si_3(Cu) and TiN. Since there was complete spreading with the Cu-50Ti alloy, it is likely that the presence of CuSiTi is critical to the spreading. Thus, the thickness of the reaction layer probably is not relevant to the spreading; it is the phase assemblage of the solids and their relative proportions that control the spreading.

Through this type of phase equilibria analysis, the proportions of the solid phases present at equilibrium can be estimated. Since spreading depends on the composition of the interface, these proportions can provide guidance concerning optimal alloy compositions. In this case, if CuSiTi is critical to the spreading, then an alloy containing 43 wt.% Ti (50 mol.% Ti) would be optimal since the corresponding alloy-ceramic experimental join would approach closest to CuSiTi.

Reaction Layer Thickness: Kim et al.[16] noted that the reaction layer thickness decreased with increasing Ti levels. This was interpreted in terms of diffusional (dynamic) processes. Again, these results can be explained with the use of the phase diagram, which describes an equilibrium (static) process. It is well known that every subsystem (binary, ternary, quaternary, etc.) has an invariant point associated with it. It is the (liquid) homogeneity region corresponding to the quaternary invariant point liquid that wets the ceramic interface at equilibrium; it is not a liquid defined by the starting alloy composition. This liquid must wet an interface that is compatible with it or else a reaction will occur. At equilibrium, the liquid will coexist with up to all four of the solid phases located on the corners of the tetrahedron. Thus, for the CuSiTi-Cu-Ti_5Si_3(Cu)-TiN subsystem, there is a eutectic or peritectic temperature and composition. The position of the overall experimental composition at the interface can be located relative to this composition. The closer the interface composition is to the invariant point composition, the greater is the amount of liquid. Since diffusivity in liquids is greater than in solids, it can be generalised that the thickness of the reaction layer can be proportional to the amount of liquid since reaction products form through diffusional processes. However, the specific volume of the reaction products, the degree of microstructural disturbance and differential thermal expansion also are important factors. It is clear that- with a fixed liquid composition, alteration of the Ti level will cause a gradual shift toward or away from the invariant point liquid. The lever rule[6] demonstrates that the relative amounts of (unique) liquid and changing solid compositions can be estimated using the phase diagram. Since the reaction layer became thinner as the Ti level increased, this suggests that the invariant point liquid composition is near the Cu portion of the CuSiTi-Cu-Ti_5Si_3(Cu)-TiN tetrahedron; as the solid composition shifts away from Cu, the relative amount of liquid decreases. Thus, it can be seen that the data can be interpreted specifically in terms of simple phase equilibria rather than generalizing about diffusion.

Through this type of phase equilibria analysis, trends in reaction layer thickness can be explained. Although dynamic diffusional processes are important to the growth rate of the reaction layer, the establishment of static equilibrium negates their influence.

6.0 SUMMARY

The present work represents the second part of a series of instructional papers on methods of derivation of phase diagrams, particularly quaternary systems relevant to ceramic-metal bonding. These diagrams are generated using reported phase equilibria data, geometric requirements of the phase rule,[6] and some assumptions. The precise steps to generate and confirm these diagrams are relatively straightforward and they require no thermodynamic calculations or model building.

Many aspects of experimental data can be interpreted in a straightforward fashion with a knowledge of the phase equilibria. These conclusions often clarify or refute the explanations offered by the authors of published wetting and bonding studies. In the three cases discussed in the present work, it is clear that alloy-ceramic joins equilibrate to form interfaces in quaternary subsystems with defined solid phases and unique liquids in proportions that can be estimated. This is a critical point that receives little consideration in the literature. The equilibria between these phases can validate experimental observations such as the formation of reaction products, the relative amounts of these phases, the failure to detect expected phases, the absence of phases that are thermodynamically stable, the direction of diffusion, potential reaction paths, the degree of spreading, the reaction layer thickness, and other features.

Ti-Ag-B-N System at 1000°C: The quaternary phase diagram for this system is relatively simple, containing only 10 quaternary subsystems. It was generated through knowledge of 3 of the 6 binary edges and 2 of the 4 ternary faces of the tetrahedron This diagram required less than 10 min. to derive.

Analysis of this system reveals that the interface at 1000°C between a Ti-1.5Ag alloy and cubic BN crosses 8 quaternary subsystems and equilibrates in the

Ag-BN-TiB_2-TiN quaternary subsystem, thus indicating diffusion of the metal toward the interface rather than diffusion of the ceramic. This conclusion is justified by the observation of an interface rich in Ti, B and N.[7]

Ti-Sn-Si-N System at 950°C: The quaternary phase diagram for this system is somewhat more complicated, containing 22 quaternary subsystems. It was generated through knowledge of 5 of the 6 binary edges and 3 of the 4 ternary faces of the tetrahedron. This diagram required ~30 min. to derive.

Analysis of this system reveals that the interfaces at 950°C between Sn-based alloys (containing up to 7 wt.% Ti) and Si_3N_4 cross only 3 quaternary subsystems, pass over most of the subsystems associated with the thermodynamically stable Nowotny phase Ti_5Si_3(N) and equilibrate in the TiN-Sn-βTi_6Sn_5-Ti_5Si_3(N) quaternary subsystem. The observation of only Sn and Ti_6Sn_5 in the final phase assemblage can be explained by lever rule[6] considerations, which suggest that the levels of the thermodynamically stable phases Ti_5Si_3(N) and TiN probably are below the limits of detection of the energy dispersive spectrometer used for analysis.[10]

Ti-Cu-Si-N System at 800°C: The quaternary phase diagram for this system is quite complicated, containing 43 quaternary subsystems. It was generated through knowledge of 5 of the 6 binary edges and all 4 of the ternary faces of the tetrahedron. This diagram required more than 8 hrs. to derive.

Analysis of this system reveals that the interfaces at 960-1140°C between Cu-based alloys with low Ti levels (up to 19 mol.% Ti) and Si_3N_4 cross 10 quaternary subsystems, the interface between a Cu-based alloy with high Ti level (57 mol.% Ti) and Si_3N_4 crosses 12 quaternary subsystems. Although CuSiTi was observed at the interface in the latter sample only, it is probable that all of the samples equilibrate in the CuSiTi-Cu-Ti_5Si_3(Cu)-TiN quaternary subsystem. As the alloy compositions shift from low-Ti to high-Ti levels, the amounts of these phases, in particular, CuSiTi, increase according to lever rule[6] considerations. This change in solid composition affects the degree of spreading in a gradual fashion because the composition also changes gradually. As the alloy compositions shift from low-Ti to high-Ti levels, the overall phase assemblage also shifts away from the invariant point associated with the CuSiTi-Cu-Ti_5Si_3(Cu)-TiN quaternary subsystem. This change in solid-liquid proportions happens gradually as the Ti level in the alloys increases. If it is accepted that the high-diffusivity liquid enhances reaction product formation at the interface, then it can be assumed that the reaction layer thickness is proportional to the amount of liquid present. Thus, the reaction layer thickness should decrease as the Ti level increases. These interpretations, which are based only on phase equilibria considerations, explain the experimental observations.[16]

Additional Observations: First, inspection of Figures 1-3 shows that each diagram is dominated by tie lines associated with certain phases. For the Ti-Ag-B-N system, these are Ag and TiN, for the Ti-Sn-Si-N system, these are Sn, Ti_5Si_3(N) and TiN; for the system Ti-Cu-Si-N, these are Ti_5Si_3(Cu), TiN and CuSiTi. The absence of Cu at the interface is notable because this shows that these Cu-based alloys are intrinsically incompatible with Si_3N_4. These observations are confirmed by the frequency of listing of these phases in the previous tables. Most of these phases are known to be thermodynamically stable and so would be expected to form at the interfaces. It is their thermodynamic stability that causes them to dominate the compatibilities in these systems.

Second, it has been noted that there are phases that are present at temperatures other than those assessed in the present work. The geometric requirements of these systems usually are such that their presence or absence is not important because these phases represent a fifth component within an established tetrahedron. For instance, $Cu_{15}Si_4$ decomposes at 800°C, so it is not present in Figure 5. However, this is a commonly observed phase and its presence at subsolidus temperatures is very likely. Inspection of Figure 5 shows that $Cu_{15}S_4$ would be located within the Si_3N_4-Cu_3Si-$Cu_{17}Si_4$-TiN tetrahedron. This would separate this tetrahedron into two separate tetrahedra, these being Si_3N_4-Cu_3Si-$Cu_{15}Si_4$-TiN and Si_3N_4-$Cu_{15}Si_4$-$Cu_{17}Si_4$-TiN. This situation generally is the case unless the appearance or disappearance of tie lines (as the temperature changes) alters other compatibilities in the region of the phase diagram where the phase of interest is located.

Third, since these derivations are geometrically derived and not confirmed experimentally, they must remain speculative. The validity of any derivation depends on the quality of the experimental data used and the absence of any unknown compounds. However, a further issue is the likelihood of the compatibilities. For instance, Figure 5 shows that $TiSi_2$ is compatible with Ti_5Si_3(N) but not Ti_5Si_3(Cu). This is not a likely tie line because $TiSi_2$ would be expected to be compatible with both Ti_5Si_3 based phases. Further, $CuSi_{19}Ti_{10}$, is compatible with Ti_5Si_3(N) but not Ti_5Si_3(Cu) which also is unlikely. The same is true of Ti_3CuN, which is compatible with Ti_5Si_3(Cu) but not Ti_5Si_3(N). However since Ti_5Si_3(Cu) and Ti_5Si_3(N) really are the quaternary solid solution Ti_5Si_3(Cu,N), this probably is not an important issue. In the final analysis, these compatibilities are correct or the phase diagram is incorrect, at least in this section of the phase diagram.

7.0 ACKNOWLEDGEMENTS

The author wishes to express his appreciation to Prof. J. C. Schuster (University of Vienna, Austria) for the kind provision of several essential publications, Prof. F. J. J van Loo (Eindhoven University of Technology, The Netherlands) for helpful comments and Dr. O. C. Standard (University of New South Wales, Australia) for the excellent and time-consuming graphical and computer work.

8.0 REFERENCES

1. C. C. Sorrell: Phase Equilibria Considerations in Ceramic-Metal Bonding, K. Uematsu, Y. Moriyoshi, Y. Saito and J. Nowotny, eds., *Interfaces of Ceramic Materials: Impact on Properties and Applications*, Trans Tech Publications, Aedermannsdorf, Switzerland, 1995, pp.127-156.
2. T. B. Massalski, H. Okamoto, P. R. Subramanian and L. Kacprzak: *Binary Alloy Phase Diagrams*, 2nd Edition, ASM, Metals Park, OH., 1990.

3. P. Villars, A. Prince and H. Okamoto: *Handbook of Ternary Alloy Phase Diagrams*, ASM, Materials Park, OH., 1995.
4. Various Editors, *Phase Diagrams for Ceramists*, American Ceramic Society, Westerville, OH., **I-XI**, 1964-1995.
5. O. Madelung: Landolt-Börnstein, *Numerical Data and Functional Relationships in Science and Technology, Substance Index 1993*, Subvolume a: Elements and Binary Substances, Subvolume b: Ternary Substances, Subvolume c: Polynary Substances, Springer-Verlag, Berlin, 1993.
6. S. T. Bowden: *The Phase Rule and Phase Reactions: Theoretical and Practical*, Macmillan, London, U.K.
7. E. Benko: Wettability Studies of Cubic Boron Nitride by Silver-Titanium, *Ceramics International*, **21**, 1995, pp.303-307.
8. P. Rogl and J. C. Schuster: *Phase Diagrams of Ternary Boron Nitride Systems*, ASM, Materials Park, OH, 1992.
9. H. J. Goldschmidt: *Interstitial Alloys*, Butterworths, London, U.K., 1967.
10. R. R. Kapoor and T. W. Eagar: TiN-Based Reactive Solders for Ceramic/Metal Joints, *Metallurgical Transactions B*, **20B**, 1989, pp.919-924.
11. M. A. Pietzka and J. C. Schuster: Phase Equilibria of the Quaternary System Ti-Al-Sn-N at 900°C, *Journal of Alloys Compounds*, **247**, 1997, pp.198-201.
12. R. Kieffer, F. Benesovsky and B. Lux: Investigation of Nowotny Phases, *Planseeber Pulvermet.*, **4**, 1956, pp.30-36.
13. E. Parthé: Contributions to the Nowotny Phases, *Acta Crystallographic.*, **8**, 1957, pp.768-769.
14. H. Nowotny: Crystal Structure and Stability of Refractory Phases, W. M. Mueller, ed., *Advances in X-Ray Analysis*, Sir Isaac Pitman and Sons Ltd., London, U.K., **5**, 1962.
15. H. Nowotny: Alloy Chemistry of Transition Element Borides, Carbides, Nitrides, Aluminides and Silicides, P. A. Beck, ed., *Electronic Structure and Alloy Chemistry of the Transition Elements*, Interscience Publishers, New York, 1963.
16. D. H. Kim, S. H. Hwang and S. S. Chun: The Wetting, Reaction and Bonding of Silicon Nitride by Cu-Ti Alloys, *Journal of Materials Science*, **26**, 1991, pp.3223-3234.
17. M. G. Nicholas and R. M. Crispin: Brazing Ceramics with Alloys Containing Titanium, *Ceramic Engineering Science and Proceedings*, **10**, 1989, pp.1602-1612.
18. R. R. Kapoor and T. W. Eagar: Brazing Alloy Design for Metal/Ceramic Joints, *Ceramic Engineering Science and Proceedings*, **10**, 1989, pp.1613-1630.
19. A. P. Tomsia, J. A. Pask and R. E. Loehman: Joining Nitride Ceramics, *Ceramic Engineering Science and Proceedings*, **10**, 1989, pp.1631-1654.

New Approaches to Metal Ceramic and Bioceramic Interfacial Bonding

Roger St. C. Smart, Pawittar S. Arora, Michael Steveson, Nobuyuki Kawashima, Giuseppe P. Cavallaro, Hui Ming and William M. Skinner
Ian Wark Research Institute, University of South Australia, The Levels Campus Mawson Lakes, South Australia 5095

1.0 INTRODUCTION

Metal-ceramic bonding and its more recent application to bioceramic bonding to metal substrates for medical and dental implants is strongly dependent on the behaviour of the boundary region. Research on this subject has a long and distinguished history.[1-7] Improvements in material performance have been shown to depend strongly on the nature of the metal and ceramic components, fabrication parameters and reaction at the interface. The strength or mechanical integrity of the metal-ceramic bond can be improved with an understanding of the influence of a number of variables including phase chemistry, interfacial chemistry and morphology, thermal and elastic mismatch, wettability, metal yield strength and processing parameters. The interfacial fracture energy is strongly influenced by residual stress distributions, flaw concentrations in the ceramics and at the interface, surface segregation of impurities and surface roughness.

Many of these limitations on the interfacial fracture resistance can be overcome using highly reactive interfaces, at which inter-diffusion and/or chemical reaction produces a functionally graded interfacial region in which the chemical and physical properties are more gradually altered from the metal substrate to the ceramic overlayer without phase discontinuity. The production of these interfacial regions has been the target of much of the recent work on metal-ceramic bonding.[3, 8-10] These methods have analogs in strategies for joining ceramics for high-temperature applications[11] such as diffusion bonding, which generally relies on the joining temperature approaching the melting temperature of the metallic layer, and reactive brazing, in which the joining temperature generally exceeds the melting temperature of a metallic interlayer between two ceramic materials. More recently, new joining approaches have used thin transient liquid phase layers to facilitate bonding, including glassy interlayers in which there is a reaction involving ion exchange and solid state diffusion.[12]

All of these methods require relatively high temperature treatment. The new strategies described in this chapter have utilised low temperature (<200°C), low density plasma reactions, combined with sol-gel coatings, to produce a variety of ceramic surface layers on metal substrates and interfacial layers between metals and oxides or other ceramics. The processes occurring at the surface of the substrate during low temperature plasma deposition of a ceramic layer, however, are similar to those involved in reactive metal-ceramic bonding at high temperatures. The merits of the low temperature plasma are that the surfaces of the substrates can be cleaned in situ, relatively low power (i.e. <100 W) is required, and the reacting species can be controlled directly by the vapour input to the plasma. We will show that layers can be designed to grade compositionally and mechanically from a metal substrate to a bulk ceramic material with the graded layers applied uniformly, bonded into the substrate structure and fully deformable without disbonding. Functionally graded layers produced in this way can be applied to corrosion protection, device fabrication and bioceramic titanium-based layers for deposition of bioactive hydroxyapatite (HAP).

2.0 REVIEW

The issues in metal-ceramic bonding are by now well known and have been extensively studied. Many of these issues relate to the fundamental solid state properties and processes including the different nature of the bonding in metals compared with ceramics causing contrasts in chemical and physical behaviour.[3, 4] If no chemical reaction occurs at the metal/ceramic interface, the driving force to form a metal-ceramic bond, characterised by the work of adhesion (W_{ad}), (as in the Young Dupré equation) can be related to the thermodynamic reduction in free energy when intimate contact is made, represented by the metal surface tension γ_{m}, and, hence, to the wettability (or contact angle θ) between the phases.[5-7]

$$W_{ad} = \gamma_m \, (1+\cos\theta)$$

However, the interfacial fracture energy is generally much larger due to additional dissipative processes including polarisation at the surfaces discouraging wetting, metal plasticity and reaction. In earlier work, it was assumed (and demonstrated) that a low metal contact angle, measured in sessile drop experiments, was a prerequisite for the formation of a strong bond but more recent experience has shown that these bonds can be very strong despite contact angles >90°.[7] There has been some evidence that the improved bonding may be due to improved removal of pores, a higher fraction of bonded area and low stress concentrations associated with low values of contact angle and lower stress shapes for residual pores.[7] Hence, the objective for optimised metal-ceramic bonding may well lie in continuous, relatively uniform, pore-free interfacial bonding in which the surface tension is lowered as far as possible by reactive mixing across the face boundaries.

Similarly, the disparate thermal and elastic properties of the two phases can lead to disbonding under thermal gradients and under shear conditions. Functionally graded materials (FGM) and metal-ceramic composites have been designed to alleviate thermoelastic and thermoplastic deformation using a variety of approaches including continuum finite element formulation of the composite phases, mean field theories, crystal plasticity models and discrete dislocation models. The link scale for compositional gradients is critical to these models and, hence, requires detailed analysis of the size, shape, continuity and spatial dispersion of the constituent phases of the graded metal-ceramic composites. It is experimentally demonstrated that these interfacial gradients can be designed to optimise bonding, optimise distribution of thermal stresses, supress the onset of plastic yielding, reduce the driving force for fracture and arrest cracks [refs. 10 and references therein].

In the approach taken here, the functionally graded material is in the form of a surface composite[13] in that this engineered material has the second phase spatially distributed in the near-surface regions such that the phase composition is linearly graded as a function of distance from the surface. One form of this functionally graded surface composite has been obtained using morphological surface modification of the bulk phase, induced by multiple-pulse laser irradiation to give micro-roughened structures with much higher surface area. The second phase is then deposited using thick or thin film deposition methods to form a three-dimensional compositionally and thermally graded interface, e.g. diamond/steel, silica/SiC.[13] In our approach, the layers are formed by low temperature plasma deposition and reaction compositionally and functionally graded from the metal through its oxide (partially reacted to ceramic nanostructures) to the ceramic overlayer, e.g. nickel/nickel oxide/silicate/silica.[8] For the same reasons as in the functionally graded materials and surface composites, there is then a gradation of composition, thermoelastic and thermoplastic properties from the metal to the ceramic across the reacted layer.

A particular area of application of FGM and surface composite engineering which is recently in focus has been the fabrication of medical and dental implants. Here, the additional requirement of biocompatibility with bone and tissue *in vivo* must be considered together with the thermal, elastic and plastic deformation properties. Dental implants with functionally graded structures derived from titanium and ceramic HAP have been fabricated to meet both mechanical and biocompatibility property requirements.[14] The Ti/HAP FGM samples showed improved biocompatibility in bone formation around the implant compared with the oxidised titanium implant in tests on rabbit femoras. An extensive series of studies of the surface modification of titanium implant substrates for improved bonding by bone growth has been described by T. Kokubo, T. Nakamura et al.[9, 15-27] They have used surface modification by oxidation, alkaline hydrolysis of the surface oxide, silica deposition, titania and sodium titanate deposition together with HAP layer deposition by sol-gel and precipitation. Very significant improvements in bond strength of the apatite layer to the substrates and in load bearing after bone growth have been achieved in these studies. In their biomimetic process, these improvements have been associated with the requirement of the coatings to be thin, deformable and achieve strong adhesion at the metal-coating interface. In their methods, reactions in aqueous solution are required, followed in many cases by high temperature treatments for consolidation of the gel-like surface layers and crystallisation of the hydroxyapatite.

In our alternative processes, the low temperature plasma reaction can be used to produce the graded interface from the titanium substrate to a thin (i.e. <50 nm) functionally graded layer with silica in anhydrous or hydrated form before hydroxyapatite growth in simulated body fluid or from sol-gel deposition. The advantages of this method lie in automatic control of surface impurities, layer thickness and composition and in the ability to produce specialised overlayers.

3.0 SUMMARY OF PREVIOUS RESEARCH

We have previously reported the low temperature plasma reaction of air:water:tetraethoxysilane (TEOS) mixtures with oxidised nickel surfaces producing silicate structures in the oxide surface layer on the metal.[28] The original objective of these studies was corrosion protection of the metal. The plasma reactions were carried out in a Harrick plasma unit PDC-32 G operated at 100 W and pressures between 1-5 $\times$ 10^{-1} torr. A three-stage sequence of reactions were used to prepare the ceramic surface layers, namely: surface cleaning using an argon plasma followed by air plasma to remove hydrocarbon contamination of the nickel surface; controlled surface oxidation of nickel using an air plasma saturated with water vapour (i.e. 20 min reaction giving an oxide depth of ~20 nm); and controlled silicate reaction using air saturated with a mixture of water and TEOS (40:1 vol.) for different lengths of time to give different concentrations of silicon (as silicate) in the surface layers. Mass spectrometric study of the plasma has shown that reactive SiO species are formed in the plasma together with a variety of other ionised species (e.g. O, OH,OCH_2CH_3) and energetic electrons. The mechanism of reaction of the plasma species with the oxide has been revealed using a combination of XPS spectra and modified Auger parameter studies.

The Auger parameter can be expressed in the modified form as

$$\alpha' = KE_A + BE_{PE}$$

where KE_A is the kinetic energy of the most intense Auger electron emission and BE_{PE} is the binding energy of the main photoelectron from the same element in the same XPS spectrum. The modified Auger parameter has the advantage that, surface charging is automatically removed by the equal and opposite shifts in each of the two contributing energies. Importantly, it can also be shown[29] that the modified Auger parameter is a measure of the energy changes due to extra-atomic relaxation and is therefore often more sensitive to the stoichiometry and strength of bonding to next nearest neighbours for the emitting atom than the XPS BE shifts alone. The information is usually presented in a chemical state plot, as developed by Wagner et al.,[29] giving values to each of the KE_A and BE_{PE} components as well as the Auger parameter. For this reason, the chemical state plot for silicon in the silicate/silica coatings provides improved information about the changes in silicon-oxygen bonding as the layer is formed by plasma reaction.[28]

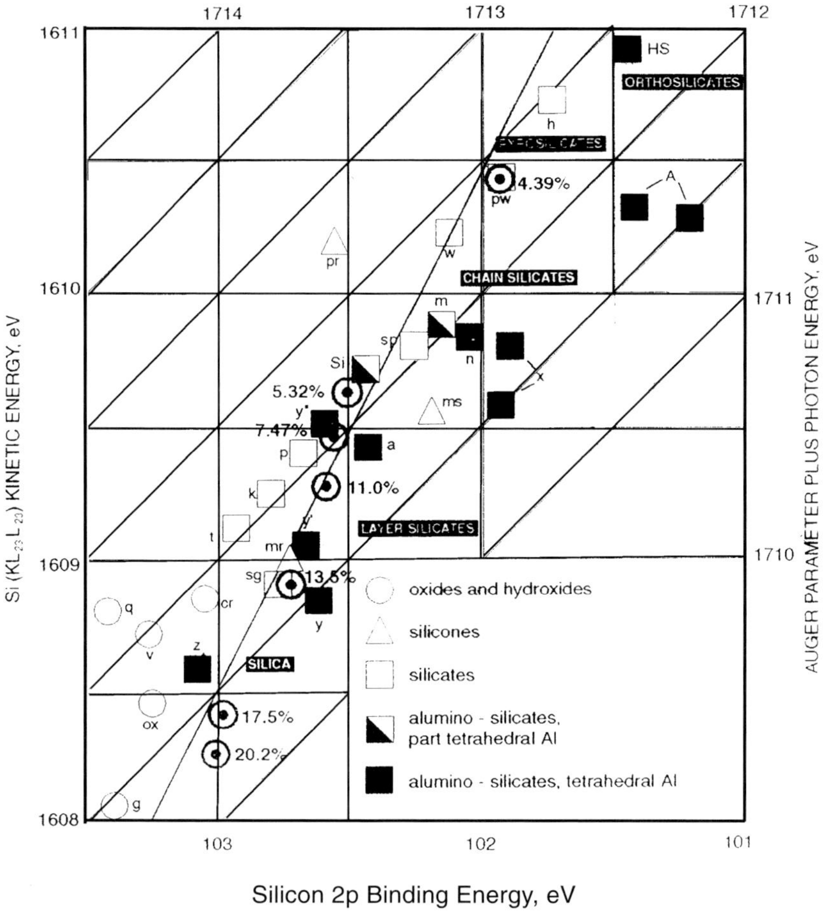

(•) % → indicates atomic concentration of silicon as measured by XPS.

Fig. 1 Modified Auger parameter plot for a variety of silicates[29] and TEOS: H_2O plasma-reacted nickel surfaces.

The chemical state plot as a function of silicon atomic concentration in the surface is shown in Figure 1. The changing values demonstrate that reactive SiO species are initially incorporated into the oxide layer to form orthosilicate groups (SiO_4^{4-}) at low silicon concentrations. As the plasma reaction proceeds, polymeric silicate structures such as pyrosilicates (e.g. $S_2O_7^{6-}$), chain and layer silicates form with increasing concentrations of Si and decreasing concentrations of Ni at the exposed surface. Increasingly-linked siloxane structures are indicated by the changing Auger parameter component values as in Figure 1 with increasing Si percent values. A complete coverage of bulk silica is achieved with no evidence of exposed nickel.[28] In

Figure 1, the value of the modified Auger parameter moves from 1712.5 eV at the lowest Si at.% (i.e. 4%) to a final value of 1711.3 eV with a full silica coating (20%). This has been shown,[28] by comparison with the work of Alfonsetti et al.[31-33] on surface stoichiometry of SiO_x thin films, to represent a stoichiometry change from $x<1.5$ (i.e. 4% Si) to $x = 2.0$ (i.e. 20% Si) during reaction.

The complete layer reduces dissolution rates of the oxidised nickel substrate in 10^{-2} M acid solution by two orders of magnitude over a period of 700 h.[28] It is noted that full coverage by silica is required for the slowest corrosion rate, with any defects in the structure of the surface layer contributing to sites of increased reaction and finally breakthrough in small conical pits observed developing progressively during acid attack beyond 4000 h. The gradation in composition and polymeric silicate structures during the TEOS reaction may appear to imply that the silicate groups are dispersed in the oxide structure rather than forming a new phase, but TEM evidence (described below) will add a more complete picture of the functionally graded surface composite.

Extension of the plasma reaction to oxidised surface layers on substrates of iron, mild steel and brass have also been reported,[34] with essentially the same mechanism of reaction producing protective surface layers on these oxidised metal surfaces. Application of the same three-stage plasma reaction sequence produces a closely similar modified Auger parameter chemical state plot for the oxidised iron surface (Figure 2) with α' varying between 1712.5 eV and 1711.1 eV corresponding to a similar change in stoichiometry from $x = 1.5$ to 2.0. Dissolution rates for the fully covered silica surface in 10^{-2} M acid again demonstrate passivation (i.e. reduction by more than two orders of magnitude in dissolution rate) of the iron surface up to times beyond 600 h.[34]

Oxidised mild steel surfaces (Figure 3) show a similar trend but there are some significant differences in this reaction. There is a smaller change in the Auger parameter across the reaction from 1712.0 eV (i.e. $x = 1.8$) to 1711.2 eV (i.e. $x = 2.0$) corresponding to an initial Auger parameter (3.5 at.% Si) representing a chain silicate structure more extensively linked by comparison with orthosilicate and pyrosilicate structures found on oxidised nickel and iron surfaces. There is a corresponding larger variation in dissolution rates with different extent of plasma reaction[34] but the fully coated mild steel surface is passivated in the same acid dissolution kinetics for periods longer than 400 h.

The oxidised brass surface presents different behaviour again with the least reacted surface corresponding to orthosilicate species at much lower stoichiometry (i.e. $x = 1.4$). The final stoichiometry of $x = 1.9$ suggests that the surface may not be fully covered by silica. The dissolution rates[34] support this contention with a reduction of one order of magnitude compared with more than two orders of magnitude on nickel, iron and mild steel surfaces under the same conditions of measurement. Extensive breakthrough of the surface silicate/silica layer occurs after 150 h compared with the much longer periods of passivation on the other surfaces.

Hence, in relation to the original objective of corrosion protection, the metal-ceramic bonding produced by reaction in the interfacial layer using a low temperature plasma has provided a mechanism for functionally grading the composition and properties from the metal through its oxide to silicate and silica overlayers. The

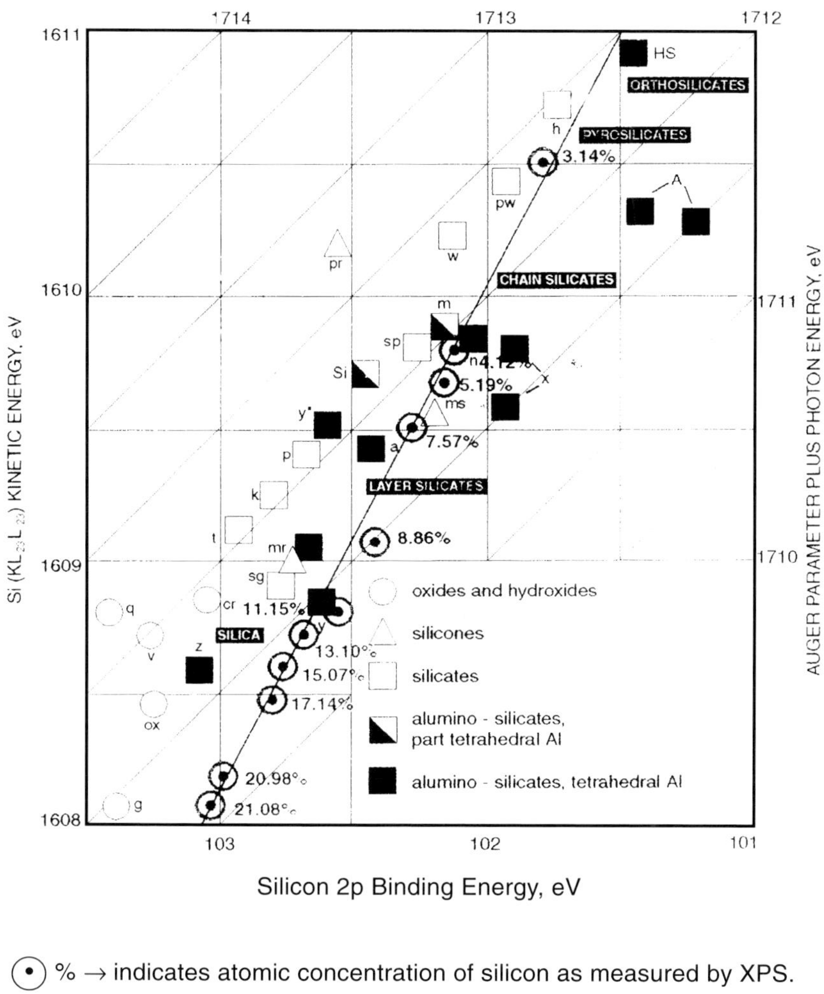

Fig. 2 Modified Auger parameter plot for TEOS: H_2O plasma reaction on oxidised iron.

silica overlayer can be extended using sol-gel deposition and relatively low temperature heat treatments.[8]

4.0 STRUCTURE AND COMPOSITION IN THE FUNCTIONALLY GRADED SURFACE LAYER

The structure of the interfacial layers has been examined using electron microscopy. Field emission scanning electron microscopy (FESEM) images of the developing

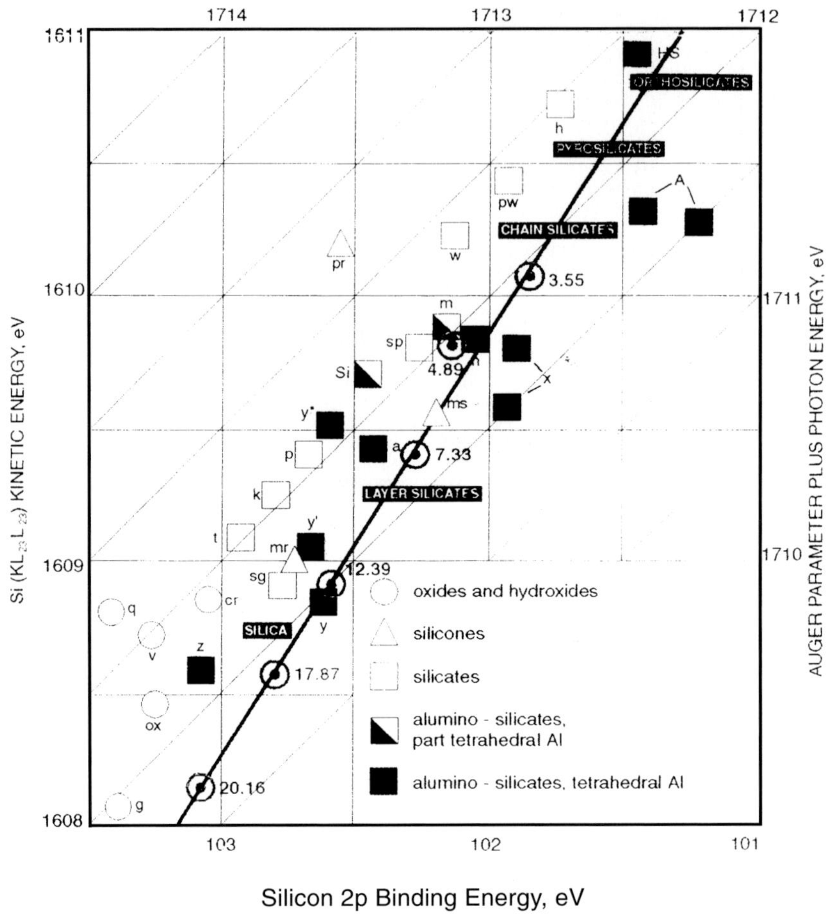

⊙ % → indicates atomic concentration of silicon as measured by XPS.

Fig. 3 Modified Auger parmeter plot for TEOS: H_2O plasma reaction on oxidised mild steel.

silicate/silica layers simply reveal uniform development of the surface layer progressively obscuring surface features of the underlying nickel metal, e.g. grain boundaries, steps, pits, protrusions on the scale of tens of nms.

For high resolution examination of the structure, TEM specimens were prepared from a nickel disc using a new procedure.[35] One side of the disc was dimpled using a conventional disc grinder followed by chemical polishing with a nickel cleaning solution used previously.[28, 36] The three-stage plasma assisted chemical vapour deposition (PACVD) method was then applied to the flat non-dimpled side of the disc effectively coating both the side and edges of the disc exposed to the plasma.

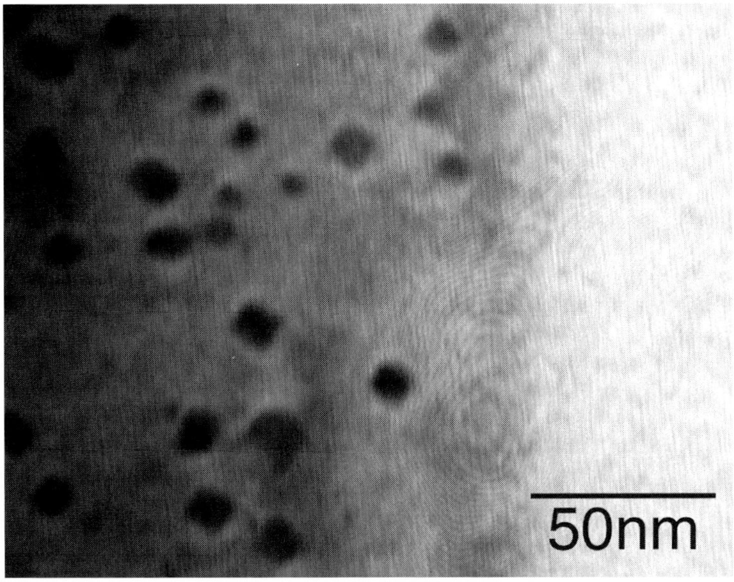

Fig. 4 TEM photograph of nickel coated with silica by PACVD (Etched Type Specimen).

The treated disc was then acid etched until a hole of ~500 μm appeared. The edges of the hole were examined using TEM in combination with EDS analysis.

Figure 4 shows a typical area at the edge of the hole, i.e. across the interfacial layers. The outer layer 50 nm in width (right hand side, Figure 4) gives an EDS analysis for silica with only a minor signal for nickel in this region. We have been unable to obtain an electron diffraction pattern from the silica in this region suggesting that it is amorphous. The darker grey region with the embedded distinct crystallites (~10 nm) produces EDS analyses with strong signals for both nickel and silicon. The nanocrystallites give strong indications of a cubic structure corresponding to either NiO or Ni_2SiO_4. The intersection between the nickel substrate and this reaction layer is shown in further detail in Figure 5. Here, it is clear that there has been interpenetration of the oxide and silicate structures between the grains of the original nickel substrate. The overlying reaction layer (top half, Figure 5) shows the functionally graded composite structure of the silicate with embedded crystallites of either nickel oxide or nickel silicate. The reacted surface of the nickel (lower half, Figure 5) contains residual grains of nickel identified (in electron diffraction) with regions giving EDS analyses for both nickel and silicon between the residual grains. The result is a reaction layer linked into the nickel substrate by this interpenetrating reaction front. There is some evidence that the crystallites in the reaction layer may be partly surface silicate-reacted nickel oxide in Figure 6. This image appears to expose the dark crystallites embedded in the predominantly silica matrix up to the

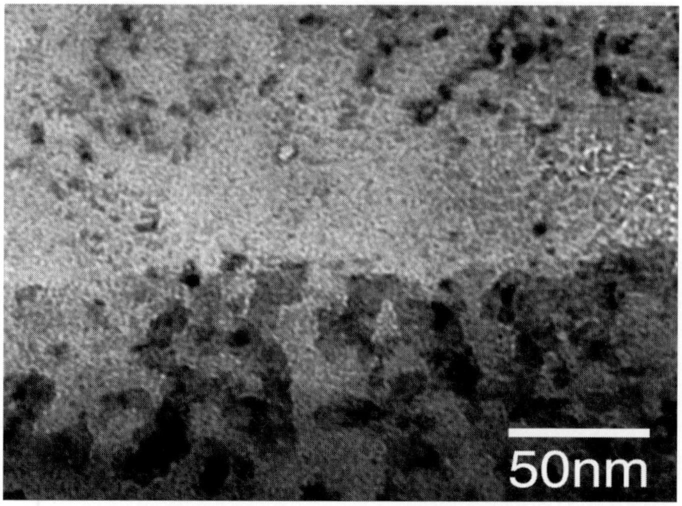

Fig. 5 TEM photograph of intersection between nickel and silica thin film produced on nickel by PACVD (Etched Type Specimen).

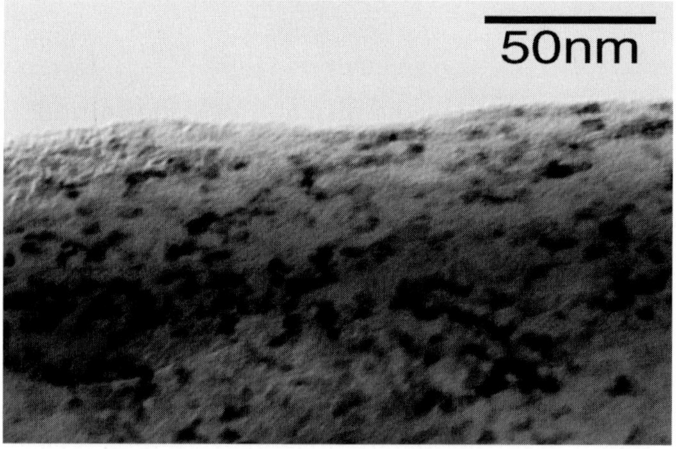

Fig. 6 TEM photograph of edge of silica thin film produced on nickel by PACVD (Etched Type Specimen).

outer edge. An electron diffraction pattern for NiO is obtained from this region. We have previously shown that reacting NiO in vacuum with quartz produces silicate surface layers in the oxide making it much less acid-soluble than the original nickel oxide.[37] This has also been found with silica-supported nickel oxide catalysts.[38, 39] However, we cannot rule out the possibility that the nanocrystallites are nickel oxide protected by an overlying silica film. In either case, the structure is quite similar to functionally graded metal-ceramic materials, in this case as surface composites.[10, 12]

The adhesion of these coatings to the substrate has been tested by adapting a scratch tester. The scratch test is the generally accepted method of assessing the interface bond strength (adhesion) between a coating and substrate, especially in the case of coatings which are less than 20 μm in thickness.[40] The test involves a hemispherical diamond indenter, usually 200 μm in radius, driven over a coated surface. The load on this indenter is linearly increased as the tip is dragged on and eventually into the coating. During this type of scratch test, the resultant tangential force, generated by the friction between the diamond tip and the coating, is monitored as it describes the dynamics of the diamond tip/coating/substrate interaction. Under these conditions, the coating detaches from the substrate because of the development of shear stresses derived from both the contact stresses developed by the hemispherical indenter and the friction generated between the diamond point and the surface of the coating. This condition is generally known as a tangentially loaded Hertzian contact. As the friction coefficient increases, the depth at which these shear stresses reach the maximum amplitude approaches the surface and, importantly, the coating/substrate interface.[41]

Failure of the coating can usually be identified by a significant change in the frictional force. In some cases, where the change might not be so noticeable, mathematical means are adopted to highlight the transitions zone/s identifying the failure of the coating.[42] It must be stressed that this test is usually conducted in conjunction with FESEM, in order to clarify the failure mechanism/s.

As expected, no differences in the frictional behaviour could be detected between the uncoated nickel surface and the nickel surface after TEOS plasma reaction to produce the 50 nm thick silicate/silica film. In order to test the effect of the graded layer on metal-ceramic bonding, a titanium nitride layer (3-5 μm thick) was deposited on to uncoated nickel and TEOS plasma-coated nickel in the same deposition run. Figure 7 shows some representative examples of frictional load testing of the two types of samples. The plasma treatment has significantly affected the friction behaviour of the coating/substrate system. In the untreated Ni sample, it is clear that the frictional load is increasing at a much greater rate than the treated sample. The higher frictional load measured on the untreated nickel surface indicates that the substrate system is less able to support the stylus under the same increasing applied normal load. The higher rate of increase of the frictional load for the untreated surface supports this difference in behaviour with the stylus experiencing a deteriorating substrate/coating interfacial adhesion, i.e. digging into the substrate. The TEOS plasma-treated nickel surface continues to show less frictional resistance to the increasing normal load. There is some evidence of possible breakthrough of the plasma coating layers near 50 N with a higher rate of increase of the frictional load above this normal loading. These reproducible results support the contention that the functionally graded metal-ceramic reaction layer produces a harder substrate and improved adhesion of the ceramic coating. Closely similar results have been measured on the titanium substrate (see below).

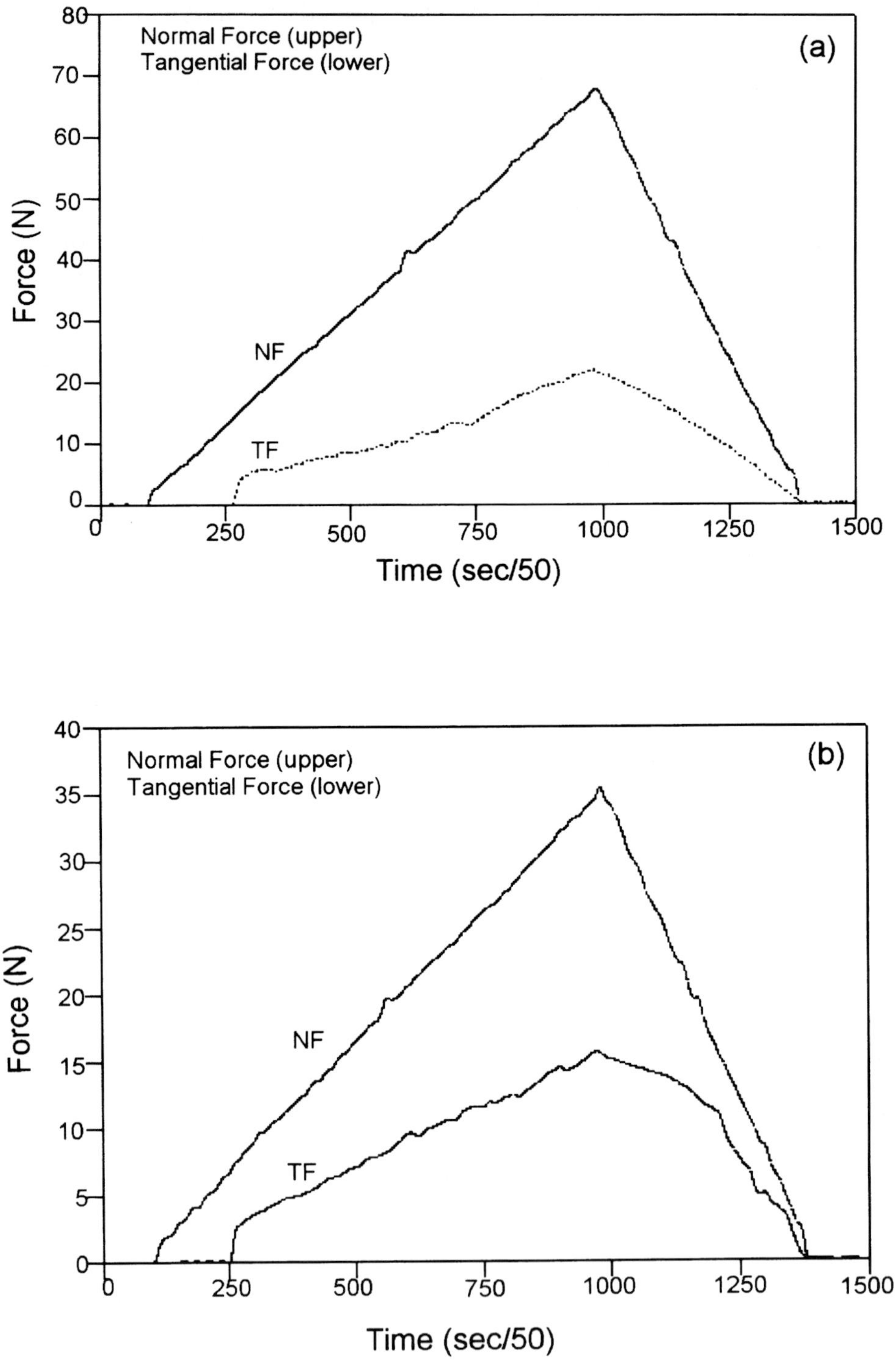

Fig. 7 Nickel sample, (a) TiN with TEOS plasma silica treatment and (b) TiN without TEOS silica treatment.

5.0 BIOCERAMIC BONDING: TITANIUM SUBSTRATES

In bioceramic applications, metals such as cobalt-chromium alloys, stainless steel and, more recently, titanium, are the materials of choice for their strength and biocompatibility in stress-bearing medical and dental implants. After an implant is fitted, the gap between the implant and the bone is filled by a layer of fibrous tissue or growing bone depending on the type of material used.[43] Bio-tolerant material, such as stainless steel, becomes surrounded by fibrous tissue whereas biocompatible material, such as titanium, can produce bone growth to fill the gap and provide a more rigid joint.[43] Bioactive materials, e.g. hydrous silica, titania, bioglasses and HAP, have been developed as surface coatings for implant use.[9, 27, 44-53] With these surface coatings, osseointegration occurs resulting in the strongest possible interface.[43]

The lower mechanical strength of these ceramic or glass surface coatings, when compared with the development of high strength at the bone/implant interface, has dictated an increasing need for high strength adhesion between the coating and the metal substrate of the implant. Failure at this interface has been a major limitation on the lifetime of implants coated by high temperature plasma spraying of hydroxyapatite layers resulting in relatively thick, brittle coatings with poor adhesion and some phase differentiation.[43] Much of the recent research internationally, particularly in groups like those of T. Kokubo et al (Kyoto University, Japan),[9,15-27, 45-53] and W. Bonfield et al (Queen Mary and Westfield College, University of London)[54] has been directed to the improvement of interfacial bonding between this substrate and the bioactive coatings. In this context, our method appears to offer some advantage in providing improved structural and chemical linkage due to the interfacial reactions.[28, 34]

The same three-stage reaction sequence applied to a titanium substrate has been shown[55] to produce a coherent, continuous coating with strong metal-ceramic bonding. The silicate/silica coated titanium can be used as a substrate for ceramic coatings of HAP or bioactive glass.

An immediate consequence of the *in situ* treatment of the titanium surface before coating is the demonstrated effects of surface cleaning and surface smoothing using an air or argon plasma (15 min). Surface hydrocarbon contamination after this treatment, measured in XPS atomic concentrations and spectra, is reduced from initial values of 60-80 at.% C to <8 at.% C (after a brief <30 s exposure to air during transfer to the spectrometer). Hence, almost all of the surface contamination is removed in the initial cleaning stage and recontamination is avoided by subsequent coating without exposure to air. The surface of the titanium is usually irregular and pitted after exposure to air before plasma treatment (Figure 8a). The initial plasma treatment removes these features to leave a relatively smooth, uniform surface (Figure 8b) with an oxide layer of controlled thickness determined by the duration of air/water plasma reaction. The response of tissue to a metal implant surface is known to be influenced by the chemical and structural properties of the surface[56] including hydrocarbon contamination and microstructure, with these variations leading to variable bone-implant bond strength. The application of subsequent plasma treatments to this cleaned, more uniform, reactive surface without exposure to air should lead to better implant/coating performance.

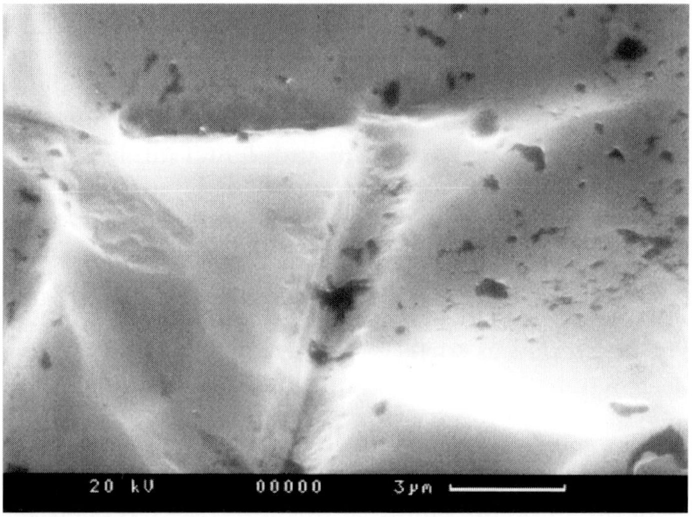

(a)

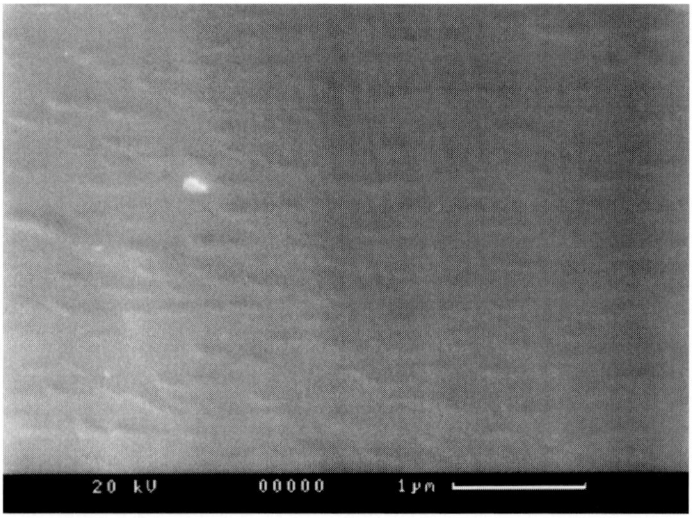

(b)

Fig. 8 SEM micrographs of (a) The naturally oxidised titanium surface showing an irregular surface and pitting and (b) Naturally oxidised titanium surface after 10 mins air plasma cleaning showing no pitting.

Figure 9 presents the silicon Auger parameter chemical state plot similar to those in Figures 1, 2, 3, for different surface silicon concentrations, at different TEOS plasma reaction times of the oxidised titanium substrate. As with the nickel substrate, there is a shift from orthosilicate-like structure to silica-like structure, via the

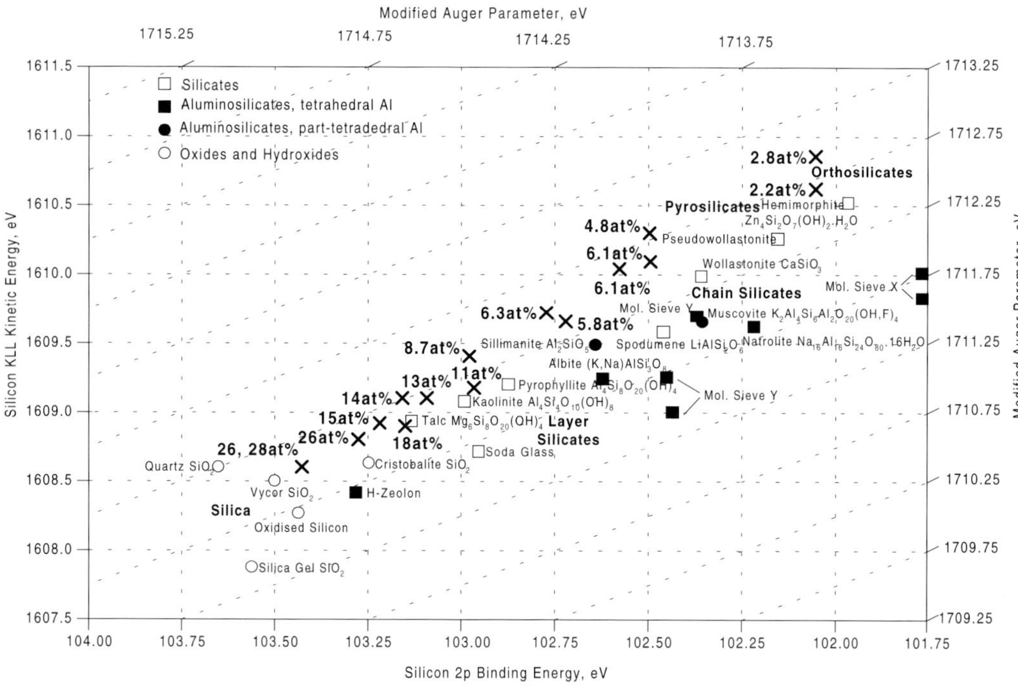

Fig. 9 Silicon Auger parameter plot for plasma reacted titanium showing spread of silicon-oxygen structures with varying surface silicon concentrations, shown next to points. TEOS plasma reaction times ranged from 0-12 mins.

intermediate pyro-, chain and layer silicates, as the silicon concentration rises. The stoichiometry of the SiO_x oxide, derived from the β parameter quoted previously,[31-33] changes from $x = 1.4$ at 2.8% Si to $x = 1.8$ at 26% Si. Depth profiles using ion beam etching in XPS were performed on these samples to determine the depth of the silicon-containing layer. Two examples are shown in Figures 10a and b for surfaces with 18% Si and 26% Si respectively. Calibration of the etch rate using a standard Ta_2O_5 layer on a Ta substrate gave 1.2 nm/min. Hence, silicon concentrations were determined for surface silicon of 6, 13, 18 and 26% down to depths of ~0.8, 2.5, 6 and 18 nm respectively. Figures 10a and b illustrate the last two of these profiles. As the TEOS plasma reaction proceeds, the silicon-containing layer becomes thicker and more silica-like, with compositional grading of the Si and Ti across the interfacial reaction layer, i.e. 5-20 nm depth, as found for the plasma reacted nickel surfaces.

This compositional grading is supported by measurements of the modified Auger parameter in depth profiles using ion beam etching through a completely coated (i.e. 28 at.% Si) surface. Figure 11 shows that there is a systematic shift of both the

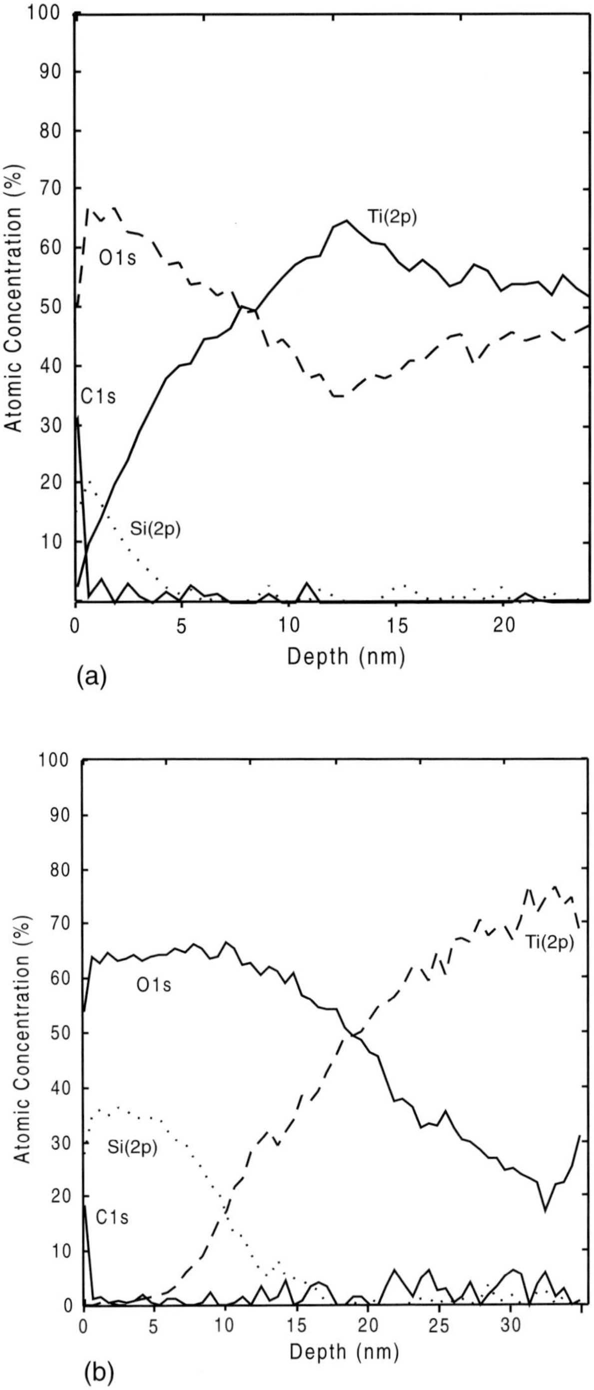

Fig. 10 (a) XPS ion etched depth profile of TEOS plasma treated titanium with 18 atomic % surface silicon and (b) XPS ion etched depth profile of TEOS plasma treated titanium with 26 atomic % surface silicon.

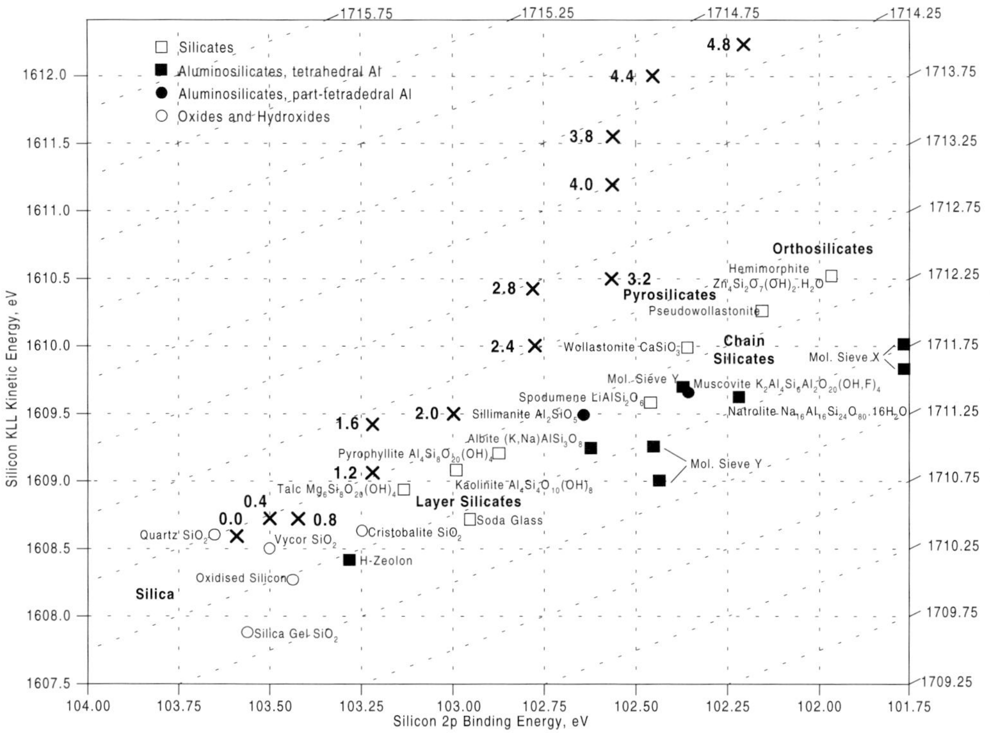

Fig. 11 Silicon Auger parameter plot for plasma reacted titanium showing the change in silicon-oxygen stoichiometry and structure within a 27. 7 at.% surface silicon coating during an XPS ion etched depth profile. Sputter rate was 1.2 nm/min; sputter times, in minutes, shown next to points.

Si-O structure and the SiO_x stoichiometry as the ion etch sputter time increases through the layer. The initial, outer surface corresponds to bulk silica moving through underlayers of layer silicates, chain silicates and pyrosilicates before there is a sharp change in stoichiometry from x = 1.3–1.4 to x = 0.7 at the limit of detection of Si near the titanium metal interface.

The 30-50 nm silicate/silica layers are completely deformable to repeated 180° bending without evidence of disbonding or cracking at the interface in SEM examination. Similar testing of tangential (frictional) resistance to an increasing normal force in scratch testing has been carried out on thick samples of titanium metal uncoated, TEOS plasma coated, and with both samples overcoated with a titanium nitride layer 3-5 μm thick. As with the nickel samples, no significant differences could be detected with the thin TEOS plasma coating compared with the uncoated titanium. The results with the TiN outer layer were, however, significantly and reproducibly different for the titanium substrate uncoated and with TEOS plasma coating. The results are compared in Figures 12a and b respectively.

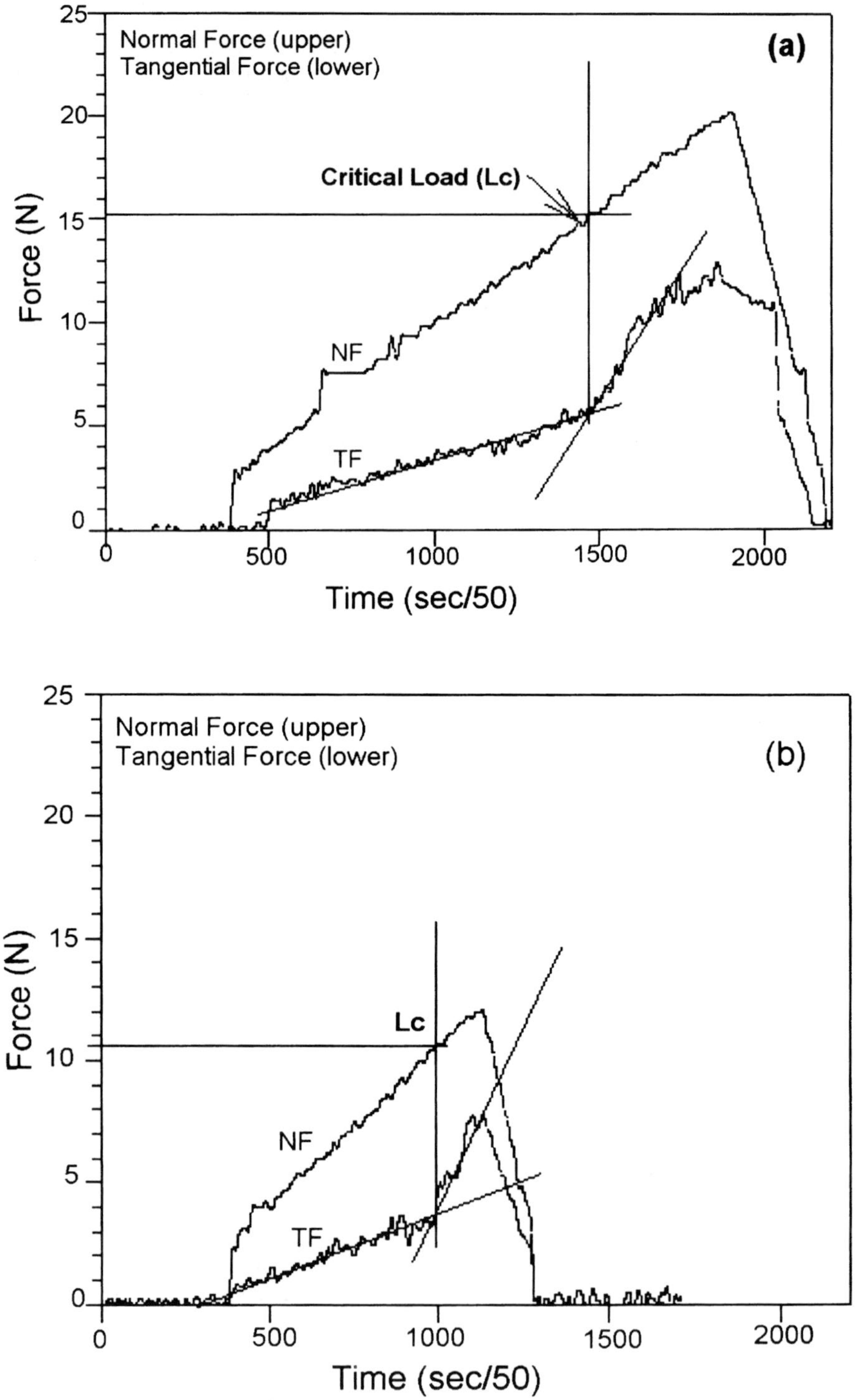

Fig. 12 Titanium (a) TiN with TEOS plasma silica treatment and (b) TiN without TEOS silica treatment.

Table 1 Surface analyses and atomic concentrations (XPS) for titanium screws T1 and T3

Element	Plasma Treated	No Plasma Treatment		Used in Bone		15 min Etch	
		As Provided	15 min Etch				
	T1	T3	T3	T1	T3	T1	T3
C(1s)	15.7	78.5	46.7	78.1	73.6	40.5	31.6
O(1s)	56.2	16.3	26.0	13.1	18.7	34.7	41.3
Ca(2p)	-	0.27	0.15	1.5	2.1	3.8	5.1
Si(2p)	26.4	1.2	0.47	4.0	1.1	16.2	-
P(2p)	-	-	-	1.3	1.0	1.2	2.8
Ti(2p)	-	2.2	22.9	-	2.0	-	17.3
F(1s)	1.7	-	-	-	-	-	-
N(1s)	-	0.87	2.5	2.0	1.5	3.5	1.9
Cl(1s)	-	0.3	1.1	-	-	-	-
Al(2p)	-	0.41	0.14	-	-	-	-

T1 Plasma treated Ti screws put into bone and removed

T3 Ti screw put into bone and removed without any plasma treatment

In this case the critical load (*Lc*) at which the coating/substrate fails is obvious in both plots. There is a sudden increase in the tangential (frictional) force at these points, and the rate at which the frictional force increases is much higher than before the *Lc* and is similar in both samples. The significant difference is in the TEOS plasma-treated TiN-coated titanium displaying an *Lc* value of 15.3 N as opposed to 10.6 N for the untreated TiN titanium substrate. As with the nickel sample, these results show that the plasma-coated titanium substrate is able to support a higher normal load due to either or both of stronger bonding of the coatings to the substrate or a hardening of the interface.

Some limited clinical testing of mechanical properties of the plasma coatings have been carried out with further work under way. In the first tests, titanium screws from a standard dental prosthesis supplied by Wingate Laboratories (South Australia) with and without the plasma silicate/silica coating were compared after insertion into bone under shear stress and load conditions simulating installation of the prosthesis. In this case the screws were removed immediately after installation. (Longer term installation tests are currently underway). Surface analysis of the plasma treated (T1) and untreated (T3) samples using XPS are summarised in Table 1. The initial surface of the three-stage plasma treated sample shows full coverage by the silica with a minor impurity of fluoride. The surface of the screw without plasma treatment shows very high levels of hydrocarbon contamination (i.e. 78% C) with correspondingly low exposure of Ti and minor impurities of Si, Ca, N, Cl and Al. The hydrocarbon contamination is sufficiently extensive and thick that it is only partly removed by 15 min. argon ion-etch with increased exposure of the underlying

TiO_2 surface layer and partial removal of the Si impurity. The screws after implantation and removal from bone implacement, have outer surfaces dominated by hydrocarbon from fibrous tissue. For the uncoated titanium implants, the surface analysis is closely similar to the initial surface before implantation with low exposure of Ti, O and the Si impurity. It is significant, however, that both the T3 and T1 samples have acquired additional Ca and P (as phosphate) resulting from bone implantation. For the plasma coated sample T3, there is reduced exposure of silica but no exposure of Ti from the metal substrate. The adherence of the silicate/silica coating is confirmed by 15 min. argon ion etching. For the plasma coated sample, the silicon concentration is increased 4-fold but there is no evidence of exposure of Ti from the substrate. Whereas, for the uncoated T3 sample, exposure of Ti is increased more than 8-fold. For both samples, the removal of some of the hydrocarbon signal increases the exposure of both calcium and phosphate indicating the initiation of some ion exchange in bone growth.

6.0 HYDROLYSED SURFACE LAYERS

A remaining issue in the use of the plasma coating technique is the desirability of producing bonded, functionally graded interfacial layers with the outer surface of the ceramic in hydrolysed form. It is now well recognised that hydrated layers of titania and silica are considerably more active than the dehydrated layers of the same material in inducing growth of hydroxyapatite on implants in simulated body fluids (SBF) and *in vivo* as well as in producing more uniform and adherent hydroxyapatite layers by sol-gel deposition. Kokubo et al. have shown[9, 15, 22, 26, 27, 46, 48, 50, 51] that hydrolysed layers of titania and silica or sol-gel prepared titania or silica show increased bioactivity as surface layers or as composite coatings in inducing hydroxyapatite growth. Improvements in bone growth have been demonstrated in clinical studies. Similarly, caustic treatment of oxidised titanium surfaces to produce hydrolysed and/or gel-like TiOH layers without coatings have produced bioactivity in *in vitro* and *in vivo* studies. The treated surfaces have shown rapid HAP growth in SBF.[9, 15-21, 24, 50] The process for production of hydrolysed titania surfaces using the caustic treatment has been patented.[57]

It has been established that the oxide/silicate/silica surfaces produced in the three - stage plasma treatment described to this point results in a substantially anhydrous silica surface layer despite the presence of water vapour in the air/water/TEOS plasma. This somewhat surprising result has been obtained using measurement of the SiOH surface concentration in DRIFT spectra and with the well-established reaction of these groups with tetrafluoroacetic anhydride (TFAA),[58] i.e. with XPS measurement of %F. Figure 13a shows the DRIFT spectrum of the normal three-stage plasma treated titanium surface with the SiO absorptions of the silicate and silica vibrations in the 1000-1200 cm^{-1} range. There is no evidence of hydrogen-bonded OH vibrations expected in the 3500- 3200 cm^{-1} region. (Vibrations of some free water vapour molecules in the DRIFT beam are observed in the 3500- 3800 cm^{-1} region.) No significant reaction of TFAA with SiOH groups was observable in XPS spectra of the

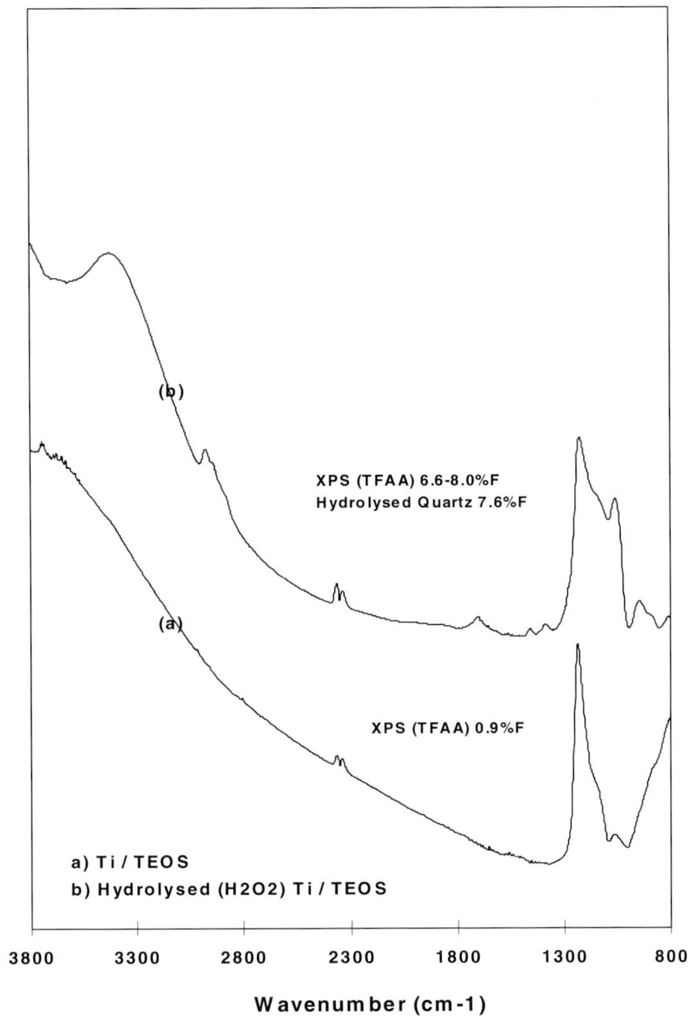

Fig. 13 Comparison (FTIR and TFAA) between (b) hydrolysed and (a) unhydrolysed silica coatings on titanium plates.

standard Ti:TEOS:H_2O plasma layer, i.e. <0.9 at.% F signal detected. The silica surface therefore appears to be substantially anhydrous and similar to high temperature (i.e. >900°C) preparation of silica or quartz.

An additional objective in the plasma treatment has, for these reasons, been to prepare a functionally graded layer with an outer surface of hydrolysed silica containing a high concentration of SiOH groups for increased reactivity. This has been achieved using a modification of the three-stage plasma treatment. The first two stages, namely air (or argon) plasma treatment for surface cleaning and air/

water plasma treatment for surface oxidation, are unchanged. In the third stage, a combined H_2O/H_2O_2/TEOS treatment is used. This stage is initiated using vapour from a 30 vol.% H_2O_2 aqueous solution in one tube with TEOS:H_2O vapour introduced from a second tube to the already-established H_2O/H_2O_2 plasma.

This modified procedure produces substantial change in the DRIFT spectra as exemplified in Figure 13b. An intense, broad hydrogen bonded OH peak centred near 3300 cm^{-1} is now observed together with new absorption structure in the 1000-1100 cm^{-1} region. Both additional features are characteristic of hydrated silica in sols or gels.[59] They appear to be superimposed on the anhydrous silica or quartz-like spectrum of Figure 13a suggesting a partially hydrated structure or anhydrous framework with hydrolysed surface layer. The spectra (Figure 13b) also show some additional uptake of hydrocarbon, i.e. absorptions in the 2800-3000 cm^{-1} CH stretching region possibly due to some additional incorporation of the alkyl groups from the TEOS structure. Reaction with TFAA now gives strong representation of F1s signals in the XPS spectra with atomic concentrations in the range 6.6-8.0%. Fully hydrolysed quartz surfaces reacted with TFAA gave 7.6% F showing that this procedure produces the equivalent of 2.6 active OH groups nm^{-2}.[60]

Hence, modification of the plasma reactants can produce a hydrated silica surface. It is likely that the level of hydration can be controlled by the vol.% H_2O_2 and this experimentation is continuing.

We have confirmed that the treatment of oxidised titanium metal surfaces in caustic solution, as developed by Kokubo, Nakamura et al.,[9, 16-21, 24] does result in alkali titanate aero-gel formation on the surface of the metal. Subsequent heating to 600°C stabilises the alkali titanate surface layer for subsequent coating, growth or precipitation of HAP. TEM imaging of titanium reacted in 10M NaOH solution at 75°C for periods up to 24 h shows extended regions of reaction 100-200 nm wide with finger-like pits penetrating into the titanium substrate to depths of order 100 nm containing gel-like TiOH reaction products. The hydrated TiOH products on the outer surface appear as sheets folded and multiply stacked protruding in leaf-like structures. An example is shown in Figure 14. It is clear that the reacted surface layer of titanium produced by this procedure, even after subsequent heating, has substantially different characteristics from the functionally graded composite layers described from the plasma reaction.

7.0 HYDROXYAPATITE COATINGS

It has been established from a large number of examples in both medical and dental implantation[15, 43] that the interface of bioactive ceramics with living bone generates an hydroxyapatite layer creating a bond between the ceramic and the living bone. The hydroxyapatite contains significant carbonate concentration and defective structure in small crystallites similar in composition and structure to the apatite in bone material.[61] The bone producing cell, the osteoblast, preferentially proliferates and differentiates to produce further apatite and collagen on the apatite layer.[46] This growth mechanism prevents the intervention of the fibrous tissue and allows

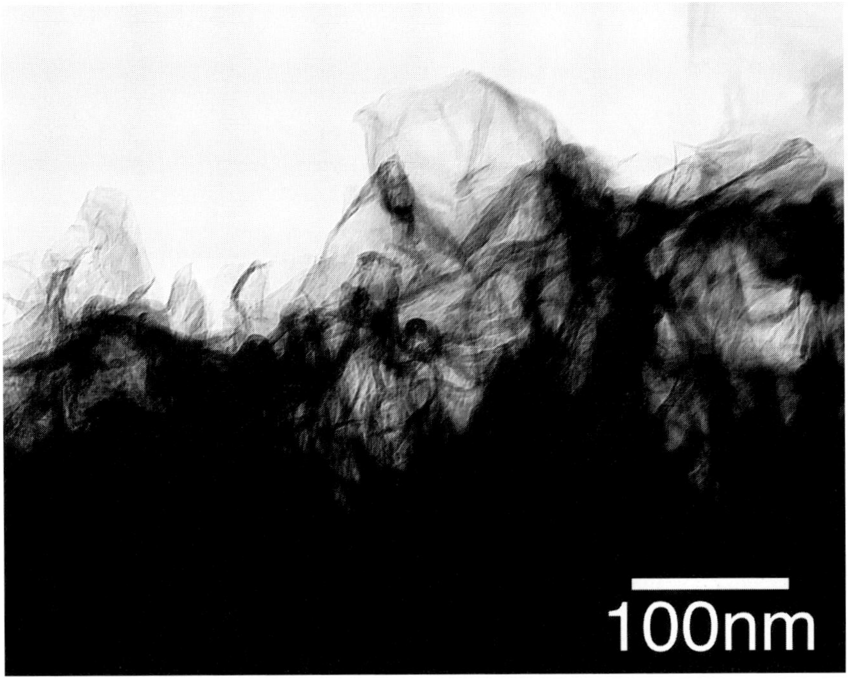

Fig. 14 TEM photograph of titanium soaked in 10M-NaOH aqueous solution at 75°C for 24 hours.

direct contact between the surrounding bone and the surface apatite layer on the ceramic so that a continuous chemical bond is formed between the bone apatite and the surface apatite reducing or removing the interfacial energy. It has been shown that apatite formation on these bioactive ceramics can be produced in SBF[15] as well as *in vivo* and *in vitro*.[43]

Bioactive hydroxyapatite coatings have been applied to the plasma-generated oxide/silicate/silica layers, with and without hydrolysis and to the hydrolysed oxidised titanium layers using three different methods, namely: sol-gel reaction; colloidal precipitation; and growth from SBF. In the first method, the hydroxyapatite sol was prepared from calcium nitrate tetrahydrate ($Ca(NO_3)_2.4H_2O$) dissolved in ethanol. Water, nitric acid and triethylphosphite ($((C_2H_5O)_3P)$) were added to the solution and the solution stirred at 50°C for 2 days under reflux. Finally, butylamine was added to the solution to produce the hydroxyapatite sol.[62] Deposition of this sol on to hydrated silica layers and hydrolysed oxidised titanium produced a uniform distribution of platey, colloidal particles (~1-2 μm) completely covering the surface with strongly adhered particles. The development of crystalline hydroxyapatite from the initially amorphous films or colloidal layers can be followed as a function of heating temperature using x-ray (Figure 15) and electron diffraction. These studies have

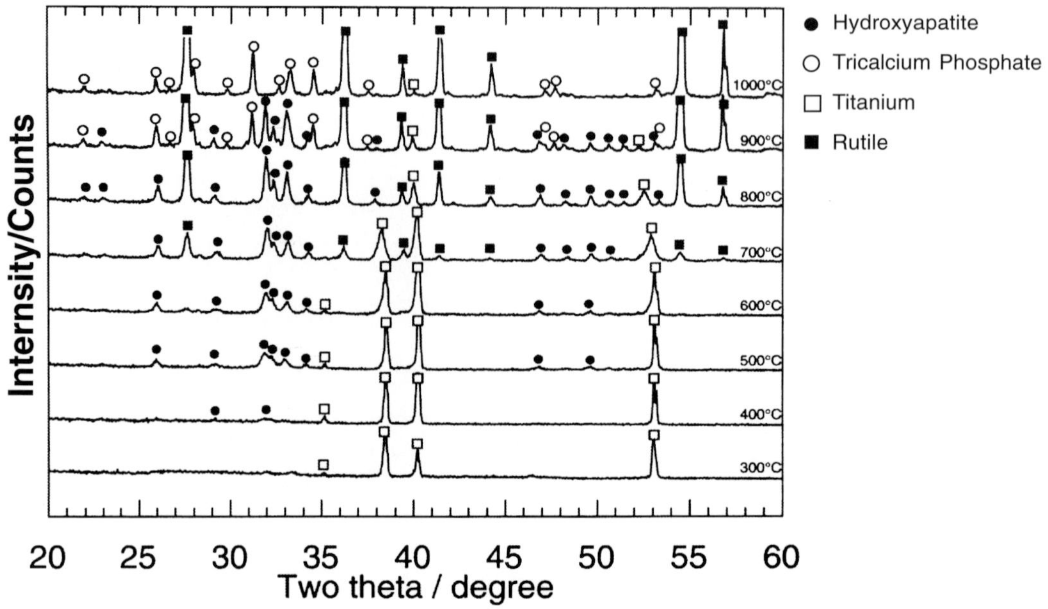

Fig. 15 XRD patterns of hydroxyapatite coating produced on plasma-silica-treated titanium substrate by sol-gel route. The coating was heated to various temperatures (300-1000°C) in air.

shown that a distinct diffraction pattern for the crystalline phase starts to appear at 400°C, is distinct by 500°C (Figure 16) and fully developed by 600°C. Some decomposition to calcium phosphate and calcium oxide begins above 700°C. This amorphous to crystalline conversion is accompanied by some shrinkage of the film, as in Figure 17, but there is no evidence for detachment of the layer from the hydrolysed substrate surface.

The second method was used as an alternative to the hydroxyapatite sol for production of a hydroxyapatite layer on the hydrolysed oxidised titanium surface. After the extended caustic treatment, the plate was soaked in saturated $Ca(OH)_2$ aqueous solution at 75°C for 24 h followed by further soaking in 10 wt.% K_2HPO_4 aqueous solution at 75°C for 24 h. This solution produced a precipitated layer of amorphous calcium hydroxyphosphate carbonate material transformed into crystalline hydroxyapatite by heating at 500°C.

In the third method, the hydrolysed surfaces of the plasma-generated silica or the caustic-generated titania were exposed to simulated body fluid.[27, 48] The growth of hydroxyapatite amorphous material and crystallites on these surfaces was confirmed with XPS and SEM measurements. Colloidal hemispherical apatite structures have been imaged similar to those described by Kokubo[15] with needle (e.g. Figure 16b) and sheet-like apatite microcrystals (0.1-0.5 μm) after heating the layers to higher temperature.

(a)

(b)

Fig. 16 SEM micrographs (a, b) Hydroxyapatite coating produced by sol-gel method. The coating was heated to 500°C for 3 hours.

The layer is not detached by repeated bending through 180° or by manual abrasion. More systematic testing of scratch and abrasion resistance and adhesion at the hydrolysed oxide/hydroxyapatite interface is currently underway. Based on the extensive testing by Kokubo et al.[27, 49, 51, 53] of similarly applied hydroxyapatite coatings on hydrolysed titania and silica, the ability to withstand loads comparable with those experienced in implantation should be obtained. Bio-mechanical testing including

(a)

(b)

Fig. 17 Hydroxyapatite coating produced by sol-gel method. The coating was heated to 600°C for 3 hours.

detachment tests, undecalcified histological examination, tension failure load tests[16, 17, 22] revealed that the hydroxyapatite-coated hydrolysed titanium implants formed direct bonds to bone tissue during early implantation whereas untreated titanium implants only showed direct contact with the bone after 16 weeks. The tension failure loading of the apatite layer-coated titanium implants was also significantly higher than that of uncoated controlled samples. The combination of

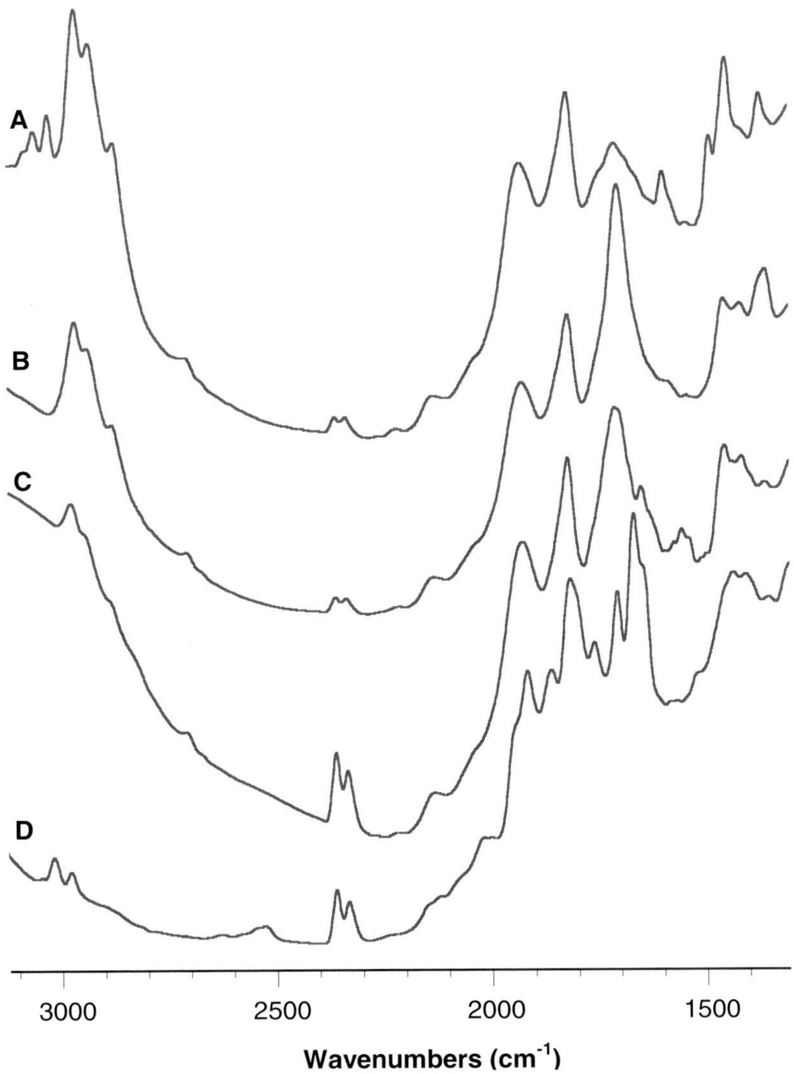

A. Styrene/Kaolinite B. Methyl Vinyl Ketone/Kaolinite
C. Acrylic Acid/Kaolinite D. Acrylic Acid/Talc

Fig. 18 Plasma assisted chemical vapour deposition on some clays.

improved interfacial titanium/hydrolysed oxide and hydrolysed oxide/HAP bonding produced by the chemical treatments appears to consistently accelerate the bone-bonding behaviour of titanium implants and enhance the strength of the bone/implant bonding.

8.0 POLYMER COATINGS ON CERAMIC SUBSTRATES

In a parallel project, we have been using PACVD to modify the surface layers of ceramic materials for incorporation into polymer matrices as fillers, paper coating materials and additives to improve hardness, brightness and tensile strength. In particular, as described in a previous review,[8] the clay minerals kaolinite and talc have been modified in this way. The objective in both cases was to improve bonding of the polymer material to the silicate or aluminosilicate surface so that, when incorporated into the polymer matrix, the particle/polymer interface did not introduce mechanical or chemical weakness. With unmodified clay particles, this interface can be subjected to disbonding (i.e. pullout), preferential fracture and hydrolysis resulting in weaker bonding, lower brightness and discolouration.

In the plasma process, it is possible to polymerise a variety of monomeric species by grafting to the plasma-activated mineral surface. A two-stage procedure is followed in which the powdered sample is either spread evenly on a glass plate inside the plasma chamber or dispersed in the plasma using an internal rotating glass cylinder. In the first stage, the sample is treated in an argon plasma for periods varying from a few minutes to 1 h at 50 W. The surface hydrocarbon contamination is removed in the first few minutes and the subsequent argon plasma treatment is used to activate the surface by breaking bonds, with some surface reconstruction in the first one or two unit cells,[8] to produce a higher surface concentration of defect and low-coordination sites. In the second stage, the vapour of the monomer is introduced to the plasma for a relatively short period of time (i.e. a few minutes). Two processes appear to occur. The monomer is adsorbed on to the active sites of the surface layer with subsequent polymerisation under the plasma (i.e. high *KE* electrons, ions, radicals) bombardment. There is also some polymerisation of the monomer within the plasma itself contributing to deposition of additional polymer film on the adsorbed and polymerised surface layer. This second process can be confirmed by the deposition of similar polymer film on to the glass surface of the plasma chamber. The presence of the polymer films on the mineral surfaces is confirmed using both XPS and FTIR. Figure 18 gives some examples of polymer coatings on kaolinite and talc. The polystyrene υCH vibrations in the range 2800-3150 cm^{-1} with corresponding chain vibrations below 1800 cm^{-1} confirm the presence of this surface layer and can be correlated with the C1s XPS spectra from this layer. Similarly, polymethylvinylketone and polyacrylate can be bonded and coated on to kaolinite to provide chemical compatibility with the corresponding polymer matrix. One example of polyacrylate reaction on to talc surfaces confirms that this procedure is also effective on silicate clay minerals.

Current research is extending the use of this procedure to polymer coatings grafted on to silicate/silica coatings on metal substrates and, in particular, to polymer coatings on titanium oxide/silicate/silica for compatibility with polymethylmethacrylate often used as the initial binding material in hip and other implants to confer early load bearing before bone growth and macro-locking of bone to the implant. The prior treatment of the metal implant with the polymer is expected to add strength to the implant/polymer interfacial bonding. This proposition will be tested using mechanical and chemical procedures already developed in the literature and in our current test programme.

9.0 CONCLUSIONS

Low temperature plasma reactions, combined with sol-gel coatings, have been used to produce a variety of ceramic surface layers on metal substrates and interfacial layers between metals and oxides or other ceramics. These layers can be designed to be compositionally and functionally graded through the metal/metal oxide/orthosilicate/pyrosilicate/chain silicates/layer silicates/silica to bulk ceramic material, eg. silica, alumina, hydroxyapatite. The graded layers are generally <50 nm thick, continuous, fully bonded to the substrate and deformable without disbonding. The objectives in design of these layers have been to produce: metal surfaces protected from oxidation, corrosion and acid attack; improved metal-ceramic bonding; and bioceramic titanium-based interfaces to bioactive hydroxyapatite for improved dental and medical implants.

Using a sequence of plasma reactions with Ar/H_2O/tetraethoxysilane, we have been able to reduce the dissolution rates in pH1 nitric acid for nickel, iron, mild steel and brass by 2-3 orders of magnitude producing passivation for periods of weeks in concentrated acid. In dilute acids, passivation can be achieved for periods of months. Modified Auger parameter studies for Si in XPS spectra have been examined in both angle resolved and ion beam etching modes as depth profiles. The results show that the structure on the metal surfaces grades from amorphous, dehydroxylated silica on the outer surface through layer silicates, chain silicates, pyrosilicates to orthosilicates close to the metal interface. TEM evidence from a dimpled nickel surface reacted on one side with this sequence of plasma reactions reveals an outer layer of pure silica (5-10 nm) overlying an amorphous, compositionally graded underlayer containing embedded, dispersed nanocrystallites with nickel oxide (or silicate) structures (10-15 nm). At the metal interface, detached grains of the metal are imaged with interpenetration of the oxide and silicate species linking the layer to the oxidised metal surface. Scratch testing with an increasing normal force across the surface of the specimen and recording of the resultant tangential (frictional) force shows that the ~30 nm layer has a substantially increased frictional load compared with the untreated oxidised metal. This behaviour is consistent with either stronger adhesion of the coating to the substrate or a harder surface.

The composition, structure and thickness of these layers can be controlled by the duration of each plasma reaction and the choice of the final reagent. The mechanisms of reaction in each process step have been elucidated with a combination of XPS, TOF-SIMS, TEM, SEM and FTIR. Thicker (i.e. μm) overlayers of ceramic materials can be added to the graded surface layers using sol-gel deposition (e.g. silica, alumina, titania, zirconia and coprecipitated mixtures).

Similar, graded titanium/oxide/silicate/silica ceramic surface layers have been shown to form using the low temperature plasma reactions on titanium alloys used in medical and dental implants. These titanium-based surface layers have been successfully coated with porous HAP sol-gel coatings. An improved chemical treatment of the dehydroxylated outer silica surface was required to produce a continuous, coherent coating of HAP but this has now been achieved. Limited initial clinical testing has shown that the plasma-induced surface layer is not measurably altered in composition, structure or thickness by insertion and removal from a bone under stress and load conditions similar to implantation in dental application.

Further variation of the low temperature plasma reaction has been achieved using argon activation of an oxide substrate or the oxide/silicate/silica surface layer followed by introduction of reactive organic monomers. It is possible to coat particle surfaces of oxides (e.g. fillers), clays (e.g. kaolinite, talc) with thin bonded polymer overlayers (e.g. polyacrylate, polyurethane). These layers have also been characterised using the same methodology. The results show improved bonding of the surface-modified particles with polymer matrices and higher resistance to hydrolysis reactions at the particle/matrix interface. This is a promising new research direction with the potential for composite metal-ceramic-polymer bonded interfacial layers.

10.0 ACKNOWLEDGEMENTS

Thanks are due to Mr. Paul Wingate (Wingate Laboratories), Dr. John Abbott and Dr. Hien Ngo (Aus Systems) for supply of samples and chemical testing. Dr. Peter Self has greatly assisted with TEM preparation and measurements.

11.0 REFERENCES

1. W. D. Kingery (ed.): Ceramic Fabrication Processes, MIT Press, 1958.
2. J. T. Klomp: American Ceramic Society Bulletin, **59**, 1980, pp.794-799.
3. H. J. De Bruin: Reactions at Phase Boundaries During Metal/Ceramic Bonding, J. Nowotny and L. C. Dufour, eds., *Surface and Near-Surface Chemistry of Oxide Materials*, Elsevier, Amsterdam, 1988, pp.507-526.
4. M. G. Nicholas: Ceramic -Metal Interfaces, J. Nowotny, ed., *Science of Ceramic Interfaces*, Elsevier, Amsterdam, 1991, pp.411-438.
5. E. Gillet: Crystallographic and Electronic Structure of Metal-Ceramic Interfaces, J. Nowotny, ed., *Science of Ceramic Interfaces*, Elsevier, Amsterdam, pp.439-456.
6. R. E. Loehman: *Scripta Metallurgica et Materialia*, **31**, 1994, pp.965-970.
7. B. J. Dalgleish, E. Saiz, A. P. Tomsia, R. M. Cannon and R. O. Ritchie: *Scripta Metallurgica et Materialia*, **31**,1994, pp.1109-1114.
8. R. St. C. Smart, P. Arora, B. Braggs, H. M. Fagerholm, T. J. Horr, D. C. Kehoe, J. G. Matisons, J. Ralston and J. B. Rosenholm: Modification of Ceramic Surfaces Using Adsorption, Plasma and Sol-Gel Reactions, K. Uematsu, Y. Moriyoshi, Y. Saito and J. Nowotny, eds., *Interfaces of Ceramic Materials,*

Engineering Materials, Trans. Tech. Publications, Switzerland, **111-112**, 1995, pp.361-404.

9. T. Kokubo: *Anales De Quimica*, **93**, 1997, pp.S49-S55.
10. S. Suresh and A. Mortensen: *International Materials Reviews*, **42**, 1997, pp.85-116.
11. M. R. Locatelli, A. P. Tomsia, K. Nakashima, B. J. Dalgleish and A. M. Glaeser: New Strategies for Joining Ceramics for High-Temperature Applications, *Interfaces of Ceramic Materials,* Trans. Tech. Publications, Switzerland, 1995, pp.157-189.
12. A. M. Glaeser: *Composites Part B: Engineering*, **28**, 1997, pp.71-84.
13. R. Singh, J. Fitzgerald: *Journal of Materials Research*, **12**, 1997, pp.769-773.
14. F. Watari, A. Yokoyama, F. Saso, M. Uo and T. Kawasaki: *Composites Part B: Engineering*, **28**, 1997, pp.5-11.
15. T. Kokubo: *Acta Materialia*, **46**, 1998, pp.2519-2527.
16. W. Q. Yan, T. Nakamura, M. Kobayashi, H. M. Kim, F. Mayaji and T. Kokubo: *Journal of Biomedical Materials Research*, **37**, 1997, pp.267-275.
17. W. Q. Yan, T. Nakamura, K. Kawanabe, S. Nishigochi, M. Oka and T. Kokubo: *Biomaterials*, **18**, 1997, pp.1185-1190.
18. H. M. Kim, F. Miyaji, T. Kokubo and T. Nakamura: *Journal of Materials Science-Materials in Medicine*, **8**, 1997, pp.341-347.
19. H. M. Kim, F. Miyaji, T. Kokubo and T. Nakamura: *Journal of Biomedical Materials Research*, **38**, 1997, pp.121-127.
20. H. M. Kim, F. Miyaji, T. Kokubo and T. Nakamura: Nippon Seramikkusu Kyokai Gakujutsu Ronbunshi, *Journal of the Ceramic Society of Japan*, **105**, 1997, pp.111-116.
21. H. M. Kim, F. Miyaji, T. Kokubo and T. Nakamura: *Journal of Biomedical Materials Research*, **32**(3), 1996, pp.409-417.
22. T. Kitsugi, T. Nakamura, M. Oka, W. Q. Yan, T. Goto, T. Shibuya, T. Kokubo and S. Miyaji: *Journal of Biomedical Materials Research*, **32**(2), 1996, pp.149-156.
23. T. Kokubo: *Thermochimica Acta*, **280**, 1996, pp.479-490.
24. T. Kokubo, F. Miyaji, H. M. Kim and T. Nakamura: *Journal of the American Ceramic Society*, **79**, 1996, pp.1127-1129.
25. K. Takatsuka, T. Yamamuro, T. Nakamura and T. Kokubo: *Journal of Biomedical Materials Research*, **29**, 1995, pp.157-163.
26. P. J. Li, I. Kanasniemi, K. Degroot and T. Kokubo: *Journal of the American Ceramic Society*, **77**(5), 1994, pp.1307-1312.
27. P. J. Li, C. Ohtsuki, T. Kokubo, K. Nakanishi, N. Soga and K. Degroot: *Journal of Biomedical Materials Research*, **28**(1), 1994, pp.7-15.
28. P. S. Arora and R. St. C. Smart: *Surface and Interface Analysis*, **24**, 1996, pp.539-548.
29. C. D. Wagner, D. E. Passoja, H. F. Hillery, T. G. Kinisky, H. A. Six, W. T. Jansen and J. A. Taylor: *Journal of Vacuum Science and Technology*, **21**, 1982, p.933.
30. C. D. Wagner: Auger and Photoelectron Energies and the Auger Parameter: A Data Set, D. Briggs and M. P. Seah, eds., *Practical Surface Analysis by Auger and X-Ray Photoelectron Spectroscopy*, 1983, pp.477-509.

31. R. Alfonsetti, G. DeSimone, L. Lozzi, M. Passacantando, P. Picozzi and S. Santucci: *Surface Interface Analysis*, **22**, 1994, p.89.
32. R. Alfonsetti, L. Lozzi, M. Passacantando, P. Picozzi and S. Santucci: *Applied Surface Science*, **70/71**, 1993, p.222.
33. R. Alfonsetti, L. Lozzi, M. Passacantando, P. Picozzi and S. Santucci: *Thin Solid Films*, **213**, 1992, p.158.
34. R. St. C. Smart and P. Arora: Ceramic Surface Layers on Oxidised Metal Surfaces Using Plasma and Sol-Gel Technology, *Proceedings Second International Meeting of Pacific Rim Ceramic Societies*, P. Walls, C. C. Sorrell and A. Ruys, eds., CD (ISBN 1440-4192), Symp., **21**, 1996, pp.35-41.
35. N. Kawashima, P. Self, P. S. Arora and R. St. C. Smart: TEM Observation of Silica Thin Films Produced on Nickel by Low Temperature PACVD and Sol-Gel Method, *Proceedings 3rd International meeting of Pacific Rim Ceramic Societies*, Korea Ceramic Society, Kyongju, South Korea, 1998, p.104.
36. W. J. McTtegart: *The Electrolytic and Chemical Polishing of Metals*, Pergamon Press, Oxford, 1956.
37. W. R. Pease, R. L. Segall, R. St. C. Smart and P. S. Turner: *Journal Chem. Soc. Faraday*, **76**, 1980, p.1510.
38. J. W. E. Coenen: Scientific Basis for Preparation of Heterogeneous Catalysts, Conference Review, Louvain, Belgium, 1978.
39. J. C. Vedrine, G. Hollinger and T. M. Duc: *Journal of Physical Chemistry*, **82**, 1978, p.1515.
40. S. J. Bull: Scratch Adhesion Testing and its Significance for Coating Performance, *Quality Control and Assurance in Advanced Surface Engineering*, K. N. Strafford and C. Subramanian, eds., Institute of Materials Press, U.K., 1997, pp.161-182.
41. A. Sackfield and D. A. Hills: *Journal of Strain Analysis*, **18**, 1983, p.107.
42. V. Bellido-Gonzalez, N. Stefanopoulos and F. Deguilhen: *Surface and Coatings Technology*, **74-75**, 1995, p.884.
43. K. Soballe: *Acta Orthopaedica Scandinavica Supplementum*, **64**, 1993, p.1.
44. L. L. Hench and H. A. Paschall: *Journal of Biomedical Materials Research Symposium*, **4**, 1973, p.25.
45. P. J. Li, K. Nakanishi, T. Kokubo and K. Degroot: *Biomaterials*, **14**, 1993, pp.963-968.
46. M. Neo, T. Nakamura, C. Ohtsuki, T. Kokubo and T. Yamamuro: *Journal of Biomedical Materials Research*, **27**, 1993, pp.999-1006.
47. S. B. Cho, K. Nakanishi, T. Kokubo, N. Soga, C. Ohtsuki, T. Nakamura, T. Kitsugi and T. Yamamuro: *Journal of the American Ceramic Society*, **78**, 1995, pp.1769-1774.
48. H. M. Kim, F. Miyaji, T. Kokubo, C. Ohtsuki and T. Nakamura: *Journal of the American Ceramic Society*, **78**, 1995, pp.2405-2411.
49. S. B. Cho, F. Miyaji, T. Kokubo, K. Nakanishi, N. Soga and T. Nakamura: Nippon Seramikkusu Kyokai Gakujutsu Ronbunshi, *Journal of the Ceramic Society of Japan*, **104**, 1996, pp.399-404.
50. S. B. Cho, K. Nakanishi, T. Kokubo, N. Soga, C. Ohtsuki and T. Nakamura:

Journal of Biomedical Materials Research, **33**, 1996, pp.145-151.

51. M. Nagano, T. Nakamura, T. Kokubo, M. Tanahashi and M. Ogawa: *Biomaterials*, **17**, 1996, pp.1771-1777.
52. H. M. Kim, F. Miyaji, T. Kokubo,M. Kobayashi and T. Nakamura: *Bulletin of the Chemical Society of Japan*, **69**, 1996, pp.2387-2394.
53. P. Li, K. Degroot and T. Kokubo: *Journal of Sol-Gel Science and Technology*, **7**, 1996, pp.27-34.
54. W. Bonfield, M. Wang and K. E. Tanner: *Acta Mater.*, **46**, 1998, pp.2509-2518.
55. M. Steveson, P. S. Arora and R. St. C. Smart: *Surface and Interface Analysis*, **26**, 1998, pp.1027-1034.
56. J. Lausmaa and B. Kasemo: *Applied Surface Science*, **44**, 1990, p.133.
57. T. Kokubo: Titanium-Based Bone -Bonding Composites Having Inverted Concentration Gradients of Alkali and Titanium Ions in a Surface Layer, US Patent Number 5,609,633, 1997.
58. I. Sutherland, E. Sheng, D. M. Brewis and R. J. Heath: *Journal of Materials Chemistry*, **4**, 1994, pp.683-687.
59. C. J. Brinker and G. W. Scherer: *Sol-Gel Science*, Academic Press Inc., 1990.
60. R. K. Iler: *The Chemistry of Silica*, Wiley and Sons, N.Y., 1979.
61. T. Kokubo, S. Ito, Z. T. Huang, T. Hayashi, S. Sakka, T. Kitsugi and T. Yamamuro: *Journal of Biomedical Materials Res.*, **24**, 1990, p.331.
62. Y. Masuda, K. Matubara and S. Sakka: *Journal Ceramic Society of Japan*, **98**, 1990, pp.1255-1266.

A Novel Optical Microscopic Approach for Understanding the Formation of Large Defects in Processing of Alumina Ceramics and its Relevance to Fracture

KEIZO UEMATSU, YAO ZHANG and NOZOMU UCHIDA
Department of Chemistry
Nagaoka University of Technology
Kamitomioka, Nagaoka, Niigata, Japan

ABSTRACT

An outline of two novel characterisation methods is presented, one for observing microstructure of sintered ceramics and the other of green compacts. They show detailed optical images of their internal structures through observation using transmission modes. The origins of large defects were identified as pores in sintered powder granules and low-density regions at the boundaries of granules in compacts. The analysis of the micrographs provided direct information on the defect size distribution in the ceramics. Strength variation was calculated with the Baratta model (spherical pore with circumferential crack) correlated with the defect size population. Comparison of the result to that of measured strength showed that both results agreed well when the crack size around a spherical pore is assumed to be about 4 times the grain size. These results demonstrate the successful and direct establishment of a processing-structure-property relationship for the first time.

1.0 INTRODUCTION

The uncontrolled variation of strength is one of the most serious problems in the application of ceramics. It makes the minimum strength of the material uncertain, and causes sudden failure of ceramic components below a designed stress level in normal use. The problem is only partially solved even with the recent design methodology based on a statistical approach. Currently, all the important components for critical applications need to be tested before application. This introduces another problem. The tedious inspection and testing, as well as high ratio of rejects, drastically increases the price of some ceramics. These problems must be solved urgently for wider application of ceramics.

The origin of strength variation is clearly the processing flaws in the microstructure of ceramics, which are developed in the processing stage of ceramics.[1] Their formation appears to be very sensitive to the processing conditions, and even a minor variation in processing parameters, affects strength properties significantly. Production under nominally controlled processing conditions is insufficient for consistent properties of ceramics. A typical example is found in a recent paper, which reports a systematic variation of properties with the climatic season of manufacturing.[2] Changes of temperature and/or humidity with season are clearly responsible for the problem. The origin of the variation should be identified directly for consistent production of ceramics.

Large pores and grains are very important flaws in the microstructure of ceramics.[1] They behave as sites of stress concentration. When a ceramic is loaded, the stress is concentrated around them.[3] The local stress value can be tens of times higher than the average stress. The ceramics fails catastrophically when the stress exceeds a certain critical value at one of these flaws. The stress concentration depends on the characteristics of the flaw, such as size, shape, location, etc. Usually, the single worst flaw, governs the strength of the ceramic.[1] A full understanding of the characteristics of detrimental flaws is clearly the starting point for examining the strength and its variation in ceramics with processing.

Direct characterisation of these flaws and non-homogenieties in microstructure requires a special technique. They are part of the microstructure, albeit unusual examples, and their detection is very difficult with conventional methods such as SEM observation. The most important requirement for a potential technique is that it should be able to find extremely rare features. A large volume must be examined to find an unusual feature of microstructure. In this respect, an examination of microstructure through transmission mode is crucial to characterise large pores and grains. High resolution may not be required, since the subject feature normally has a large size.

Simple optical microscopy is well suited for this purpose.[4] Many ceramics including alumina, silicon nitride and silicon carbide are characteristically transparent materials. They appear opaque due to the internal scattering of light from residual pores and/or second phases in thick sections. Made into thin sections, they become transparent, and their internal structure can be directly inspected with an optical microscope in the transmission mode.

Once large defects are characterised, they should be directly related to the strength characteristics of ceramics. The strength of ceramics has been modelled in terms of large defects in microstructure, but the models and tests all apply to artificial defects of spherical shape.[5] There appears to be no direct model or test for ceramics with natural defects. The detailed characteristics of real defects will provide a better opportunity for examining the basic theory of fracture for ceramics.

The structure of green compacts also needs to be examined to understand the origin of defects in ceramics. The structure of ceramics is developed from the structure of the green body in the densification process.[6] However, the green body can not be made transparent by simple thinning. The transparency of the green body needs to be improved to examine its structure with optical microscopy. This can be achieved with a designed immersion liquid,[7] in which the transparency of the green body increased drastically by suppressing the scattering of light at the surfaces of powder particles. Full understanding of defect structures of green and sintered bodies provides the missing link in the processing-structure-property relationship.

This paper presents a novel optical microscopic approach for understanding the formation of large defects in processing of alumina ceramic.[8] Their possible origins will be discussed to suggest the approach for improvement. The relation of these large defects to the fracture strength will be also discussed to directly show the effect of defect characteristics on the strength and its variation in ceramics.[9,10] Other typical non-uniformities in powder granules and compacts will also be presented.

2.0 METHOD OF OBSERVATION

The microstructures of green and sintered bodies can be best examined by observing their interior directly with a transmission optical microscope.[4,7,8] Both normal and cross polarized light microscopes are used for observation. Features successfully characterised with the former microscope include pores, cracks, inclusions and organic additives. Those characterised with the latter microscope are aggregates, orientation of particles and large particles and grains. The details of methods and results have been presented elsewhere.[11-13] The outline of the methods is presented here.

In the observation of sintered ceramics, a thin specimen is needed with both surfaces mirror polished using diamond paste. The allowed thickness for observation varies with the relative density and material. A typical thickness is 50 μm for industrial grade alumina with 98% of theoretical density. It can be over 1 mm for hot isostatically pressed alumina with theoretical density above 99.9%,[4] since it has high transparency.

The structure of green bodies can also be examined after they are made transparent with an adequate immersion liquid as shown in Figure 1. The principle for transparency is as follows. For normal incidence of light, the fraction of light reflected at an interface R is given by,

$$R=(n'-1)^2/(n'+1)^2 \quad . \tag{1}$$

where n' is the relative refractive index given by the ratio of refractive index of the powder particle (n_1) to that of the surrounding medium(n_2). An adequate immersion liquid with $n'=1$ diminishes the light reflection at the interface and realizes the complete transmission of light through the body.

The optimum relative refractive index varies with the object of interest. The best value is unity in the observation of structures relevant to foreign materials in the matrix, such as inclusions, binders and other additives, for which the refractive index is different from that of the matrix powder. This value makes the matrix highly transparent and the structure of interest less transparent. In the observation of structures associated with non-uniform packing density in the structure, such as pores and cracks, the relative refractive index should be slightly off unity. A slight residual reflection of light is needed to generate optical contrast for these structures. Excessively transparent specimen shows no structure in the microscopic observation. In the observation of aggregates, particle orientation and large particles and grains with a crossed polarized light, the relative refractive index should be close to unity. The optical contrast is generated by the anisotropic optical property of the material in this observation, not by the residual reflection of light at the interface.

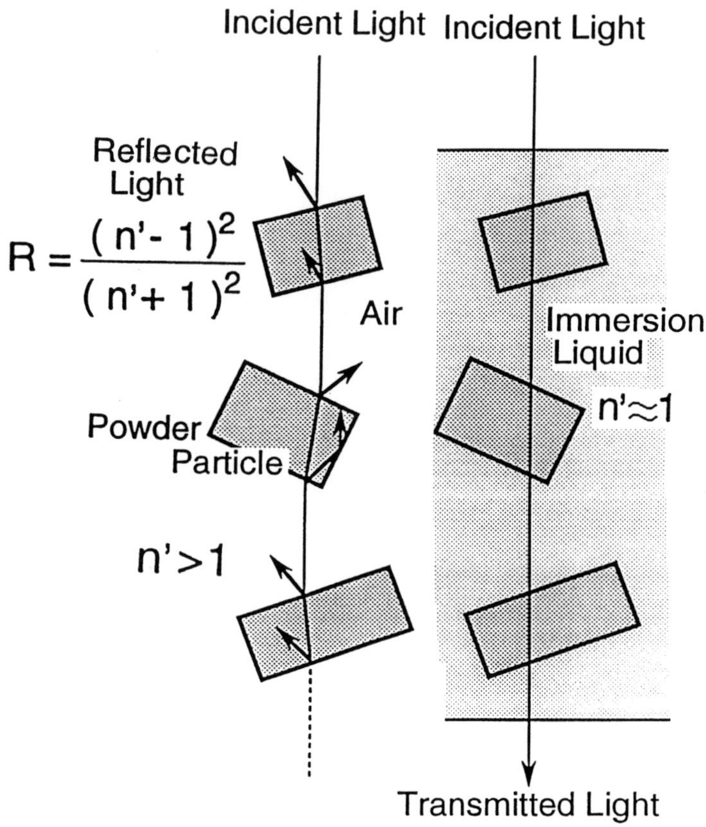

Fig. 1 Principle of the immersion liquid method.

3.0 SPECIMENS

An alumina ceramic distributed by Japan Fine Ceramic Centre and its raw materials were used in this study. The ceramic is called Referceram, and is often applied for domestic and international round-robin tests. It has consistent characteristics and is regarded as a semi-reference material. Details of the specimen and its preparation procedure are presented elsewhere.[6]

4.0 RESULTS OF OBSERVATION

Figure 2 shows spray-dried granules observed by the liquid immersion technique with normal light. They all contain internal defects, and have irregular shapes.[6] In

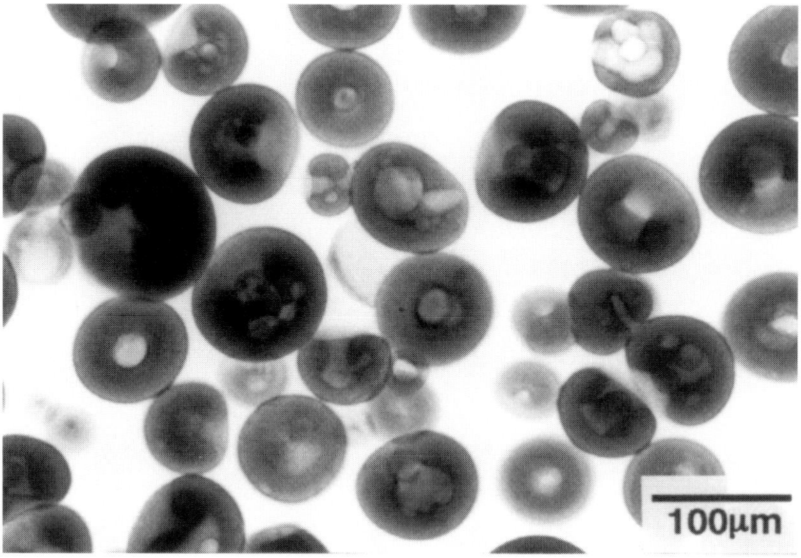

Fig. 2 Immersion liquid optical micrograph of spray dried granules.

this observation, the internal structure of pores can be better examined with the present method than with SEM. With SEM, surface exposed on dimples will not be seen if they are placed downward. Both slurry characteristics and the drying conditions appear to affect the shape of granules and the formation of these pores. The strong effect of slurry characteristics on the shape and structure of silicon nitride granules has been reported.[14] The granules made from the dispersed slurry have dimples without exception. Those made from flocculated slurry have round shape without dimples.

Figure 3 shows the microstructure of an alumina compact made of the above granules with dimples and internal pores.[6] The compact has a non-homogeneous structure. The boundaries of granules are clearly visible. The interfaces of granules contain cracks and low-density regions. Cracks are also often noticed inside the granules. Those in the granules are clearly formed from the dimples in the granules.

Figure 4 shows the microstructure for a sintered ceramic of alumina. It contains many features and their close correlation to the green body structure is clear. Large round pores are clearly developed from the dimples in granules. They survive the forming process of compaction and the subsequent sintering process. There are also sharp narrow pores in the microstructure. They clearly are formed from the low-density regions at the granule boundaries. These large defects must be very detrimental to strength characteristics since each one of them is potentially a site of fracture initiation.

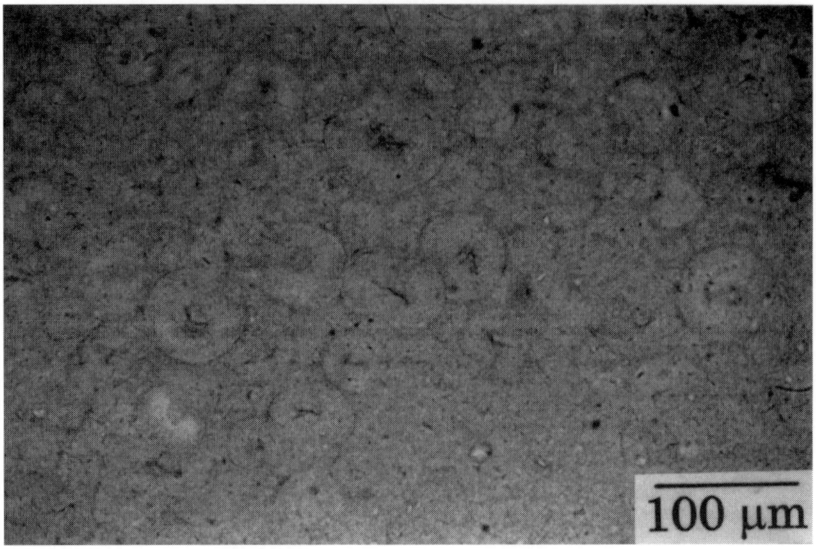

Fig. 3 Packing structure of powder particles in alumina green compact.

Figure 5 shows a logarithmic plot of the pore size distribution.[10] The maximum defect found in the specimen was 52 μm in the total observation volume of 0.225 mm^3. The pores larger than 14 μm are examined for the subsequent evaluation. It shows two linear regions over the size region examined, suggesting that the following equation can represent the result for the large pore sizes >28 μm.

$$y = ax^{-b} \tag{2}$$

Small pores < 28 μm also follow a linear relation for the region examined. However, there should be a limit for this relation. There should be a maximum in the relation. Otherwise, number of small pores would be too large to account for the relative density of the ceramic.

The data of defect size larger than 28 μm were used to fit a defect distribution function, since these large defect are critical to the material strength. Other functions were also examined, but the exponential function in Eqn. 2 was found to provide the best fit. For the present material, the values of a and b are 4.5×10^{10} and 5.97, respectively. It should be of great interest to directly examine the relevance of these large pores to the strength characteristics of the ceramics from first principles.

5.0 STRENGTH CALCULATION BASED ON LARGE DEFECTS

Based on the experimental results, simulated specimens with a given size distribution of flaws were generated in a computer, and the strength was calculated for each of

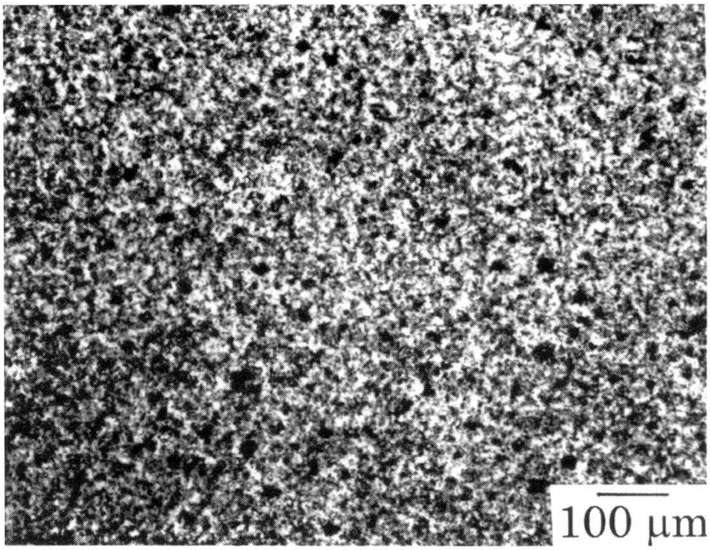

Fig. 4 Microstructure of sintered ceramics.

them with the fracture mechanics model.[9, 10] Figure 6 illustrates an example of the model specimen for a four-point bending test by computer. A given number of defects with various sizes were placed randomly in a specimen with size 4 × 1.5 × 10 mm, which corresponds to a volume under tension in the strength measurement. The shape of the defect was represented by the Baratta's model,[15,16] i.e. a spherical void with a circumferential crack. This was justified since a majority of defects in the material were spherical. The interaction between defects was neglected.

In the calculation, any defect in the above volume of tensile stress was regarded as a potential fracture origin. The stress intensity for each flaw was calculated at a given load from their position and size. The load was increased until the stress intensity for any one defect reached the critical stress intensity K_{IC} of the material. The strength of this specific model specimen is determined with this load. This procedure was repeated to determine the strength for a large number of specimens.

The following relations were used in this study. The fracture stress for *i*th defect can be calculated from its size and the location as follows:

$$\sigma_{fi} = \frac{1}{Y}\frac{K_{IC}}{\sqrt{\pi L}} f(x,z) \tag{3}$$

where K_{IC} is the critical stress intensity factor, which is assumed to be a constant in the present calculation (4.0 MPa · $m^{1/2}$ for the material). *Y* is a parameter which depends on the specimen and defect geometry, and is given for this model by[10]

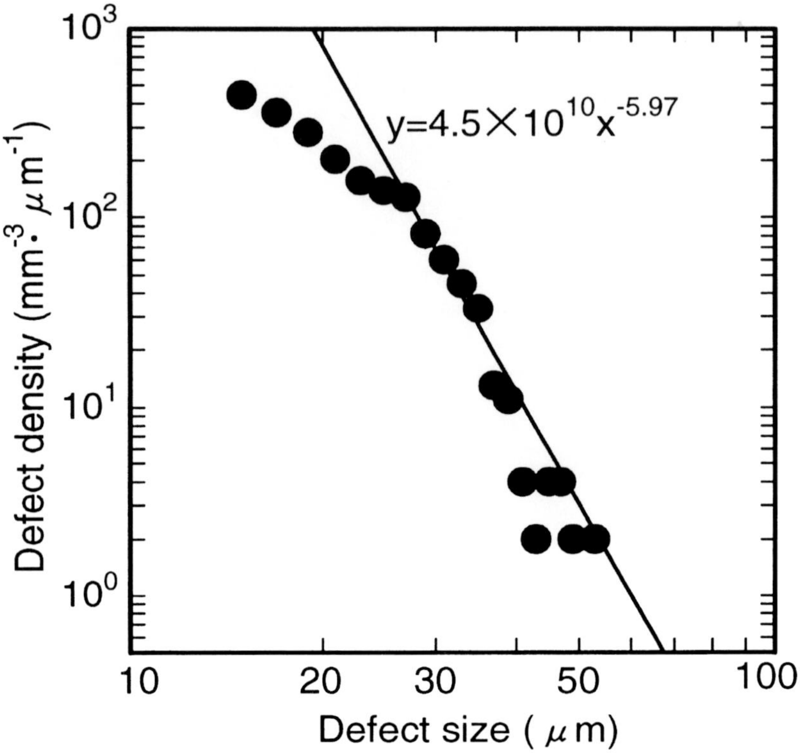

Fig. 5 Pore size distribution in alumina ceramics.

$$Y = [c - k \cdot \arctan(\alpha)]\left\{\frac{1}{2(7-5v)}\left[(4-5v)\left(\frac{1}{\alpha+1}\right)^3 + 9\left(\frac{1}{\alpha+1}\right)^5\right]+1\right\} \quad (4)$$

where c and k are constants. c is 1.12 and k is 0.3 and 0.22 for the internal defect and the surface defect, respectively. ν is Poisson's ratio of the material, and is taken as 0.25 here. α is

$$\alpha = \frac{L}{R} \quad (5)$$

where R is the radius of the void. L is the crack length. L is assumed to be in the range 1-10 time the grain size.

The function $f(x, z)$ in Eqn. 3 is the term defining the position of the defect, where x and z are coordinates of the centre of the defect. For $|x|<S_1$, $0 < z \leq h$, $f(x, z)$ is given by,

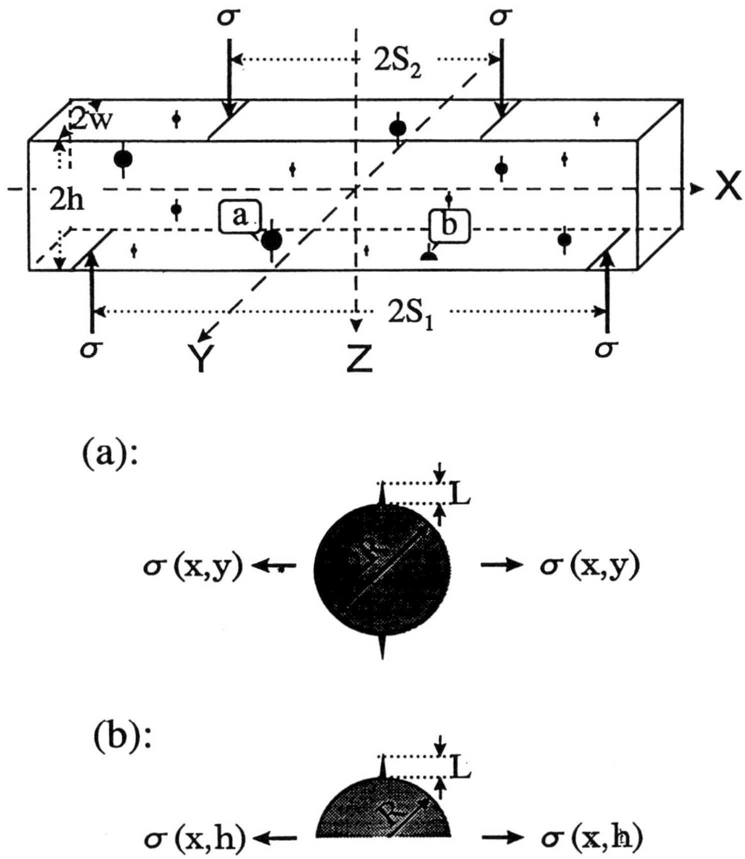

Fig. 6 Model specimen used for the calculation of strength, in a personal computer, a bending specimen was generated, in which defects of various sizes were placed randomly by following the eqn. (2).

$$f(x,z)=\begin{cases} \dfrac{h}{z} & |x|\le S_2 \\ \dfrac{h}{z}\cdot\dfrac{1}{S_1-|x|} & S_2<|x|<S_1 \end{cases} \tag{6}$$

where h = 1.5 mm, S_1 = 15 mm, S_2 = 5 mm.

Of all fracture stresses σ_{fi} calculated by the above equations, the lowest is taken as the fracture strength of the specimen. To remove the effect of random seeding, 500 specimens were calculated.

Figure 7 shows the Weibull plot of the experimentally measured and the calculated strengths for the material. Good agreement was found for the measured and the calculated strengths when the crack size is taken as 4 times the grain size.

6.0 DISCUSSION

Defects in ceramic materials could be easily characterised by the present technique. The results provide us with better understanding than any other conventional methods on the characteristics of flaws in ceramics. They also deepen our understanding of fracture mechanics. This is an important result, which shows the relation between the characteristics of natural flaws in a material before fracture and the strength of the material.

The characteristics of the flaws even suggest the formation mechanisms of these defects as discussed above. In brief, the large pores of approximately spherical shape were developed from dimples in granules used for the production of this ceramic through the powder compaction process. Pores of more or less irregular shape are formed from the low-density regions at the boundaries of granules.

The size of flaws responsible for fracture seems to be much larger than those actually found in a small volume, i.e., the size of the largest fracture origin was 170 μm in 500 specimens generated by computer. This is much larger than the largest defect, which was actually found in 0.225 mm^3 in the microscopic observation, about 50 μm. This size is acceptable, although it may appear unrealistic in the first impression. Defects of this size have actually been found in a HIPed alumina, where the high transparency associated with the high density allows one to examine flaws in much larger volume, i.e. 15 mm^3.[4]

In the calculation of strength with the defect size, the parameter L was adjusted to be 1-10 times the grain size. Good agreement was found between the experiment and the calculation at crack lengths 4 times the grain size for the tested material. The presence of cracks of this size have been suggested in other studies.[12,13,17] Although the real significance of parameter L is still unclear, two effects should be considered. One is the presence of real cracks, the size of which is nearly equal to the grain size. The other is the factors derived from the effects of the finite specimen, R-curve behaviour and sub-critical crack growth. In a bending test, the fracture occurs often from defects on or near the tension surface where the maximum stress is applied to the material in the specimen. The stress concentration around a defect near the surface of a finite specimen is several times higher than that for the internal defect far from the surface[15] This requires a large value of L in the calculation, since Baratta's model is valid for the internal defects in an infinite specimen or the surface defects on the semi-infinite specimen. K_{IC} was taken as a constant in the calculation, however it can change with the defect size because of the R-curve behaviour.

In a recent paper, Rodel et al. showed that the value of K_{IC} for small cracks was around 2.5 MPa $m^{1/2}$ for all alumina ceramics for all grain sizes.[17, 18] If this value is correct, the cracks around the pores can be much shorter, about 2 times of the grain size. This value also appears realistic. At the present stage of study, it is difficult to clarify the true length L and also its physical meaning. Further study is needed to clarify them.

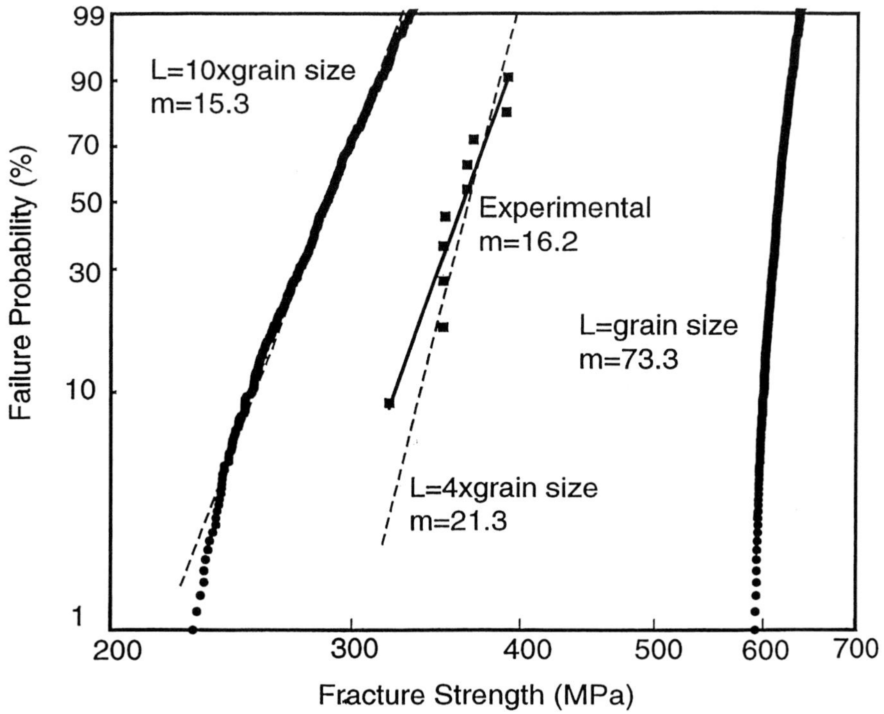

Fig. 7 Comparison of measured and calculated strengths, solid squares show the measured strength variation. Solid circles show strengths calculated for two conditions, with the crack size, L, equal to the grain size and ten times the grain size. The dotted line show the strength calculated with $L = 4$ times the grain size.

7.0 REFERENCES

1. F. F. LANGE: *Journal of American Ceramics Society*, **72**, 1989, p.3.
2. N. SHINOHARA, M. OKUMIYA, T. HOTTA, K. NAKAHIRA, M. NAITO and K. UEMATSU: *American Ceramics Society Bulletin*, **78**, 1999, p.81.
3. B. R. LAWN: *Fracture Mechanics of Ceramics*, R. C. Bradt, A. G. Evans, D. P. H. Hasselman and F. F. Lange, eds., Plenum, New York, **5**, 1983, p.1.
4. K. UEMATSU, M. SEKIGUCHI, J. Y. KIM, K. SAITO, Y. MUTOH, M. INOUE, Y. FUJINO and A. MIYAMOTO: *Journal of Materials Society*, **28**, 1993, p.1788.
5. A. ZIMMERMANN, M. HOFFMAN, B. D. FLINN, R. K. BORDIA, T. J. CHUANG, E. R. FULLER and J. RODEL: *Journal of American Ceramics Society*, **81**(9), 1998, pp.2449-2457.
6. N. SHINOHARA, M. OKUMIYA, T. HOTTA, K. NAKAHIRA, M. NAITO and K. UEMATSU: *Journal of Materials Science*, **34**, 1999, p.4271.

7. K. Uematsu, J. Y. Kim, M. Miyashita, N. Uchida and K. Saito: *Journal of American Ceramics Society,* **73**, 1990, p.2555.
8. K. Uematsu: *Powder Technology*, **88**, 1996, pp.291-298.
9. Y. Zhang, M. Inoue, N. Uchida and K. Uematsu: *Journal of Materials Research,* **14**, 1999, p.3370.
10. K. Uematsu, Y. Zhang, N. Uchida, T. Hotta, M. Naito, N. Shinohara and M. Okumiya: *Key Engineering Materials*, **161-163**, 1999, pp.145-150.
11. Y. Iwamoto, H. Nomura, I. Sugiura, J. Tsubaki, H. Takahashi, K. Ishikawa, N. Shinohara, M. Okumiya, Y. Yamada, H. Kamiya and K. Uematsu: *Journal of Materials Research*, **97**, 1994, p.1208.
12. K. Uematsu, H. Ito, Y. Zhang and N. Uchida: *Ceramic Transactions*, **54**, 1995, p.83.
13. K. Uematsu, H. Ito, S. Ohsaka, H. Takahashi, N. Shinohara and M. Okumiya: *Journal of American Ceramics Society,* **78**, 1995, p.3107.
14. H. Takahashi, N. Shinohara, M. Okumiya, K. Uematsu, J. Tsubaki, Y. Iwamoto, and H. Kamiya: *Journal of American Ceramics Society,* **78**, 1995, p.903.
15. F. I. Baratta: *Journal of American Ceramics Society,* **61**(11-12), 1978, p.490.
16. F. I. Baratta: *Journal of American Ceramics Society,* **64**(1) 1981, p.C3.
17. A. Zimmermann, M. Hoffman, B. D. Flinn, R. K. Bordia, T. J. Chuang, E. R. Fuller Jr, and J. Rodel: *Journal of American Ceramics Society*, **81**, 1998, p.2449.
18. J. Seidel and J. Rodel: *Journal of American Ceramics Society*, **80**(2), 1997, p.433.

Formation Mechanisms of the Grain Boundary Layer of Pr-Doped ZnO Varistor Through Eutectic Phenomena

SUNG-YONG CHUN, KAZUO SHINOZAKI and NOBUYASU MIZUTANI
Department of Inorganic Materials
Faculty of Engineering
Tokyo Institute of Technology
2-12-1 O-okayama, Meguro-Ku, Tokyo 152, Japan

ABSTRACT

The morphology of grain boundary and the behaviour of mass transport in praseodymium(Pr) doped ZnO ceramics are strongly affected by firing atmosphere and temperature. The system ZnO-PrO_x, showed a simple eutectic type phase diagram with eutectic point of c.a. 50 mol.% of Pr at 1382 ± 5°C. The drastic changes in grain boundary morphology and mass transport behaviour between 1350 and 1400°C were ascribed to the increase of the vaporisation contents with the formation of liquid phase. The penetration of liquid (ZnO-PrO_x) into the grain boundaries of sintered Co-doped ZnO pellets resulted in varistors with breakdown voltages per grain boundary in the range of 1-3 V and non-linearity coefficients of 15-58.

1.0 INTRODUCTION

Zinc oxide varistors containing rare-earth metal oxides[1-6] have been widely used for protection of electronic devices from transient over voltage. The microstructure of the ZnO varistors consists of highly conductive n-type ZnO grains and electrically insulating Schottky barriers of grain boundaries.

Bismuth oxide doped ZnO varistors were devised in the early 1970s by Matsuoka.[7] It has been known that Bi_2O_3 melts at relatively low temperature (860°C) and forms grain boundaries after sintering between 1100-1300°C. From these studies, it is now recognized that liquid phase has been considered to be The Schottky barrier at the grain boundary of the ZnO microstructure. A variety of works have been presented describing the dependence of varistor properties on processing conditions about ZnO varistors but few of them were studied about Pr-doped ZnO varistors, especially in the field of grain boundary formation. To understand the structure, composition, and properties of the boundaries of a solid or liquid, it is very important to understand the interfaces between phases, for they have a strong influence on many mechanical properties, chemical phenomena and electrical properties.

In Pr-doped ZnO varistor systems. Mukae et al. suggested that oxidation of interstitial zinc donors in the grain boundary region caused the reduction of Pr_6O_{11} to Pr_2O_3 during sintering, resulting in abrupt changes in the electrical characteristics of ceramic varistors.[8] Alles and Burdick reported that abrupt changes in donor concentration and equilibrium barrier results arose from a marked increase in the fraction of electrically active grain boundaries.[4]

In this study, the author prepared Pr-doped ZnO varistors and Pr, Co-doped ZnO varistors respectively and analysed their grain growth behaviour and grain boundary formation mechanism. In addition, phase-diagrams and electrical properties were also investigated to confirm the mechanism of the formation of grain boundary phase, which is closely related to liquid-phase.

2.0 EXPERIMENTAL

2.1 Preparation and Characterisation of Sample

Reagent-grade raw materials of ZnO and Pr_6O_{11} were used for preparation of ZnO varistor samples. Raw materials were mixed in wet condition, by ball milling for 24 hrs. using alcohol dried at 200°C. The dried powder was ground and sifted through a 325-mesh screen and packed at a pressure of 100 MPa. The disks were sintered at 1200 to 1500°C in N_2, air and O_2 for 2 hrs. and furnace cooled. Microstructures were observed with a scanning electron microscope, while the measurement of mean grain size and grain size distribution was carried out.

2.2 Differential Thermal Analysis (DTA)

DTA analysis was carried out on laboratory-made DTA apparatus with a Pt/30% RhPt/6% Rh thermocouple for detecting the DTA signal in air. The DTA signal was detected by a digital microvolt meter (R6871-DC, Advantest, Japan) and the amplified signal was transferred to the microcomputer through an IEEE-488 interface.

2.3 Measurement of Depth Profiles of Pr Concentration

The Pr-concentration distribution of sintered bodies, as well as the composition, was analysed using an ICP atomic emission analyser (ICP-AES; SPS1500VR; Seiko instruments Inc.) In this case, the Pr-concentration distribution measurement of the sintered bodies was performed by compositional analysis of samples. For this experiment, the powders of the pellets were ground from surface into the bulk at intervals of 50 μm. The ground powders were collected and dissolved in nitric acid for ICP analysis.

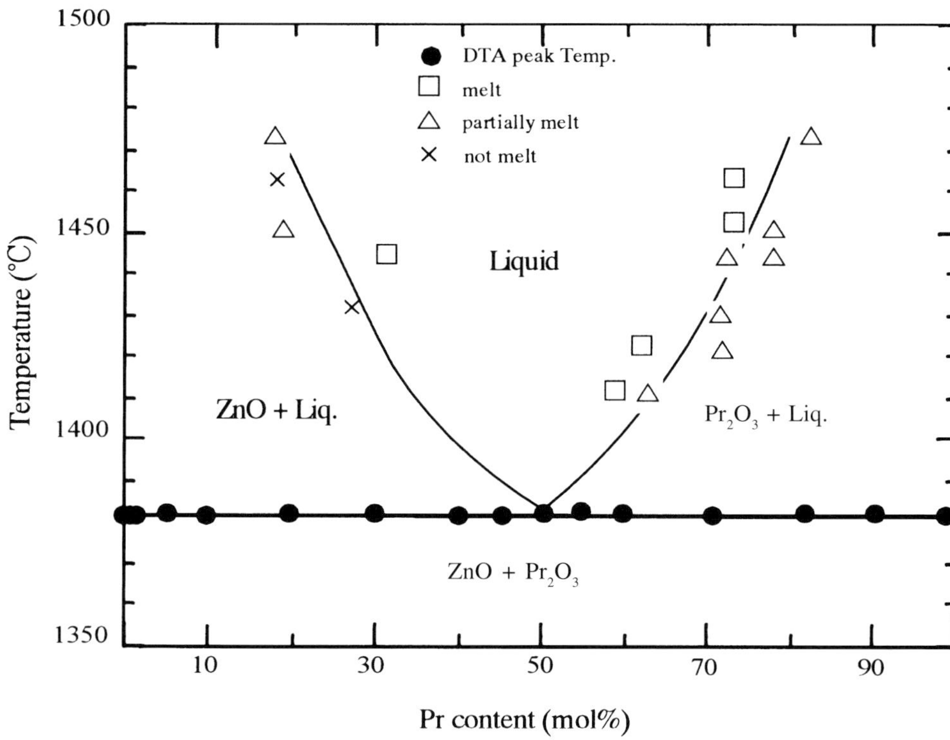

Fig. 1 Experimentally determined ZnO solidus and liquidus boundary for the ZnO-PrO_x system.

2.4 Measurements of Varistor Properties

The current-voltage characteristics and the capacitance voltage characteristics of each sample were determined at room temperature by stepping up the voltage in increments of either 5 or 10 V using programmable impedance analyser (Model 4192A LF impedance analyser. Hewlett-Packard) at 100 kHz.

3.0 RESULTS AND DISCUSSION

3.1 Phase Diagram of ZnO-PrO_x System

DTA measurements and microstructural observation elucidated the binary phase diagram of ZnO-PrO_x system shown in Figure 1.[9] Obtained data were smoothed and the liquid phase formation temperature was determined at the point of departure

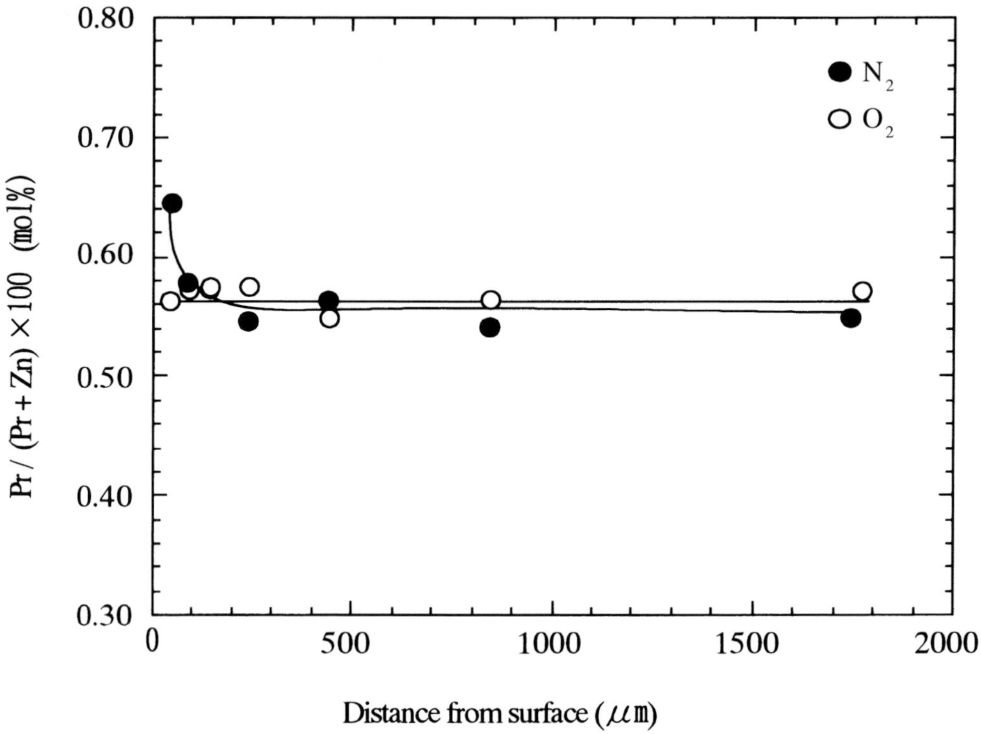

Fig. 2 Depth profiles of Pr concentration in 0.5 mol.% Pr-doped ZnO sintered at 1350°C for 2 hrs. in N_2 and O_2.

from the baseline of the DTA profiles. The thermocouple was calibrated every 10 measurements using the melting temperature of metal palladium i.e. 1552°C.

It is clarified that the liquid formation temperature is constant at 1382 ± 5°C which indicates the solidus boundary in this system. The results of microstructure and the changes integrated value of endothermic peak on DTA indicated the eutectic point to be c.a. 50 mol.% of Pr at 1382°C.

3.2 Depth Profiles of Pr Concentration for the 0.5 mol.% Doped ZnO Samples

Figure 2 shows the depth profiles of the Pr concentrations from the sample surfaces of sintered bodies fired for 2 hrs. at 1350°C in N_2 and O_2, using the ICP atomic emission analysis. Figure 2 indicates that at a firing temperature of 1350°C, the Pr concentration of the samples, fired in both N_2 and O_2 has a constant value of about 0.56 mol.% in the region deeper than 100 μm from the surface. The higher value of 0.64 mol.% Pr in N_2 compared to 0.5 mol.% Pr in O_2 can be ascribed to the higher

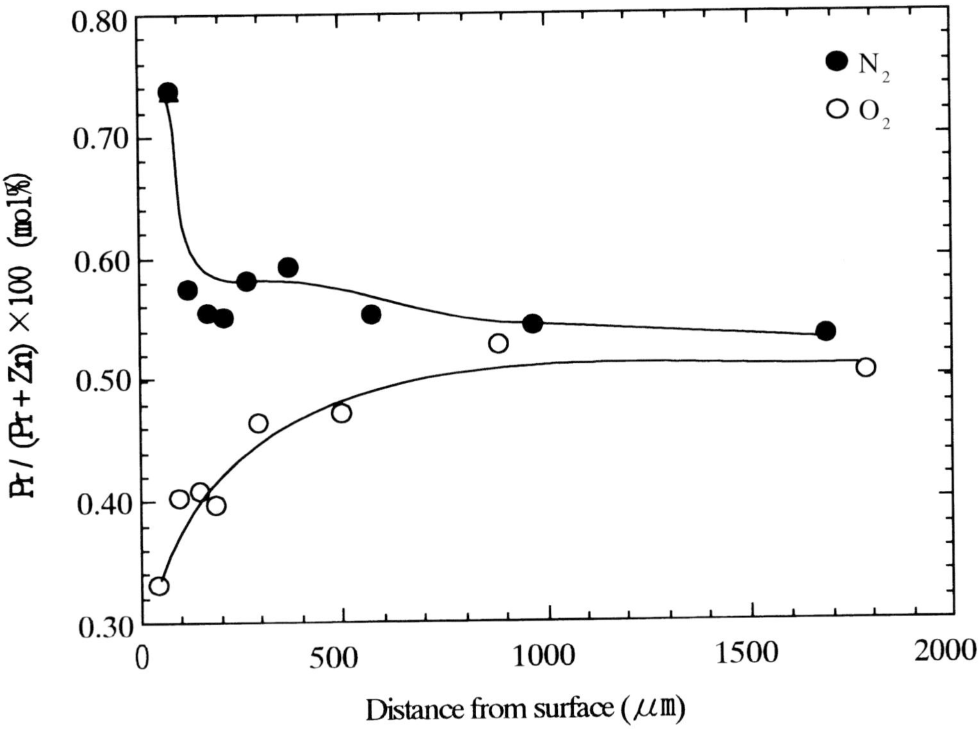

Fig. 3 Depth profiles of Pr concentration in 0.5 mol.% Pr-doped ZnO sintered at 1500°C for 2 hrs. in N_2 and O_2.

vapour pressure during the dissociation of ZnO in N_2. The equilibrium vapour pressure of ZnO is approximately 10^8 times higher than that of Pr_2O_3.[10,11]

On the other hand, Figure 3 shows the depth profiles of the Pr concentrations from the sample surfaces of sintered bodies fired for 2 hrs. at a firing temperature of 1500°C, in N_2 and O_2. The Pr concentration is highest (0.74 mol.%) at the surfaces of the samples fired in N_2, as in the case of 1350°C. However, it was found that in the case of firing in O_2, the Pr concentration is lowest (0.32 mol.%) at the surfaces of the samples and forms an increasing curve in the region 300 μm from the surface, and constant in the region deeper than 300 μm.

These results shown in Figures 2 and 3 indicate that after firing the pellets prepared by mixing ZnO and $PrO_{1.83}$(Pr_6O_{11}) uniformly, the concentration gradient depends on the atmosphere at the time of firing. The cause of such a different concentration gradient according to firing atmosphere is explained as follows. The weight decrease accompanying the evaporation of the elements revealed that the weight decrease of samples is larger in N_2 compared to O_2. Especially in the case of firing at 1500°C in N_2, the weight decrease (20.2 wt.%) is about 10 times higher compared to 2.9 wt.% in O_2.

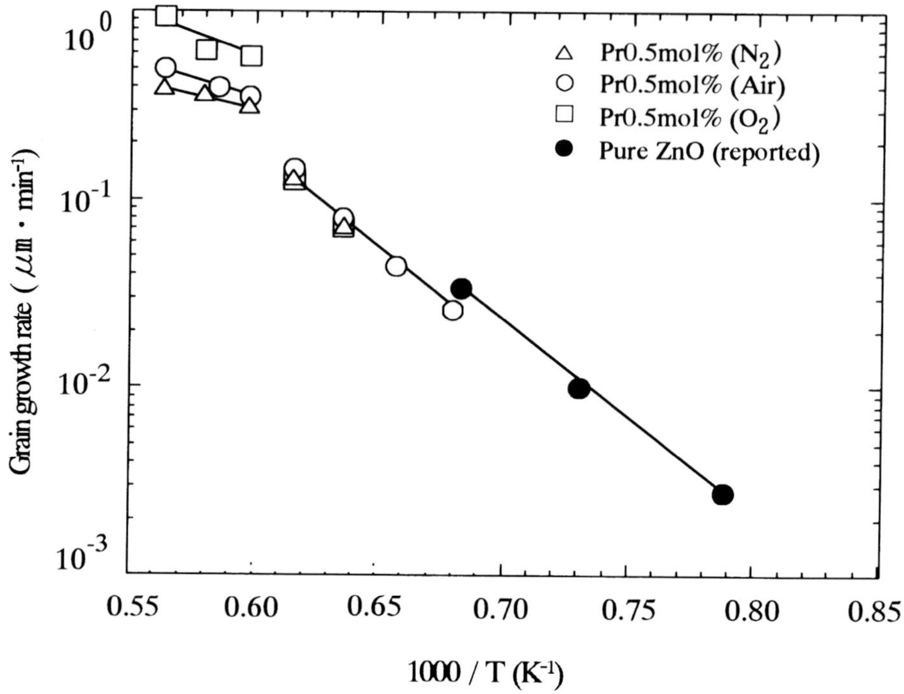

Fig. 4 Arrhenius plot of grain growth rate of 0.5 mol.% Pr-doped ZnO sintered in N_2, air, and O_2 for 2 hrs.

3.3 Grain Growth Behaviour

Figure 4 shows the Arrhenius plot of grain growth rate of samples containing 0.5 mol.% of Pr fired for 2 hrs. With respect to the gradient of each plot of grain growth rate, a large change is encountered at 1350°C. The calculated activation energy below 1350°C is about 60 ± 10 kcal which agrees quite well with the activation energy[12] of pure ZnO reported conventionally. However, over 1350°C, it was found that the atmosphere dependence that appears in the grain growth is considered to change between 1350°C and 1400°C. The calculated activation energy for grain growth above 1350°C is about 25.1 ± 10 kcal. This value is about 40% of that below 1350°C. The fact that the activation energy for grain growth below 1350°C equals that of pure ZnO means that the grain boundary migration in solid-state reaction is the main grain growth mechanism below 1350°C. A large reduction of activation energy above 1350°C irrespective of firing atmosphere can be ascribed to solution-precipitation process in the liquid-phase sintering.

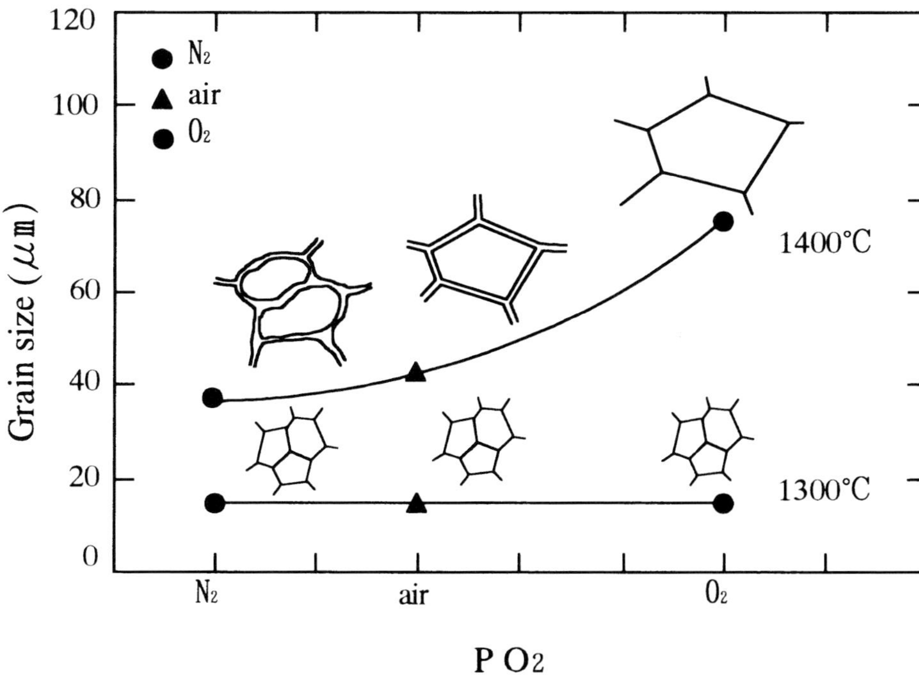

Fig. 5 Effect of firing atmosphere (oxygen partial pressure) on the grain size and grain boundary morphology at the surface of 0.5 mol.% Pr-doped ZnO samples.

3.4 Microstructure

Figure 5 shows the schematic microstructure of the surfaces of ZnO samples containing 0.5 mol.% Pr fired at 1300 and 1400°C in each of N_2, air and O_2. As we reported in the earlier study,[13] it was found that the mean particle size for the specimens fired at 1300°C (below the eutectic temperature) is about 15 μm with no dependence on firing atmosphere. The change of both grain size and grain boundary morphology by atmosphere is remarkable on sample fired at 1400°C (above the eutectic temperature). As the grain size increased in the order of N_2, air and O_2, the distinct precipitation of grain boundary phases was encountered on the surfaces of samples fired in N_2. Furthermore, it was found that the existence state of grain boundary phases differs with firing atmosphere along sample surfaces and inside. In other words, a large amount of grain boundary phases precipitate on the surfaces or samples fired in N_2, while the grain boundary phase is not encountered on the surfaces of samples fired in O_2. On the other hand, when microstructures inside the samples fired in O_2 are observed, it is found that a large volume fraction of grain boundary phases segregate on the triple point in contrast to surfaces.

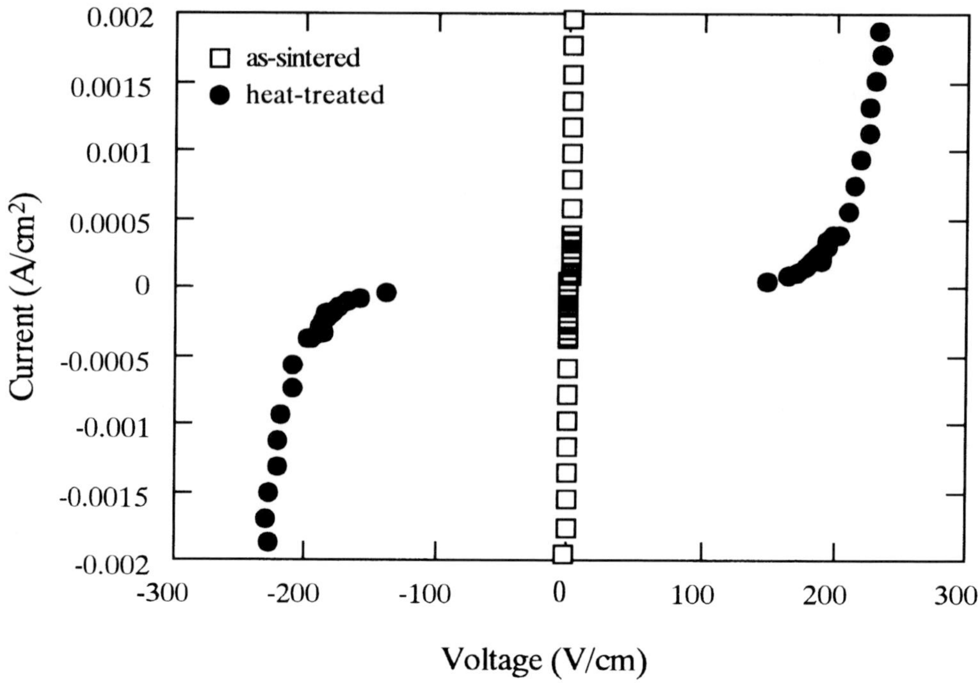

Fig. 6 I-V characteristics of Co-doped ZnO sample before and after the heat-treatment with Pr_6O_{11} at 1300°C for 30 min.

3.5 Varistor Properties

To elucidate the role or liquid-phase on the formation of electrically "active" grain boundaries in ZnO varistor, the possible technique of varistor-formation via infiltration of molten liquid ZnO-PrO_x into the grain boundaries of pre-sintered ZnO ceramics and their influence on the resulting properties were studied. Figure 6 compares the I-V response of the as-sintered 7 mol.%. Co-doped ZnO disks with those that were heat-treated with Pr_6O_{11} at 1300°C for 30 min. The marked nonlinear I-V response of the Pr_6O_{11} heat-treated samples is obvious. The breakdown voltage at a current density 0.1 mA/cm^2 determined to be 220 V/cm, is very similar to the value of conventionally fabricated bismuth-doped low-voltage varistors.

The change of nonlinearity (α) vs. heat-treatment temperature is shown in Figure 7. From these results, nonlinearity was one for the sample heat-treated at 1250°C, which means it has no varistor-property. However, the value of nonlinearity begins to increase above 1300°C and the maximum value of α =47 was obtained at 1350°C. Further higher temperature of heat-treatment decreased the nonlinearities.[14]

In addition, the change of grain sizes of 7 mol.% Co-doped ZnO with heat-treatment temperature is shown in Figure 8. As we can see, the grain size shows, no

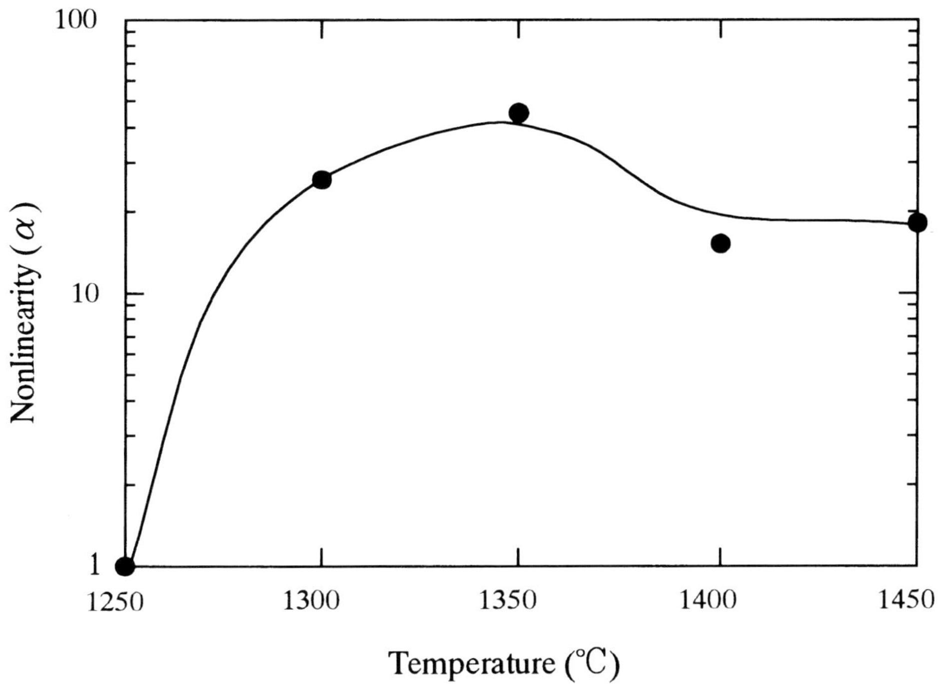

Fig. 7 Log of non-linearity vs. heat-treatment temperature of 7 mol.% Co-doped ZnO samples after heat-treated with Pr_6O_{11} at elevated temperature for 30 min.

noticeable changes for the samples heat-treated between 1200 and 1250°C. However grain size begins to increase between 1250 and 1300°C because the samples heat-treated above 1300°C were infiltrated with ZnO-PrO_x, liquid. We reported that the formation temperature of liquid-phase decreases to 1275 ± 5°C with addition of Co in previous work.[15] Thus, the increase of ZnO grain size between 1250 - 1300°C can be ascribed to the formation of liquid-phase at these temperatures. The results of Figures 7 and 8 indicate that the varistor properties of ZnO originate from the liquid phase.

4.0 SUMMARY

The formation of different morphologies of grain boundary according to firing temperature (below and above liquid-phase formation temperature) and atmosphere (oxygen partial pressure) can be explained by different vaporisation of components in Figure 9. Added Pr component existed between the grain boundaries of ZnO below liquid-phase formation temperature (1380°C). This Pr component begins to

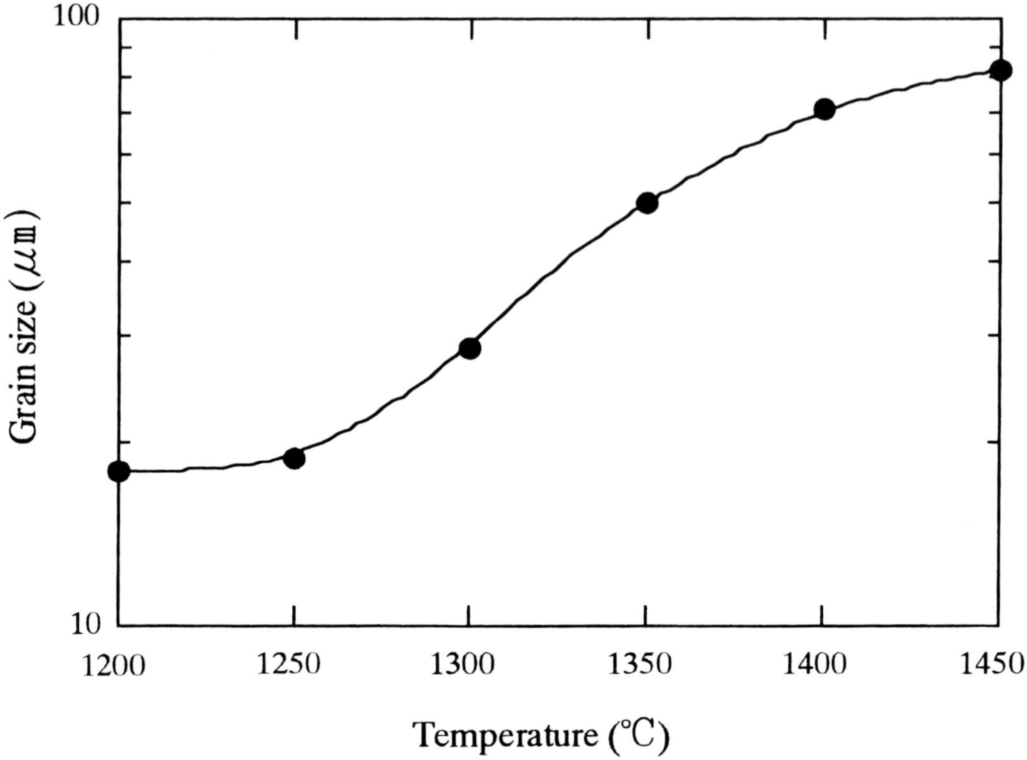

Fig. 8 Log grain size of 7 mol.% Co-doped ZnO samples with heat-treatment temperature for 30 min.

produce liquid-phase by reacting with ZnO above 1380°C. In N_2, large Pr segregates are observed on the surface of fired ZnO, because large vaporisation of ZnO occurred at the surface. On the other hand, in O_2 small Pr segregate is observed on the surface of fired ZnO because, large vaporisation of Pr occurred at the surface.

From above results, Pr concentration gradient of samples sintered at 1500°C in Figure 3 can suggest that two kinds of mechanism have occurred simultaneously. One is the higher vaporisation pressure during dissociation of ZnO compared to $PrO_{1.83}$ and this is the reason of higher Pr concentration at the surface of sample sintered in N_2. Another is the formation of a liquid accelerates the evaporation of Pr oxide at 1500°C and this is the reason for lower Pr concentration at the surface of samples sintered in O_2.

On the basis of the formation of varistor-properties per grain boundary (e.g. threshold voltage, nonlinear coefficient) the number of active grain boundaries are believed to be increased when the samples were heat-treated above liquid-phase temperature.

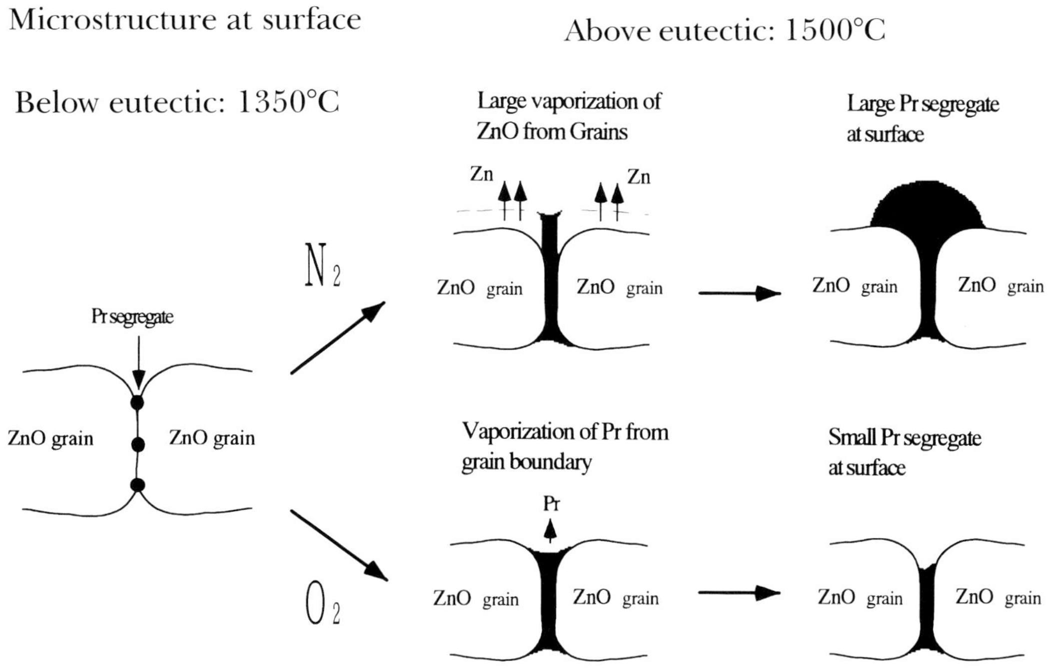

Fig. 9 Formation of different morphologies of grain boundary at surface according to firing atmosphere.

5.0 REFERENCES

1. K. MUKAE, K. TSUDA and I. NAGASAWA: Non-Ohmic Properties of ZnO-Rare Earth Metal Oxide-Co_3O_4 Ceramics, *Japanese Journal of Applied Physics,* **16**(8), 1977, pp.1361-1368.
2. K. MUKAE and I. NAGASAWA: Effect of Praseodymium Oxide and Donor Concentration in the Grain Boundary Region of ZnO Varistors, *Advances in Ceramics, Additives and Interfaces in Electronic Ceramics,* M.F. Yan and A.H. Heuer, eds., American Ceramic Society, Columbus, OH, **7**, 1983, pp.331-342.
3. K. MUKAE: Zine Oxide Varistors with Praseodymium Oxide, *American Ceramics Society Bulletin,* **66**(9), 1987, p.1329-1331.
4. A. B. ALLES and V. L. BURDICK: The Effect of Liquid-Phase Sintering on the Properties of Pr_6O_{11}-Based ZnO Ceramics, *Journal of Applied Physics,* **70**, 1991, pp.6883-6890.
5. A. B. ALLES, R. PUSKAS. G. CALLAHAN and V. L. BURDICK: Compositional Effects on the Liquid-Phase Sintering of Praseodymium Oxide-Based Zinc Oxide Varistors, *Journal of American Ceramics Society,* **76**, 1993, pp.2098-2102.
6. H. K. VERMG K. P. KUMAR, K. G. WARRIER and A. D. DAMODARAN: Effect of K_2O on the Sintered Microstructure of Praseodymium-doped ZnO Varistors, *Journal of Materials Science Letters*, **8**, 1989, pp.974-976.

7. M. Matsuoka: Progress in Research and Development of Zinc Oxide Varistors, in Advances in Ceramics, *Grain Boundary Phenomena in Electronic Ceramics*, L. M. Levinson, ed., The American Ceramic Society, Columbus OH, **1**, 1981, pp.290-308.
8. K. Mukae, K. Tsuda and I. Nagasawa: Capacitance-vs-Voltage Characteristics of ZnO Varistors, *Journal of Applied Physics*, **50**, 1979, pp.4475-4476.
9. S. Y. Chun, N. Wakiya. K. Shinozaki and N. Mizutani: Phase Diagram and Microstructure in the ZnO-Pr_2O_3 System, *Journal of American Ceramics Society*, **80**(4),1997, pp.995-998.
10. I. S. Kulikov: Thermodinamika Oksidov Spravochik, Nisso Tsushinsha, Moscow, USSR, 1986, pp.263-266.
11. I. S. Kulikov: Ibid, 1986, pp.241-242.
12. M. Drofenik: Oxygen Partial Pressure and Grain Growth in Donor-Doped $BaTiO_3$, *Journal of American Ceramics Society*, **70**(5), 1987, pp.311-314.
13. M. Sida, S. Y. Chun, N. Wakiya, K. Shinozaki and N. Mizutani: Effect of the Sintering Temperature and Atmosphere on the Grain Growth and Grain Boundary Phase Formation of Pr-doped ZnO Varistor, *Journal of Ceramics Society*, Japan, **104**, 1996, pp.44-48.
14. Y. M. Chiang and W. D. Kingery: Compositional Changes Adjacent to Grain Boundaries During Electrical Degradation of a ZnO Varistor, *Journal of Applied Physics*, **53**, 1982 pp.1765-1768.
15. S. Y. Chun, N. Wakiya. K. Shinozaki and N. Mizutani: Investigation of the Solidus Boundaries and Microstructure in the ZnO-$PrO_{1.5}$-CoO System, *Journal of Materials Research*, **13**(8), 1998, pp.2110-2116.

Ceramic Precursors: A Theoretical Study and Review of Experimental Data Relating to the Formation of $Al(OH)_3$

T. J. SOAR and A. R. GERSON
Ian Wark Research Institute
University of South Australia
Mawson Lakes, South Australia 5095

ABSTRACT

Understanding of nucleation and growth mechanisms enables the physical characteristics of ceramic precursors to be tailored to specific outcomes. Gibbsite ($Al(OH)_3$) crystallises under conditions of supersaturation of Al(III) in caustic solutions. On calcination of gibbsite, Al_2O_3 is formed.

Numerous hydroxo-aluminate monomeric species have been proposed including, $Al(H_2O)_6{}^{3+}$, $Al(OH)(H_2O)_5{}^{2+}$, $Al(OH)_2(H_2O)_4^+$, $Al(OH)_3^0$ and $Al(OH)_4^-$. The latter species, the aluminate ion, has generally been accepted as being the dominant monomeric species present under caustic conditions. In addition, a multitude of dimeric and polymeric species have been proposed including the Keggin ion ($(Al_{13}O_4(OH)_x(H_2O)_{36-x})^{(31-x)}$), $Al_2(OH)_2(H_2O)_8^{4+}$, $Al_2(OH)^{5+}$, $Al_2O(OH)_6{}^{2-}$ and $Al_5(OH)_{13}{}^{2+}$.

The relative stability of monomeric, dimeric and trimeric species and possible reaction pathways between these indicates that the $Al(OH)_4^-$ ion may not be directly involved in the nucleation process but rather nuclei growth may occur through the addition of minor, less stable monomeric species. This offers a possible explanation for the apparently contradictory slow nucleation rate of gibbsite and its relatively low activation energy for growth. The most stable dimer formed is found to be $Al_2O(OH)_6^{2-}$ which readily forms from unstable monomers. Further growth of this dimer may be through the process of addition of non $Al(OH)_4^-$ species for instance $Al(OH)_5^{2-}$ or $Al(OH)_6^{3-}$.

This manuscript reviews the literature relevant to aluminium speciation in caustic liquors in conjunction with calculations carried out using semi-empirical quantum chemical modelling to assess relative stabilities and the reaction pathways between aluminium hydroxide species. A nucleation mechanism for aluminium hydroxide is proposed.

1.0 INTRODUCTION

Few conceptual advances have been made in the area of nucleation and pre-crystalline particle growth in the last 40 years.[1] Crystalline growth models generally rely on classical ideas of monomeric unit addition at step sites on the crystal surface which are not appropriate for very small non-crystalline particles.[1] Fundamental models that have been developed to predict nucleation[2] generally rely on unobtainable 'perfect' solutions neglecting hetero-particles or substrate nucleation sites. These models do not accurately predict real crystallisation behaviour.[3, 4]

A considerable body of research on the ideal particle size distribution for dense ceramic composites is already available in refs. 5-7 and references therein. These distributions require the addition of fine, colloidal particles in different ratios with coarse particles to form matrix structures with different packing densities. Large (>1 μm) ceramic precursor particles lead to less effective packing and therefore lower ultimate density and flexural strength in the ceramic product. By adjusting the proportion of fine to coarse particles, it is possible to achieve a maximum green density although this is normally less than the ideal calculated value derived from the model structures.[6,7] Hence, there is a considerable technological interest in the controlled formation of colloidal ceramic precursors through the control of nucleation and precipitation and subsequent thermal decomposition. The precursor forms of the dissolved species, nucleating species and colloidal phases are critical to the form of the final ceramic product.

Colloidal aluminium hydroxide and silica are both important ceramic precursors. It has become apparent that the mechanism of formation of these particles is unusual in that they both undergo a polymeric nucleation step, which is in contrast to the established view of precipitation mechanisms. In this regard, in order to fully understand the nucleation mechanism of aluminium hydroxide, it is useful to briefly review in parallel the known nucleation processes undergone by both aluminium hydroxide and silica. However, the particle nucleation and growth behaviour of these two ceramic precursors display some distinct differences.

The ability to control silica particle size (up to 800 nm) and, to a more limited extent, size dispersion through the manipulation of solution concentration and temperature has already been demonstrated by Stober and Fink[8] and Bogush, Tracy and Zukoski.[9] Three contrasting mechanisms for silica growth are shown in Table 1. The proposed mechanisms have a common initial growth step from monomer to polymer by the addition of monomers. After this, the growth paths differ essentially in the relative emphasis placed on growth by agglomeration, monomer addition and internal cross-linking.

Cryovitrification through super fast cooling offers a mechanism whereby structures existing within solutions can be vitrified, i.e., held within a glassy solid state and subsequently imaged using transmission electron microscopy (TEM) or scanning electron microscopy. Slow cooling results in crystallisation and in aqueous solutions the formation of ice and hence the original solution structures can be destroyed. This technique has been used previously to study the nucleation mechanism of silica from silicon alkoxide.[14] The initial particles ($\approx$10 nm) appear to be polymeric and of low density. These species undergo a sudden collapse to form high density species. These may then grow by the subsequent attachment of further low-density species. The extent of polymerisation of silicon alkoxides has been modelled as a function of time through the use of cryovitrification - TEM (cryo-TEM).[10]

The effect of water concentration on the mechanism of SiO_2 nucleation has been studied by Bailey and Mecartney[14] who examined colloidal SiO_2 growth from tetraethoxysilane in a water, alcohol and ammonium solution. It was proposed that

Table 1 Contrasting proposed mechanisms for the nucleation and growth of silica.

SiO_2[11]	SiO_2[12,13]	SiO_2[14]
Monomer ↓ Polymers ↓ Nuclei ↓ Aggregation to form colloidally stable particles ↓ Further aggregation of the colloidally stable particles with nuclei	Monomer ↓ Polymers ↓ Colloidally stable particles ↓ Further addition of monomers	Monomers ↓ Polymers ↓ Addition of monomers and internal cross linking ↓ Continued growth until insoluble* ↓ Collapse as enthalpy over comes entropy ↓ Growth by further monomer addition

* solubility of polymers decreases with increasing size and cross linking[15]

a high water concentration leads to a higher degree of hydrolysis (Reaction 1) and thus a higher degree of cross linking through condensation (Reactions 2 and 3). Hence, at high water concentrations more dense colloidal particles may be predicted.

≡Si-OR + HOH	→	≡Si-OH +RH	Hydrolysis	Reaction 1
≡Si-OR + HO-Si≡	→	≡Si-O-Si≡ + ROH	Condensation	Reaction 2
≡Si-OH + HO-Si≡	→	≡Si-O-Si≡ +HOH	Condensation	Reaction 3

Alumina can be formed by the thermal decomposition of aluminium hydroxide. Coarse aluminium hydroxide (5-40 μm) is generally produced by the Bayer processing of gibbsite ($Al(OH)_3$) from bauxite ores. A recent review of the nucleation and precipitation mechanisms of precursors to colloidal alumina concluded that in principle it is possible to prepare colloidal alumina in specific size ranges (e.g. 50, 100, 150, 250 and 500 nm) by the controlled precipitation of aluminium hydroxide.[16]

During the growth processes of both silica and aluminium hydroxide a chemical reaction occurs. For silica these reactions appear to be hydrolysis and condensation

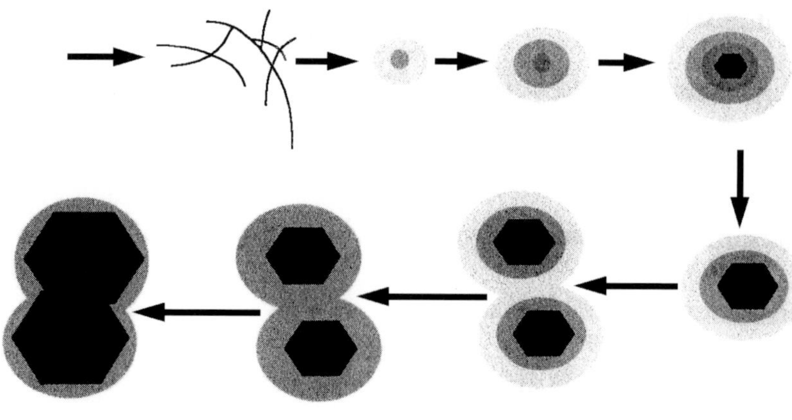

Fig. 1 A schematic mechanism of gibbsite nucleation and particle growth of aluminium hydroxide from caustic aluminate liquors.

reactions (Reactions 1, 2 and 3). There is substantial experimental evidence to suggest that $Al(OH)_4^-$ is the dominant Al-containing species in caustic environments.[17] If this species plays a role in the formation of dimers, polymers and eventually solid $Al(OH)_3$ then the ejection of an OH^- from each monomer must occur at some stage of the reaction process. The final structures of silica and aluminium hydroxide differ in that the Si in silica is bridged by -O-, however, the Al in aluminium hydroxide is bridged by $-OH^-$. Therefore a condensation reaction of the type shown in Reactions 2 and 3 is not possible during the formation of aluminium hydroxide.

A speculative growth path for aluminium hydroxide in highly caustic liquors has been proposed (Figure 1).[18] This mechanism does not propose a series of detailed molecular reactions. In this mechanism it is postulated that initially, in a fresh caustic aluminate solution (pH > 14), $NaAl(OH)_4$ ion pairs are present. Whether these ion pairs are solvent separated or in close contact is not discussed. As the solution ages regions of loose polymeric networks form. These polymeric networks slowly densify to form regions of low phase contrast to the background solution.[18]

This study[18] reported the results from a number of experiments carried out with techniques that rely on phase contrast for nuclei or species detection, e.g., small angle x-ray and neutron scattering (SATS and SANS respectively), dynamic light scattering and cryo-TEM. Neither small angle x-ray nor neutron scattering gave rise to significant scattering above the solvent even when visible precipitated particles were present. It has been estimated that a contrast in x-ray scattering length density of as little as 5% is sufficient to give rise to x-ray scattering.[19] Similar observations have been made by Bottero et al. in OH/Al = 2.5 solutions using SAXS.[55] These were reported to have weak x-ray scattering power that they presumed to be due to lack of phase contrast of the species with the background solution. After 24 hours the scattering power increased indicating an increase in density of the solution species. Dynamic light scattering gave rise to poor scattered light intensity until the particles were approximately 150 nm in diameter.[18] The minimum particle size observed was 10 nm using cryo-TEM. Cryo-TEM indicated that particles simultaneously grew and densified with aging. Further densification leads to a crystalline $Al(OH)_3$ core.[20]

Rheology, colloid stability and AFM colloid probe measurements have shown non-DLVO repulsive particle interaction at high pH and high ionic strength.[21] Particle interactions between colloidal $Al(OH)_3$ particles in synthetic, supersaturated Bayer liquors are initially repulsive but become increasingly adhesive with time as indicated by both rheological and AFM measurements. The repulsive forces have a shorter-range, non-DLVO steric and a longer-range hydrodynamic component. A steric layer thickness of about 20 nm is indicated by AFM measurements at the gibbsite-solution interface. AFM measurements also indicate that on aging, the steric forces decrease followed by a systematic increase of inter-particulate adhesion. From a mechanistic point of view, these observations have been interpreted as being due to a diffuse polymeric layer initially causing steric repulsion, then densifying and becoming adhesive (Figure 1).

Although the mechanism of aluminium hydroxide nucleation is now quite well developed for highly caustic liquors no real understanding of the rate determining step in the nucleation mechanism yet exists. It has been shown that the rate determining step is not the coordination transformation required to go from $Al(OH)_4^-$ in solution to octahedrally coordinated Al in gibbsite.[22] In addition it is known that water has a peculiar effect on supersaturation in caustic aluminate solutions with dilution often increasing supersaturation.[23] It is possible, although no direct evidence has yet been obtained that the rate determining step in the nucleation and growth process of aluminium hydroxide is the formation of bridging -OH groups. This reaction may be closely related to the availability of free water in these highly caustic solutions.

2.0 METHODOLOGY

Molecular modelling provides a means to discriminate between particular reaction pathways on the basis of calculated Gibbs free energy of formation of the participating and resulting species and hence calculation of Gibbs free energy of reaction.[24] Moreover, the process of energy minimisation results in optimised species geometry. In all calculations described in this text all atomic positions were allowed to refine. Final species geometries are described as unstable where one or more Al to O distances are greater than 2 Å. A normal species bond length as found from optimization of an $Al(OH)_4^-$ species has been found to be 1.76 Å. In the crystalline form, where the coordination of the Al is greater, the average Al-O bond length is 1.90 Å.[25]

The calculations described herein employed the restricted Hartree-Fock method with self consistent field calculations using AM1.[26] Electrostatic repulsion and exchange stabilisation integrals are evaluated by approximate means, a restricted basis set of one *s* and three *p* Slater-type orbitals per atom is used, and overlap integrals *S* are ignored in the secular equation,

$$|H - ES| = 0 \tag{1}$$

Molecular orbitals and *s* and *p* bond strengths are generated from overlap integrals. AM1 parameters for Al were optimised by Dewar and Holder.[27] This method has been found to produce more accurate heats of formation for Al-containing species than the comparable semi-empirical quantum mechanical methods of PM3 and MNDO. Heats of formation are calculated by the following method:

$$\Delta H_f^o = E_{elect} + E_{nuc} + \sum_A E_{isol}(A) + \sum_A E_{atom}(A) \tag{2}$$

where E_{elect} is the electronic energy of the final species, E_{nuc} is the internuclear repulsion energy of the final species, $E_{isol}(A)$ is the energy required to remove all the valence electrons from all the atoms in the system while in an isolated state and $E_{atom}(A)$ is the energy required to isolate the atoms from their standard state. As AMl is parameterised to reproduce experimental values at 298 K, the heat of formation is, by default, calculated at this temperature. This definition of heat of formation is equivalent to the standard definition. A comprehensive comparison of experimental and calculated heats of formation ΔH_f^o values has been carried out.[28] The experimentally measured ΔH_f^o of some relevant aluminium oxide and hydroxide species as well as the differences between these values and the calculated values are listed in Table 2. The root mean square (RMS) of the error indicates the magnitude of the errors while the unsigned average error indicates systematic deviation of the calculated from experimentally derived heats of formation. The RMS for the 50 Al containing species tested was 57 kJ mol.$^{-1}$ and the unsigned error was 84 kJ mol.$^{-1}$.[28]

Solvent effects are modelled using COSMO, Conductor-like Screening Model.[29] COSMO belongs to the class of dielectric continuum models. In these models the solute molecule is embedded in a dielectric continuum of permittivity ε, i.e. the solute molecule forms a cavity within the dielectric. From electrostatics it is known that the response of a homogeneous dielectric continuum to any charge distribution of solute consists of a surface charge distribution on the interface arising from polarisation of the dielectric medium. It is a nontrivial problem to calculate the screening charge densities $\sigma(r)$ given by

$$4\pi\xi\sigma(r) = (\varepsilon - 1)n(r)E^{-}(r) \tag{3}$$

where $n(r)$ is the surface normal vector at the point r and $E^{-}(r)$ is the total electric field at the inner side of the cavity surface. This cannot, in general, be solved analytically for arbitrarily shaped surfaces. Numerical approaches involve the segmentation of the cavity surface with a constant charge density on each segment.

It is known that the relative permittivity or dielectric constant for concentrated sodium aluminate solutions is difficult to measure experimentally. Experimental dielectric constants have been reported for lower electrolyte concentrations by Rao et al.[30] NaOH solutions of 0.25 M and 2.00 M (at 298 K) produced dielectric constants of 73 ($\pm$2) and 59.5, respectively. Similarly, the electrolyte $MgCl_2$ at concentrations 1 M, 3 M and 4 M (at 298 K) gave rise to dielectric constants of 57.1, 44.3 and 42.0, respectively. From these experimental results, it is apparent that the dielectric constant decreases with increasing electrolyte concentration. In

Table 2 Experimental, and the difference between the calculated and experimental ΔH_f^o of aluminium-containing species.[28]

Species	Experimental ΔH^o_f (kJ mol.$^{-1}$)	Difference Between Calculated and Experimental ΔH^o_f (kJ mol.$^{-1}$)
AlO	66.99	-31.82
AlO^+	993.53	-11.26
AlO^-	-269.53	48.32
AlOH	-180.03	12.27
$AlOH^+$	544.28	117.02
$AlOH^-$	-230.27	42.96
H-Al = O	-33.49	-58.15
AlO_2	-86.25	69.92
AlO_2^-	-485.67	89.96
AlO_2H	-460.55	127.57

addition, electrolyte double-layer theories reported in literature utilise various dielectric constant values.[31] These range from 1 through to 20 where the average is approximately 10. Therefore, to model the low dielectric conditions expected within the solution double layers or solvation shells of a caustic aluminate liquor, a dielectric constant of 10.0 was selected for these computational investigations.

The heat of reaction ΔH^o_r can be calculated from the ΔH^o_f of the reactants (A and B) in isolation from each other, and the ΔH^o_r of the final configuration of the products (A + B):

$$\Delta H_r^o = \Delta H_f^o (A + B) - [\Delta H_f^o (A) + \Delta H_f^o (B)] \tag{4}$$

The more negative the reaction energy, the more favourable the reaction. This methodology has been successfully applied to a limited study aimed at defining the initial polymerisation reaction steps, in caustic aluminate liquors, involved in the formation of $Al(OH)_3$.[24] These calculations, in contrast to the majority of those presented here, were carried out in a vacuum environment. A comparison of the effect of the vacuum and dielectric continuum environments on stability and reaction energy will be made.

In addition it is possible to calculate the entropy of a given species taking into account vibrational, rotational and transformational contributions. It is not possible to calculate the conformational contribution to the entropy. However, for small species this can safely be assumed to be a relatively minor constituent. Given the entropy and enthalpy it is then possible to calculate Gibbs free energy of formation (ΔG^o_f) and therefore Gibbs free energy of reaction (ΔG^o_r) using the same methodology as for the calculation of ΔH^o_r. It is possible that even though the ΔG^o_r, for a particular reaction is favourable, in actuality, the reaction will not readily take place. This may be due to exceptionally high activation energies.

It is also possible to calculate transition states and therefore activation energies. This involves locating and following the reaction coordinate connecting the two stable systems, reactants and products, on a (3n-6)-dimensional geometry surface.[32] The transition state is associated with the highest energy point located along this reaction coordinate. Energy of activation is defined as the difference in energy between the transition and reactant compounds. Transition state location and, hence, energy of activation calculations reported here were performed using the SADDLE technique within MOPAC93. This technique involves specifying the geometries of the reactants and also the products, which are located on either side of the transition state. The reaction coordinate is then identified by locating geometries of progressively higher energy whilst progressively shortening the distances between the reactants and products.[28] These calculations were also performed using a dielectric constant environment of 10.0.

3.0 MONOMERS

Numerous Al-speciation studies have been carried out under slightly alkali or near neutral conditions. Baes and Mesmer,[23] using solutions made by the dissolution of $Al(ClO_4)_3$ from pH 2 to 12, predicted the stability constants of a range of species including $Al(H_2O)_6^{3+}$, $Al(OH)(H_2O)^{2+}$, $Al(OH)_2(H_2O)^+$ and $Al(OH)_3$. Several studies using ^{27}Al NMR have been carried out under 'aqueous' conditions.[34-37,53] These have generally concluded that the major monomeric species is $Al(H_2O)_6^{3+}$.

The Al speciation under conditions of moderate pH is in contrast to that observed at high pH. It is the high pH speciation that we are focusing on within this study. Several Raman and infrared spectroscopic studies have been carried out on concentrated caustic aluminate solutions. In dilute solutions ([Al] < 1 M) most studies have interpreted the spectra obtained as being due solely to the tetrahedrally coordinated aluminate ion.[17,38,39,40,43] Conductance, acid-titration and freezing-point measurements have been used to show that the aluminate ion is a mono-valent ion.[41] The specific gravity for a range of aluminium concentrations with varying OH/Al ratios was measured and from the change in volume due to the coordination of H_2O with Al it was concluded that the aluminate ion probably existed as $Al(OH)_4^-$.[41] This monomer has characteristic vibrational frequencies, 725-740 cm^{-1} using infrared and 625 cm^{-1} using Raman spectroscopies.[17,43,38,39,40]

In an early study of caustic aluminate solutions Raman bands were observed at 615 cm^{-1} and 310 cm^{-1}.[43] These were interpreted as arising from a tetrahedral $Al(OH)_4^-$ species. A further study[42] did not agree with this assignment of these bands. The argument proposed was based on the lack of observation of a 720 cm^{-1} ν_3 $Al(OH)_4^-$ band which should be present if the 615 cm^{-1} ν_1 and 310 cm^{-1} ν_2 bands[43] were correctly assigned. It was suggested on the basis of selection rules and polarisation that the main species from pH 8 to 12 is either square planar or a polymeric octahedral

Table 3 ΔH^o_f for monomeric aluminium species using either vacuum or dielectric continuum environment. The monomers are listed in order of the most stable to least stable species for the species environment using dielectric constant of 10.0. n.s. indicates the Al-containing species was sterically unstable.

Monomer	ΔH^o_f (kJ mol.$^{-1}$)	
	Vacuum	Dielectric Constant = 10.0
$Al(OH)_6^{3-}$	-397	-2355
$Al(OH)_5^{2-}$	-1131	-2073
$Al(OH)_4(H_2O)^-_2$	n.s.	-2062
$Al(OH)_4(H_2O)^-$	n.s.	-1838
$Al(OH)_3(H_2O)_3$	-1639	-1771
$Al(OH)_4^-$	-1308	-1613
$Al(OH)_3(H_2O)_2$	-1408	-1528
$Al(OH)_3(H_2O)$	-1142	-1246

structure. This species did under go vibrational shifts on deuteration. A 628 cm^{-1} band was observed above pH 13 (4 M Al). This was interpreted as being due to AlO_2^- with $D_{\infty h}$ symmetry. It was also observed that this frequency did not under go any shift on analysis of an equivalent deuterated solution.

Subsequently, this interpretation was brought into doubt by the spectroscopic study carried out by Moolenaar et a1.[17] A 1 M Al and 5 M NaOH solution gave rise to an intense polarised Raman band at 625 cm^{-1}. On deuteration this band shifted to 605 cm^{-1}. Depolarised bands at 325 and 725 cm^{-1} were also observed in the hydrogenated liquor. The band at 725 cm^{-1} was also observed by IR. This band shifted to 745 cm^{-1} on deuteration (IR and Raman). A broad IR band at 950 cm^{-1} was also observed.[17] All these bands were assigned to $Al(OH)_4^-$.

On the basis of Moolenaar et al.'s study it is generally accepted that $Al(OH)_4^-$ comprises the vast majority of the Al-containing species in fresh (un-aged) concentrated caustic aluminate solutions. It is therefore reasonable to accept that any theoretical study should predict this as the most stable aluminium hydroxide monomer. In order to test this we have optimised the geometry and thus the ΔH^o_f of all possible negatively charged or neutral Al monomers with 4, 5 and 6 coordination with all possible stoichiometries and geometries of water and/or OH^-. Only negatively charged or neutral species were assessed as it was felt that there was no possibility of a positively charged species existing in a solution containing a high concentration of OH^-. The resulting ΔH^o_f values for the eight most stable species are shown in Table 3 for both a vacuum environment and a dielectric continuum environment. In neither case is the $Al(OH)_4^-$ species the most stable species. In the vacuum

Table 4 ΔH^o_r for monomeric species with respect to $Al(OH)_4^-$ in a species environment of either vacuum or dielectric constant of 10.0. These are listed in order of the most likely to least likely reaction for the species environment using dielectric constant of 10.0. n.s. indicates the Al-containing species was sterically unstable.

Monomer	ΔH^o_r (kJ mol.$^{-1}$)	
	Vacuum	Dielectric Constant = 10.0
$Al(OH)_6^{3-}$	-1029	-276
$Al(OH)_3(H_2O)_3$	-354	-176
$Al(OH)_3(H_2O)$	-355	-139
$Al(OH)_3(H_2O)_2$	-337	-138
$Al(OH)_4(H_2O)_2^-$	n.s.	-113
$Al(OH)_4(H_2O)^-$	n.s.	-56
$Al(OH)_5^{2-}$	-236	-49
$Al(OH)_4^-$	0	0

environment the neutral species tend to have greater stability. In the dielectric continuum environment where the charge can be partially stabilised by the local polarisability of the dielectric continuum the species with more negative charge tend to be more stable.

However, to properly assess the relative stability of species it is necessary to calculate the ΔH^o_f required to convert one monomeric species to another. In order to do this stoichiometric (and charged) balanced equations must be defined and the ΔH^o_f of all the species involved in the reaction (including OH^- and H_2O) must be calculated. It was found on carrying out this process, that $Al(OH)_4^-$ gives rise to the most stable system in both the vacuum and dielectric continuum environment. ΔH^o_r energies for the conversion of the 8 most stable species (by ΔH^o_f listed in Table 3) to $Al(OH)_4^-$ are given in Table 4. The more negative the ΔH^o_r energy the more favourable the forward reaction, i.e. the more likely the reaction is to give rise to an $Al(OH)_4^-$ product. It is apparent from Table 4 that all the reactions listed are favourable. This confirms the experimental result that $Al(OH)_4^-$ is the most abundant monomer in caustic aluminate solutions.

The formation of ion pairs, for instance $NaAl(OH)_4$, is still a matter of some discussion. Analysis of potassium and sodium aluminate solutions (1.21 M KOH, 0.71 M $Al(OH)_3$ and 2.42 M KOH or NaOH and 1.42 M $Al(OH)_3$) gave rise to one Raman band at 636 cm^{-1} which was interpreted as being due to the presence of $Al(OH)_4^-$ thus indicating the speciation was not a function of the cation present.[39] ^{27}Al NMR shows considerable broadening at high solution concentrations but no

Table 5 ΔH^o_f of monomeric Al-containing species ion-paired with Na^+ or K^+ in a vacuum species environment and ΔH^o_r for the formation of the ion pair. These are listed in order of the most stable to least stable ΔH^o_f for the Na^+ ion paired species. n.s. indicates the Al-containing species was sterically unstable.

Monomer	ΔH^o_f (kJ mol.$^{-1}$)		ΔH^o_r for the Formation of an Ion Pair (kJ mol.$^{-1}$)	
	Na^+	K^+	Na^+	K^+
$Al(OH)_5^{2-}$	-1822	-1726	-799	-685
$Al(OH)_4^-$	-1622	-1568	-422	-350
$Al(OH)_3(H_2O)_2$	-1391	-1369	-91	-51
$Al(OH)_6^{3-}$	n.s.	-1353	n.s.	-1046
$Al(OH)_3(H_2O)$	-1080	-1078	-46	-26
$Al(OH)_3(H_2O)_3$	n.s.	n.s.	n.s.	n.s.

change in chemical shift is observed.[17] The line broadening observed is commensurate with the formation of contact ion pairing. However, the concentration of ion paired species versus non-ion paired species is unknown as is the species that may be involved in dimerisation reactions. In Table 5, the calculated ΔH^o_f of a range of K^+ and Na^+ ion paired species are given. These have been calculated using a vacuum environment. Unfortunately, software limitations currently restrict the calculation of species involving Na^+ or K^+ in a dielectric continuum environment. It is apparent from Table 5 that Na^+ stabilises species more than K^+. This is expected due to the higher charge density on Na^+ as compared to K^+. However, as in the previous case with non-ion paired species, it is necessary to assess the ΔH^o_r required to convert each species to the ion paired $Al(OH)_4^-$.

ΔH^o_r energies for the conversion of ion paired species to ion paired $Al(OH)_4^-$ are given in Table 6. The results shown in Table 6 indicate that in a vacuum environment ion paired $Al(OH)_4^-$ does not give rise to the most stable system. The positive ΔH^o_r calculated for the transformation from ion paired $Al(OH)_5^{2-}$ to ion paired $Al(OH)_4^-$ indicates that this reaction is not favourable and that therefore ion paired $Al(OH)_5^{2-}$ is more likely to form than ion paired $Al(OH)_4^-$. This is a result of the increasing stabilising effect of ion pairing on species as the species charge increases (columns 4 and 5 Table 5). For instance the ΔH^o_r of ion pairing of $Al(OH)_5^{2-}$ with Na^+ is -799 kJ mol^{-1} as compared to the comparable reaction for $Al(OH)_4^-$ which gives rise to a ΔH^o_r of –422 kJ mol^{-1}.

It is possible that if a dielectric continuum model could be used then the stabilisation effect of ion pairing on species with larger charges would be less

Table 6 ΔH^o_r for the reaction from Na^+ or K^+ ion-paired monomeric species to Na^+ or K^+ ion paired $Al(OH)_4^-$ in a vacuum environment. These are listed in order of the most likely, to least likely reaction, for the calculations using Na^+. n.s. indicates the Al-containing species was sterically unstable.

Monomer	ΔH^o_r (kJ mol.$^{-1}$)	
	Na^+	K^+
$Al(OH)_3(H_2O)$	-731	-679
$Al(OH)_3(H_2O)_2$	-668	-636
$Al(OH)_6^{3-}$	n.s.	-333
$Al(OH)_5^{2-}$	141	99

pronounced than in vacuum. Hence, $Al(OH)_4^-$ may in this case, still give rise to the most stable species system. An indication of this is given in Table 4 where the more highly charged species are unstable with respect to $Al(OH)_4^-$ but are more stable in the dielectric continuum rather than the vacuum environment. We know from experimental evidence that $Al(OH)_5^{2-}$, whether ion paired or not may only exist in very minor quantities and that $Al(OH)_4^-$ is the major species. Therefore, for our purposes, non-ion paired models in a dielectric continuum environment give rise to more realistic results than ion paired species in a vacuum environment. Hence, it is the non-ion paired species in the dielectric continuum environment that we will focus on for the remainder of this text.

In order to be able to critically assess the likelihood of conformational exchange between different monomeric species it is necessary to calculate the activation energy (E_a) required for conversion of one species to another. An example of the reaction pathway for the conversion of $Al(OH)_5^{2-}$ to $Al(OH)_4^-$ is given in Figure 2. The E_a for a range of reactant monomers going to $Al(OH)_4^-$ are given in Table 7. The negative E_a figures given in Table 7 indicate that the reaction occurs spontaneously and requires no energy input.

The reverse reaction E_a (column 3 Table 7) from the $Al(OH)_4^-$ monomer to give another monomeric species are all positive. However, some of these are sufficiently small that it is possible that, at any one time, in solution a very small concentration of these other monomers exist even though they are less energetically favourable than $Al(OH)_4^-$.

In order to calculate these concentrations the entropic contribution (ΔS^o_f) to the monomeric Gibbs free energy of formation (ΔG^o_f) must be calculated where

$$\Delta G^o_f = \Delta H^o_f - T(\Delta S^o_f) \qquad (5)$$

ΔG^o_r can then be calculated by applying the same reaction scheme as in eqn. (4). The calculated ΔG^o_r of the most energetically favourable monomers to give $Al(OH)_4^-$

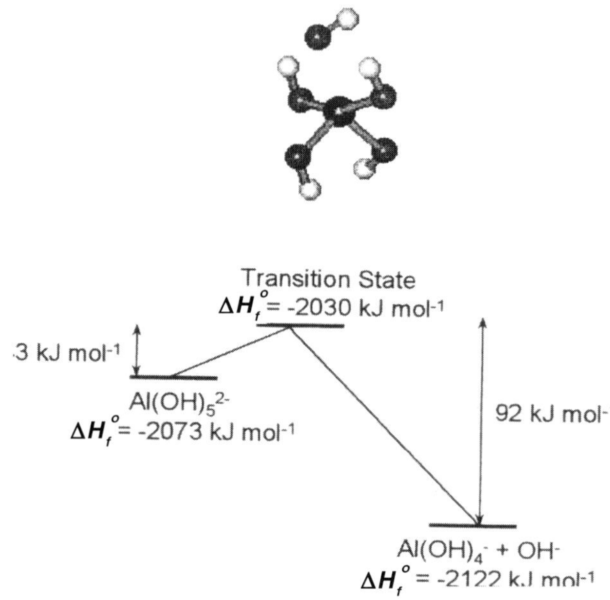

Fig. 2 The activation energy required for the conversion of $Al(OH)_5^{2-}$ to $Al(OH)_4^-$ and the transition state structure.

Table 7 E_a for monomeric species transforming to $Al(OH)_4^-$ in a dielectric continuum environment of dielectic constant of 10.0.

Monomer	E_a for the Forward Reaction to $Al(OH)_4^-$ (kJ mol.$^{-1}$)	E_a for the Backward Reaction from $Al(OH)_4^-$ (kJ mol.$^{-1}$)
$Al(OH)_4(H_2O)_2^-$	-52	61
$Al(OH)_4(H_2O)^-$	-49	7
$Al(OH)_3(H_2O)_3$	-21	155
$Al(OH)_6^{3-}$	11	287
$Al(OH)_3(H_2O)_2$	26	164
$Al(OH)_5^{2-}$	43	92
$Al(OH)_3(H_2O)$	52	191

is given in Table 8. Having obtained the ΔG_r^o it is then possible to calculate species concentration in solution. Assuming a reaction

$$A + B \rightarrow C + D \tag{6}$$

the equilibrium constant, K, can be defined

$$K = [(a)C(a)D]/[(a)A(a)B] \tag{7}$$

Table 8 ΔG^o_r for monomeric species reacting to form $Al(OH)_4^-$ and resulting calculated solution species concentration.

Monomer	ΔG^o_r (kJ mol.$^{-1}$)	Molality (mol.kg^{-1})
$Al(OH)_4(H_2O)_2^-$	-210	3.1×10^{-33}
$Al(OH)_4(H_2O)^-$	-111	7.1×10^{-17}
$Al(OH)_3(H_2O)_3$	-242	23.5×10^{-43}
$Al(OH)_6^{3-}$	-268	1.8×10^{-47}
$Al(OH)_3(H_2O)_2$	-164	1.8×10^{-30}
$Al(OH)_5^{2-}$	-77	4.6×10^{-17}
$Al(OH)_3(H_2O)$	-116	1.3×10^{-24}

where $a = \gamma m$ and γ is the activity coefficient and m is the species molality (mol. Kg^{-1}). The equilibrium constant may be calculated from

$$\Delta G^o_r = -RTln(K) \tag{8}$$

where R is the gas constant (8.3145 J K^{-1} mol^{-1}) and T is the temperature in Kelvin (298.0 K). Hence by assuming an activity coefficient of 1 and some idealised concentrations, $[Al(OH)_4^-]$ = 2.7 M, $[OH^-]$ = 1.1 M and $[H_2O]$ = 30 M, the concentration of the minor monomeric species can be calculated. These are given in Table 8, column 3. The most major species after $Al(OH)_4^-$ is calculated to be $Al(OH)_5^{2-}$ at a concentration of 9.9×10^{-14} mol. Kg^{-1}. It should be noted that all monomeric species, other than $Al(OH)_4^-$, are predicted to be present in insufficient concentrations to be detectable by spectroscopic techniques.

4.0 DIMERS

Concentrated aluminate ([Al] > 1 M) solutions have been found to give rise to more complicated spectra than more dilute solutions. These spectra include the spectral characteristics of the $Al(OH)_4^-$ ion. The additional absorptions are generally interpreted as being due to polymerisation of aluminium-containing species. It has been proposed that the $Al(OH)_4^-$ species, surrounded by a hydrated water layer, would undergo dehydration as the caustic or aluminium concentration is increased, to form possibly AlO_2^- or $Al_2O(OH)_6^{2-}$.[17,65,67] Study of the vapour pressure of aqueous aluminate solutions has indicated the formation of $Al(OH)_4^-$ at low aluminium concentrations followed by gradual dehydration to AlO_2^- as the alumina concentration was increased.[44]

The concentrations of the species that give rise to these extra vibrational absorptions increase with increasing caustic concentration.[17,39] There is contradictory evidence concerning the effect of increasing aluminium concentration. Moolenaar et al. report that increasing aluminium concentration (in an excess of OH^-) results in increasing concentration of the non $Al(OH)_4^-$ species.[17] However, Kopylova et al. found that increasing aluminium concentration has little effect on the concentration of these extra species.[39]

The Raman and IR band at 625 cm^{-1} due to the $Al(OH)_4^-$ species was observed on analysis of a 6 M Al, 24.3 M NaOH solution.[17] In addition, medium Raman polarised bands at 705 and 540 cm^{-1} and a weak depolarised band at 330 cm^{-1} were observed. A new IR band at 900 cm^{-1} was also observed. Interpretation of the extra bands was made by the direct comparison of their frequencies with the spectra of the $K_2[(HO)_3AlOAl(OH)_3]$ crystalline structure[46] and its deuterated analogue. It was concluded that the additional bands were likely to be due to the $(OH)_3AlOAl(OH)_3^{2-}$ species. The three bands characteristic of the new solution species at 900 (IR), 705 (Raman) and 540 cm^{-1} (Raman) compare favourably to the Al-O stretching bands at 920 cm^{-1} $\nu(AlO_3)$, an average of 725, 700, 685 and 628 $\nu(AlO_3)$ and 546 cm^{-1} ν_s(AlOAl) bands for the crystalline $K_2[(HO)_3AlOAl(OH)_3]$ structure.

Calculations of vibrational frequencies of hydroaluminate anions[45] were carried out and compared to those observed by Moolenaar et al.[17] Force constants were derived from the frequencies observed for $Al(OH)_4^-$. The crystal structure of $K_2[(HO)_3AlOAl(OH)_3]$[46] was used to derive the structural model of the proposed species. Symmetry C_{2v} was assumed. This model was then used to calculate a set of theoretical frequencies which were then compared to the vibrational frequencies measured for the crystalline $K_2[(HO)_3AlOAl(OH)_3]$ structure.[47,48] The calculated vibrational frequencies gave a good fit to the solid state vibrational spectra. These vibrational frequencies measured for the solid state ion in turn compare well to the vibrational frequencies measured for the proposed solution dimer.

Analyses have been carried out on a range of solution concentrations; 4.84 M KOH and 2.85 M $Al(OH)_3$, 12.10 M KOH and 7.12 M $Al(OH)_3$ and 16.13 M NaOH and 3.23 M $Al(OH)_3$, 20.97 M NaOH and 1.75 M $Al(OH)_3$.[39] Absorption bands were observed at 900-920 and 740 cm^{-1} by IR and 550 cm^{-1} by IR and Raman. These were interpreted as being due to ν_{as}Al-O-Al, superposition of ν_s AlO(H) of the non-bridging fragment, (H)OAlO(H) of the polymeric complex and ν_2 $Al(OH)_4^-$ and ν_sAl-O-Al respectively. It was proposed that the species giving rise to the frequencies might be dimers and infinite linear polymers formed by linking of tetrahedral hydroxy complexes through Al-O-Al.

In order to test the relative stability of various dimers a range of molecular models were developed. A number of assumptions were made regarding the structure of these models to limit the enormous range of possibilities. All dimers modelled were assumed to be symmetrical, i.e. each Al had identical coordination. In addition only OH^- coordination was considered. However, bridging groups could be either -O- or –OH^-- with either one or two bridges being present The coordination of the Al atoms in the dimer was varied from 3 to 6-fold. Table 9 shows the 11 dimers with the most stable (negative) ΔH_f^o.

It was found on calculation of ΔH^o_r for transformations between dimers that the dimer $[(OH)_3Al\text{-}O\text{-}Al(OH)_3]^{2-}$ gives rise to the most stable system. This dimer is the structure suggested in Moolenaar's spectroscopic study.[17] In Table 10 ΔH^o_r with respect

Table 9 ΔH_f^o of dimeric Al-containing species using a dielectric constant environment of 10.0. These are listed in order of the most stable to least stable species. n.s. indicates the Al containing species was sterically unstable.

Dimer	ΔH_o^f (kJ mol.$^{-1}$)
$[(OH)_4Al\text{-}(OH)_2\text{-}Al(OH)_4]^{4-}$	-3801
$[(OH)_4Al\text{-}O\text{-}Al(OH)_4]^{4-}$	-3587
$[(OH)_4Al\text{-}(OH)\text{-}Al(OH)_4]^{3-}$	-3577
$[(OH)_4Al\text{-}O_2\text{-}Al(OH)_4]^{6-}$	-3303
$[(OH)_3Al\text{-}O_2\text{-}Al(OH)_3]^{4-}$	-3193
$[(OH)_3Al\text{-}(OH)_2\text{-}Al(OH)_3]^{2-}$	-3147
$[(OH)_3Al\text{-}O\text{-}Al(OH)_3]^{2-}$	-2877
$[(OH)_3Al\text{-}(OH)\text{-}Al(OH)_3]^{-}$	-2535
$[(OH)_2Al\text{-}(O)_2\text{-}Al(OH)_2]^{2-}$	-2521
$(OH)_2Al\text{-}(OH)_2\text{-}Al(OH)_2$	-1977
$(OH)_2Al\text{-}O\text{-}Al(OH)_2$	-1562

to (i.e. as the product) $[(OH)_3Al\text{-}O\text{-}Al(OH)_3]^{2-}$ are given. In all cases the ΔH^o_r is negative indicating the reaction is favourable. The second most favourable dimer is $[(OH)_3Al\text{-}(OH)_2\text{-}Al(OH)_3]^{2-}$ with a ΔH^o_r to give 'Moolenaar's' dimer of only -11 kJ mol^{-1}. These two dimers are structurally closely related by

$[(OH)_3\ Al\text{-}O\text{-}Al(OH)_3]^{2-} + H_2O \rightarrow [(OH)_3\ Al\text{-}(OH)_2\text{-}Al(OH)_3]^{2-}$.

The four most energetically favourable dimers are predicted to be

$[(OH)_3Al\text{-}O\text{-}Al(OH)_3]^{2-}$
$[(OH)_3Al\text{-}(OH)_2\text{-}Al(OH)_3]^{2-}$
$[(OH)_2Al\text{-}(O)_2\text{-}Al(OH)_2]^{2-}$
$[(OH)_4Al\text{-}(OH)\text{-}Al(OH)_4]^{3-}$

in order of decreasing stability. These dimers show the full range of possible bridging structures, one and two -O- bridges and one and two -OH- bridges Spectroscopic discussion of dimeric structures in the literature has centred on single -O- bridges. The evidence for the conclusion that this is the only bridging structure and that it is distinguishable from other possibilities is flimsy. It is generally agreed that $Al(OH)_4^-$ is the predominant monomeric solution species under caustic conditions. This monomer has a theoretical point group of S_4, however its IR and Raman vibrations are assigned successfully with the point group T_d. This implies that the hydrogens are sufficiently diffuse to be effectively ignored This in turn implies that it may not be possible to distinguish -O- and -OH-. Absorption frequencies very similar to those attributed to -O- have been observed in hydroxyaluminate compounds which

Table 10 ΔH^o_r for the dimeric species listed below reacting to form $[(OH)_3Al\text{-}O\text{-}Al(OH)_3]^{2-}$ in an environment of dielectric constant of 10.0. These are listed in order of the most likely to least likely reaction.

Dimer	ΔH_r^o (kJ mol.$^{-1}$)
$[(OH)_4Al\text{-}O_2\text{-}Al(OH)_4]^{6-}$	-1329
$[(OH)_3Al\text{-}O_2\text{-}Al(OH)_3]^{4-}$	-421
$[(OH)_4Al\text{-}(OH)_2\text{-}Al(OH)_4]^{4-}$	-375
$[(OH)_4Al\text{-}O\text{-}Al(OH)_4]^{4-}$	-308
$(OH)_2Al\text{-}O\text{-}Al(OH)_2$	-297
$(OH)_2Al\text{-}(OH)_2\text{-}Al(OH)_2$	-163
$[(OH)_3Al\text{-}(OH)\text{-}Al(OH)_3]^-$	-114
$[(OH)_4Al\text{-}(OH)\text{-}Al(OH)_4]^{3-}$	-90
$[(OH)_2Al\text{-}(O)_2\text{-}Al(OH)_2]^{2-}$	-75
$[(OH)_3Al\text{-}(OH)_2\text{-}Al(OH)_3]^{2-}$	-11

are known to contain no oxygen bridging, e.g. gibbsite 585, 559 and 532 cm^{-1} [49] and bayerite, approximately 540 cm^{-1}.[50]

5.0 THE DIMERISATION REACTION

Calculations of ΔH^o_r of monomers to form dimers has indicated that two $Al(OH)_4^-$ monomers will not react to form any dimer, i.e. ΔH^o_r is always positive. On analysis this is not a surprising finding given that the nucleation process to form solid $Al(OH)_3$ is very slow. The E_a for the growth of gibbsite has been measured to be relatively low; 54 kJ mol.$^{-1}$ [51] and 79 kJ mo1.$^{-1}$.[52] If the dominant monomeric species reacted to form a dimer it would be expected that this process, and possibly subsequent polymerisation reactions, would occur quite rapidly particularly if the E_a is as low as that indicated by experiment. In addition, all reactions between the two most likely monomers, $Al(OH)_4^-$ and $Al(OH)_5^{2-}$ also give rise to positive ΔH^o_r on formation of any of the dimers.

We have calculated the ΔH^o_r of all possible combinations of all possible monomeric reactants to give all the dimers modelled in this study (approximately 400 reactions). However, we will limit our description here to the reactions between the four most energetically favourable monomers ($Al(OH)_4^-$, $Al(OH)_5^{2-}$, $Al(OH)_4(H_2O)^-$, $Al(OH)_3(H_2O)$) to give the two most energetically favourable dimers ($[(OH)_3Al\text{-}O\text{-}$

Table 11 ΔH^o_r for monomer + monomer → dimer reactions between the four most energetically favourable monomers to give the two most energetically favourable dimers. Only the favourable, i.e. negative, heats of reaction are given. All calculations have been carried out using a dielectric continuum environment with a dielectric constant of 10.0.

Monomer Reactant	Monomer Reactant	Dimer Product	ΔH^o_r (kJ mol.$^{-1}$)	ΔG^o_f (kJ mol.$^{-1}$)
$Al(OH)_4^-$	$Al(OH)_3(H_2O)$	$[(OH)_3Al\text{-}(OH)_2\text{-}Al(OH)_3]^{2-}$	-60	7
$Al(OH)_4^-$	$Al(OH)_3(H_2O)$	$[(OH)_3Al\text{-}O\text{-}Al(OH)_3]^{2-}$	-71	-56
$Al(OH)_5^{2-}$	$Al(OH)_5^{2-}$	$[(OH)_3Al\text{-}(OH)_2\text{-}Al(OH)_3]^{2-}$	-19	-38
$Al(OH)_5^{2-}$	$Al(OH)_5^{2-}$	$[(OH)_3Al\text{-}(O)\text{-}Al(OH)_3]^{2-}$	-30	-101
$Al(OH)_5^{2-}$	$Al(OH)_4(H_2O)^-$	$[(OH)_3Al\text{-}(OH)_2\text{-}Al(OH)_3]^{2-}$	-26	-45
$Al(OH)_5^{2-}$	$Al(OH)_4(H_2O)^-$	$[(OH)_3Al\text{-}O\text{-}Al(OH)_3]^{2-}$	-37	-108
$Al(OH)_5^{2-}$	$Al(OH)_3(H_2O)$	$[(OH)_3Al\text{-}(OH)_2\text{-}Al(OH)_3]^{2-}$	-109	-89
$Al(OH)_5^{2-}$	$Al(OH)_3(H_2O)$	$[(OH)_3Al\text{-}O\text{-}Al(OH)_3]^{2-}$	-120	-152
$Al(OH)_4(H_2O)^-$	$Al(OH)_4(H_2O)^-$	$[(OH)_3Al\text{-}(OH)_2\text{-}Al(OH)_3]^{2-}$	-33	-52
$Al(OH)_4(H_2O)^-$	$Al(OH)_4(H_2O)^-$	$[(OH)_3Al\text{-}O\text{-}Al(OH)_3]^{2-}$	-44	-115
$Al(OH)_4(H_2O)^-$	$Al(OH)_3(H_2O)$	$[(OH)_3Al\text{-}(OH)_2\text{-}Al(OH)_3]^{2-}$	-116	-96
$Al(OH)_4(H_2O)^-$	$Al(OH)_3(H_2O)$	$[(OH)_3Al\text{-}O\text{-}Al(OH)_3]^{2-}$	-127	-159
$Al(OH)_3(H_2O)$	$Al(OH)_3(H_2O)$	$[(OH)_3Al\text{-}(OH)_2\text{-}Al(OH)_3]^{2-}$	-199	7
$Al(OH)_3(H_2O)$	$Al(OH)_3(H_2O)$	$[(OH)_3Al\text{-}O\text{-}Al(OH)_3]^{2-}$	-210	-56

$Al(OH)_3]^{2-}$, $[(OH)_3Al\text{-}(OH)_2\text{-}Al(OH)_3]^{2-}$) which give rise to a negative ΔH^o_r on formation of a dimer (Table 11).

It can be seen in Table 11 that the least energetically favourable monomer, $Al(OH)_3(H_2O)$, is the most likely to react to form a dimer. However, the least energetically favourable monomer, while being the most reactive, will also be present

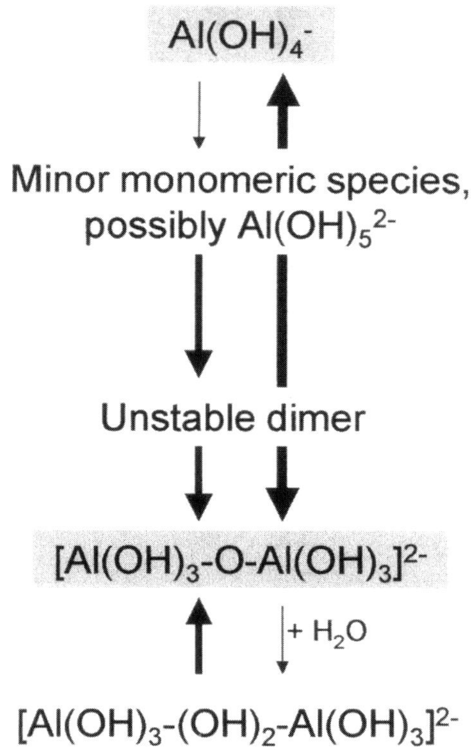

Fig. 3 The proposed dimerisation mechanism. The species shown within gray shaded boxes are those which are proposed on the basis of both experimental and theoretical evidence. The thickness of the arrows indicates the extent of the reaction in the direction.

in smaller concentration than the other three more stable monomers (Table 8). $Al(OH)_5^{2-}$ will react with any other monomer except $Al(OH)_4^-$. It is likely that the dimerisation reaction may follow many paths and proceed through both unstable monomers and unstable dimers before reaching the most stable dimer. The proposed reaction scheme is shown schematically in Figure 3. This reaction scheme is shown to finish with the dimer $[(OH)_3Al\text{-}(OH)_2\text{-}Al(OH)_3]^{2-}$. In order to build up an $Al(OH)_3$ structure, which possess 6-fold octahedral -OH- coordination with pairs of Al atoms being bridged by pairs of -OH- bridges, it is necessary at some stage of the nucleation process for -OH- bridges to form between Al atoms. This -OH- bridged dimer is shown at this stage of the nucleation mechanism to indicate that these bridges may viably form during the dimerisation process even though it would be expected that this dimer would be present in smaller concentrations than $[(OH)_3Al\text{-}O\text{-}Al(OH)_3]^{2-}$.

6.0 TRIMMERS, TETRAMERES AND OLIGOMERS

A large number of complementary studies have been carried out on dilute aluminate salt solutions in acidic or mildly alkaline conditions. It has generally been accepted that under these conditions that the rate-determining step in precipitation of aluminium hydroxide is the formation of Al-OH-Al bridges, hence, many of the studies carried out have focused on the ratio of OH/Al in solution.

The $Al_{13}O_4(OH)_x(H_2O)_{36-x}{}^{(31-x)}$ (Keggin ion) structure was deduced by Bottero, Cases, Fiessinger and Poirer[53] in acidified aluminium nitrate solutions on the basis of ^{27}Al NMR. The Keggin ion comprises a central tetrahedrally coordinated Al surrounded by 12 octahedrally coordinated Al. SAXS investigations by Rausch and Bale[54] in similar solutions observed plate like structures with a radius of gyration, $R_g = 4.3$ Å. It was concluded that these structures were Keggin ions.

Bottero, Tchouber, Cases and Fiessingerss[35] performed SAXS experiments on aged and fresh aluminium chloride solutions over a range of solution concentrations. Solutions with OH/Al = 2.0 (0.1 M $AlCl_3$) showed no change in speciation as a function of time. The principal component of this solution was identified as $Al_{13}O_4(OH)_{28}{}^{3+}$. The radius of gyration was calculated from small angle x-ray scattering data to be 9.8 Å corresponding to an ionic radius of 12.6 Å. In the solid state this ion has been found to have a radius of 5.4 Å.[56] The extra size of the solution Keggin ion was assumed to be due to hydration. No explanation has been proposed as to why one species could give rise to such widely varying R_g measurements in solution (4.3 Å to 9.8 Å) although molecular modelling has indicated that the smaller R_g is more realistic.[57]

Solutions of greater hydroxide to aluminium ratio (OH/Al = 2.5 and 0.1 M $AlCl_3$) were also examined by Bottero et al.[55] The solutions speciation was interpreted as being made up of Keggin ions and larger polymeric species. These latter species were observed to change shape with time from cylindrical particles of $R_g = 15$ Å and length 310 Å after 1.5 hrs. to platelets of $R_g = 250$ Å and thickness, $R_t = 60$ Å after 24 hrs. During this time the intensity of scattered x-rays increased by 1.5-fold indicating a densification of the colloidal species

Similar studies have been performed by Schaefer et al., with similar results.[58] X-ray, static and dynamic light scattering studies measurements were carried out on caustic aluminate solutions of 0.5 M Al. It was found that the speciation observed was dependent on the concentration of caustic. Below OH/Al = 2.0 the Keggin ion was again observed. At OH/Al = 2.5, large clusters with a radius of gyration of approximately 1400 Å were observed. Below OH/Al = 2.5, the solution was found not to gel. Above OH/Al = 2.5, gelation was observed. The value of OH/Al of 2.5, is very close to the vector percolation threshold of 2.4 which represents the minimum number of cross-links required for a 3-D structure to attain rigidity.[59] Static light scattering analysis indicated that on a four-fold dilution the molecular weight decreases by a factor of 3 but the radius of gyration remained constant, Thus the polymers formed, swelled and broke upon dilution. In addition, the fractal dimension decreased from 2 to 1.4. A value of 2 suggests a reaction limited cluster-cluster and monomer-cluster aggregation whereas a value of 1.4 may suggest diffusion limited cluster-cluster growth.[60] Rausch and Bale[54] also observed this behaviour in solutions with OH/Al < 2.5. If has been concluded that for dilute aluminium containing salt solutions the nature of the Al-containing species in solution is very sensitive to the OH/Al ratio which, to a large extent, governs the likelihood of precipitation. In addition the size of the complexes that form depend a great deal on the degree of hydrolysis.[61] At OH/Al < 2.5 the observations made are consistent with a series of equilibria involving various polymeric and monomeric species.[62]

Ruff and Tyree[61] have performed static light scattering measurements on acidified aluminium nitrate solutions. It was found that turbidity increased with increasing aluminium concentration and the second virial coefficient decreased with increasing OH/Al ratio. This latter observation can be interpreted to mean that either the species became less repulsive or that they had less affinity for the surrounding solvent on increasing OH/Al ratios. The weight average molecular weight was found to be equal to approximately 1000 g mol.$^{-1}$. This corresponds closely to the molecular weight of $[Al_{13}O_4(OH)_{24}(H_2O)_{12}]^{7+}$ of 1039 g mol.$^{-1}$.

Hsu[63] has investigated the mechanism of gibbsite crystallisation from partially neutralised aluminium chloride solutions. The solutions were analysed for pH, turbidity and changes in Al-containing species as a function of aging. Speciation was investigated using the kinetics of Al-Ferron colour development.[64] It was concluded that the initial polymers present evolved on aging to either gibbsite or stable polymers and that this evolution occurred by two different pathways.

Concentrated caustic aluminate solutions have been described as having a considerable degree of hydrogen bonding, causing clustering of the aluminate ions to form complex structures.[65, 66] The results of a number of experiments carried out on concentrated caustic aluminate liquors have concluded that the aluminate ion has a tendency to polymerise and that the formation of aluminium-containing polymers in solution occurs by the release of bound H_2O during the period before the visual detection of any crystals.[67] No observation, which the authors are aware of Keggin ion formation in concentrated caustic aluminate liquors has been made. There have been no additional reliable observation of vibrational absorptions that increase on aging of solutions other than those already attributed to $Al(OH)_4^-$ and $[(OH)_3Al\text{-}O\text{-}Al(OH)_3]^{2-}$ thus the bonding within any polymeric species that form must be indistinguishable from the monomeric and dimeric species that have been identified.

It is not within the scope of this study to model large polymeric systems. However, by probing the thermodynamics of some small oligomers we may gain insight into the formation mechanisms of larger species. To this end we have modelled some trimer and tetramer species. These species were all symmetric, were either neutral or negatively charged, contained aluminium atoms with 4, 5 or 6-fold OH^- coordination and contained Al pairs bridged by two -OH- groups. OH coordination was assumed due to the necessity to render the model more gibbsite-like on increase in oligomeric size. There is no particular reason to assume that trimers or tetramers will be symmetrical. This has been done solely to reduce the number of possibilities for modelling. It is also entirely possible that trimer or larger species may exist with a mixture of 1 or 2-OH- or -O- bridging structures.

Table 12 ΔG^o_f of Al trimers and ΔH^o_f and ΔG^o_f to give $[(3)Al\text{-}(2)\text{-}Al(l)\text{-}(2)\text{-}Al(3)]^{2-}$. The number in () indicate the number of attached OH^- group or -()- bridging hydroxyl groups. All calculations are carried out in a dielectric continuum environment with dielectric constant of 10.0.

Timer	Al Coordination and Charge	ΔG^o_f (KJ mol.$^{-1}$)	ΔH^o_r (kJ mol.$^{-1}$)	ΔG^o_r (kJ mol.$^{-1}$)
$[(4)Al\text{-}(2)\text{-}Al(2)\text{-}(2)\text{-}Al(4)]^{5-}$	$[6\text{-}6\text{-}6]^{5-}$	-5834	-513	-639
$[(4)Al\text{-}(2)\text{-}Al(1)\text{-}(2)\text{-}Al(4)]^{4-}$	$[6\text{-}5\text{-}6]^{4-}$	-5065	-316	-398
$[(3)Al\text{-}(2)\text{-}Al(2)\text{-}(2)\text{-}Al(3)]^{3-}$	$[5\text{-}6\text{-}5]^{3-}$	-4758	-103	-145
$[(3)Al\text{-}(2)\text{-}Al(1)\text{-}(2)\text{-}Al(3)]^{2-}$	$[5\text{-}5\text{-}5]^{2-}$	-4343	0	0
$[(3)Al\text{-}(2)\text{-}Al(2)\text{-}Al(3)]^{-}$	$[5\text{-}4\text{-}5]^{-}$	-3685	-146	-98
[(2)Al-(2)-Al(1)-(2)-Al(2)]	4-5-4	-3127	-193	-96
$[(3)Al\text{-}(O)\text{-}Al(1)\text{-}(O)\text{-}Al(3)]^{2-}$	$[4\text{-}3\text{-}4]^{2-}$			55

The ΔG^o_f of the trimer species are given in Table 12. The most energetically favourable trimer is $[(3)Al\text{-}(2)\text{-}Al(l)\text{-}(2)\text{-}Al(3)]^{2-}$ where the number in () indicates the number of OH^- groups. The ΔH^o_r and ΔG^o_r of transformation of the other trimers modelled, to give this trimer are given in Table 12. A full range of -O- bridges species have not been examined. However, the possibility that these species may exist has not been discounted, particularly as calculations have indicated that $[(OH)_3Al\text{-}O\text{-}Al(OH)_3]^{2-}$ is the most stable dimer formed. To assess this possibility the trimer $[(3)Al\text{-}O\text{-}Al(1)\text{-}O\text{-}Al(3)]^{2-}$ was also investigated. This trimer was found to be more stable than $[(3)Al\text{-}(2)\text{-}Al(l)\text{-}(2)\text{-}Al(3)]^{2-}$ with a ΔG^o_r, to form $[(3)Al\text{-}(2)\text{-}Al(l)\text{-}(2)\text{-}Al(3)]^{2-}$ of 55 kJ mol.$^{-1}$ (bottom line of Table 12). Thus the energetic relationship between the -OH- bridged timer and its -O- bridged analogue (i.e. minus one H_2O per pair of -OH- bridges) appears to be the same as for dimers. This suggests that one of the possible rate determining steps in the formation of the aluminium hydroxide crystalline structure may be the formation of the -OH- bridges.

The combinations of monomeric and dimeric reactants which will give rise to an energetically favourable reaction to form the most energetically favourable trimer, $[(3)Al\text{-}(2)\text{-}Al(l)\text{-}(2)\text{-}Al(3)]^{2-}$, are limited. In Table 13, the first row shows the only energetically favourable ΔG^o_r to form the most stable trimer from a combination of the four most energetically favourable monomers and the two most energetically favourable dimers. The remaining rows give the favourable ΔG^o_r, of combinations of the four most favourable monomers with the third and fourth most favourable dimers. It is apparent from Table 13 that the more unstable the reacting monomer and/or dimer and hence the more minor concentration of this species in solution, the more likely the trimerisation reaction will occur. Consequently, it appears that, as for the dimerisation reaction, the trimerisation reaction will primarily occur

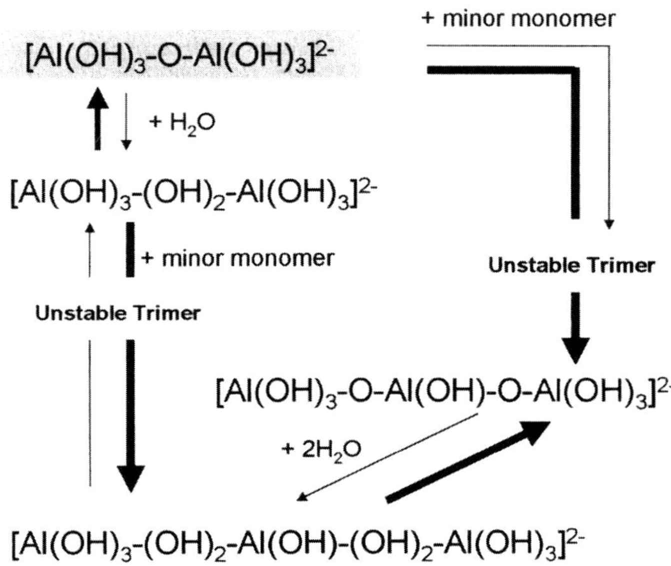

Fig. 4 Possible reaction pathways from the most energetically favourable dimers to the most energetically favourable trimers.

Table 13 ΔH^o_r and ΔG^o_r of the four most stable monomers with the four most stable dimers to give $[(3)Al\text{-}(2)\text{-}Al(1)\text{-}(2)\text{-}Al(3)]^{2-}$. All calculations are carried out in a dielectric continuum environment with a dielectric constant of 10.0.

Monomer	Dimer	ΔH^o_r (kJ mol.$^{-1}$)	ΔG^o_r (kJ mol.$^{-1}$)
$Al(OH)_3(H_2O)$	$[(OH)_3Al\text{-}(OH)_2\text{-}Al(OH)_3]^{2-}$	-79	-44
$Al(OH)_3(H_2O)$	$[(OH)_2Al\text{-}(O)_2\text{-}Al(OH)_2]^{2-}$	-143	-14
$Al(OH)_4^-$	$[(OH)_4Al\text{-}(OH)\text{-}Al(OH)_4]^{3-}$	-19	-18
$Al(OH)_5^{2-}$	$[(OH)_4Al\text{-}(OH)\text{-}Al(OH)_4]^{3-}$	-68	-114
$Al(OH)_4(H_2O)^-$	$[(OH)_4Al\text{-}(OH)\text{-}Al(OH)_4]^{3-}$	-75	-121
$Al(OH)_3(H_2O)$	$[(OH)_4Al\text{-}(OH)\text{-}Al(OH)_4]^{3-}$	-158	-165

through the reaction of minor solution constituents. A schematic mechanism for the formation of trimer from a dimer is shown in Figure 4.

The mechanism of formation of tetramers is highly significant, as this is the first opportunity for a branched structure to form. The most energetically favourable tetramer of those modelled was $[(3)Al\text{-}(2)\text{-}Al(2)\text{-}(2)\text{-}Al(2)\text{-}(2)\text{-}Al(3)]^{4-}$ (Table 14). The middle two aluminium atoms in this tetramer have 6-fold coordination to OH^- groups and thus resemble the coordination of Al within the gibbsite structure. It can be seen from the ΔG^o_r values given in Table 14 that the branched tetramer

Table 14 ΔG^o_f of Al tetramers and ΔH^o_r and ΔG^o_r to give [(3)Al-(2)-Al(2)-(2)-Al(2)-(2)-Al(3)]$^{4-}$ The number in ( ) indicate the number of attached OH group or -( )- bridging hydroxyl groups.

Tetramer	Al Coordination and Charge	ΔG^o_f (kJ mol.$^{-1}$)	ΔH^o_r (kJ mol.$^{-1}$)	ΔG^o_f (kJ mol.$^{-1}$)
[(4)Al-(2)-Al(2)-(2)-Al(2)-(2)-Al(4)]$^{6-}$	[6-6-6-6]$^{6-}$	-6787	-1038	-1234
[(3)Al-(2)-Al(2)-(2)-Al(2)-(2)-Al(3)]$^{4-}$	[5-6-6-5]$^{4-}$	-6227	0	0
[(2)Al-(2)-Al(2)-(2)-Al(2)-Al(2)]$^{2-}$	[4-6-6-4]$^{2-}$	-5257	-514	-524
[(3)Al-(2)-Al(1)-(2)-Al(1)-(2)-Al(3)]$^{2-}$	[5-5-5-5]$^{2-}$	-5380	-386	-401
[(2)Al-(2)-Al(1)-(2)-Al(1)-(2)-Al(2)]	[4-5-5-4]^{9}	-4198	-536	-463
[(4)Al-(2)-Al-(2)-Al(4)]$^{6-}$ | (2) | Al(4)	[6-6-6]$^{6-}$ | 6	-6739	-1081	-1282
[(3)Al-(2)-Al-(2)-Al(3)]$^{3-}$ | (2) | Al(3)	[5-6-5]$^{3-}$ | 5	-5856	-420	-485
(2)Al-(2)-Al(2)-(2)-Al(2) | (2) | Al(2)	[4-6-4]° | 4	-4161	-575	-500

[(3)Al-(2)-Al{-(2)-Al(3)}-(2)-Al(3)]$^{3-}$ (second from bottom) is the 4th most likely tetramer to form. This may indicate that branching is not likely to occur in significant quantities in this stage of polymerisation. It is possible that polymeric branching is entropically driven. Thus, when a linear polymer reaches a critical length the entropy associated with such a polymer may be sufficiently unfavourable to induce the formation of branches in preference to increasing polymer length.

7.0 SUMMARY AND CONCLUSIONS

The computational results and analysis thereof in the light of extensive publications based on experimental studies have helped to clarify a number of issues relating the nucleation mechanism of aluminium trihydroxide from caustic solutions.

It appears from both experimental studies and the computational study reviewed here that $Al(OH)_4^-$ is the dominant aluminium containing monomeric species found in conditions of moderate to high pH. Other monomeric species are likely to be present but in concentrations less than that observable by current spectroscopic means. $[(OH)_3Al\text{-}O\text{-}Al(OH)_3]^{2-}$ is indicated, by both computational and experimental studies to be the most prolific dimer found in caustic aluminate solutions. However, there is some ambiguity in the assignment of the presence of the -O- bridge in this species as spectroscopically it may be difficult to distinguish between dimeric species undergoing rapid protonation and deprotonation. Semi-empirical quantum mechanical calculations show that $[(OH)_3Al\text{-}(OH)_2\text{-}Al(OH)_3]^{2-}$ is almost as likely to form as $[(OH)_3Al\text{-}O\text{-}Al(OH)_3]^{2-}$.

The mechanism of dimerisation has not been elucidated by experimental means although it has been shown that the concentration of dimers in solution increases with increasing caustic concentration and probably also increase with increasing aluminium solution concentration. Semi-empirical quantum mechanical calculations indicate that the $Al(OH)_4^-$ species will not dimerise. Hence, dimerisation must be a process undergone on the collision of minor, less stable species within caustic aluminate solutions. It is highly likely that dimerisation can occur by a number of different pathways. The most likely pathway appears to involve the dimerisation of $Al(OH)_5^{2-}$ species as these have been calculated to be the second most common monomeric species in solution, albeit at a concentration of less than 10^{-10} that of $Al(OH)_4^-$. A dimerisation reaction involving $Al(OH)_5^{2-}$ is consistent with the experimental observations of increasing dimer concentration on increasing caustic and Al solution concentration. Thus, the rate limiting step for dimerisation and possible further polymerisation steps may be the collision frequency of minor unstable monomeric species, either with each other or growing oligomers.

The stability and mechanism of formation of trimers has been found by computational means to be similar to that of dimers. The trimeric species $[(OH)_3Al\text{-}O\text{-}Al\text{-}O\text{-}Al(OH)_3]^{2-}$ has been found to be slightly more likely to form than $[(OH)_3Al\text{-}(OH)_2\text{-}Al\text{-}(OH)_2\text{-}Al(OH)_3]^{2-}$. In addition, the reaction mechanisms to form trimers appears to involve either minor monomeric species or minor dimeric species or both. The involvement of $Al(OH)_4^-$ in trimer formation appears to be limited to reactions with reasonably unstable dimers. It is unlikely that spectroscopic means could distinguish between the -OH- bridged and -O- bridged trimers or between these trimers and their dimeric analogues. Hence, it is possible that extensive Al polymerisation in caustic solutions occurs but is not detected spectroscopically.

In solid aluminium trihydroxide, each Al is 6-fold coordinated through 3 pairs of -OH- bridges to 3 Al atoms. In order to form this structure, therefore, at some stage of nucleation -OH- bridges must form. Calculations on the relative stabilities of dimers and trimers indicate that -O- bridged species are slightly more stable than their -OH- bridged analogues i.e. with addition of H_2O. It is possible therefore that the formation of these bridging groups may be rate determining and may be dependent on the concentration of free water in solution. That is, it may be possible for dimerisation and subsequent oligomerisation to occur rapidly in highly concentrated solution but the bridging required for the formation of solid aluminium hydroxide is inhibited in these solutions by the lack of free water. Thus, the rate of formation of aluminium hydroxide nuclei is dependent on two factors, the collision frequency of minor species and the ability to form -OH- bridges. The former is retarded on dilution and the latter is enhanced on dilution. This provides a possible explanation for the non-linear gibbsite solubility curve observed as a function of solution dilution.

The 4-fold $Al(OH)_4^-$ to 6-fold (gibbsite) Al coordination transformation must be accounted for in an aluminium trihydroxide nucleation mechanism. The calculations carried out indicate that tetrameric Al containing species may contain 6-fold coordinated Al. This increase in coordination results from increasing bridging to Al atoms via -OH- bridges. It appears unlikely that substantially branching will occur at the tetrameric stage of oligomerization. This may be induced on further polymerisation due to a growing unfavourable entropic component to ΔG^o_f. If this is the case then the formation of branching in the Al-containing species is analogous to the collapse and densification of silica spheres as discussed in the Introduction of this review.

8.0 REFERENCES

1. J. W. Mullins: *Crystallisation*, Butterworth-Heinemann Ltd., 1993.
2. G. H. Nancollas and N. Purdie: The Kinetics of Crystal Growth, *Quart. Review*, London, **18**(1), 1964, pp.1-20.
3. J. P. Hirth and G. M. Pound: *Progress in Materials Science*, **2**, 1963, pp.1-4.
4. V. K. Lamer and R. H. Dineger: *Journal of American Chemical Society*, **72**, 1950, pp.4847-4854.
5. E. Matijevic: Monodispersed Inorganic Colloids Have Become Reality, *Proceeding of Materials Research Society International Meeting on Advance Materials, Symposium-X Frontiers of Materials Research*, 1988, pp.178-204.
6. G. L. Messing and G. Y. Onoda Jr.: *Journal of American Ceramics Society*, **61**, **1978**, p.1.
7. S. Tarutg Y. Itou, K. Kawashima, K. Kitajima, N. Takusagawa, K. Okada and N. Otsuka: Effect of Additives on Packing and Sintering of Bimodeal Size-Distributed Alumina Powder Mixtures, *Key Engineering Material Transactions Technology Publication*, **111-112**, 1995, pp.321-329.
8. W. Stober, A. Fink and E. J. Bohn: *Collection of International Science*, **26**, 1968, p.62.
9. G. H. Bogush, M. A. Tracy and C F. Zukoski: Preparation of Monodisperse Silica Particles: Control of Size and Mass Fraction, *Journal of Non-Crystalline Solids*, **104**, 1988, pp.95-106.

10. J. K. Bailey, C. W. Macosko and M. L. Mecaatney: Modeling the Gelation of Silicon Alkoxides, *Journal of Non-Crystalline Solids,* **125**, 1990, pp.208-223.
11. G. H. Bogush and C F. Zukoski: Studies on the Formation of Monodisperse Silica Powders, *Ultrastructure Processing of Advanced Ceramics,* J.D. Mackenzie and D.R. Ulrich, eds., Wiley, New York, 1988, pp.477-486.
12. C. H. Byers and M. T. Harris: A Dynamic Laser-Light Scattering Study of Solvent Effects in Silica Synthesis by Alkoxide Hydrolysis; *Ultrastructure Processing of Advanced Ceramics,* J.D. Mackenzie and D.R. Ulrich eds., Wiley, New York, 1988, pp.843-853.
13. M. T. Harris, R. Rbrunson and C. H. Byers: *Journal of Non-Crystalline Solids,* **121**, 1990, p.397.
14. J. K.Bailey and M. L. Mecartney: Formation of Colloidal Silica Particles from Alkoxide, *Colloids and Surfaces,* **63**, 1992, pp.151-161.
15. P. C. Heimenz: *Polymer Chemistry,* Marcel Dekker, New York, 1984.
16. R. St Smart, J. A. Counter, J. Addai-mensah, A. R. Gerson and J. Ralston: Nucleation and Precipitation Mechanisms in Precursors to Colloidal Alumina, *Ceramic Interfaces; Properties and Applications,* Chapman and Hall, 1989.
17. R. J. Moolenaar, J. C. Evans and L. D. Mckeever: The Structure of the Aluminate Ion in Solutions at High pH, *Journal of Physical Chemistry,* **74**(20), 1970, pp.3629-3636.
18. A. R. Gerson, J. Addai-mensah, J. Counter, T. Soar and J. Watson: The Nucleation Mechanism of Gibbsite from Bayer Liquors, *Light Melals,* 1998, pp.167-172.
19. H. H. Paradies: *Colloids and Surfaces,* **6**(4), 1983, p.405.
20. A. R. Gerson, J. A. Counter and D. J. Cookson: Influence of Solution Constituents, Solution Conditioning and Seeding of the Crystalline Phase of Aluminium Hydroxide using In Situ X-ray Diffraction, *Journal of Crystalline Growth,* **160**, 1996, pp.346-354.
21. J. Addai-mensah, A. R. Gerson, C. A. Prestidge, I. Ametov and J. Ralston: Interactions Between Gibbsite Crystals in Supersaturated Caustic Aluminate Solutions, *Light Metals,* 1998, pp.159-166.
22. P. M. May, G. T. Hefter and P. Sipos: *Private Correspondence,* 1997.
23. C. Misra and E. T. White: The Precipitation of Bayer Aluminium Trihydroxide, *Journal of Crystalline Growth,* 1970, pp.172-178.
24. A. R. Gerson, J. Ralston and R. St. C. Smart: An Investigation of the Mechanism of Gibbsite Nucleation using Molecular Modelling, *Colloids and Surfaces,* **110**, 1996, pp.105-117.
25. H. Saalaeld and M. Wedde: Refinement of the Crystal Structure of Gibbsite, $Al(OH)_3$, *Zeitschrift für Kristallographie,* **139**, 1974, pp.129-135.
26. M. J. S. Dewar, E. G. Zoebisch, E. F. Healy and J. J. P. Stewart: AMI: A New General Purpose Quantum Mechanical Molecular Model, *Journal of American Chemical Society,* **107**(13), 1985, pp.3902-3909.
27. M. J. S. Dewar and A. J. Holder: AMI Parameters for Aluminum, *Organometallics,* **9**, 1990, pp.508-511.
28. J. J. P. Stewart: *Mopac Manual 1993,* Fujitsu Limited Tokyo, Japan, 1993.

29. A. Klamt and G. Shüürmann: COSMO: A New Approach to Dielectric Screening in Solvents with Explicit Expressions for the Screening Energy and its Gradient, *Journal of the Chemical Society*: *Perkin Transactions*, **2**, 1993, pp.779-805.
30. P. S. K. M. Rao and D. Premaswarup: Dielectric Constants of Aqueous Electrolytic Solutions, *Indian Journal of Pure and Applied Physics,* **4**(8), 1996, pp.332-333.
31. J. R. Macdonald and C. A. Barlqw, Jr.: Equilibrium Double Layer Theory, *Proceedings of the First Australian Conference on Electrochemistry*, J.A. Friend, F. Gutmann and J.W. Hayes, eds., J. W., Pergamon Press Ltd., 1965.
32. M. J. S. Dewar, E. F. Healy and J. J. P. Stewart: Location of Transition States in Reaction Mechanisms, *Journal of Chemical Society Faraday Transactions 2,* 1984, 80, pp.227-233.
33. C. F. Baes and R. E. Mesmer: Chapter 6, *The Hydrolysis of Cations*, John Wiley and Sons, New York, 1976.
34. J. W. Akitt, N. N. Greenwood, B. L. Khandelwal and G. D. Lester: Aluminium-27 Nuclear Magnetic Resonance Studies of the Hydrolysis and Polymerisation of the Hexa-aquo-aluminum(III) Cation, *Journal of Chemical Society Dalton Transaction,* **57**, 1972, pp.604-610.
35. J. Y. Bottero, J. E. Poirer and F. Fiessinger: Study of Partially Neutralised Aqueous Aluminium Chloride Solutions: Identification of Aluminium Species and Relation between the Composition of the Solutions and their Efficiency as a Coagulant, *Water Science Technology*, **13**(1), 1981, pp.601- 612.
36. J. W. Akitt and A. Farthing: *Journal of Chemical Society Dalton Transactions,* 1981, p.1606.
37. J. Y. Bottero, J. P. Marchal, J. M. Cases, J. E. Poirier and F. Flessinger: Aluminium-27 NMR Study of Dilute Partially Neutralized Solutions of Aluminium Chloride, *Bulletin. Society Chim.* Fr., **11-12**(1), 1982, pp.439-444.
38. N. Y. Chen, J. C. Evans and L. D. Mckeever: *Science in China (Series B),* **36**, 1993, pp.32-38.
39. E. A. Kopylova, L. P. Ni, M. V. Zakharova and Y. F. Lkuchnikov: Technique and Results of Investigation of the Structure of Aluminate Solutions with the Aid of IR and Raman, *Journal of Applied Chemistry (USSR),* **47**(10), 1974, pp.2396-2399.
40. T. P. Porotnikova, V. A. Derevyankin, S. O. Kuznetsov, M. P. Kuznetsova and A. K. Chirkov: Technique and Results of Investigation of the Structure of Aluminate Solutions with the Aid of IR and Raman, *Journal of Applied Chemistry (USSR),* **46**, 1973, pp.479-480.
41. K. Sakamoto: Mechanism of the Decomposition of Sodium Aluminate Solutions, *Extractive Metallurgy of Aluminium, International Symposium,* **1**, 1963, pp.175-189.
42. L. A. Carriera, V. A. Maroni, J. W. Swaine, Jr and R. C. Plumb: Raman and Infrared Spectra and Structures of the Aluminate Ions, *Journal of Chemical Physics,* **45**(6), 1966, pp.2216-2220.
43. E. R. Lippincott, J. E. Psellas and M. C. Tobin: The Raman Spectra and Structures of Aluminate and Zincate Ions, *Journal of Chemical Physics,* **20**, 1952, p.536.

44. I. A. Dibrov, G. Z. Mal'tsev and V. P. Mashovets: *Zhurnal Prikladnoi Khimii*, **37**, 1964, pp.1920-1929.
45. I. S. Ignat'ev, A. N. Lazarev and V. A. Kolesova: Normal Vibrations of Aluminate with Tetrahedrally Coordinated Aluminium Atoms', *Izvesteya Akademii Nauk SSSR, Neoganisheskie Materialy*, **12**(7), 1976, pp.1230-1239.
46. G. Johansson: The Crystal Structure of the Potassium Aluminate $K_2[Al_2O(OH)_6]$, *Acta Chemica Scandinavica*, **20**(2), 1966, pp.505-515.
47. J. Haladjian and Rozlere: Vibrational Spectra of Potassium Hexahydroxyaluminate and Digallate, *Journal of Inorganic and Nuclear Chemistry*, **55**, 1973, pp.3821-3829.
48. V. A. Kolesova, Y. I. Ryskin and I. V. Guseva: *Izvesteya Akademii Nauk SSSR, Neoganisheskie Materialy*, **4**, 1968, p.1822.
49. H. V. Olphen and J. J. Fripiat: *Data Handbook for Clay Materials and other NonMetallic Materials*, Pergemon Press, 1979.
50. S. M. Bradley, R. A. Kydd and R. F. Howe: The Structure of Al-gels Formed through Base Hydrolysis of Al^{3+} Aqueous Solutions, *Journal of Colloid and Interface Science*, **9**, 1993, p.405.
51. J. Mordini and B. M. Cristol: Mathematical Model of Alumina Trihydrate Precipitation from Bayer Liquors, Presented at the *4th Yugoslav International Symposium on Aluminium*, 1982.
52. W. R. King: Some Studies in Alumina Trihydrate Precipitation Kinetics, *Light Metals*, **2**, 1973, pp.551-563.
53. J. Y. Bottero, J. M. Cases, F. Fiessinger and J. E. Poirier: Studies of Hydrolyzed Aluminium Chloride Solutions I. Nature of Aluminium Species and Composition of Aqueous Solutions, *Journal of Physical Chemistry*, **84**(22), 1980, pp.2933-2939.
54. W. V. Rausch and H. S. D. Bale: Small Angle X-ray Scattering from Hydroyzed Aluminium Nitrate Solutions, *Journal of Chemical Physics*, **40**(11), 1964, pp.3391-3394.
55. J. Y. Bottero, D. Tchouber, J. M. Cases and F. Fiessinger: Investigation of the Hydrolysis of Aqueous Solutions of Aluminium Chloride, *Journal of Physical Chemical*, **86**,1982, pp.3667-3671.
56. G. Johansson: On the Crystal Structure of some Basic Aluminium Salts, *Acta Chemica Scandinavica*, **14**(3), 1960, pp.771-773.
57. A. R. Gerson: Unpublished Research Results, 1996.
58. D. W. Schaefer, R. A. Shelleman, K. D. Keefer and J. E. Martin: Equilibrium Structure and Rigidity of Alumina Polymers, *Physica*, **140a**, 1986, pp.105-113.
59. M. F . Thorpe: Continuous Deformations in Random Networks, *Journal of Non-Crystalline Solids*, **57**(3), 1983, pp.355-70.
60. R. Pynn: Sludge Physics and the Liberty Bell, *Los Alamos Science, Summer*, 1990, pp.90-113.
61. J. K. Ruff and S. Y. Tyree: Light Scattering Studies on Aqueous Aluminium Nitrate Solutions, *Journal of American Chemical Society*, **80**(7), 1958, 1523-1526.
62 R. J. Stol, A. K. vanHelden and P. L. deBruyn: Hydrolysis-Precipitation Studies

of Aluminium (III) Solutions, *Journal of Colloid Interface Science*, **57**, 1976, pp.115-131.

63. P. H. Hsu: Mechanism of Gibbsite Crystallisation From Partially Neutralized Aluminum Chloride Solutions 2. A Kinetic Study and Model, *Clays and Clay Minerals*, **36**, 1988, pp.25-30.
64. P. Ptsai and P. H. Hsu: Expansion of Smectites as a Function of Vapour Pressure of Ethylene Glycol, *Solid Science Society American Journal*, **48**(4), 1984, pp.935-959.
65. J. R. Glastonbury: Nature of Sodium Aluminate Solutions, *Chemistry and Industry*, 1969, pp.121-125.
66. N. Brown: Secondary Nucleation of Aluminium Trihydroxide - Kinetics and Mechanism, *Light Metals*, 1977, **2**, pp.1-14.
67. J. Zambo: Structure of Sodium Aluminate Liquors; Molecular Model of the Mechanism of their Decomposition, *Light Metals*, 1986, pp.199-215.

Molecular-Assembly Mediated Crystallization as a Novel Synthetic Route for Ceramic Thin Films

K. KOUMOTO, S. SEO and W.S. SEO
Department of Applied Chemistry, Nagoya University
Furo-cho, Chikusa-ku, Nagoya 464-8603, Japan

ABSTRACT

Thin films of γ-FeO(OH) and Fe_3O_4 were successfully grown by supplying NH_3 gas to $FeCl_2$ solution through an appropriate Langmuir monolayer. A (010) preferentially oriented γ-FeO(OH) film was deposited under a stearyl alcohol monolayer at the beginning of the crystal growth, and later turned into a non-oriented Fe_3O_4 particulate film. The molecular arrangement of a stearyl alcohol monolayer and the ionic arrangement of the (010) plane of γ-FeO(OH) match geometrically well with each other, indicating that a stearyl alcohol monolayer worked as a template for epitaxial growth of a (010) oriented γ-FeO(OH) thin film.

We have also succeeded in fabricating micropatterned TiO_2 films on self-assembled monolayers (SAM) of phenyltrichlorosilane (PTCS) through the precipitation from an aqueous solution containing TiF_6^{2-} ions. A PTCS monolayer was first formed on a silicon wafer and UV light was irradiated onto the masked monolayer to obtain the SAM with two-dimensionally patterned hydrophilic/hydrophobic surfaces. This patterned substrate was immersed in an aqueous solution to directly deposit TiO_2 (anatase). Sonication treatment after the deposition resulted in a micropatterned TiO_2 thin film.

1.0 INTRODUCTION

The so-called biomimetic materials chemistry has recently become a very attractive field as it is expected to open new pathways to synthesize inorganic materials at low temperatures through low-energy processes,[1] control of the crystal-axis orientations and morphologies of inorganic materials through the choice of appropriate organic templating agents represents an important goal and critical theme of such research. A number of organic systems including Langmuir monolayers, Langmuir-Blodgett (LB) films, self-assembled monolayers (SAM), reverse micelles, vesicles, liquid crystals, etc. have been shown to be effective as templates for this purpose.[2]

Among those systems Langmuir monolayers have simple two-dimensional structures and can easily be applied as substrates for thin film growth out of aqueous solutions. A variety of inorganic crystals such as NaCl,[3] $CaCO_3$,[4, 5] $BaSO_4$,[4] CdS,[6] PbS,[7] hydroxyapatite,[8] have been successfully grown with certain crystal-axis orientations under Langmuir monolayers with various functional groups lying at the air/water interfaces serving as nucleation sites.

SAM is another kind of organic monomolecular layer that can be spontaneously formed by immersing a solid substrate into a precursor solution,[9] and can also serve as a template for inorganic crystal growth.[10] SAMs formed through chemisorption of organosilanes of general formula $RSiX_3$ (where "R" is an organic functional group and "X" is an alkoxide or halide) to surface hydroxyl groups have long been popular as templating agents[9] due to their favourable film properties. For example, the spontaneous self-assembly of $RSiX_3$ species that occurs at hydroxylated surfaces generally yields homogeneous, robust SAMs, strongly chemisorbed to the substrate. In addition, chemical and physical properties of the SAM are readily controlled through judicious choice of "R". These features have prompted the use of organosiloxane SAMs as templates for a variety of advanced applications, including fabrication of non-linear optics multilayer assemblies,[11-14] alignment of liquid crystals[15-17] and conducting polymers,[18,19] and growth of oriented mineral structures[20-22] on surfaces.

The ability to alter the chemical reactivity of the "R" pendant groups through exposure to UV light,[23-27] electron[28] or ion[29-31] beams, or plasmas or low energy electrons generated using proximal probes[32,33] comprises an additional advantage in that, it provides a means to create spatially well-defined reactivity templates in the organosiloxane SAM. Si-C bond cleavage predominates during exposure for many,[24,26,27,31] though not all,[23,25] organosiloxane SAMs studied to date, leading to formation of reactivity templates comprised of separate regions of intact pendant "R" groups and surface silanol species. Such simple templates have proven effective for the fabrication of biosensors[34-36] and microelectronic circuitry.[27]

The main feature of biomimetic materials chemistry is the control over crystal nucleation and growth events at the inorganic/organic interfaces through the process of templating or molecular recognition. This review reports two such examples of inorganic crystallization control; firstly the thin film growth of iron compounds under Langmuir monolayers, and secondly, micropatterning of TiO_2 thin films on SAM's, both of which have been recently carried out successfully in our laboratory.

2.0 THIN FILM GROWTH OF IRON COMPOUNDS UNDER LANGMUIR MONOLAYERS[37]

2.1 PRECIPITATION OF γ-FeO(OH) AND Fe_3O_4

The crystal growth was carried out with a Langmuir trough in a closed system. A 10 mM $FeCl_2$ (99.9%, Kojundo Chemical Laboratory, Ltd.) solution was prepared for the subphase. Surfactant molecules, such as stearyl alcohol or stearyl amine, dissolved in chloroform (5×10^{-3} mol/l) were spread onto the subphase and compressed to the constant pressure of 25 mN/m to form a stable monolayer at 30°C. After the monolayer was formed, 5.6 wt.% ammonia solution was placed outside the subphase as a source of NH_3 gas for the reaction. NH_3 gas evaporated from the ammonia solution was transported to the Fe-containing subphase, and was dissolved into the subphase increasing its pH for the precipitation reaction (crystal growth) to take place at the air/water interface.

A glass substrate (1 × 1 cm) was placed at the bottom of the subphase before starting the crystal growth. The crystalline films grown under the monolayer were transferred onto the glass substrate by lowering the subphase level by sucking the subphase with an aspirator. Crystals grown in the subphase in the absence of a monolayer were gathered in a powder form by evaporating water from the subphase.

In the absence of a monolayer, yellow precipitates were formed inside the subphase 3 min after the reaction started. The amount of black particles formed from the yellow precipitates gradually increased with increasing reaction time, and they were obviously deposited in 6 hours at the bottom of the subphase. No preferential precipitation was observed at the air/water interface and the upper part of the subphase became transparent in 6 hours. The yellow precipitates formed in 3 min were x-ray amorphous, as shown in Figure 1a. They are believed to be amorphous phases of α, β or γ-FeO(OH), since yellow solids of α, β and γ-FeO(OH) are known to precipitate usually out of an aqueous solution containing Fe(II) ions.[38] The black particles deposited at the bottom of the subphase were determined to be of Fe_3O_4 as in Figure 1b.

In the presence of a stearyl alcohol monolayer, on the other hand, a considerably different crystal growth manner was observed. Brown thin film was formed at the air/water interface in 5 min, which was assigned to be composed of γ–FeO(OH) crystals (Figure 2a). As shown in Figure 2b, the formed γ–FeO(OH) film showed a preferred orientation of (010) planes parallel to the interface, i.e. the monolayer. The morphology of the formed film was observed with TEM and its photograph is shown in Figure 3a. Very thin crystals overlapping with each other to form a thin film can be observed in the photograph. Figure 3b shows the electron diffraction pattern which coincides with that of γ-FeO(OH) and a small amount of Fe_3O_4. Slightly diffused diffraction spots generated from γ-FeO(OH) crystals resemble a single-crystal like pattern and suggest that the crystal axes of each thin crystal are almost all oriented in the same direction. This observation firmly verifies that γ-FeO(OH) has grown epitaxially under a stearyl alcohol monolayer. The intensity of XRD peak corresponding to the Fe_3O_4 phase increased and that of γ-FeO(OH) (020) peak decreased with increasing reaction time (Figures 2c and d). Finally, Fe_3O_4 phase was only observed in 6 hours (Figure 2e).

When a stearyl amine monolayer was used in the experiment, the crystal growth manner was similar to the case in which no monolayer was employed.

2.2 Roles of a Monolayer in the Precipitation of γ-FeO(OH)

The formation of iron compounds through the oxidation of ferrous salt solutions depends on the oxidation rate, the concentration, the pH and the temperature of the solution.[39,40] It is known that γ-FeO(OH) could be obtained by a slow oxidation of an aqueous solution of a Fe(II) salt if pH is kept about 6-7.[40]

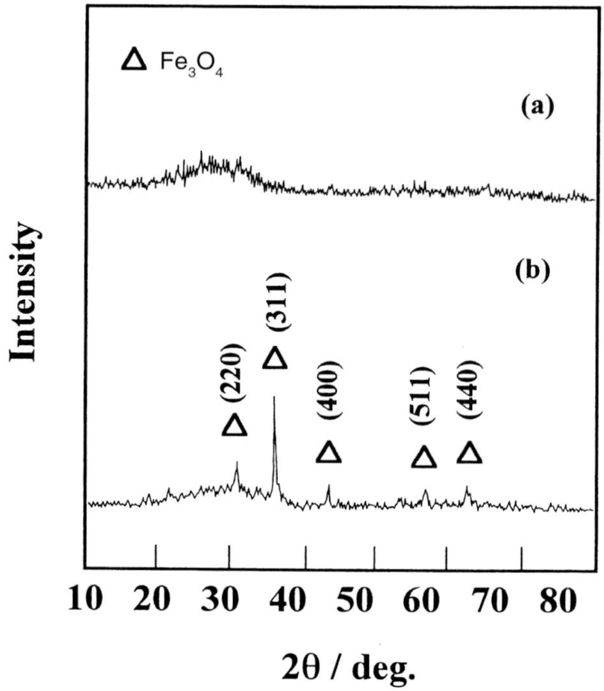

Fig. 1 XRD patterns of the precipitates formed without a monolayer, (a) at the air/water interface after 10 min and (b) at the bottom of the subphase after 6 hours.

The precipitation reaction for γ-FeO(OH) can be expressed by the following equation:

$$2Fe^{2+} + 3H_2O + 1/2\ O_2 \rightarrow 2\ \gamma\text{-}FeO(OH) + 4H^+ \qquad (1)$$

where γ-FeO(OH) is produced by the oxidation and hydration of Fe(II) ions, accompanying the generation of protons. Therefore, the formation of γ-FeO(OH) would be prevented if alkali is not added to the solution to keep the pH value between 6-7.

In the present study, NH_3 gas was used for the crystal growth at the air/water interface, so that γ-FeO(OH) might be the main product in the presence of a stearyl alcohol monolayer, although the pH value could not be controlled. However, the presence of a monolayer regulated somehow the pH of the subphase. In fact, pH of the subphase remained almost unchanged for ~15 min after the reaction started, indicating that the dissolved ammonia was used to neutralize the protons generated together with the precipitation of γ–FeO(OH). That is, in the beginning of the reaction, the amount of OH^- ions formed by dissolution of NH_3 was almost equal to that used for the neutralization of the generated protons and must have given rise to the proper conditions for the growth of γ–FeO(OH). In this case, hydroxyl groups of a stearyl alcohol monolayer, which attracted ferrous ions preferentially, must

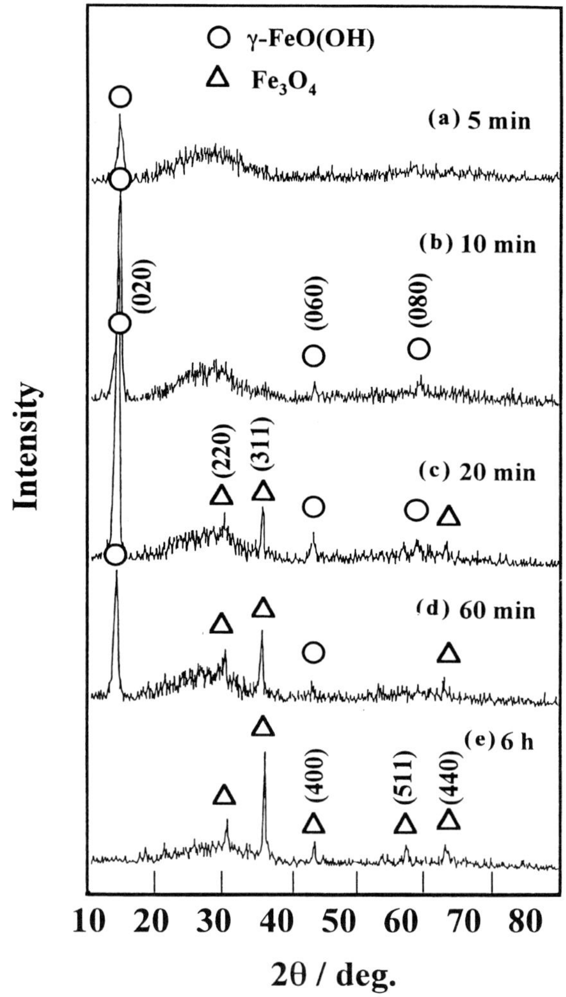

Fig. 2 XRD patterns of the films formed under the stearyl alcohol monolayers after different reaction times.

have acted as the heterogeneous nucleation sites to enhance the crystal growth under the monolayer.

On the other hand, the air/water interface in the absence of a monolayer cannot regulate the dissolution of ammonia into the subphase and cannot attract ferrous ions. A large amount of ammonia, therefore, dissolved into the subphase and the OH^- formed must have quickly diffused away from the air/water interface down to the lower part of the subphase. Consequently, precipitation and crystal growth proceeded everywhere in the subphase, where the solid phases precipitated are considered to have been different at different depth due to the gradient in the pH

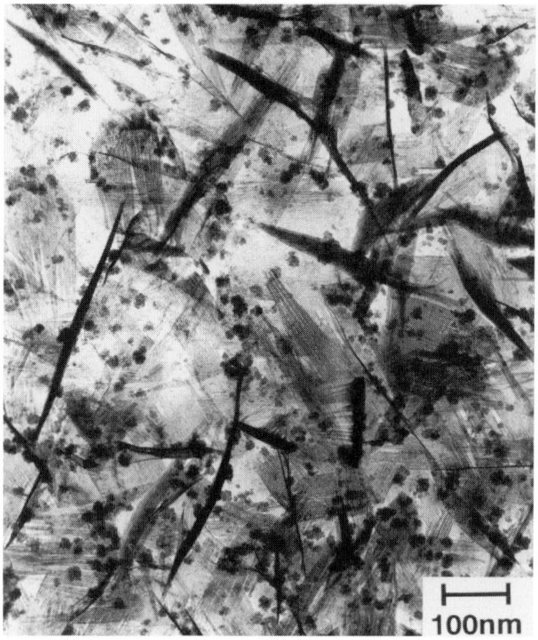

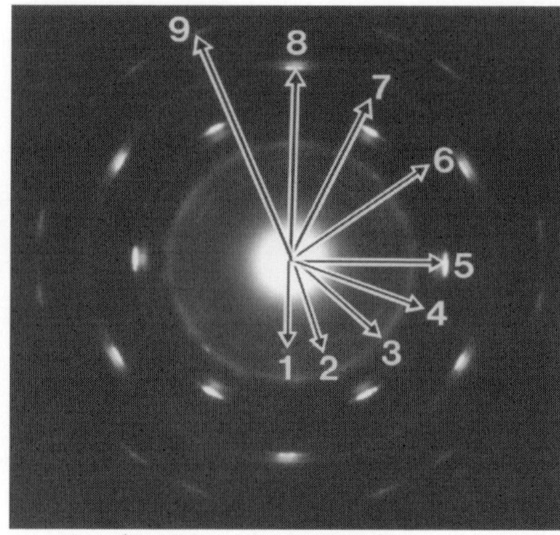

No.	d (Å)	(hkl)
1	2.97	Fe_3O_4 (220)
2	2.63	γ-FeOOH (011)
3	2.23	γ-FeOOH (111)
4	1.85	γ-FeOOH (220)
5	1.75	γ-FeOOH (151)
6	1.52	Fe_3O_4 (440) or γ-FeOOH (231)
7	1.43	γ-FeOOH (171) or (260) or (180)
8	1.34	γ-FeOOH (251)
9	1.06	γ-FeOOH

Fig. 3 Transmission electron micrograph (a) and the electron diffraction pattern (b) of the film formed under the stearyl alcohol monolayer after 5 min.

value. This situation may have prevented the crystal growth of γ–FeO(OH) and gave amorphous precipitates in the beginning of the reaction.

2.3 Crystallographic Matching

It should be noted that a thin film of a (010) preferentially oriented γ–FeO(OH) has epitaxially grown under a stearyl alcohol monolayer at the initial stage of the reaction. The epitaxial growth of a thin film can be attributed to its geometric matching with the substrate, namely a monolayer in the present case. In general, stearyl alcohol molecules take hexagonal packing structure when they form a monolayer on an aqueous subphase, as shown in Figure 4a. The black circles indicate hydroxyl groups, which are arranged hexagonally in a uniform manner. The average nearest-neighbor distance between the two hydroxyl groups was calculated to be 0.48 nm from the isothermal pressure-area (π–A) curve of the monolayer, which showed that the area per molecule was about 0.23 nm^2 in the solid state.

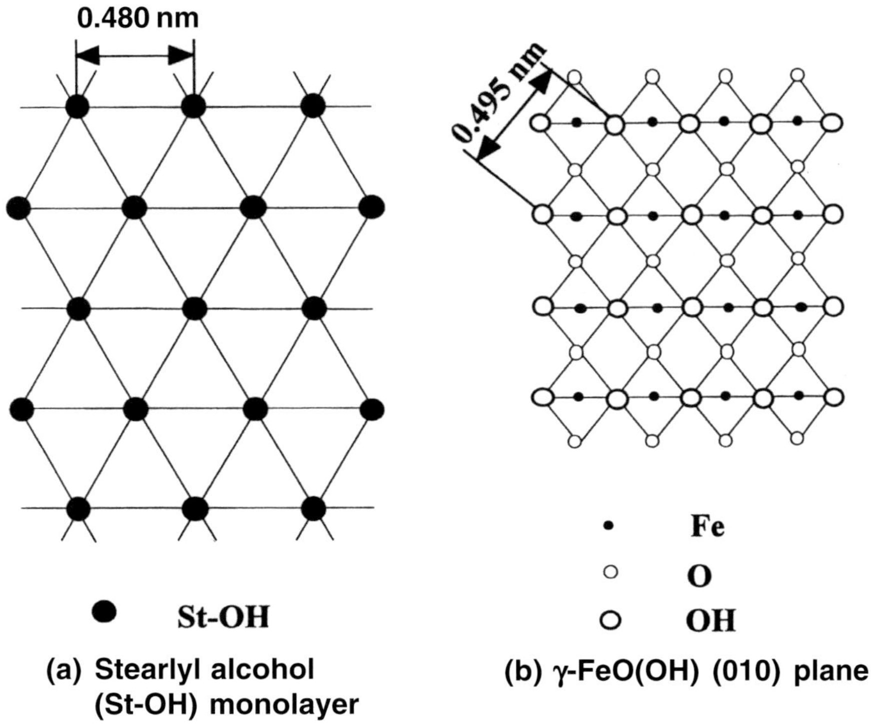

Fig. 4 The two-dimensional structures of a stearyl alcohol monolayer and the (010) plane of γ–FeO(OH).

It is known that γ–FeO(OH) takes a layer structure shown as a perspective drawing in Figure 4b.[41] Oxygen ions are situated at the vertices of each octahedron with an Fe ion at its centre. The O-H•••O hydrogen bonds only exist between the hydroxyl groups belonging to different layers, which in turn connect the layer structure. The distance between the oxygen ions of different layers is only 0.272 nm, so that there are OH groups on the outer surface of each layer. The lattice parameters of the orthorhombic crystal according to the JCPDS card (No.080098) are $a = 0.388$ nm, $b = 1.254$ nm, and $c = 0.307$ nm. Thus, the distance of the two hydroxyl groups on γ–FeO(OH) (010) plane is 0.495 nm, as shown in Figure 4b. This coincides with the two-dimensional lattice parameter of a stearyl alcohol monolayer within only 3% mismatch. The overlapping structure of Figures 4a and b shows their well-matching situation. Consequently, the stearyl alcohol monolayer must have worked as a template inducing the epitaxial growth of a (010) preferentially oriented γ–FeO(OH) film.

On the other hand, a stearyl amine monolayer did not enhance the nucleation of γ–FeO(OH), although the lattice parameter of the monolayer is similar to that of γ–FeO(OH). Since amine is too weak to attract ferrous ions to form a ferrous ion layer under the amine monolayer, the concentration of ferrous ions under the stearyl amine monolayer was very low at the initial stage of the reaction, resulting in a crystal growth manner similar to that observed in the absence of a monolayer. This result would firmly indicate that anionic surfactant molecules (e.g. stearyl alcohol) can be effective to enhance the nucleation of γ–FeO(OH), while cationic surfactant molecules (e.g. stearyl amine) cannot serve as nucleation sites.

3.0 MICROPATTERNING OF TiO_2 THIN FILM ON SELF-ASSEMBLED MONOLAYER (SAM)

3.1 Micropatterning Technique

Micropatterning techniques are becoming increasingly important in electronics. Micromachining of semiconductor devices has made steady progress, and the degree of device integration is increasing at an amazing speed along with its miniaturization.[42] In this context, the device structure has become complex owing to the enlargement of the surface area of the insulating dielectric film, such as SiO_2 and SiN_x, to achieve higher capacitance. However, it is technically difficult to fabricate a thin film with a large surface area and with no pin holes. Accordingly, substitution of the other high-permitivity materials, such as TiO_2, Ta_2O_5 and (Ba,Sr)TiO_3, for the current Si-based materials is regarded as promising for the future of microelectronics,[43] and a sophisticated micropatterning technique for these dielectric oxide ceramics is urgently required.

Payne et al. first reported the micropatterned thin films of ferroelectric oxides[44] and Ta_2O_5[45] fabricated by employing SAMs for a micro-contact printing process[46] based on the usual sol-gel method. Their process is attractive because it can easily produce micrometer resolution patterns of various oxide films. However, it usually requires post heat-treatment for complete crystallization of the film at high temperatures since an as-deposited film is amorphous and does not show the properties expected for the crystalline state. Other reported patterning techniques based on the sol-gel method also always need a post heat-treatment process.[47-49]

Our novel patterning technique developed recently is based on the so-called liquid-phase deposition instead of the sol-gel method, and does not need post heat-treatment as a crystalline oxide film is directly deposited near room temperature. This technique incorporates a patterned SAM composed of hydrophilic and hydrophobic surfaces, and utilizes the difference in their reactivity with metal-containing complex ions in an aqueous solution for the fabrication of patterned oxide films.

3.2 Liquid-phase Deposition on SAM

Recently, liquid-phase deposition has been developed as a new synthetic process for metal oxide thin films.[50-53] In this process, it is possible to form a thin film of metal oxide directly on the substrate which is immersed in the solution. A metal oxide thin film is formed by means of a ligand-exchange equilibrium reaction of metal-fluoro complex ions and a F^--consuming reaction of boric acid which acts as a scavenger for F^-, as expressed by the following equations:

$$MF_x^{(x-2n)-} + nH_2O \Leftrightarrow MO_n + xF^- + 2nH^+ \tag{1}$$

$$H_3BO_3 + 4HF \Leftrightarrow BF_4^- + H_3O^+ + 2H_2O \tag{2}$$

By adding boric acid which readily reacts with F^- ions to form stable BF_4^- ions according to Eqn. 2, the equilibrium reaction, Eqn. 1, is shifted towards the right-hand side leading to the deposition of metal oxide. This process does not require any special apparatus and does not need high-temperature treatment after deposition. This process can also be applied to synthesize various complex metal oxides.[53]

It has been reported that the SAM formed by phenyltrichlorosilane (PTCS) precursor changed its surface functionality from hydrophobic (phenyl group) to hydrophilic (hydroxyl group) by UV irradiation in air.[54,55] Therefore, by fabricating a phenyl/hydroxyl group-patterned SAM (Figure 5) and subsequently using it as a substrate for a liquid-phase deposition process, different reactivity of the two different functional groups can be expected to lead to selective deposition of metal oxide resulting in its micropatterned thin film. In fact, micropatterned circuits of Ni metal have been successfully fabricated on various SAMs of a similar kind by electroless plating.[24,27,55]

3.3 Method of SAM Patterning and Thin Film Growth of TiO_2

P-type Si (100) wafers were employed as substrates. They were cleaned ultrasonically in deionized water (>17.6 MΩ cm), immersed in a 1:1 (vol.) $HCl:CH_3OH$ solution for 30 min, and cleaned again in deionized water. They were further immersed in conc. H_2SO_4 for 30 min and then in boiling water for 5 min, and were cleaned with acetone. Self-assembled monolayers of phenyl-trichlorosilane (Aldrich; PTCS-SAM) were prepared by immersing the substrate into an anhydrous toluene solution containing 1 vol.% PTCS for 5 min under a N_2 atmosphere. The substrate with a PTCS-SAM was then baked at 120°C for 5 min to remove residual solvent and to promote chemisorption of the SAM.

PTCS-SAM was exposed for 2 hours to UV light (220 - 400 nm) from a Hg lamp through a photo mask. The UV-irradiated regions became hydrophilic due to Si-OH group formation, while the non-irradiated part remained unchanged, *i.e.* it was composed of hydrophobic phenyl groups, which gave rise to a patterned PTCS-SAM. In order to check for successful film formation and functional group change, water drop contact angle was measured for both irradiated and non-irradiated surfaces. Initially deposited PTCS-SAM showed a water contact angle of 74°, but the UV-irradiated surface was wetted completely (contact angle < 5°).

Ammoniumhexafluorotitanate $(NH_4)_2TiF_6$ (Mitsuwa Chemical Co., AHFT) and boric acid H_3BO_3 (Kishida Chemical Co.) were separately dissolved into deionized water, respectively, and mixed with each other to form a solution to which an

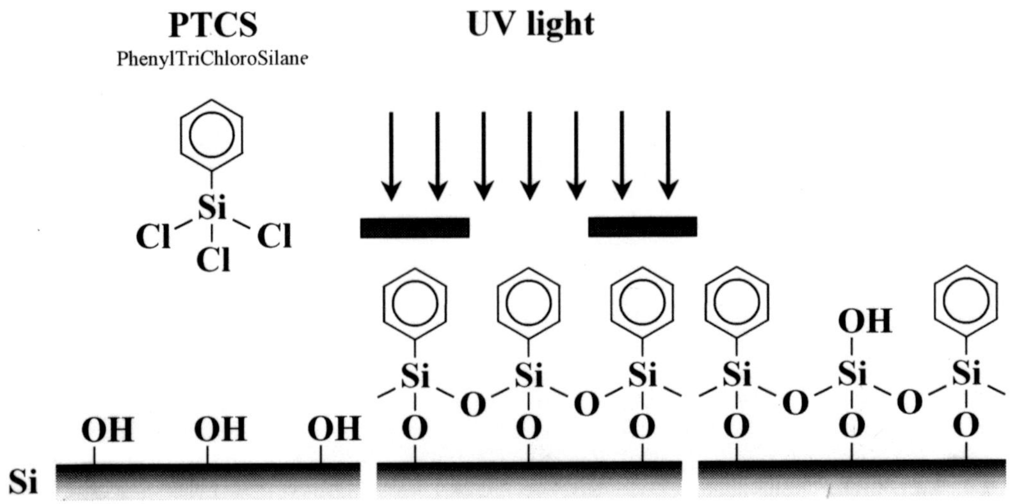

Fig. 5 Schematic illustration of the patterning process for a PTCS-SAM.

appropriate amount of HCl was added to control the pH just before the reaction started at a prescribed temperature. Distilled water was further added to this solution to finally yield a 0.05M AHFT: 0.15M H_3BO_3 precursor solution. The substrate with patterned PTCS-SAM was immersed into the precursor solution at 50°C for 12 hours. It was held vertically on the bottom of a reaction vessel to prevent the precipitate particles formed in the solution from accumulating on the surface. The substrate was then sonicated for 1 min, rinsed with distilled water, and dried at room temperature.

3.4 Thin Film Growth Process and Micropatterning

Figure 6 shows the variation of the morphology of the film deposited on the patterned PTCS-SAM with deposition time. It is clearly seen that the contrast of a SEM image changes with the change in the microstructure of the film, and the contrast for hydrophilic and hydrophobic surfaces becomes reversed during deposition. On the hydrophobic surface, the nucleation takes place rapidly and extensively in the initial stage of deposition, and a large number of particles are formed (Stage I). Particles then grow larger and coalesce to form a thin film in the later stages (Stages II and III). In contrast, on the hydrophilic surface, the nucleation takes place only very slowly in the initial stage (Stage I), and particle growth proceeds in the second stage (Stage II), followed by the coalescence of particles to form a film (Stage III). The SEM image contrast for hydrophilic and hydrophobic surfaces becomes similar finally in Stage III, *i.e.* the films on the two surfaces become indistinguishable.

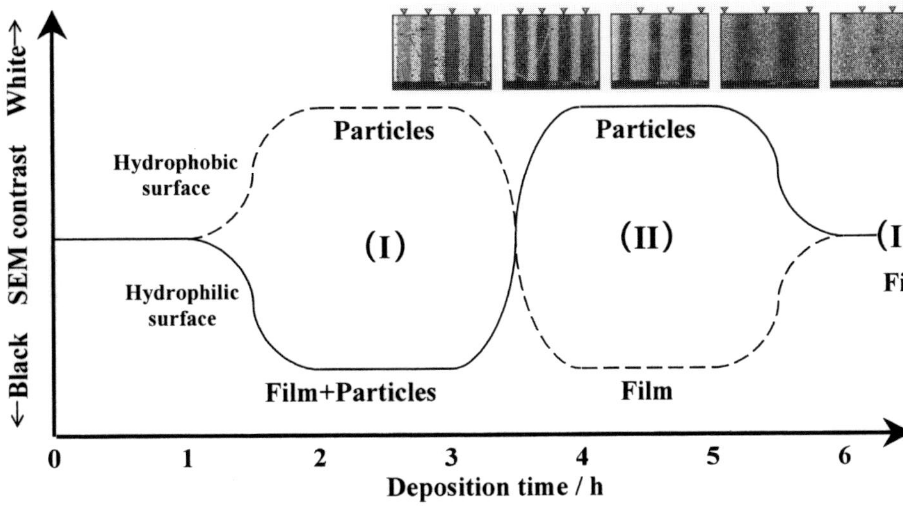

Fig. 6 SEM photographs of the TiO_2 film showing the morphology change with deposition time (▼ indicates the film on the hydrophilic surface); qualitative SEM contrast vs. deposition time is shown together with the morphology change in the films on hydrophilic (—) and hydrophobic (····) surfaces.

The fact that a film on the hydrophilic surface shows black contrast in the Stage I firmly indicates that an extremely thin film of TiO_2 has already been formed, while only particles could be observed on the hydrophobic surface. This observation suggests that the growth mechanism may be different on the two surfaces. It can be easily imagined that the reactivity of $[TiF_6]^{2-}$ ions with hydrophilic Si-OH groups differs greatly from that with hydrophobic phenyl groups, and this difference must have caused the difference in the growth mechanism of a film.

On the hydrophilic surface, $[TiF_6]^{2-}$ ions would react with Si-OH groups to form strong Si-O-Ti bonds, and subsequent hydrolysis and polymerization would yield a network of -Ti-O-Ti-O- (anatase). This reaction would have promoted the epitaxial growth resulting in the film growth in the very initial stage of deposition (Stage I). However, lattice mismatch between anatase and a two-dimensional network of Si-OH groups is too large for epitaxial growth to continue, and hence, film growth became suspended when certain critical thickness of an anatase film was reached, and the growth mode changed into that of particle growth (Stage II).

In contrast, on the hydrophobic surface, common heterogeneous nucleation must have caused the particle growth in Stage I, possibly because no specific interactions take place between $[TiF_6]^{2-}$ ions and surface phenyl groups. This film growth mode appears in the final stage on both hydrophilic and hydrophobic surfaces. However, the reason the growth mode changes from particle to film growth may be associated with the fact that the deposition rate decreases with increasing deposition time as the $[TiF_6]^{2-}$ ion concentration decreases.

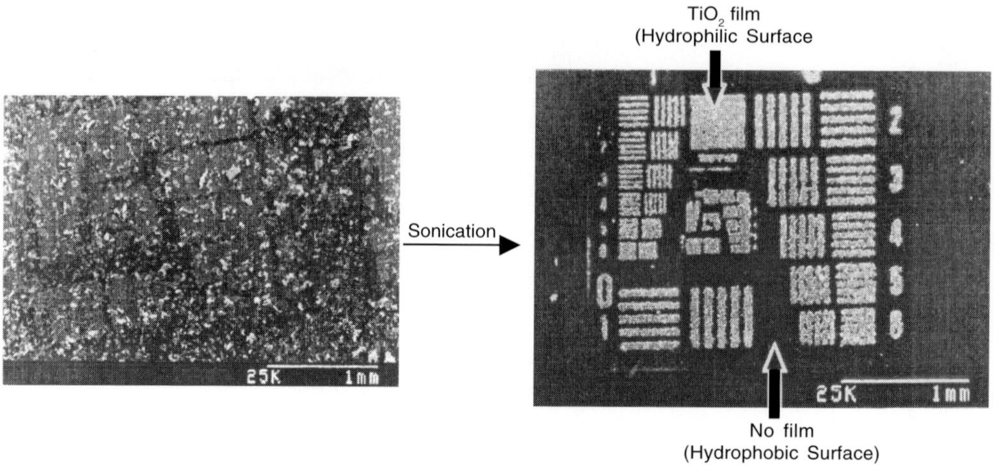

Fig. 7 SEM photographs of (a) the film deposited after 12 h and (b) the micro-patterned film remaining only on the hydrophilic part after sonication.

In Figure 7 are shown SEM photographs of an as-deposited thin film (50°C, 12 h) and a micropatterned thin film which was obtained after 10 min sonication of an as-deposited film. It is obvious that the TiO_2 film formed on the hydrophobic surface completely peeled off during sonication, while the film formed on the hydrophilic surface stayed on the substrate resulting in a micropatterned TiO_2 film. This phenomenon indicates that the chemical bonding states between the PTCS-SAM and the TiO_2 film on the two surfaces differ significantly. Although the film is held very weakly to the hydrophobic regions of the PTCS-SAM due to physisorptive bonding, strong bonding of Si-O-Ti formed on the hydrophilic surface would give rise to strong adhesion of the film to the substrate.

4.0 SUMMARY

We have demonstrated that two-dimensionally assembled molecular layers can control over the crystallization of inorganic solids precipitating out from aqueous solutions; crystal-plane oriented epitaxial growth of lepidocrocite took place under a Langmuir monolayer of stearyl alcohol due to a crystallographic matching between a monolayer and a growing crystal. Selective bonding of TiO_2 to silanol surface gave rise to a micropattern of TiO_2 deposited on a patterned self-assembled monolayer.

Living creatures have developed their own techniques to synthesize inorganic materials at specific sites of molecular assemblies with required size and shape since ancient times. The future microelectronics and photonics would require exquisite techniques to fabricate necessary functional materials at specific regions in a small length scale with required size and shape, and this process should be similar to what living creatures are doing. The findings presented in this paper are expected to provide an important step towards the fabrication of next-generation microdevices and to trigger new motivation to study inorganic materials from a biomimetic point of view.

5.0 REFERENCES

1. S. Mann: "Biomineralization and Biomimetic Materials Chemistry", *Biomimetic Materials Chemistry*, S. Mann, ed., VCH Publishers, New York, 1996.
2. J. H. Fendler and F. C. Meldrum: *Advanced Materials*, **7**, 1995, p.7.
3. E. M. Laudau, R. Popovitz-Biro, M. Levanon, L. Leiserowitz, M. Lahar and J. Sagiv: *Mol. Cryst. Liq. Cryst.*, **134**, 1986, p.323.
4. S. Mann, D. D. Archibald, J. M. Didymus, T. Douglus, B. R. Heywood, F. C. Meldrum and N. J. Reeves: *Science*, **261**, 1993, p.1286.
5. J. Lahiri, G. Xu, D. M. Dabbs, N. Yao, I. A. Aksay and J. T. Groves: *Journal of American Chemical Society,* in press.
6. J. Yang, F. C. Meldrum and J. H. Fendler: *Journal of Physical Chemistry*, **99**, 1995, p.5500.
7. X. K. Zhao, J. Yang, L. D. Mccormick and J. H. Fendler: *ibid.*, **96**, 1992, p.9933.
8. H. Lin, W. S. Seo, K. Kuwabara and K. Koumoto: *Journal of Ceramic Society of Japan*, **104**, 1996, p.291.
9. A. Ulman: *An Introduction to Ultra Thin Organic Films from Langmuir-Blodgett to Self-Assembly*, Academic Press, New York, 1991.
10. B. C. Bunker, P. C. Rieke, B. J. Tarasevich, A. A. Campbell, G. E. Fryxell, G. L. Graff, L. Song, J. Liu, J. W. Virde, G. L. Mcvay: *Science*, **264**, 1994, p.48.
11. W. Schrof, S. Rozouvan, E. Van Keuren, D. Horn, J. Schmitt, G. Decher: *Advanced Materials*, **10**, 1998, p.338.
12. W. B. Lin, W. P. Lin, G. K. Wong and T. J. Marks: *Journal of American Chemical Society*, **118**, 1996, p.8034.
13. S. B. Roscoe, A. K. Kakkar, T. J. Marks, A. Malik, M. K. Durbin, W. P. Lin, G. K. Wong and P. Dutta: *Langmuir*, **12**, 1996, p.4218.
14. H. W. Jiang, A. K. Kakkar, A. M. Lebuis, H. T. Zhou and G. K. Wong: *Journal of Materials Chemistry*, **6**, 1996, p.1075.
15. D. Shenoy, K. Grueneberg, J. Naciri and R. Shashidhar: *Japanese Journal of Applied Physics-2,* **37**(11A), 1998, p.L1326.
16. S. D. Evans, H. Allinson, N. Boden, T. M. Flynn and J. R. Henderson: *Journal of Physical Chemistry B*, **101**, 1997, p.2143.
17. K. Ichimura, H. Tomita and K. Kudo: *Liquid Cryst.*, **20**, 1996, p.171.
18. C. G. Wu and J. Y. Chen: *Chemistry of Materials*, **9**, 1997, p.399.
19. Z. Huang, P. C. Wang, A. G. MacDiarmid, Y. Xia and G. Whitesides: *Langmuir*, **13**, 1997, p.6480.

20. C. B. Mao, H. D. Li, E. Z. Cui, G. G. Feng, H. Wang and C. L. Ma: *Journal of Materials Chemistry*, **8**, 1998, p.2795.
21. D. D. Archibald, S. B. Qadri and B. P. Gaber: *Langmuir*, **12**, 1996, p.538.
22. B. R. Heywood and S. Mann: *Advanced Materials*, **6**, 1994, p.9.
23. R. J. Collins, H. Shin, M. R. DeGuire, A. H. Heuer and C. N. Sukenik: *Applied Physics Letters*, **69**, 1996, p.860.
24. W. J. Dressick, C. S. Dulcey, M. S. Chen and J. M. Calvert: *Thin Solid Films*, **284/285,** 1996, p.568.
25. J. Jennane, T. Boutros and R. Giasson: *Canadian Journal of Chemistry*, **74**, 1996, p.2509.
26. W. J. Dressick, C. S. Dulcey, J. H. Georger and J. M. Calvert: *Chemistry of Materials*, **5**, 1993, p.148.
27. W. J. Dressick and J. M. Calvert: *Japanese Journal of Applied Physics*, **32**, 1993, p.5829, and references therein.
28. N. Mino, S. Ozaki, K. Ogawa and M. Hatada: *Thin Solid Films*, **243**, 1994, p.374.
29. T. Schenkel, M. Schneider, M. Hattass, M. W. Newman, A. V. Barnes, A. V. Hamza, D. H. Schneider, R. L. Cicero, C. E. D. Chidsey: *Journal of Vacuum Science and Technology B*, **16**, 1998, p.3298.
30. R. Younkin, K. K. Berggren, K. S. Johnson, M. Prentiss, D. C. Ralph and G. M. Whitesides: *Applied Physics Letters*, **71**, 1997, p.1261.
31. E. T. Ada, L. Hanley, S. Etchin, J. Melngailis, W. J. Dressick, M. S. Chen and J. M. Calvert: *Journal of Vacuum Science and Technology B*, **12**, 1995, p.2189.
32. S. L. Brandow, J. M. Calvert, E. S. Snow and P. M. Campbell: *Journal of Vacuum Science and Technology B*, **15**, 1997, p.1455, and references therein.
33. H. Sugimura and N. Nakagiri: *Journal of American Chemical Society*, **119**, 1997, p.9226.
34. L. A. Chrisey, C. E. O'Ferrall, B. J. Spargo, C. S. Dulcey, and J. M. Calvert: *Nucleic Acids Res.*, **24**, 1996, p.3040.
35. M. Yan, S. X. Cai, M. N. Wybourne and J. E. W. Keana: *Journal of American Chemical Society*, **115**, 1993, p.814.
36. S. Bhatia, J. L. Teixeira, M. Anderson, L. C. Shriver-Lake, J. M. Calvert, J. H. Georger, J. J. Hickman, C. S. Dulcey, P. E. Schoen and E. S. Ligler: *Anal. Biochem.*, **208**, 1992, p.197.
37. H. Lin, H. Sakamoto, W. S. Seo, K. Kuwabara and K. Koumoto: *Journal of Crystal Growth*, **192**, 1998, p.250.
38. U. Schwertmann and R. M. Corneil: *Iron Oxides in the Laboratory*, VCH Publishers, New York, 1991.
39. M. Kiyama and T. Tanaka: *Bulletin Chem. Soc. Jpn.*, **45**, 1972, 1923.
40. Y. Tamaura, K. Ito and T. Katsura: *Journal of Chemical Society Dalton Trans.*, **1983**, 1983, p.189.
41. A. F. Wells: *Structural Inorganic Chemistry*, Clarendon Press, Oxford, 1954.
42. A. Yoshida, H. Tsutsu and T. Hirao: *Oyo Buturi*, **67**(6), 1998, p.306.
43. Y. Ohji, S. Iijima, N. Nakanishi and I. Asano: *Oyo Buturi*, **66**(11), 1997, 1210.
44. N. L. Jeon, P. G. Clem, R. G. Nuzzo, D. A. Payne: *Journal of Materials Research*, **10**, 1995, p.2996.

45. P. G. Clem, N. L. Jeon, R. G. Nuzzo and D. A. Payne: *Journal of American Ceramic Society*, **80**, 1997, p.2821.
46. Y. Xia and G. M. Whitesides: *Angew. Chem. Int. Ed.*, **37**, 1998, p.550.
47. N. Tohge, T. Minami, A. Matsuda, Y. Matsuno, S. Katayama and Y. Ikeda: *Journal of Ceramic Society of Japan*, **96**, 1988, p.1127.
48. K. Shinmou, N. Tohge and T. Minami: *Japanese Journal of Applied Physics*, **33**, 1994, p.L1181.
49. T. Ohishi and A. Katoh: *Journal of British Ceram. Trans.*, **92**(2), 1993, 79.
50. N. Nagayama, H. Honda, H. Kawahara: *Journal of Electrochem. Society*, **135**, 1988, 2013.
51. S. Deki, Y. Aoi, O. Hiroi and A. Kajinami: *Chemistry Letters*, **1996**, 1996, p.433.
52. S. Deki and Y. Aoi: *Journal of Materials Research*, **13**(4), 1998, p.883.
53. T. Yao: *Journal of Materials Research*, **13**(5), 1998, 1091.
54. C. S. Dulcey, J. H. Georger, Jr., V. Krauthamer, D. A. Stenger, T. L. Fare, J. M. Calvert: *Science*, **252**, 1991, p.551.
55. J. M. Calvert, G. S. Calabrese, J. F. Bohland, M. S. Chen, W. J. Dressick, C. S. Dulcey, J. H. Georger, Jr., J. Kosakowski, E. K. Pavelcheck, K. W. Rhee, L. M. Shirley: *Journal of Vacuum Science Technology*, **B12**(6), 1994, 3884.

The Local Dielectric Function in Ceramics: Influence of Chemistry and Anisotropy in the Grain Boundary Vicinity

A. THOREL
Centre des Matériaux
Ecole des Mines de Paris
BP87, 91003 Evry Cedex, France

L. NORMAND
Institut Français du Pétrole
1 and 4 avenue du Bois-Préau
BP311, 92506 Rueil-Malmaison Cedex, France

B. DELCROIX
Centre des Matériaux
Ecole des Mines de Paris
BP87, 91003 Evry Cedex, France

C. ECHER
National Center for Electron Microscopy
Lawrence Berkeley National Laboratory
University of California
Berkeley CA 94720, U.S.A.

P. CHEMINANT-COATANLEM
CEA-Saclay, DRN/DMT/SEMI/LE2M
91191 Gif sur Yvette, France

Y. MONTARDI
RHODIA, Centre de Recherches d'Aubervilliers
52 rue de la Haie-Coq
93308 Aubervilliers, France

ABSTRACT

Using the low loss region of Electron Energy Loss (EELS) spectra obtained via a $\vec{k}$-controlled TEM experiment, we measured locally the complex permittivity in some ceramics with the spatial resolution typical of the conventional Transmission Electron Microscope. Real and imaginary parts of the dielectric function were obtained after extracting the zero loss peak and multiple scattering from the EELS spectra which were then analysed via the classical

Kramers-Krönig formulation. The possible artifacts of the methodology as well as its major difficulties are described in detail. To make the typical relaxation times of the different polarization mechanisms clearly appear, results are presented in Cole-Cole diagrams where the imaginary part of the dielectric function is given as a function of the real part. Results of such an analysis are shown and discussed for $BaTiO_3$ sintered with different amount of dopants segregated near the grain boundaries, for HfB_2 in which HfO_2 precipitates are present, and for Y_2O_3. Since the method is $\vec{K}$-dependent, it is possible to detect the local anisotropy of the dielectric function as it is presented for Si_3N_4. In that case, the implication on the stability of the interfacial vitreous film thickness is discussed.

1.0 INTRODUCTION

1.1 General Considerations

Materials, such as ceramics, are more and more reliable and sophisticated, with less and less defects due to the now well controlled sintering processes. As a consequence, empirical approaches to improve their performances are less efficient since a real knowledge of the elementary mechanisms is required. In lots of cases the properties depend on the intrinsic "defects" such as grain boundaries, intergranular phases, precipitates (high temperature mechanical behaviour of Si_3N_4, varistance effect in ZnO, Josephson junctions in high T_C superconductors, thermal resistivity in AlN). Material improvement must rely on experiments that allow the detection of the local influence of such microstructural features, that is to say with a spatial resolution comparable to the extension of these features. Knowing the local properties, it is possible in some cases to reconstruct and predict the bulk properties with the aid of the statistics, and eventually drastically improve the behaviour of the material. In anyway, when such local property is known, it gives a powerful tool to theorists to identify the fundamentals of the elementary mechanisms.

Figure 1 presents a typical schematic ceramic microstructure viewed at different scales, such as those obtained by optical microscopy and Scanning Electron Microscopy; one can see that, as it is often the case with ceramics, the material consists of a major phase and a secondary phase. The major phase, which composes most of the grains, has a more or less convex morphology; it confers on the material its overall characteristics such as thermal conductivity, superconductivity. The second phase, usually distributed at grain boundaries or triple junctions, most of the time displays a concave shape; this intergranular phase is the result of a chemical reaction between the major phase and additives, which are added to improve the sintering behaviour. In that respect, if they enhance the sintering ability, for the same reason they also are the factor that limits the material performances. As an example, yttria added to aluminium nitride will react with the alumina layer coating the grains to form an yttrium-aluminium oxide that will be liquid at the sintering temperature, enhancing then strongly the sintering. But the resulting intergranular phase, which wets the interfaces, will strongly affect and limit the bulk properties, such as the thermal conductivity, for it gives rise to very efficient thermal barriers throughout the material. The same scenario occurs when silicon nitride is sintered in the presence

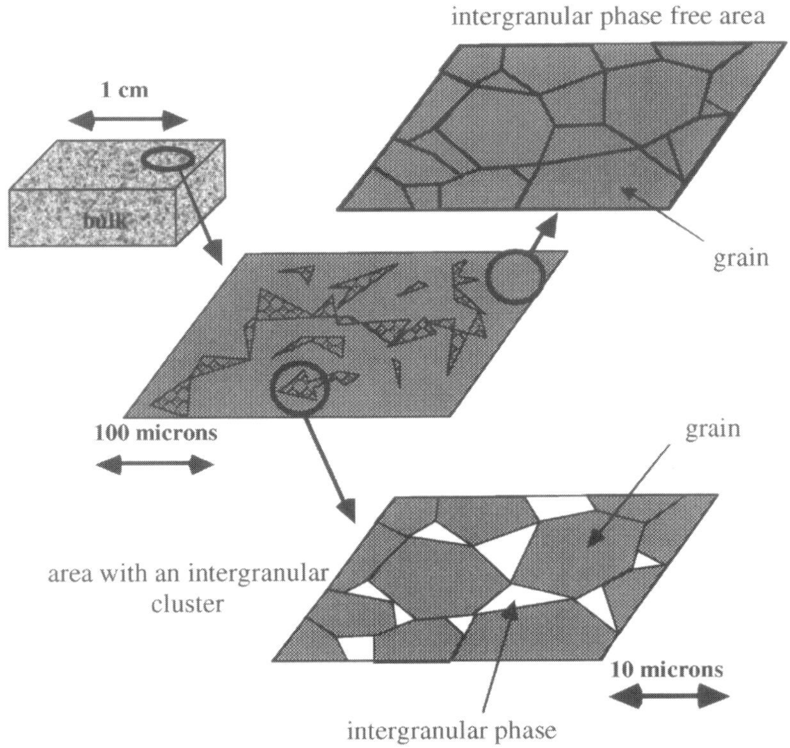

Fig. 1 Schematic view of a typical ceramic microstructure at different scales.

of alumina and yttria; sintering is significantly enhanced when the additives react with the silicon nitride grains to form the SiAlON phase and eventually an intergranular glassy silicate phase. When used at high temperature, mechanical properties of the material are strongly limited by the viscosity loss of the intergranular glassy phase.

It is commonly accepted that the chemical composition and morphology of the intergranular phase tailor the bulk properties. Therefore, if one can measure the local properties of the intergranular phase and model its morphology, it is then possible to predict the bulk behaviour of the material. This approach has been proved to be successful in different systems such as polycrystalline aluminium nitride in which the bulk thermal conductivity could be predicted with a 5% accuracy,[3] provided that the local diffusivity be measured locally at the micrometer scale via photoreflectance experiments[4] and the AlN microstructure described by random set models[5]. Dielectric properties are among the most important properties for they govern the behaviour of a large number of materials, i.e. electromagnetic absorption, optical reflectivity, capacitance. They may also play an important role in the stability and viscosity of liquid films, and control for example the thickness of interfacial films in ceramics.[6] We will present thereafter a methodology that allows the determination of the dielectric function in materials at the nanometer scale. It is based on the use of Electron Energy Loss Spectroscopy (EELS) and was applied to ceramics earlier[7-10] and to polymers more recently.[11, 12] We have applied this methodology to different ceramic systems and tested its ability to render slight chemistry or crystallography changes and to give fundamental insights on the associated elementary dielectric mechanisms.

1.2 Electron Energy Loss Spectroscopy and Dielectric Properties

Electron Energy Loss Spectroscopy (EELS) is a now well-established and powerful analytical method that allows chemical composition measurements as well as the detection of the electronic structure or the local atomic arrangement through different analyses of the EELS spectra. The main advantage of the EELS related experiments is that they are always carried out in a Transmission Electron Microscope; EELS analyses can be then achieved while observing simultaneously the sample under conventional or eventually high resolution phase contrast TEM conditions. Moreover, the sample can be tilted in the microscope so that, an interface can be viewed edge-on. We will see later that the tilting capabilities can serve other more sophisticated purposes based on the directional transfer of a momentum to the material. Furthermore, very small probe size can be used for EELS; routine experiments at the nanometer scale can now be conducted, even down to a few angstroms with VG type machines or with the more advanced, available, transmission electron microscopes.

Even if early studies have tried to use the low loss region of the EELS spectrum for chemical composition analysis purpose, in order to take advantage of a highly favourable peak to noise ratio,[13] it was proved to be very little affected by alloying effects or composition changes. The major reason is that this low loss region is directly related to the collective excitation of valence and conduction electrons, and that substituting a few atoms in the structure does not drastically change the electron density. Very early it also appeared that the low loss region of EELS spectra could be interpreted as the signature of a gas of free electrons, excited by the incident beam, then moving collectively within a network of cations. A very simple analytical expression for the intensity of the spectrum versus the energy loss can be derived from these considerations, the whole approach being named the dielectric formalism. The very interesting point is that the dielectric function, or complex permittivity, appears in this expression, giving an opportunity for its local measurement at the nanometer scale.

In lots of materials, for example those containing transition elements in which outer levels are completely filled with electrons whereas inner levels are not, low energy interband transitions can occur. The corresponding electrons, weakly bonded, participate actively to the dielectric function even if they are not truly collective and delocalized electrons.

This approach has been applied here to determine the local complex dielectric function of different materials in the case of a small addition of dopants (Nb and Co in $BaTiO_3$), a chemical substitution (HfO_2 particles in a HfB_2 matrix), Y_2O_3 close to and away from a grain boundary. Furthermore, in order to find out whether the electronic contribution to the polarization mechanisms could display a crystallography related anisotropy, EELS experiments have been carried out on a highly anisotropic cell, that is β-Si_3N_4. If proved, this anisotropy could be of a significant importance in explaining interfacial film thickness discrepancies in silicon nitride based polycrystalline ceramics, as it has already been suggested.[2]

2.0 EXPERIMENTS

2.1 Materials

2.1.1 $BaTiO_3$

Pure barium titanate is a perovskite structure ABO_3, which exhibits a strong dielectric discontinuity at the Curie temperature, which corresponds to the paraelectric/ ferroelectric transition; during cooling, the paraelectric cubic phase transforms to the tetragonal ferroelectric phase at the Curie temperature (130°C). The strong and narrow dielectric peak at the Curie temperature reflects the imminent instability of the paraelectric phase, which can then be easily polarized. It is associated to a symmetry loss (m3 m to 4 mm) and to the appearance of a domain structure in which the local polarization occurs along one of the six pseudo-cubic [001] directions by a slight displacement of ions. The tetragonal phase is stable between 130 and 0°C, temperature at which the material undergoes another transition (tetragonal to orthorhombic), again associated to a dielectric peak. Therefore, because of its highly peaked dielectric behaviour, pure barium titanate cannot be used in practical applications such as electronic devices, which require a stable capacitance over a wide range of temperature. In order to overcome this temperature dependence, two successful attempts have been made. On one hand, authors have tried to build up an ordering of the B-site cations of the perovskite structure; this leads to a high diffusion phase transition instead of the sharp ferroelectric transformation at the Curie temperature. The consequence is a very efficient broadening of the dielectric peak. On the other hand, since the Curie temperature is composition dependent, the incorporation of chemical gradients throughout the structure allows obtaining a wide distribution of Curie temperatures. Since diffusion occurs more efficiently along grain boundaries, the resulting microstructure consists of "core-shell" grain morphology where the doped shell has a lower Curie point than the pure core (see Figure 2 for example). The dielectric constant is then roughly independent of the temperature and fulfils the requirements of the "X7R" technical specification. Even though in this latter case the properties are sufficiently stable for the material to be safely used, the exact role of the dopants on the dielectric properties is not precisely known, especially at high frequencies where the electronic polarization mechanism, which accounts for 2/3 of the total polarization, remains the only active one.

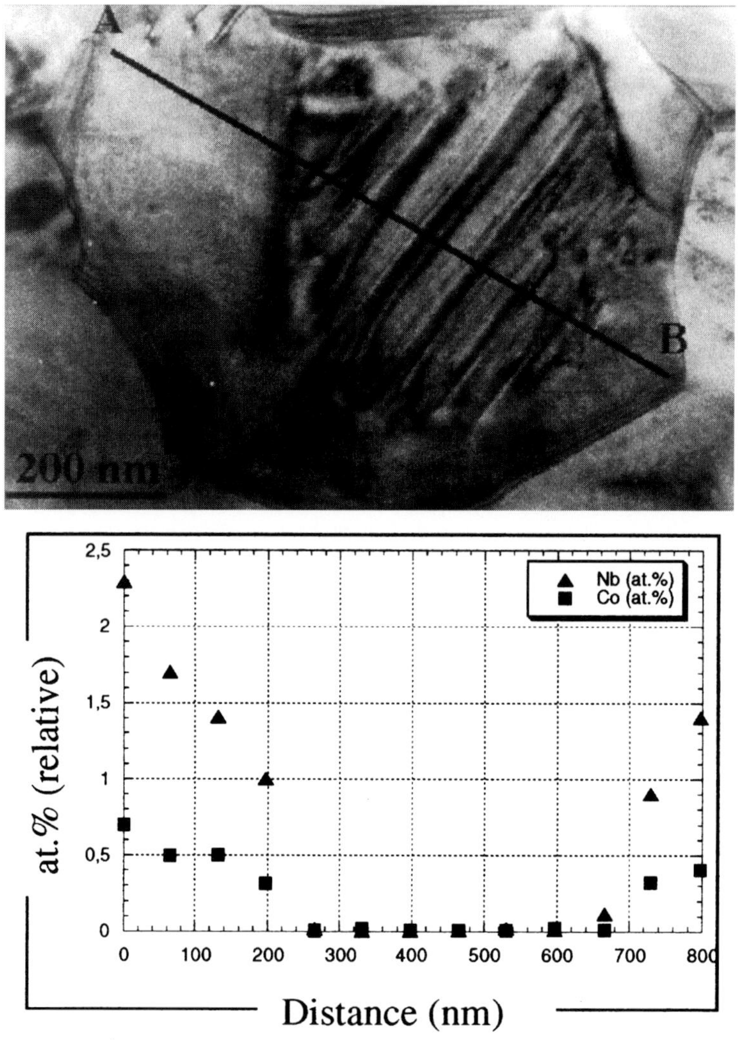

Fig. 2 "Core-Shell" structure observed in a (Nb;Co) doped $BaTiO_3$ by Transmission Electron Microscopy.

Additions of niobium and cobalt to pure barium titanate were found to be efficient enough for the material to satisfy the "X7R" specifications; a typical core-shell structure was observed (see Figure 2), the larger the addition the higher the core-shell structure density. In this work, we explored the influence of three different amounts of dopants, (Nb, Co) = (0, 0), (1.755, 0.488) and (2.4, 0.6), on the local electronic polarization mechanisms.[9] For each composition, the material was sintered at 1300°C for 2 hours. Observations and measurements were carried out on large grains in the non-doped material, and from the core to the shell in the "core-shell" structure in the doped materials.

2.1.2 HfB_2

Hafnium diboride is a material which excellent neutron absorption properties make it a serious candidate for the next generation of control-rods in pressurized water reactors. This is due to the high cross section for neutron capture displayed by ^{10}B and Hf isotopes. One of the major problems encountered in such applications is the accommodation of the fission products after the ^{10}B (n,α) reaction[17] in the material structure. Thorough studies have been carried out to characterize the damage occurring in HfB_2 when ^{10}B transforms into He and Li.[18, 19] Neither He nor Li precipitation was found; however, neutron irradiated samples observed in a TEM showed the presence of vacancy and interstitial dislocation loops respectively in basal and prismatic planes. These loops induce an anisotropic distortion of the HfB_2 cell as well as microcracking while accumulating near the grain boundaries.[20] An elegant way of inhibiting the irradiation damage is to use a material with pre-existing cavities acting as sinks for irradiation defects. Such cavities have been obtained through the precipitation of vacancies that were the result of adding 10% vol. of Hf in the starting powder of the stoichiometric HfB_2.[21] The metal rich powder was sintered at 1800°C for 0.25 hrs. under 47 MPa, and then annealed 24 hrs. at 1200°C. Analyses of the microstructure after sintering revealed that the major phase was HfB_2; we detected also the presence of minor phases such as Hf-α, HfB and fine particles of HfO_2 associated with the cavities (Figure 3). The simultaneous close presence of HfB_2 and HfO_2 was a good opportunity to attempt to characterize the influence of the nitrogen/oxygen substitution on the electronic polarization mechanisms in this particular structure that displays highly covalent B-B bonds as well as metallic Hf-B bonds.

2.1.3 Y_2O_3

Yttria displays an interesting high dielectric constant, in the range 14 to 17. Its properties are known to be strongly dependent on the presence of defects in the structure i.e. the electrical conductivity. Fracture toughness as well as thermal diffusivity are also sensitive to an oxygen deficiency. Yttria is often considered as a typical second phase-forming material and is commonly used as an additive in order to enhance the sintering of ceramics such as nitrides. It is also known to favour the ionic conductivity in oxides when substituted to a heterovalent ion (for example in ZrO_2) and, for the same basic reason, to act as a phase stabilizer (also in ZrO_2). But yttria possesses intrinsic properties related to the specific electronic structure typical of transition metal oxides, which can be used for example in the optic or laser industry.

The properties of transition metals derive from the fact that p electrons can be excited whereas all s or d electrons are not; as an example, 4p electrons in yttrium can start being excited resulting in a collective reorganisation of 4d electrons following $4p^6 4d^N \Rightarrow 4p^5 4d^{N+1}$. The situation is more complex when oxidation of yttrium occurs

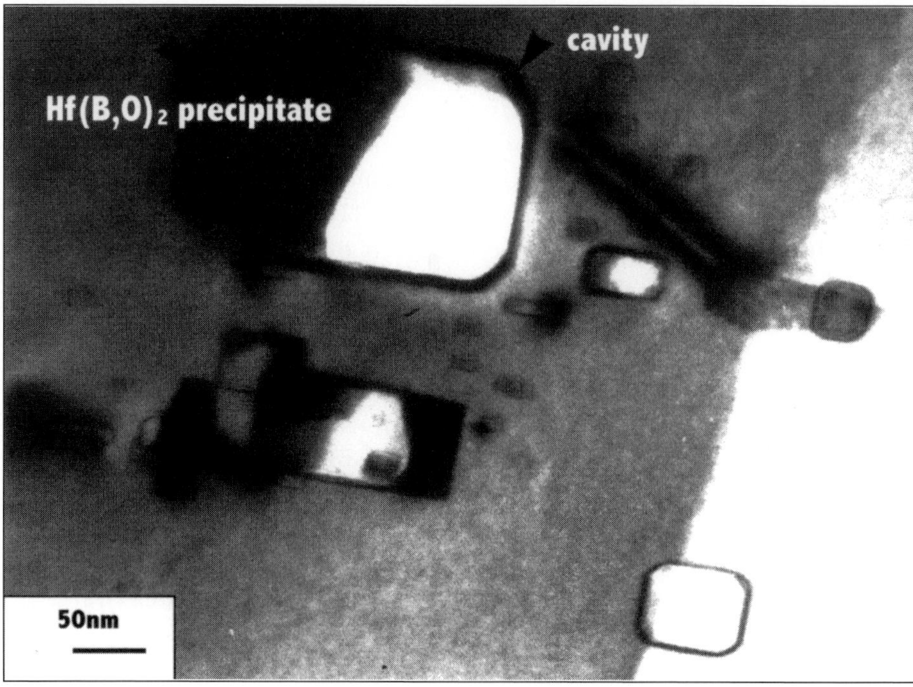

Fig. 3 TEM observation of HfO_2 precipitate associated with cavities in HfB_2+10% Vol Hf.

and affects its band structure; but the low loss region of EELS spectra recorded from a Y_2O_3 sample shows up a signature where the excitation of Y4p electrons is clearly visible at 37 eV, in addition to other peaks characteristic of valence electrons at lower energy.[22] Since yttria crystallizes in a cubic anion deficient fluorite structure, important deviations from the stoichiometry can occur,[23] eventually aggravated at grain boundaries.[24,25] In certain cases of ionic crystals, such oxygen composition gradient may give rise to an electrical activity at grain boundaries that can be strongly affected by an annealing under an oxidizing or reducing atmosphere.[26]

Ultra-pure yttria samples were provided by T. Rouxel who sintered the material under hot isostatic pressing (HIP) conditions[27] for superplasticity experiments.[28] The major impurity in the Rhône-Poulenc ultrapure Y_2O_3 starting powder was found to be SiO_2 which amounts to less than 25 ppm. After the HIP treatment at 1250°C under 150 MPa for 45 min, Rouxel et al. heated the samples in air at 1500°C for two hours in order to avoid any oxygen deficiency. Dielectric function characterizations were carried out within the pure yttria grains, close to and across grain boundaries to eventually detect an interface related oxygen deficiency. A detailed analysis of the material microstructure and dielectric properties can be found elsewhere.[29]

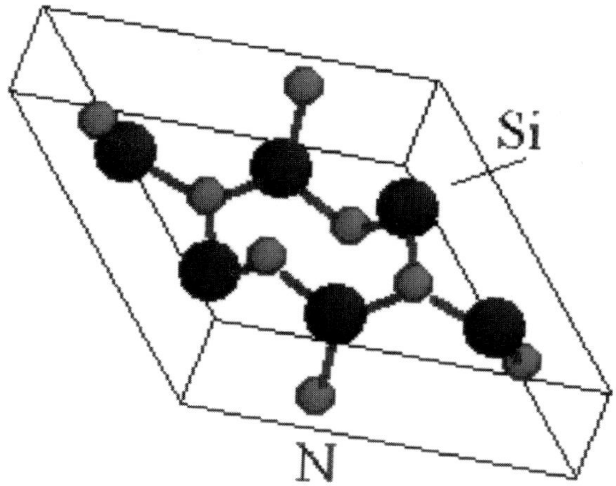

Fig. 4 β-Si_3N_4 unit cell.

2.1.4 Si_3N_4

To a certain extent, silicon nitride has been for the last three decades the best example of what is a high performance advanced ceramic; as in many structures where covalent bonding predominates, silicon nitride exhibits a high hardness, a strong refractory character, a high mechanical resistance and intrinsic toughness. Following the Pauling electronegativity difference rule between Si and N, one finds a 30% ionic character to the bonding in Si_3N_4. The significant ionicity in Si_3N_4, more generally the compatibility of the structure with ionic bonding, allows extensive solid solutions to form with different oxides such as Al_2O_3 as shown in early works in the Si-Al-O-N system.[30,31] The Si_3N_4 structure is made of SiN_4 tetrahedra where nitrogen atoms are coordinated so that sp^3 hybridization of silicon atoms is satisfied. The SiN_4 tetrahedra are joined together in order to have the nitrogen atoms in a sp^2 configuration. Such atomic arrangement where partial ionic charges are assumed, satisfies Pauling's second rule for local charge neutrality. It leads to two varieties of very open and anisotropic hexagonal structures, the polytypes α and β. α and β-Si_3N_4 structures and the α to β transformation have been extensively studied[32] as well as the corresponding α' and β'-SiAlON phases where aluminum is substituted for silicon in tetrahedral coordination, and oxygen for nitrogen simultaneously for keeping charge neutrality.[33]

β'-SiAlON is the most popular substituted polytype for it displays a higher oxidation resistance and is easier to sinter than pure β-Si_3N_4. The commonly accepted generic formula unit is $Si_{6-z}^{(24-4z)+}Al_z^{3z+}O_z^{2z-}N_{8-z}^{(24-3z)-}$ where $z \leq 4$ (see[15] for an overview) with a crystal structure remaining essentially that of β-Si_3N_4 whose symmetry is $P6_3/m$) and hexagonal parameters c_0 = 0.2909 nm and a_0=0.7753 nm (Figure 4). The material under consideration here is a low substituted β'-SiAlON where z = 0.4. It was obtained from a mixture of 80.35%^{w} Si, 5.6%^{w} Al and 14.05%^{w} Y_2O_3 heated at 1400°C under nitrogen, grounded and then sintered at 1750°C again under nitrogen.

Fig. 5 β'-$Si_{5.6}Al_{0.4}O_{0.4}N_{7.6}$ typical microstructure.

The significant amounts of impurities in this material ($0.35\%^w$ Ca and $0.6\%^w$ Fe) allow an intergranular glassy phase to develop,[34] whose composition is that of a Y, Al silicate[35] (Figure 5). If the amount of impurities is lowered, the same process leads to the crystallization of the intergranular Y, Al silicate phase. Interfacial vitreous films can be encountered in silicon nitride when the second phase is glassy, but also though less frequently at hetero-interfaces when the second phase is crystallized.

There are numerous techniques that have been employed for the detection of such interfacial films, such as dark field imaging and the Fresnel fringes techniques in TEM; but most of them can give rise to artifacts.[36] Only High Resolution Transmission Electron Microscopy (HRTEM) has proved to be non-ambiguous and to be giving access to a fundamental structural information on interfacial films.[37,38] Nevertheless, one problem associated with HRTEM for such a purpose is that the required microscope resolution must be comparable to the Si-N or Si-O distances (≈0.16 to 0.17 nm) with a high tilting capability to have the film edge-on; this is strongly favoured when the interface planes correspond to crystallographic planes, for example prismatic or semi-prismatic planes. When there is a common zone axis for both adjacent grains, condition that can be thought as the most favourable for a HRTEM image, there is a strong possibility that the crystallographic relationship at the grain boundary be then a special one.[2,36,39] In that case, there is no gain in energy for this particular grain boundary to be wetted by a vitreous phase and therefore no film will be observed (Figure 6). Nevertheless, situations can be found where a common zone axis exists and the boundary is still a general boundary as seen in Figure 7 where a film is imaged in a commercial silicon nitride (SSN from Desmarquest). We checked on that particular boundary that the crystallographic

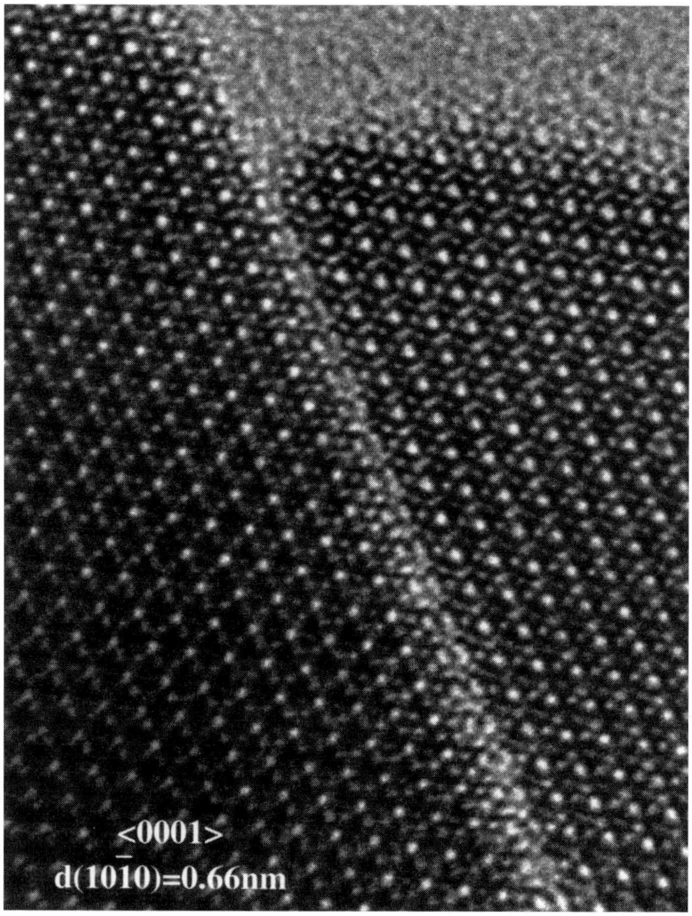

Fig. 6 HRTEM of a film free Low Angle Boundary in $\beta'\text{-}Si_{5.6}Al_{0.4}O_{0.4}N_{7.6}$.

relationship at the interface remained complex $[\langle 000\rangle_1 \,//\, \langle 5\bar{1}\bar{4}6\rangle_2;\ \{10\bar{1}0\}_1 \,//\, \{21\bar{3}0\}_2]$ and did not correspond to a low energy grain boundary.[2] Observation of this interface along grain 1 showed that the film thickness could vary by more than a factor of two, depending on the orientation of the interface planes and eventually on that of the second grain[2] (Figure 7b). Eventually, there are specific areas along the interface in Figure 7 where no film at all is present (Figure 7c). When the amount of impurities is low enough to allow the intergranular silicate to crystallize, there are still cases where such films are observed at interfaces (see Figure 7d).

These are interesting observations since the film thickness is always considered only chemistry dependent (see[40] for a thorough film thickness analysis). Such assumption is based on the fact that the equilibrium film thickness, as described by Clarke,[6] is derived from the balance between forces acting on the film when the

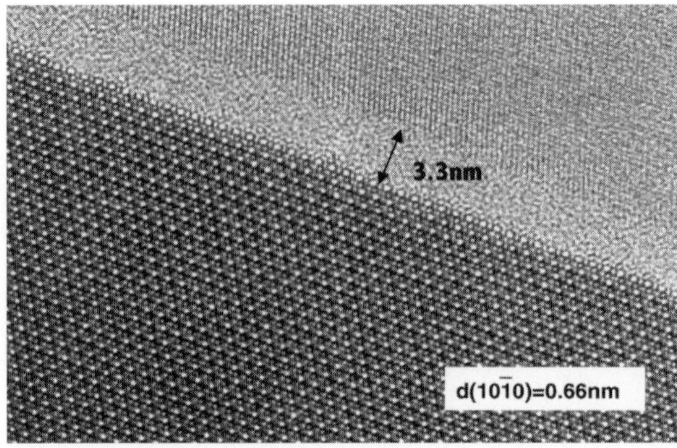

(a)

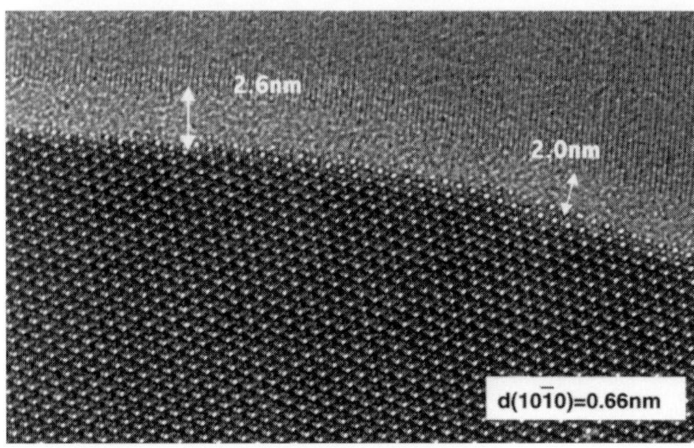

(b)

(c)

Fig. 7 (a) interfacial film in silicon nitride (SSN from Desmarquest) [$<0001>_1$ // $<5\bar{1}\bar{4}6>_2$; $\{10\bar{1}0\}_1$ // $\{21\bar{3}0\}_2$], (b) Another area of the same interface, with variation of the film thickness, (c) Another area of the same interface, with no film. (Please see (d) in next page).

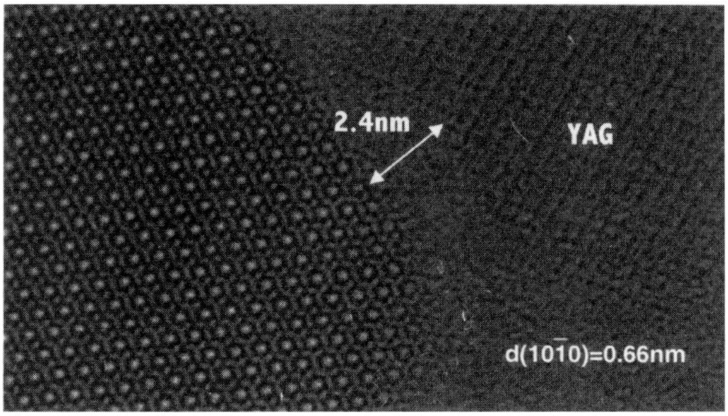

Fig. 7 (d) Vitreous film at β-Si_3N_4/ (Y,Al) silicate.

latter is liquid at the sintering temperature of the material. In particular, Clarke introduced a London type dispersion force and a structural disjoining pressure.[6] Referring to previous work,[41] he suggested that the notion of structural disjoining pressure could be applied to ceramic systems where interfacial films are liquid at the sintering temperature. Authors most generally affirm that the chemical composition of intrfacial films is identical to that of the glassy triple junctions (see[42] for example). This assumption is not necessarily exactly verified,[2] but it nevertheless seems reasonable to propose that the dominant structural unit in the film is the SiO_4 glass tetrahedron since the film has obviously a silica based composition. Then, Clarke suggests in his formalism that the SiO_4 tetrahedron in the film are ordered at the immediate vicinity of the adjacent nitride grains interfaces, the centre of the film remaining eventually vitreous. In this scheme, Clarke's repulsive disjoining force is naturally due to interactions between ordered layers of SiO_4 tetrahedra when the two adjacent grains get closer under the action of external pressure, capillary force or London dispersion forces. A few HRTEM observations where a certain ordering appears at the surface of Si_3N_4 grains tend to give an experimental support to this mechanism,[37,38] then it becomes also evident that this epitaxial ordering must be strongly related to the nature of the interface plane, and hence the structural disjoining pressure and the film equilibrium thickness must also be related to the local crystallography of the system.

The dispersion force originates from the electromagnetic interaction between fluctuating dipoles associated with the different phases in the film vicinity. In the case of an intergranular silica film between two silicon nitride grains, the dispersion force is attractive and proportional to the Hamaker constant $H_{\alpha\beta\alpha}$ of the system, where α stands in our case for Si_3N_4 and β for the intergranular liquid silica, for which a simplified expression is given below:

$$H_{\alpha\beta\alpha} = \frac{3}{4}kT\left(\frac{\varepsilon_\alpha - \varepsilon_\beta}{\varepsilon_\alpha + \varepsilon_\beta}\right)^2 + \frac{3\pi\hbar\nu_e}{8\sqrt{2}}\frac{(n_\alpha^2 - n_\beta^2)^2}{(n_\alpha^2 + n_\beta^2)^{3/2}} \quad (1)$$

In that expression, where k is the Boltzmann constant, T the temperature $\hbar$ the Plank constant and ν_e the absorption frequency in the ultraviolet, appear the static dielectric constant ε and the refractive index n for both α and β phases. The chemistry dependence of the Hamaker constant and hence of the London dispersion force, and consequently of the film equilibrium thickness is evident through changes in dielectric constant and refractive index of the film as already shown.[40] Since they essentially rely on the electronic polarization mechanisms, it seems also that there is no reason for quantities such as ε_β and n_β not being affected by the local crystallography, i.e. the interface planes, that underlie a spatial variation of the electronic structure which may strongly differ if one considers different orientations of the highly non-isotropic silicon nitride structure.

To give some support to this assumption, experimental evidences on another system, essentially obtained through spatially resolved Electron Energy Loss Spectroscopy, have already reported an abrupt change of the electronic structure at interfaces, which was related to the local atomic coordination and bonding (see[8,44] for example). Furthermore, French carried out thoroughly the full calculation of the Hamaker constant, as a function of the excitation frequency,[10] and found different values for bulk glass and films of the same composition in Y_2O_3 doped Si_3N_4. As an example he showed that the index of refraction was lower in the grain boundary film than in bulk glass, leading then to a higher Hamaker constant for a particular interfacial film; this was attributed to an increased oxygen content in the film. In addition, Turan measured on HRTEM images of the thickness of intergranular films in the system h-BN/β- Si_3N_4/ β-SiC and compared them to calculated thickness on the basis of an estimation of the Hamaker constant for films sandwiched at h-BN/β- Si_3N_4 and h-BN/β-SiC interfaces,[39] the comparison showed clear deviations from what could be expected from Clarke's model.[6]

It is suggested here that the observed discrepancies between calculated and actual film thickness on the one hand, and between the Hamaker constant for bulk glass and glassy film as well as for different films in the same material on the other hand, are both chemistry and crystallography dependent. The crystallography dependence (interface planes, coincidence site lattice type relationship) was never formally considered by the former authors, probably because of the extreme difficulty to obtain a sufficient number of experimental evidences. Nevertheless it seems a natural option, accepted for a long time in the grain boundary community, since the electronic structure at the interface is the key-point. We will show thereafter that Electron Energy Loss Spectroscopy can give evidence of such crystallography dependence, through the anisotropy of the dielectric function, when applied to silicon nitride along different zone axis.

2.2 Acquisition of EELS Spectra

A number of different microscopes in different laboratories have been used for the present study for cross-correlation purposes. Spectra recorded on $BaTiO_3$ and on Si_3N_4 have been obtained at 200 KV on a Jeol 200 CX equipped with a Parallel

Electron Energy Loss Spectrometer (Lawrence Berkeley Laboratory, National Center for Electron Microscopy, PEELS # 666 from Gatan) and at 100 KV on a Philips CM 30 (Rhodia, formerly Rhône-Poulenc-Centre de Recherche d'Aubervilliers, with the same PEELS # 666 from Gatan). Spectra on HfB_2 have been recorded in Berkeley, and those on Y_2O_3 in the Laboratoire de Physique des Solides at Orsay University on a PEELS # 666 coupled to a VG-HB501 dedicated STEM operated at 100 KV. Except for the case of Si_3N_4, experiments have been carried out in the image mode; diffraction mode was used for Si_3N_4 for reasons that will be explained latter. Optimal recording conditions (convergence angle, acceptance angle) were looked for; a precise and complete description and influence of these conditions are given elsewhere.[16] The energy resolution of the experiments performed on the Jeol 200 CX and CM 30 attained 0.8 to 1.0 eV, whereas that of the spectra recorded on the VG-HB501 reached 0.3 eV. For the two first microscopes, the spatial resolution (≈10 nm) was defined by the spectrometer entrance aperture and the magnification. The spatial resolution obtained with the VG-HB501 was given by the probe size, that is ≈1.5 nm.

2.3 Procedures for the Dielectric Function Calculation

In a Transmission Electron Microscope, more than 90% of the electrons crossing the sample do not interact with it, and then experience no energy loss. Therefore, these electrons give rise to a strong peak on the EELS spectra at zero energy loss, which is called Zero Loss Peak (ZLP). But an incident electron can also be scattered by an atom of the target and loose an energy E_1 while being deflected by an angle $\theta_1(E_1)$. Since eventually the sample is thick enough to allow the incident electrons to experience multiple scattering, the former electron having an energy E-E_1 and an incident direction $\theta_1(E_1)$ can be scattered again and loose an energy E_2 and be deflected by an angle $\theta_2(E_2)$. The maximum energy this electron can loose is $E_2 = E - E_1$. Therefore, an EELS spectra is made of several contributions as it can be seen in the following expression:

$$J_{\text{total}}(E) = \text{ZLP} \oplus J_1(E) \oplus \underbrace{J_2(E) \oplus \ldots \oplus J_N(E)}_{J_{\text{ms}}(E)} \tag{2}$$

where ZLP is the Zero Loss Peak, $J_{ms}(E) = J_2(E) \oplus \ldots \oplus J_N(E)$ is the multiple scattering signal and $J_1(E)$ the single scattering distribution for the energy E which was involved in the interaction between the incident electrons and the atoms in the sample.

In the dielectric theory, $J_1(E)$ is the useful information that can be expressed in terms of the dielectric properties of the sample:

$$J_1(E) = K.Im\left\{-\frac{1}{\varepsilon(E)}\right\}.f(\theta_E) \tag{3}$$

where $\varepsilon(E)$ is the dielectric function of the target for the energy E, K a constant proportional to the thickness of the area where the spectrum is recorded, and $f(\theta_E)$ a function of the experimental conditions. K and $f(\theta_E)$ will be detailed thereafter. Given experimentally the total energy loss distribution J_{total} (E), the following steps will describe how one can extract the pertinent information $\varepsilon(E)$.

2.3.1 Deconvolution and Removal of the Zero Loss Peak

Since the electrons corresponding to the ZLP do not bear utilizable information, it has to be removed from the spectrum. Strictly speaking, the ZLP contains information about electrons that have suffered an extremely low energy loss, i.e. phonon excitation. But the energy spreading of the incident beam, even with the new field emission guns, and the resolution of the most sophisticated spectrometers (at best 0.2 eV) do not allow to extract information down to such an energy level.

The ZLP has both a Lorentzian and a Gaussian character; different procedures have been tested for an optimum deconvolution and subtraction of this elastic signal.[9] It came out that the ZLP deconvolution available in the EL/P software from Gatan was often satisfactory and was used for HfB_2 and Y_2O_3, although in the latter case an Ax^m extrapolation was finally preferred. On the other hand, deconvolution for $BaTiO_3$ and Si_3N_4 was carried out by means of a home-made ZLP auto-deconvolution routine.[9,16] It must be emphasized here that this step is crucial since it can influence considerably the low loss region of the EELS spectra in the range ≈2 to 10 eV where much of the electronic transitions of interest are located. In most of the cases, practically, the region below 5 eV after deconvolution must be considered with extreme caution for its physical significance is questionable.

2.3.2 Deconvolution and Removal of the Multiple Scattering

Once the ZLP is removed from J_{total} (E), the multiple scattering term J_{ms} (E) has to be evaluated and subtracted following (eqn. 2). There is a strong dependence of the multiple scattering events with the thickness of the sample, the thicker the sample the higher the probability for an incident electron to experience multiple collisions with the target atoms. One can express the intensity of interaction J^t_m of order m for all energies integrated over the thickness t of the sample by a Poison distribution following:

$$J^t_m = \left(\frac{I}{m!}\right)\left(\frac{t}{\lambda}\right)^m \exp\left(-\frac{t}{\lambda}\right) \tag{4}$$

where I is the incident beam intensity, λ the mean free path of the interaction under consideration. The intensity of the scattering of order m is maximized when $(t/\lambda) = m$. Incidentally, (eqn.4) shows that J^t_m and then (t/λ), are known experimentally from the EELS spectrum; thus λ can be calculated providing that the thickness t of the sample be measured in the area under consideration, which is sometimes possible either by convergent beam diffraction, stereoscopy or morphological analysis.[45] This is a very important statement, as it will be discussed in the following sections of this paper.

Technically, the deconvolution of the multiple scattering events from the EELS spectrum can be carried out through two different techniques, one based on a Fourier-Logarithmic approach, widely detailed by Egerton,[14] the other is a purely numerical auto-deconvolution analysis of the experimental data.[9,16] The later one was used in the present work to analyse spectra recorded on $BaTiO_3$ and some on Si_3N_4, and the first one for HfB_2, Y_2O_3 and for some Si_3N_4 spectra. The Fourier-Logarithmic technique is extremely fast but requires that the spectra being deprived of any abrupt discontinuity and being carefully extrapolated to high losses with an adequate function, i.e. an AE^{-r} function. The single scattering event intensity is thus given explicitly by:

$$J_1(E) = \frac{I_0 t}{\pi a_0 m_0 v^2} . Im\left\{-\frac{1}{\varepsilon (E)}\right\} . \int_0^\beta \frac{2\theta \, d\theta}{\theta^2 + \theta_E^2} \tag{5}$$

where K and $f(\theta_E)$ of (eqn. 3) are now defined, and where I_0 is the intensity of the ZLP, v the velocity of the incident electrons, a_0 the Bohr radius, m_0 the mass of an electron at rest, β the spectrometer collection semi-angle and $\theta_E = E/(\gamma m_0 v^2)$ the characteristic diffusion angle relative to an energy loss equal to E and where γ is the relativistic factor.

This expression is valid only if the convergence angle of the incident beam is small in comparison with the collection angle of the spectrometer, conditions that can be easily fulfilled experimentally.

After integrating, (eqn. 5) can be written as follows:

$$Im\left\{-\frac{1}{\varepsilon (E)}\right\} = \frac{J_1(E)}{\ln\left[1 + (\beta / \theta_E)^2\right]} \frac{\pi a_0 m_0 v^2}{I_0 t} \tag{6}$$

As it has already been suggested, $Im\left\{-\frac{1}{\varepsilon (E)}\right\}$

can be calculated at this step provided that the thickness t is known accurately, and assuming that β and v are given experimental conditions.

2.3.3 Normalization

Since the measurement of the thickness t is only seldom possible and requires at best off-line calculations for each acquired spectrum, one often prefers to give (eqn.6) an absolute scale by introducing the optical refractive index n of the material (eqns. 7 and 8) through the Kramers-Krönig summation rule for $E = 0$:

$$1 - Re\left\{\frac{1}{\varepsilon (0)}\right\} = \frac{2}{\pi} \int_0^\infty Im\left\{-\frac{1}{\varepsilon (E)}\right\} \frac{dE}{E} \tag{7}$$

In that case, for materials that are rather insulating, that is to say for which Im $\{\varepsilon (E \rightarrow 0)\}$ is very small (often true for ceramics), one has[46]:

$$1-Re\left\{\frac{1}{\varepsilon\ (0)}\right\}=1-\frac{1}{\varepsilon\ (0)}\approx 1-\frac{1}{n_0^2}$$
$$\text{with } \varepsilon\ (0)=n_0^2 \qquad (8)$$

It can be of interest here to notice that what one uses to name ε(0) in the EELS community refers to ε(∞) in the optics or electronic community. The reason is that the exploitable low loss EELS region extends to frequencies that range roughly between 7.10^{14} Hz (for 3 eV) to 5.10^{16} Hz (for 200 eV) whereas the optical frequencies range between $3.75.10^{14}$ Hz (for red) and $7.5.10^{14}$ Hz (for violet). Then the highest optical light frequency, expressed symbolically by "∞" and regarded as the upper infinite limit for frequency for micro-onde people, corresponds to the lowest value of the frequency (or energy) attainable in our EELS spectra (violet or ultraviolet), then symbolized by "0". In our case, a calculation taking into account the value of the permittivity for an energy loss really equal to zero -the static permittivity-, or eventually any experimental value for a given frequency (as it will be shown for Y_2O_3), would be more rigorous. But the conventional formalism based on (eqn. 8) is accurate enough for ceramics to give our results a sufficient consistency in the frame of this paper, for example to discriminate the different polarization mechanisms.

Then, by combining (eqns. 6 and 7), one obtains:

$$K=\frac{2}{\pi}\frac{\displaystyle\int_0^{\infty}\frac{J_1(E)}{ln\left[1+(\beta/\theta_E)^2\right]}\frac{dE}{E}}{\left(1-\dfrac{1}{n_0^2}\right)} \qquad (9)$$

where K is the constant that appeared in (eqn. 3). One can see that the thickness of the sample disappears from the equation and that K can be fully calculated from the experimental single scattering distribution $J_1(E)$ knowing the refractive index for visible light n_0, that can be either measured or found easily in handbooks. In turn, since $K = I_0 t / (\pi a_0\ m_0 v^2)$ from (eqn. 5), the thickness t can be evaluated at this step.

Evaluation of K requires the calculation of the integral in (eqn.9) from $E=0$ to $E=\infty$; $J_1\ (E)$ has thus to be extrapolated toward higher energy losses by a AE^{-r} or $A(E/E_m)^{-r}$ function, A, E_m and r being estimated from experimental spectra, so that the function falls practically to zero when E becomes large. This type of exponential function is known and often used in Energy Loss Spectroscopy to extract a signal from the background; it has a scientific basis.[13] It must be noticed here that this extrapolation is already carried out if multiple scattering has been removed as seen in section 2.3.2. Finally, as soon as K is known, the calculation of the imaginary part of (eqn. 6) is straightforward following:

$$Im\left\{-\frac{1}{\varepsilon(E)}\right\}=\frac{1}{K}\times\frac{J_1(E)}{ln\left[1+(\beta/\theta_E)^2\right]} \qquad (10)$$

2.34 Kramers-Krönig analysis

Knowing $\mathrm{Im}\left\{-\frac{1}{\varepsilon\,(E)}\right\}$ from section 2.3.4., it is easy to determine

$\mathrm{Re}\left\{\frac{1}{\varepsilon\,(E)}\right\}$ from the so-called Kramers-Krönig transformation that is given below:

$$\mathrm{Re}\left\{\frac{1}{\varepsilon\,(E)}\right\}=1-\frac{2}{\pi}\,\mathrm{PP}\int_0^\infty \mathrm{Im}\left\{-\frac{1}{\varepsilon\,(E')}\right\}\frac{E'}{E'^2-E^2}\,dE' \tag{11}$$

where PP stands for the Cauchy principal part of the integral that must be considered to avoid the discontinuity at pole E=E'. Two methods can be envisioned to calculate this integral: the first one is based on the use of sinusoidal Fourier series[47] and was applied to $BaTiO_3$ and Si_3N_4. It has the advantage of rapidity and avoids the delicate calculation of the Cauchy principal part; but we have observed with this method slight deviations from theoretical predictions when tested for the Debye model. The second method is based on the numerical integration of (eqn. 11); it has been tested for the Debye, Drude, and ionic models and the result was in a very good agreement with the theory. Owing to the now very powerful numerical tools available on computers, this numerical technique is no more time consuming; it was applied here to the Y_2O_3, HfB_2 and some of the Si_3N_4 EELS spectra.

3.0 RESULTS AND DISCUSSION

In order to emphasize the dielectric relaxation distributions that occur in a material throughout the frequency spectrum, we will present the results under the so-called Cole-Cole plot.[48] It is obtained by plotting the imaginary part of the permittivity as a function of the real part; appearance of a perfect semicircle indicates that there exists in the material a very narrow distribution of relaxation times. In that case, this is a strong evidence that one single polarization mechanism exists and is excited at a frequency (or energy) that can easily be obtained from the plot. Multiple semicircles indicate that multiple polarization mechanisms contribute to the global permittivity of the material. If the Cole-Cole distribution displays a tail, corresponding to increasing imaginary part with increasing real part, this is a good indication that the relaxation times distribution is large, which may indicate a strong overlapping of polarization mechanisms or eventually losses due to conduction. Owing to its simplicity and pertinence in emphasizing the polarization mechanisms, the Cole-Cole plot will be used thereafter to present the dielectric properties of the materials under considerations in this paper.

3.1 Role of Dopants on the Electronic Polarization in $BaTiO_3$

Figure 8a shows the results on doped and on large grains of the non-doped materials; the Cole-Cole plot displays a succession of perfect circles that give a clear evidence of a series of distinct and narrow relaxation times. The frequency (or energy) does not appear explicitly in this graph; nevertheless the frequency (or energy) is implicitly given when one follows the curves from the right hand side ($\approx 7.10^{14}$ Hz for 3 eV) to the left hand side ($\approx 5.10^{16}$ Hz for 200 eV). It is to be noted here that, in opposition to typical Cole-Cole plots for complex impedance, i.e. RC circuits, the circles are not centred on the real axis.

The most striking feature in Figure 8a is that two major circles occur for the non-doped material whereas for the doped materials the right-hand side circle splits into two other circles which expand with increasing amounts of niobium and cobalt. Cusps occur roughly at 15 eV for the non-doped material and at 7 and 15 eV for the doped materials. For doped materials, the right-hand side circle is strongly affected by the amount of dopants; inversely, the circle at high energy on the left-hand side of the plot seems poorly affected by the dopant concentration. For all the materials, irregularities are observed for higher energies.

Comparing the band structure of $BaTiO_3$[49] and the experimental dielectric function, the common circle at higher energy than 15 eV (left hand-side) for the pure material is thought to correspond to a plasmon type excitation whereas the circle for an energy below 15 eV (right hand-side) is attributed to the excitation of valence electrons (oxygen 2p electrons) and barium 5p electrons toward titanium 3d free levels of the conduction band. A compensation mechanism between Nb^{5+} as a donor and Co^{2+} or Co^{3+} as an acceptor leads to a roughly constant carrier density and then to equivalent plasmon characteristics for doped and non-doped materials; this explains why the circle attributed to a plasmon is in a first approximation independent of the dopant concentration.

Appearance of a third circle is clearly related to the introduction of dopants in the structure. In order to accommodate the charge excess given by the donors, point defects such as barium vacancies are created for a small concentration of dopants, in addition to titanium vacancies for higher concentration of dopants. These vacancies give rise to two additional levels in the band structure respectively at 1.8 eV and 1.3 eV below the conduction band. For reasons explained in the section 2.3.2, transitions from both these levels toward the conduction band cannot be isolated here. Nevertheless they strongly participate to the relaxation time corresponding to the right hand-side circle, the consequence of the increasing appearance of vacancies with the doping being that it gives rise to numerous additional possible intermediate transitions, decipherment of which is out of the frame of this paper. It is then observed in Figure 8a that the higher the dopant concentration the larger the diameter of the right hand-side circle. We will assume here that for the doped materials this right hand-side circle is mainly the consequence of transitions from energy levels corresponding to the titanium and barium vacancies and oxygen 2p levels, toward the conduction band. The intermediate circle is then attributed to transitions of barium 5p electrons towards the titanium and barium vacancy levels and the conduction band.

More interesting is the Cole-Cole diagrams recorded locally, on a small "core-shell" grain of a highly doped material, for which (Nb;Co) = (1.750:0.488), from the core region to the shell region. Results are presented in Figure 8b where Cole-Cole plots for the core region, the intermediate zone and the shell region of the same grain are shown. Like in the case of the grain in Figure 2, we checked that the core was deprived of dopant whereas a steep dopant concentration occurs toward the grain boundary. Surprisingly, three main circles are present whatever the curve which aspect is similar to that of the Cole-Cole plot for (Nb;Co) = (1.750:0.488) in Figure 8a. The general frame of our interpretation will then be roughly the same; nevertheless, comparison between both sets of curves shows slight discrepancies, i.e. the more the concentration of dopant the smaller the right hand-side circle. Moreover, it can be seen for the shell region a faint additional fourth circle at about 15 eV, between the plasmon and the barium 5p circles. Further interpretation of Figure 8b must take the following points into account:

1. The plasmon relaxation, owing to the compensation mechanism already mentioned, is independent of the dopant concentration in small grains as well as in large grains,
2. The dopant concentration in the shell region can be considerably higher than what is expected from the starting formulation,
3. We checked by diffraction that the core region of a core-shell grain is pseudo-cubic,
4. A cell distortion occurs along the dopant concentration gradient, along with a stress gradient,
5. Vacancies appear to coalesce and form loops in the high doped shell region, as observed in Figure 9, which gives a substantial confirmation of a high amount of doping in these regions.

A full interpretation of Figure 8b is not attempted here; but it is thought that diffusion of vacancies toward the core region, as well as the presence of a stress field, may explain the difference between the dielectric behaviour of the core regions in the doped materials and that of the pure material.

3.2 Role of the N/O Substitution in HfB_2

Very little information on HfB_2 is available; some calculations and measurements on the band structure of HfO_2 can be nevertheless found in the literature (see[51] for example). It is admitted that the B-B bonds in HfB_2 have a sp^2 character; furthermore, π orbitals have been detected through NMR experiments,[52] eventually completed by delocalized electrons from hafnium since the $5d^26s^2$ configuration allows a transfer of electrons from the metal to boron. B-B bonds are highly covalent[53] whereas Hf-B ones have a strong metallic character, partly consisting of hafnium delocalized electrons.[54] It is also admitted that, like in other transition elements, low energy transitions can occur for pure Hf following the example $5p^65d^2 \Rightarrow 5p^55d^3$. Calculations[51] of the band structure of HfO_2 are based on the assumption that the 4f level belongs to the core and do not participate in the chemical bonding, and that the valence

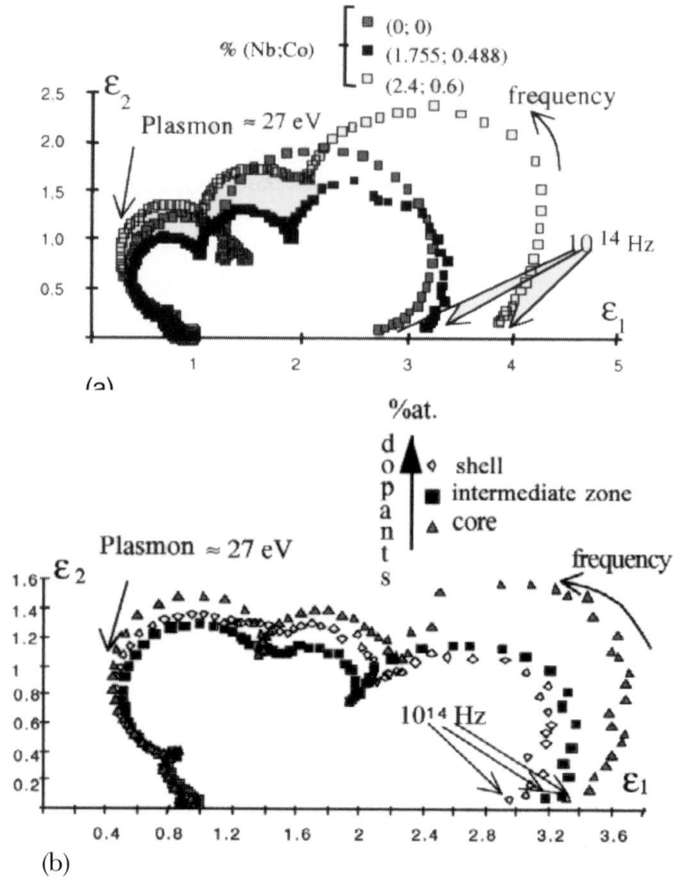

Fig. 8 **(a)** Cole-Cole plot for doped and non-doped $BaTiO_3$ (Nb and Co) and (b) Cole-Cole plot for a core-shell grain in the doped $BaTiO_3$; core region, shell region, intermediate region

states are 5d, 6s and 6p. The bonding is of the s-p-d type, involving oxygen 2s and 2p orbitals. Results of the calculations[51] of the densities of states for HfO_2 show that it consists of a low energy narrow band corresponding to the oxygen 2s states, a valence band of hybrid oxygen 2p and hafnium s-p-d states, and a conduction band bottom formed mainly by the hafnium d states. This leads to a gap of the order of 5 eV.

Calculation of the Fermi surface of HfB_2 has also been attempted;[55] it is observed that for discrete directions of the Brillouin zone the conduction and valence band may slightly intersect, the Fermi level being obligatory in that case in the conduction band; this gives evidence of the semi-metallic character of HfB_2.

As far as the chemical bonding of HfB_2 and HfO_2 is concerned, as a first and rough approximation, the same s-p-d type bonding assumption can eventually be made, with the noticeable difference that the valence state of oxygen is –2 and that of boron is +3, the p level of oxygen having a tendency to fill, whereas the electrons

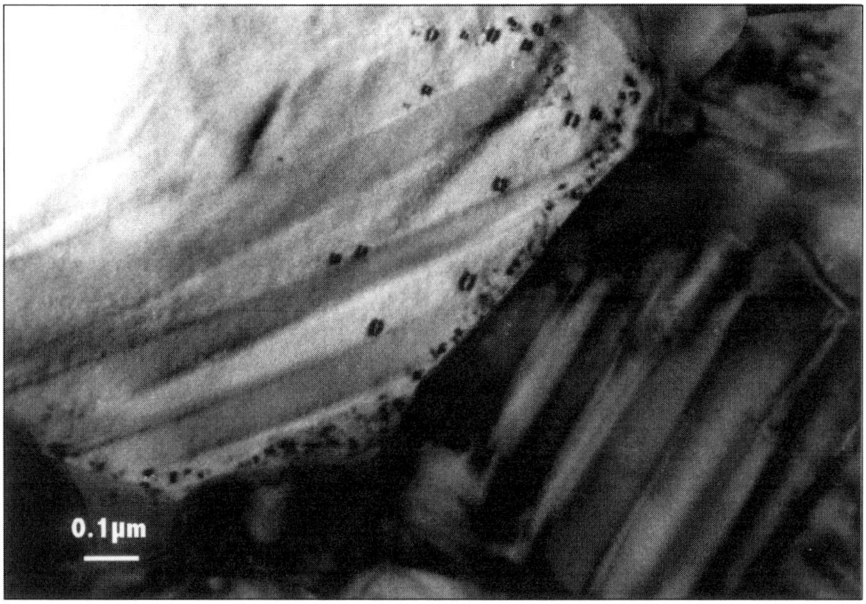

Fig. 9 Vacancy loops in the shell region of a core-shell grain in (Nb;Co) doped $BaTiO_3$

of both p and s levels of boron have a tendency to empty. In anyway, like in transition metals with more than a single d electron, the distribution of the hafnium d electrons among the available orbitals will occurs so that the preference will be given for orbitals of lower energy in the ligand field and for a maximum number of parallel spins.

Transitions elements, which have an incomplete d shell, have more or less similar chemical properties since the filling of the inner (N)d shell has little effect on the ionization potential and properties of the (N+1)s electrons. An even more pronounced effect is seen for rare earth elements in which the 4f shell is incompletely filled. Then useful information for the HfB_2 and HfO_2 analysis can be obtained from more documented materials like oxide or boride of other transitions or rare earth elements. If one compares the low loss region of the EELS spectra of transition metals in the Gatan Atlas,[22] mainly recorded on oxidized states, one finds strong similarities immediately; the spectra are always made of two major peaks corresponding to the collective excitation of valence electrons (at low energy, between roughly 10 and 20 eV), and to interband transitions (at higher energy, between roughly 30 and 50 eV) from the p levels toward available levels of lower energy, i.e. in the conduction band. The spectrum given in the atlas for HfO_2 is in that respect very similar to that of Sc_2O_3. In that HfO_2 spectrum, two major peaks are visible at about 18 eV and 47 eV, with three secondary peaks at 7 eV (which is incidentally equal to the first ionization potential for Hf in its ground state), at 26 eV and at 37 eV, the one at 26 eV being quite strong. It is admitted here that the peak at 18 eV corresponds to a plasmon and that at 47 eV to excitation of hafnium $5p^{1/2}$ and $5p^{3/2}$ electrons.

Experimental EELS spectra and single scattering distributions of HfO_2 and HfB_2, obtained after a standard Gatan ELP Zero Loss Peak and multiple scattering deconvolution, are given in Figure 10a. Although the peaks for the single scattering of HfO_2 appear at positions in a good agreement with the Gatan Atlas (shoulder at 8.2 eV, peaks at 17 eV, 26.5 eV, 38.4 eV, 47.6 eV), the relative height of peaks at 17 eV and 26.5 eV is different from that of pure HfO_2. This, in addition to a faint shoulder visible at 23.5 eV, indicates a probable mixed $Hf(B,O)_2$ stoichiometry. The single scattering distribution for HfB_2 displays a shoulder at 8.2 eV, two strong peaks at 23.5 eV and 48.2 eV, and a slight shoulder at 38.4 eV. In that respect, the EELS signature of HfB_2 is very similar to that of all the lantinide oxides, even though these have an incomplete 4f shell and a completely free 5d shell. The full analysis of the band structure of HfB_2 is out of the frame of the present paper, but it seems reasonable to assume that the peak at 23.5 eV is a plasmon of metallic character mainly consisting of electrons involved in the Hf-B π bonding, whereas the peak at 48.2 eV is the signature of interband transitions of hafnium 5p electrons.

The major difference between HfO_2 and HfB_2 consists in the splitting of the plasmon peak when boron is replaced by oxygen; according to the calculations mentioned above, it is thought here that the series of peaks below 30 eV for HfO_2 consists successively of the excitation of the oxygen 2s electrons, of electrons participating to the hybrid between oxygen 2p and hafnium s-p-d, and to the excitation of the hafnium d states in the conduction band.

The Cole-Cole plot calculated from the HfO_2 and HfB_2 single scattering distributions is presented in Figure 10b; the diagram for HfO_2 and that for HfB_2 reasonably superimpose below 10 eV and above 50 eV, which is not a surprise since the dielectric function is equal to $(\varepsilon_2,\varepsilon_1)=(0,1)$ when the frequency is infinite, and since an arbitrary value of $\varepsilon_1(0)=2$ has been chosen for both materials. Apart from the fact that no data is available in the literature, there is no formal justification for this latter value, which is probably underestimated since for example the refractive index of borate glasses range between 1.5 and 2.1, and that of oxides of elements of group 4 (TiO_2, ZrO_2) are respectively 2.5 and 2.2.

Between 10 eV and 50 eV, the Cole-Cole diagrams differ strongly, both presenting a series of well defined circles that corresponds to a series of narrow relaxation times more or less intersecting. Taking into account the interpretation of the single scattering distributions described above and given the value of the energy at which the circles intersect, the identification of the electronic polarization mechanisms are straightforward: they are directly related to the peaks in the single scattering distribution. The Cole-Cole plot allows to compare the mechanisms that participate to the permittivity of the materials, and to determine which are the most efficient. In that respect, it appears that the plasmon at 17 eV for HfO_2 and that at 23.5 eV for HfB_2 give the highest contribution to the dielectric properties of both materials.

Moreover, the Cole-Cole plot is a very sensitive and discriminative display of any fine structure of the single scattering distribution as it can be observed on Figure 10c, which gives an enlargement of the cusps region at about 40 eV of Figure 10b. As an illustration, a small circle appears clearly between 34 eV and 43 eV on the HfO_2 curve that corresponds to the peak at 38.4 eV in Figure 10a.

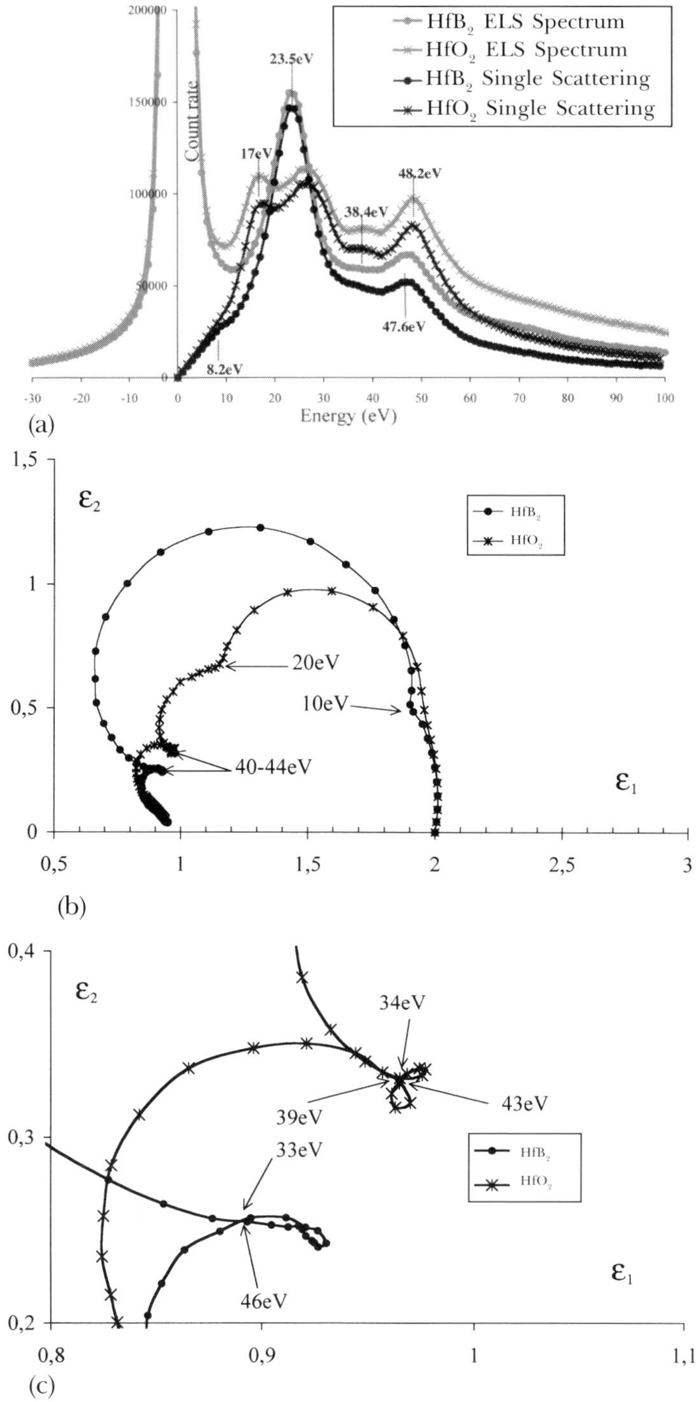

Fig. 10 (a) EELS spectra and single scattering distributions for HfO_2 and HfB_2, (b) Cole-Cole distributions for HfB_2 and HfO_2 corresponding to figure 10a and (c) Cole - Cole plot for HfB_2 and HfO_2: enlargement of the loops at low (ε_2, ε_1).

Evidently, there is lack of data concerning hafnium-based compounds. If hafnium diboride was to be extensively used as a neutron absorbent for control rod application in PWR reactor, it is suggested here that not only its nuclear properties are to be considered, but also its electronic structure in order to understand and optimize the ability of this material to absorb the point defects created by the neutron irradiation, thus strongly inhibiting the appearance of the associated damage.

3.3 Dielectric Function in Y_2O_3

The electronic structure of yttria is much more documented than that of hafnium based compounds (see[56, 57] for example). In these works, density of states calculations, based on a semi-empirical tight-binding method applied to oxygen 2s, 2p and to yttrium 4d, 5s, 5p orbitals, were coupled to X-ray Absorption Spectroscopy (XAS) and X-ray Photoemission Spectroscopy (XPS) experiments. The 5p orbital is unoccupied in yttrium, but is considered here as a final state for a dipolar transition. The calculations could reproduce the fine structure in the lower part of the conduction band, and the signature of p orbitals confined to a tetrahedral environment in the valence band. Dipolar transitions of the s $\Rightarrow$ p and p $\Rightarrow$ d type have been given evidence at K and L yttrium edges via XAS, and many-body relaxation effects were observed following the creation of the core-hole, eventually with a collective character. These authors show that the ligand-field approach is not sufficient in this case to account for the shape of the valence band, and suggest that the quasi-octahedral local environment of yttrium atoms is responsible for the splitting of the unoccupied d levels, which are involved in the hybridization of the oxygen p orbitals. They attribute the peaks of the density of states of the valence band to mainly the oxygen p orbitals, to long-range band effects and to direct-hopping terms between neighbouring oxygen atoms. Yttrium levels seem to be more involved in the conduction band through hybridization of oxygen p with unoccupied yttrium d states, probably dz^2 and $d(x^2\text{-}y^2)$ orbitals.

Figure 11a presents the EELS spectrum recorded on yttria, as well as the multiple and single scattering distributions. This spectrum has been obtained from an intragranular region; no difference between grain boundary regions and intragranular regions has been observed, indicating that the cautious post-HIP thermal treatment at 1500°C in air was sufficient to avoid any oxygen deficiency. Two major peaks are visible at 14.7 eV and 37.2 eV, in addition to more or less observable shoulders at 8 eV, 20 eV and 30 eV.

Following the same kind of reasoning as that used for $BaTiO_3$ and $Hf(B,O)_2$, we assume that the peak at 14.7 eV is a plasmon corresponding to a collective excitation of electrons from the valence band (oxygen p orbitals), with eventually the effect of oxygen hopping that could be responsible for the peak at 20 eV. The second strongest peak at 37.2 eV (and eventually the faint one at 30 eV) is due to the excitation of yttrium $4p^{1/2}$ and $4p^{3/2}$ electrons that are then available for interband transitions, and move toward the unoccupied level of the conduction band (hybrid oxygen p and yttrium d orbitals).

The Zero Loss Peak deconvolution used above to obtain the multiple scattering signal is based on an extrapolation of the Ax^m type. It has among others the interest of satisfying the continuity of the first derivative. The influence of the type of numerical treatment used is clearly visible on Figure 11b that shows the Cole-Cole distributions corresponding to a parabolic, a linear and an Ax^m ZLP deconvolution. For this particular calculation, we have used $\varepsilon(0) = 10$, value measured at room temperature and obtained at 10^6 Hz. We will assume here that the Ax^m ZLP deconvolution is the most adequate one in the case of Y_2O_3 for it allows to make appear the faint shoulder below about 8-10 eV (see Figure 11a), which is also present in the Gatan Atlas.[22] This could be the signature of an event taking place in the range 5-8 eV, which gives a maximum of absorption as it can be seen on the dielectric function (Figure 11c) derived from the EELS spectrum with an Ax^m ZLP deconvolution, and from which the Cole-Cole diagram in Figure 11b is drawn.

Figure 11d presents an enlargement of the loop region in Figure 11b; it shows that the differences between the different numerical options for the ZLP deconvolution diminish for higher energy, but are still observable (see Figure 11e, a magnified view of the intersection point in Figure 11d). The reason why a numerical treatment at essentially very low energy has still an influence at high energy comes from the calculation of the integrals in eqns. 9 and 11 for which the summation has to be considered from 0 to ∞.

Whatever the ZLP deconvolution, comparing Figures 11b and 11f, which gives an enlargement of the dielectric function above 10 eV, it is observed that the Cole-Cole distribution can be roughly divided into four energy domains bounded by 0 eV, 10 eV, 20 eV, 30 eV and ∞. A close-view of the fractal-type loop structure appearing in the (1,2) plane (see Figure 11g) suggests that it could be somehow more complicated and that numerous relaxation mechanisms could be involved; but this would be out of the frame of the present paper. The four domains arbitrarily defined above correspond to four major relaxation times, the first two (from 0 to 20 eV) probably related to the collective excitation of valence electrons at different levels, and the last two (to 20 eV to ∞) to interband transitions of yttrium $4p^{1/2}$ and $4p^{3/2}$ electrons.

As far as the dielectric properties are concerned, and at least in the range of frequency under consideration here, but there is no evident reason that the hierarchy between these different electronic polarization mechanisms not be maintained at lower frequencies-, it appears that the most efficient mechanism is that related to the excitation of valence electrons. But the example of Y_2O_3 clearly demonstrates the strong need for a non-ambiguous ZLP deconvolution and for a high resolution in energy.

3.4 Crystallography Related Dielectric Anisotropy in Si_3N_4

Figures 5 to 7 present different intergranular morphologies typical of silicon nitride ceramics. The faceted and often elongated β silicon nitride grains, particularly at heterophase interfaces (see Figure 5) suggest immediately some degree of anisotropy

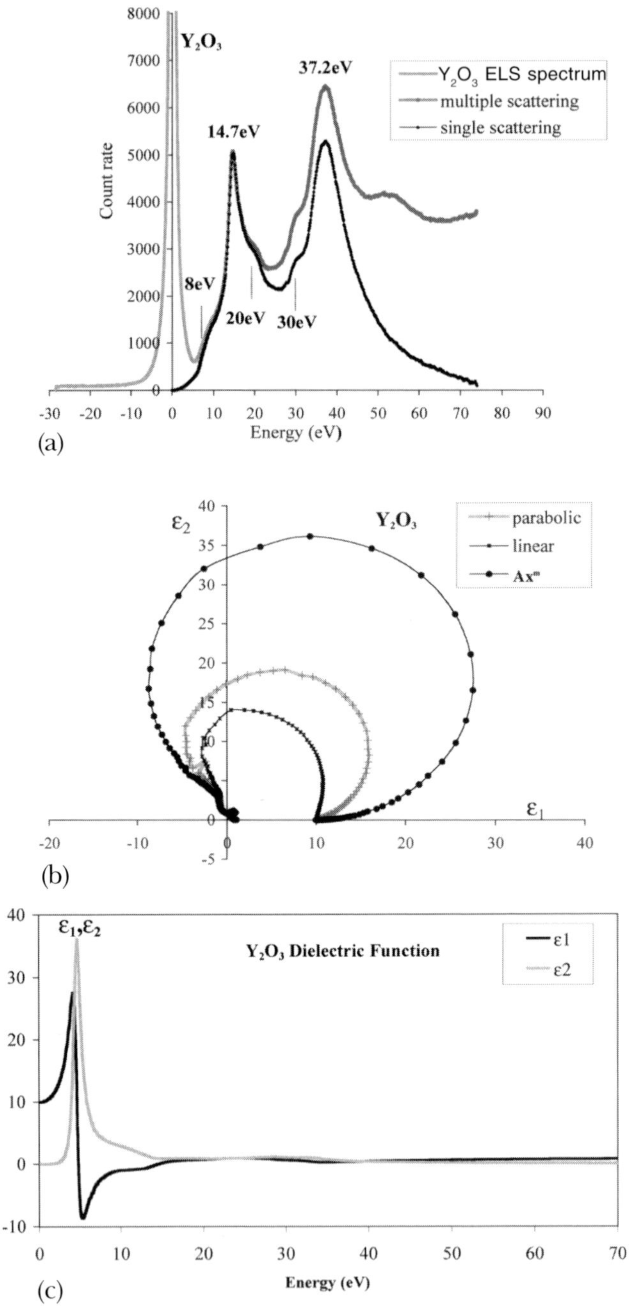

Fig. 11 (a) Experimental Y_2O_3 EELS spectrum of and corresponding multiple and single scattering distributions, (b) Cole-Cole distributions for different Zero Loss Peak deconvolutions, (parabolic, linear and Ax^m), (c) Dielectric Function of Y_2O_3, (d) Enlargement of the loop observed in figure 11b, (e) Enlargement of the intersection point region of figure 11d, (f) Enlargement of the Dielectric Function in figure 11c and (g) Structure of the loop in the (ε_1, ε_2) plane. (Please see Figs. (d-f) in next page).

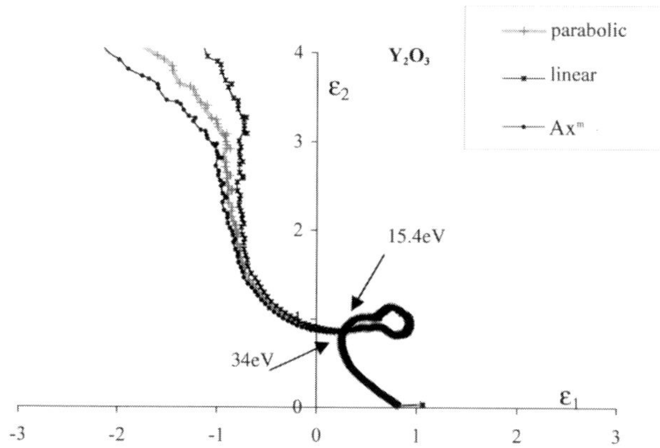

(d)

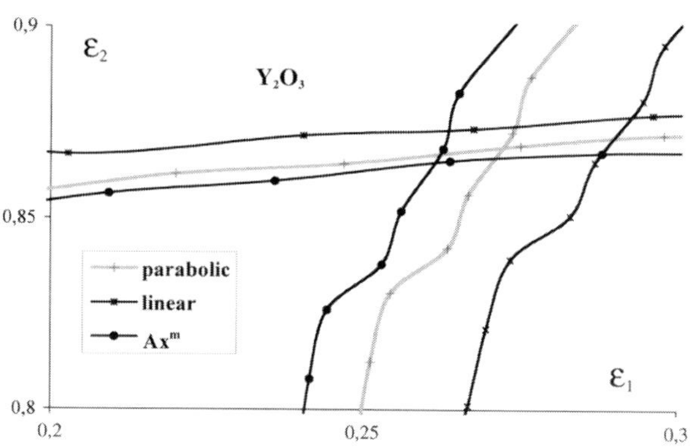

(e)

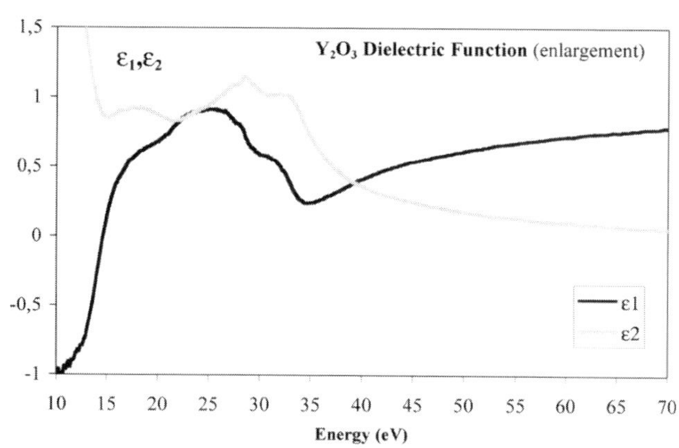

(f)

Fig. 11 continued. (Please see Fig. (g) in next page).

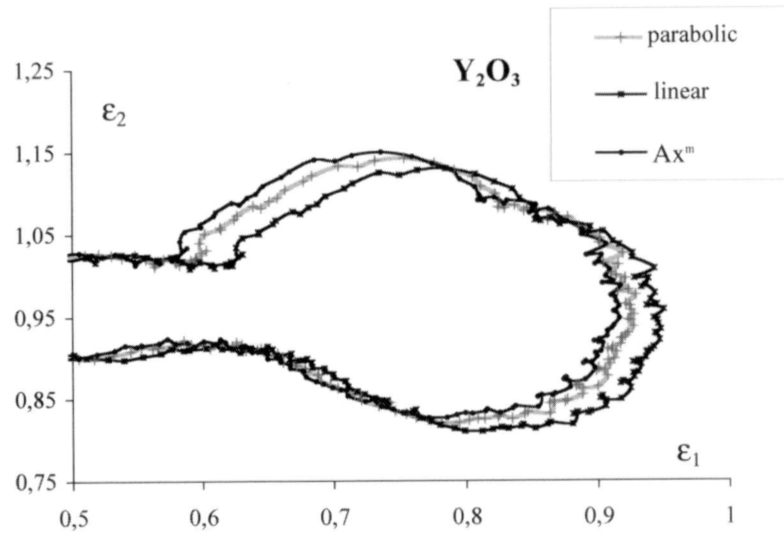

(g)

Fig. 11 continued.

in this ceramic; this is also observed locally at β-Si_3N_4/glass interfaces where $\{10\bar{1}0\}$ faceting often occurs (Figure 12), which is the remnant of solid/liquid interface energy anisotropy at the sintering temperature. We suggested in section 2.1.5 that such crystallographic anisotropy could have a significant dielectric or optical counterpart, and that this anisotropy of the dielectric function and refractive indices of silicon nitride could be experimentally detected through EELS experiments.

Since in an EELS experiment the electrons, from the valence or conduction bands or from low energy outer levels, are collectively excited by the momentum $\vec{k} = \hbar\vec{q} = \hbar(\vec{K}' - \vec{K})$ ($\vec{K}$ and $\vec{K}'$) are respectively the wave vectors associated with the transmitted and scattered beams) transferred to the material when submitted to the incident beam (see Figure 13), it is theoretically possible to excite selectively these electrons in a discrete direction if one controls the orientation of $\vec{k}$. In that case, it is intuitively understandable that the excitation of the electrons will depend on the orientation of the molecular orbitals with respect to that of the momentum $\vec{k}$. The control of the orientation of $\vec{k}$ is experimentally easily obtainable in a Transmission Electron Microscope when the specimen is thin ($\| \vec{K}' \| \approx \| \vec{K} \|$) and when it is tilted so that a Bragg condition is fulfilled; in that case, all the $\vec{k}$ vectors lay approximately in the reciprocal plane observable on the screen. A particular momentum $\vec{k}$ or scattered $\vec{K}'$ vector is thus chosen by positioning the EELS entrance aperture along a particular reciprocal direction at a distance from the transmitted spot corresponding to the plasmon characteristic angle (typically 0.010 mrd at 200 KV) (see Figure 13); this is easily achieved by adjusting the shift controls of the projector system lenses.

Two types of experiment have been attempted for the present paper, both being carried out in the diffraction mode. In the first experiment, the EELS 1 mm entrance

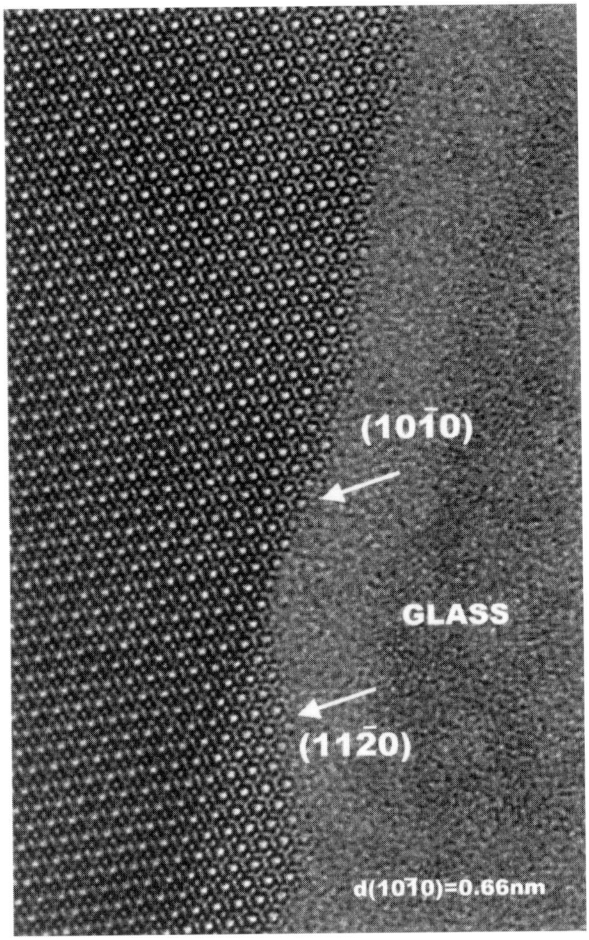

Fig. 12 Faceting at β'-SiAlOn/Glass interfaces.

aperture was centred on the transmitted beam and the TEM camera length (L = 82 cm) taken so that the scattered $\vec{K}$ vectors corresponding to the plasmon characteristic angle were included in the geometrical cone defined by the aperture, and able to be taken into account by the spectrometer. In that case, all the scattered beams corresponding to all the possible $\vec{k}$ vectors included in the aperture and laying in the reciprocal plane were considered (see[9] for a complete analysis of this experiment for $BaTiO_3$). Different crystallographic orientations have been investigated by tilting the sample so that the optical path symmetry axis of the microscope lay along different real space [uvw] directions of Si_3N_4. Figure 14a presents the Cole-Cole distributions for [uvw] equals to [001] and [233], which are respectively perpendicular to real space planes (001) and (8 32 7) (see Figure 14b for the corresponding projected views of the unit cell). The value of ε(0) has been here arbitrary chosen equal to 2. A full interpretation of these Cole-Cole diagrams is

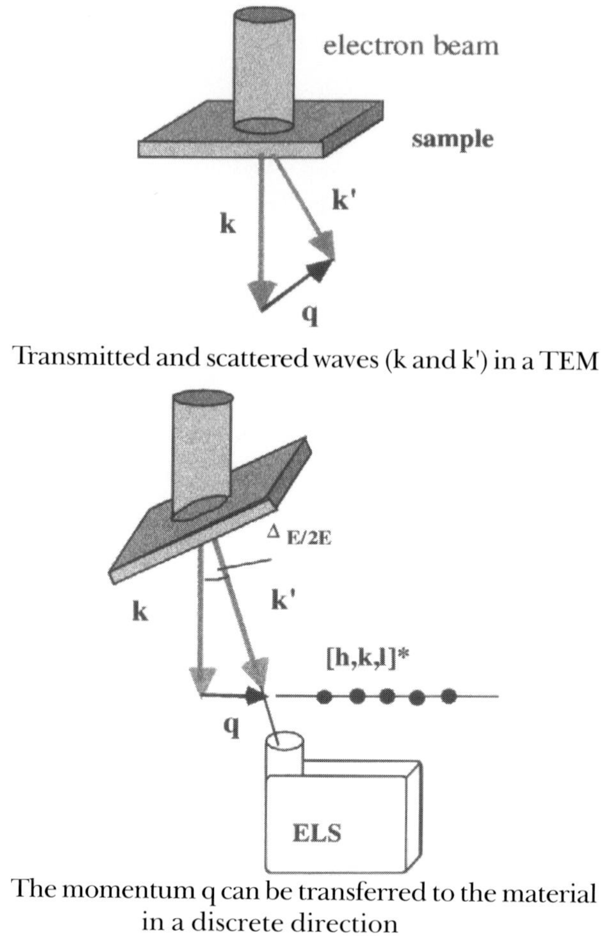

Fig. 13 Schematic of the momentum controlled EELS experiment.

not attempted here, but it is clear that a dielectric crystallography related anisotropy exists, that can be given evidence by a $\vec{k}$-controlled EELS experiment. The underlying idea is that the excitation of sp^2 and sp^3 hybrid orbitals depends on the angle between their respective directions and that of the $\vec{k}$ vectors.

In the second experiment, only one single [uvw] real space direction was investigated at a time, but the diffraction pattern was shifted so that the 1 mm EELS entrance aperture was positioned along a specific reciprocal direction at a distance of the transmitted spot corresponding approximately to the plasmon characteristic angle. In order to be more selective, a higher camera length was chosen. Figure 14c presents the Cole-Cole distributions when the material is viewed along [103] (direction perpendicular to the real space plane $(20\ \overline{1}\overline{0}\ 9)$) with the aperture centred on $(000)^*$, and with the aperture shifted along the $[010]^*$ direction; the camera length is here equal to 205 cm. In that case, $\varepsilon(0)$ has been here arbitrary chosen equal to 4; thus Figures 14a and c cannot be directly compared. Again, a significant change between the Cole-Cole plots on Figure 14c is observed.

It must be noted here that the size of the EELS entrance aperture corresponds to an angle that is (even for a large camera length) of the same order of magnitude as that of the plasmon characteristic angle. The immediate consequence is that this experiment does not allow to isolate a single $\vec{k}$ vector along $[010]^*$, but rather a distribution of $\vec{k}$ vectors whose directions are more or less aligned with $[010]^*$ within roughly a $\pm$ 30° uncertainty. But since the sample is very thin (High Resolution images have been obtained from the area under consideration, which indicates that the thickness is of the order of 10 nm), all these $\vec{k}$ vectors lay approximately in the same reciprocal plane. Increasing drastically the camera length or diminishing the EELS entrance aperture would be the solution for a better discrimination of the $\vec{k}$ vectors had the EELS signal become rapidly negligible. For this reason, the EELS standard equipment does not allow the detection of a weak dielectric anisotropy; as an example, in spite of a spontaneous polarization in $BaTiO_3$, we could not detect any dielectric anisotropy in this material via this method (see[9]).

The conclusion of these two experiments suggests strongly that there may be different electronic polarization mechanisms activated when the excitation comes from different crystallographic directions; this is not a very new idea in principle, and for example EELS results on graphite K-edge showing a crystallography dependence of the π/σ signals are long known, and eventually also on the same kind of dependence of the plasmon energy in silicon nitride[36]. But to our knowledge, no data on the local determination of the dielectric function anisotropy on polycrystalline ceramics is available in the literature. The consequence of these results on the stability of interfacial films in silicon nitride is not straightforward, in particular because the High Resolution and EELS experiments have not been carried out simultaneously due to experimental constraints. But considering that Van der Waals forces, acting at the interfacial film scale and playing a key-role for its stability, are essentially dependent on the local dielectric permittivity balance,[59] and that the electronic polarization mechanisms often account for the major contribution to the total polarizability, it should be a natural option to consider our results as a possible explanation for the film thickness variations that have been experimentally observed.

4.0 CONCLUSIONS

Electron Energy Loss Spectroscopy was already known as a powerful analytical technique that allows scientists to obtain fundamental information on materials, particularly at interfaces. It is our belief that the low loss region of EELS spectra has not been investigated enough, and that the possibility of attaining the local dielectric function to explore the electronic structure of materials remains an interesting challenge. We have tested in this work the high potentiality of this method when applied to different sintered polycrystalline ceramics.

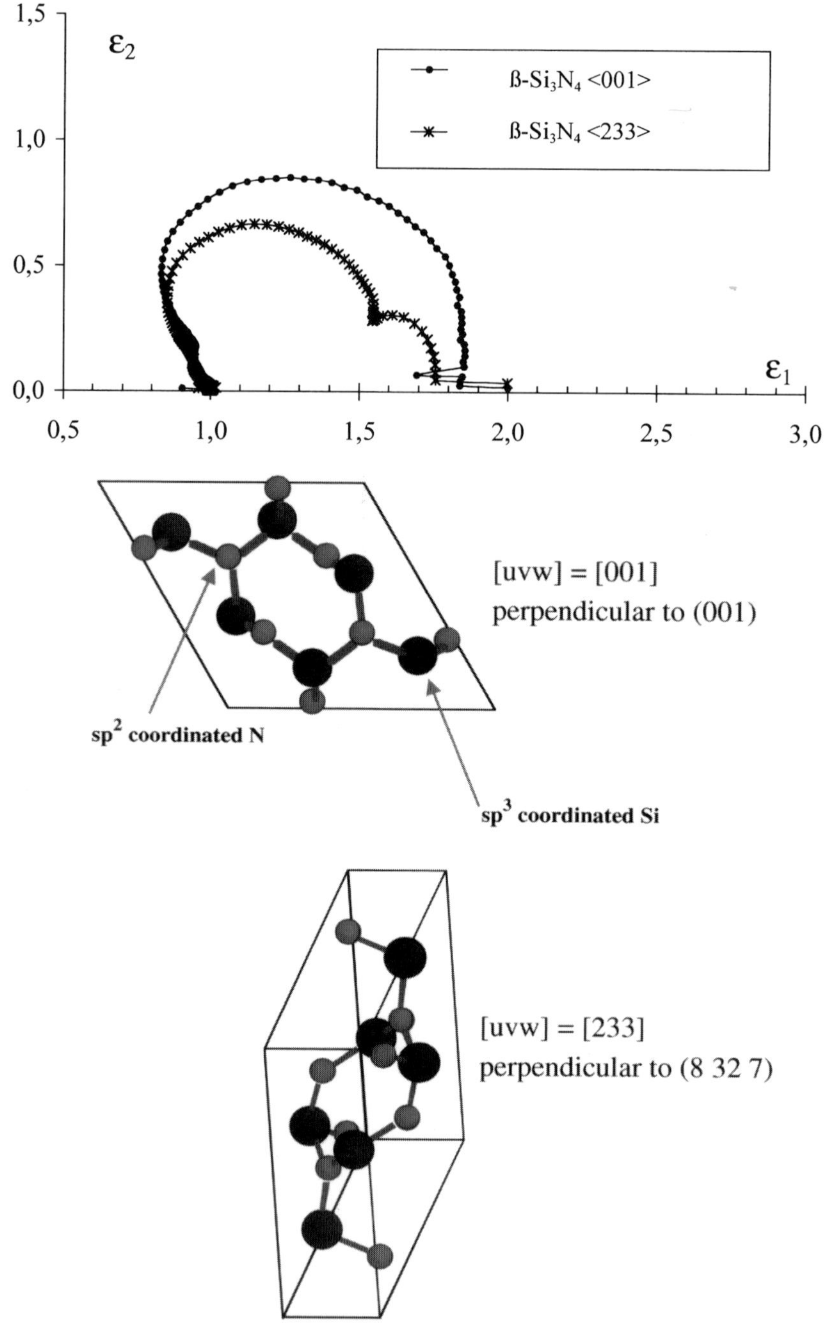

Fig. 14 (a) Cole-Cole diagrams for Si_3N_4 "viewed" along real space directions [uvw] = [001] and [233], (b) Si_3N_4 unit cell "viewed" along real space directions [uvw] = [001] and [233], (c) Cole-Cole diagrams for Si_3N_4 "viewed" along real space direction [uvw] = [103] with = $\vec{k}$ transferred along [010]*. (Please see fig. (c) in next page).

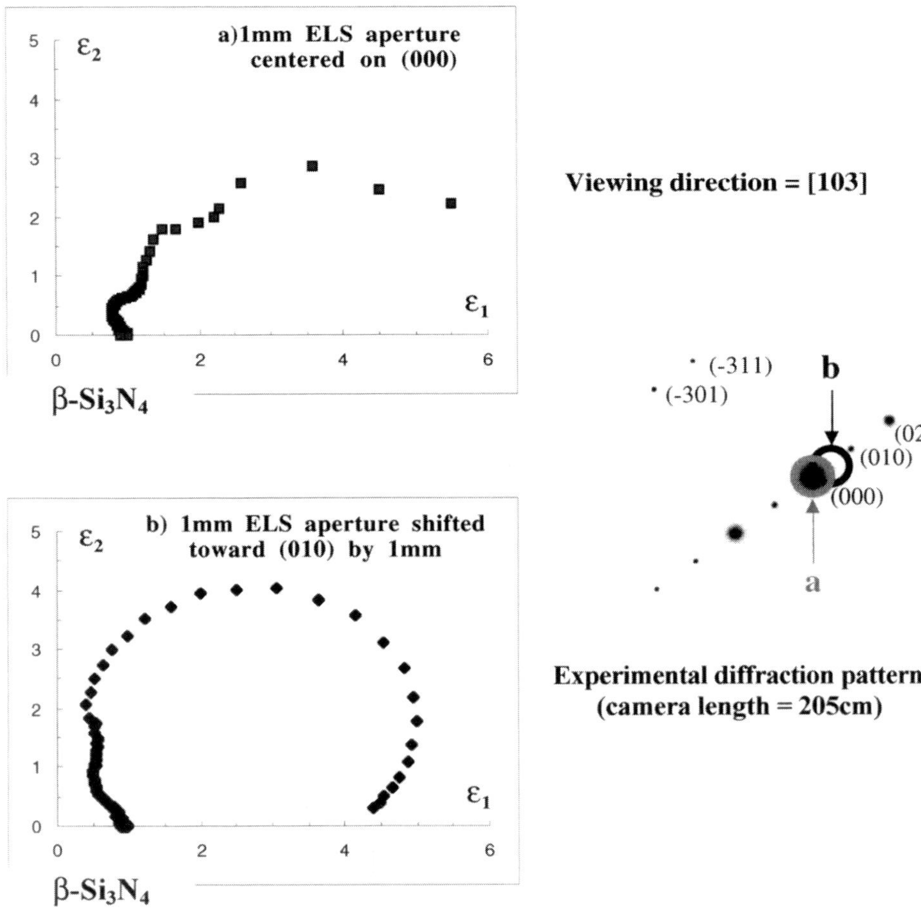

Fig. 14 continued.

We could thus determine the role of dopants on the dielectric function of barium titanate, and find that they introduce vacancies in the structure which lead to additional levels of energy in the band structure; these levels are responsible for the appearance of unexpected polarization mechanisms. We could also show a strong difference between the dielectric function of hafnium diboride and hafnium dioxide, and give an interpretation in terms of their respective band structures, the Cole-Cole plot clearly shows the difference of metallic character between both materials. The method has been also efficient in proving that no oxygen deficiency, which would have been associated with electronic effects, could be detected between trans and intergranular zones of yttria; it was particularly well suited to show the versatility of its band structure, provided that an adequate spectrum deconvolution is achieved. Finally, the local anisotropy of the dielectric function of silicon nitride has been detected through an EELS experiment in which the momentum transferred to the material was controlled. Implications of these results on the stability of vitreous interfacial films in silicon nitride have been suggested; this could provide useful informations for current theories that are based on local forces that strongly rely on the local dielectric permittivity balance, and explain why the film thickness can vary within the same sample.

5.0 ACKNOWLEDGEMENT

The authors are grateful to K.H. Westmacott (Lawrence Berkeley Laboratory, University of California) and B.Concina (Centre des Matériaux, Ecole des Mines de Paris) for enjoyable discussions and useful calculations, to D.Imhoff (Laboratoire de Physique des Solides, Université d'Orsay) for accurate EELS experiments on the Orsay VG STEM microscope, and to S. Laurent-Fontaine (Centre des Matériaux, Ecole des Mines de Paris) for help with image scanning and layout. They also thank T.Rouxel (Université de Rennes, Laboratoire de Chime des Matériaux) for providing them with the yttria sample. This work is supported by the French "Conférence des Grandes Ecoles" who provided the financial support that allowed two of us to visit the NCEM, LBNL at the University of California at Berkeley, by the French "Délégation Générale à l'Armement" (contract No. 96/075), and by the Director, Office of Energy Research, Office of Basic Energy Sciences, Materials Sciences Division of the U.S. Department of Energy under Contract No. DE-AC03-76SF00098.

6.0 REFERENCES

1. C. Pélissonnier, L. Pottier, D. Fournier and A. Thorel: *Materials Science Forum*, **207-209**, 1996, pp.813-816.
2. A. Thorel, C. Pélissonnier and S. Hagège: *Fourth Euro Ceramics*, **2**, 1995, pp.311-318
3. C. Pélissonnier: Ph.D. Thesis, Ecole Nationale Supérieure des Mines de Paris, 1996.
4. A. M. Mansanares, T. Velinov, Z. Bozoki, D. Fournier and A. C. Boccara: *Journal of Applied Physics*, **75**, 1994, p.1.
5. D. Jeulin: Probabilistic Models of Structures, Invited Key-Note Lecture, Workshop PROBAMAT 21rst Century, Perm, Russia, 10-12, G. N. Frantziskonis, ed., NATO ASI Series, **46**, 1997, pp.233-257.
6. D. Clarke: *Journal of American Ceramics Society*, **70**(1), 1987, pp.15-22
7. L. Normand, A. Thorel and Y. Montardi: *Proceedings of the " Microscopy Society of America 52nd Meeting"*, New Orleans, G. W. Bailey and A. J. Garratt, eds., Reed, 1994, pp.566-568.
8. H. Mullejans, J. Bruley, R. H. French and P. A. Morris: *Proceedings of the MRS Symposium on "Determining Nanoscale Physical Properties of Materials by Microscopy*

and Spectroscopy, M. Sarikaya, M. Isaacson and K. Wickramasighe, eds., 1994.
9. L. Normand: Ph.D. Thesis, Ecole Nationale Supérieure des Mines de Paris, 1996.
10. R. H. French, C. Scheu, G. Duscher, H. Müllejans, M. J. Hoffmann and R. M. Cannon: *Materials Res. Soc. Symposium,* **357**, 1995, pp.243-258
11. B. Delcroix, L. Savary, D. Jeulin and A. Thorel: *Proceedings of PIERS 98,* in Press.
12. K. Varlot, J. M. Martin, C. Quet, *Journal of Microscopy,* **191**, No.2, 1998, pp.187-194
13. P. Trebbia: Ph.D. Thesis, Orsay University, France, 1976.
14. R. F. Egerton: "Electron Energy Loss Spectroscopy in the Electron Microscope" Plenum Press, 1986.
15. Y. M. Chiang and D. P. Birnie: III, W. D. Kingery, *Physical Ceramics,* John Wiley and Sons, Inc., 1997.
16. L. Normand and A. Thorel: *Proceedings of the MSA 98 Meeting,* **3**, 1998, p.655.
17. V. P. Goltsev, et al.: *Proceedings of Scientific Technical Conference,* Ulyanovsk, **2**, 1970, p.535.
18. P. Cheminant, L. Boulanger, X. Deschanels, A. Thorel and Y. Serruys, EUROMAT 96, *Conference on Materials and Nuclear Power,* Bournemouth, U.K., 1996.
19. P. Cheminant, L. Boulanger, X. Deschanels and A. Thorel: *Journal of Nuc. Mat.*, **256**, 1998, pp.180-188.
20. P. Cheminant, X. Deschanels, L. Boulanger and A. Thorel: *Key Engineering Materials*, **132-136**, 1997, pp.643-646.
21. P. Cheminant-Coatanlem, Ph.D. Thesis, Université d'Orsay, 1997.
22. GATAN EELS Atlas by C. C. Ahn and O. L. Krivanek
23. A. E. Miller and A. H. Daane: *J. Inorg. Nuclear Chemicals,* **27**, 1955, 1965.
24. R. Einzinger: "Grain Boundaries in Semiconductors ", Leamy Pike and Seager, eds., North Holland, New York, 1982, p.343.
25. G. H. Jonker: "Grain Boundaries Phenomena in Electronic Ceramics ", Levinson, American Ceramic Society, Colombus, Ohio, 1981.
26. H. Thrig and M. Klerk: *Applied Physics Letters,* **35**, 1979, p.307
27. T. Rouxel, J. F. Baumard, J. L. Besson, F. Valin and M. Boncoeur: *European. Journal of Solid State Inorg. Chemistry,* **32**, 1995, pp.617-627.
28. T. Rouxel, D. Murat, J. L. Besson and M. Boncoeur: *Acta Mater.* **44**(1), 1996, pp.263-278.
29. A. Thorel, K. H. Westmacott, B. Concina, D. Imhoff and T. Rouxel: in Preparation.
30. Y. Oyama and O. Kamigaito: *Japan Journal of Applied Physics,* **10**, 1971, p.1937.
31. K. H. Jack and W. J. Wilson: *Nature Physical Science,* **283**, 1972, p.28.
32. K. H. Jack: "Progress in Nitrogen Ceramics ", F. L. Riley, ed., Martins Nijhoff Publishers, Boston, 1983, pp.45-60.
33. K. H. Jack: *Journal of Material Science,* **11**, 1976, pp.1135-1158.
34. A. Thorel, J. Y. Laval, D. Broussaud, G. Teste de Sagey and G. Schiffmacher: "Ceramic Microstructure 86; Role of Interfaces", University of California,

Berkeley, U.S.A., Plenum Press-New york, 1986, pp.263-270
35. A. Thorel, Y. Kihn, J. Y. Laval and J. Sévely: *Proceedings of ICXOM 11th,* London, Canada, 1986, pp.347-350.
36. A. Thorel: Ph.D. Thesis, Ecole des Mines de Paris, 1988.
37. A. Thorel, J. Y. Laval and D. Broussaud: "ISE87 –Lake Placid ", Les Editions de Physique, C5 49, 1988, pp.289-292.
38. J. Douin, T. Epicier, J. M. Pénisson and A. Thorel: *Materials Chemistry and Physics,* **32**, 1992, pp.77-85
39. S. Turan and K. Knowles: *Journal of European Ceramics Society,* **17**, 1997, pp.1849-1854.
40. H. J. Kleebe, M. K. Cinibulk, R. M. Cannon and M. Rühle: *Journal of American Ceramics Society,* **76**(8), 1993, pp.1969-77.
41. B. V. Derjaguin and N. V. Churaev: *Journal of Colloid Interface Science,* **49**, No.2, 1974, pp.249-255
42. H. J. Kleebe and M. Rühle: *Material Research Society. Symposium Proceedings,* **238**, 1992, pp.859-864.
43. H. C. Hamaker: *Physica,* **4**, 1937, pp.1058-1072.
44. K. M. Suenaga, D. Bouchet, C. Colliex, A. Thorel and D. G. Brandon: *Journal of the European Ceramic Society,* **18**, 1998, pp. 1453-1459
45. L. Savary, D. Jeulin and A. Thorel: *Acta Stereologica,* in Press.
46. J. Daniels, C. V. Festenberg, H. Raether and K. Zeppenfield, *Springer Tracts in Modern Physics,* Springer, Berlin, 54, 1970.
47. D. W. Johnson: *Journal of Physics A: Math.Gen.,* **8**(4), 1975.
48. K. S. Cole and R. H. Cole: *Chemical Physics,* **10**, 98, 1942.
49. T. B. Wu and J. N. Lin: *Journal of American Ceramics Society,* **77**(3), 1994, pp.212-216.
50. G. Koschek and E. Kulakek: *Journal of American Ceramics Society,* **68**, No.11, 1995, pp.582-586.
51. N. I. Medveda, V. P. Zhukov, M. Y. Khodos and V. A. Gubanov: *Physics Stat. Sol.* (b) **160**, 1990, p.517.
52. N. N. Greenwood, R. V. Parish and P. Thorton: *Quaterly Reviews,* **20**, 1966, p.441.
53. B. Post, F. W. Glaser and D. Moskowitz, *Acta Metallurgica,* **2**, 1954, p.20.
54. G. V. Samsonov, P. Hagenmüller and T. Lundstrom: " Boron and Refractory Borides ", 6th Matkovich, ed., Springer Verlag, 1977.
55. T. Tanaka and Y. Ishizawa, Albuquerque NM 1990, *AIP Conference Proceedings of 231,* D. Emin, et al. eds., 1991, p.46.
56. F. Jollet, C. Noguera, N. Thromat, M. Gautier and J. P. Durand: *Physical Review B,* **42**, No.12, 1990, pp.7587-7595.
57. N. Thromat, C. Noguera, M. Gautier, F. Jollet and J. P. Durand: *Physical Review B,* **44**, No.15, pp.7904-7911.
58. Handbook of Chemistry and Physics 76th Edition, David R.Line, 1996, pp.12-59
59. D. R. Clarke: *Materials Science Forum,* **294-296,** 1999, pp.1-8.

Interface Studies of Cu/α-Al_2O_3 (00.1) by Crystal Truncation Rod Analysis Using Synchrotron X-Ray

K. Chung and Haydn Chen
Department of Materials Science and Engineering
University of Illinois, Urbana
Illinois 61801, USA

ABSTRACT

The surface structure of the Cu/α-Al_2O_3 system was studied to obtain a systematic understanding about the relationship between the growth and the resulting structures of the system and its effect on the film properties. Discussion is focused on the crystal truncation rod (CTR) data analysis measured by synchrotron X-ray at National Synchrotron Light Source (NSLS), Brookhaven National Laboratory (BNL). Theoretical equation of the diffracted CTR intensity profile of our system based upon the kinematical approximation was fitted to the measured data. The CTR analysis method provided us with the nature of the terminating layer sequence of the substrate, the structural relaxations (interatomic distance change) of the layers near the surface, nature of interface Cu atoms, and the interfacial or the surface roughness.

1.0 INTRODUCTION

Due to the ever-demanding miniaturization of electronic devices, Cu thin films have become increasingly useful as contacts and interconnects. Consequently the interface structure and accompanied properties between Cu and a variety of device-based materials are topics of fundamental and technological interest. The electrical and the mechanical properties of the thin films are strongly dependent on the initial conditions (structure, composition and morphology) of their substrate. To study the effect of annealing on the substrate surface morphology and the film texture, Cu thin films were deposited on two different substrate surfaces at room temperature. One substrate had a high temperature annealing prior to the deposition of Cu and the other without annealing. As a reference, it is important to understand the CTR intensity profile of the bare substrate first. From the CTR intensities measured from the bare α-Al_2O_3 substrate, surface structure including the relaxed interatomic distances near the surface, roughness parameters and the appropriate termination sequence can be obtained. The model structure of the bare substrate will be used as a reference.[1] Based upon the CTR intensity from the bare substrate, any changes in the CTR intensity from the Cu/α-Al_2O_3 system can be analysed by modifying the termination with a different sequence, changing relaxed interatomic distance near the surface in the structure factor and adding a new copper layer to the surface. Then, fitting the calculated intensity to the measured one using a least-square method can derive proper values of roughness parameters.

In order to measure the agreement between the calculated CTR intensity and the measured one, a factor R is used which can be defined as

$$R = \frac{\sum | I_{measured}^{1/2} - I_{calculated}^{1/2} |}{I_{measured}^{1/2}} \quad (1)$$

The equation for the CTR intensity of a model structure is

$$I_{CTR} = S \cdot | F(\vec{q}) |^2 \cdot \frac{1}{2\sin^2(\vec{q}) \cdot \vec{a}_3 / 2)} \cdot C(\vec{q}), \quad (2)$$

where S is an overall scaling factor, $F(\vec{q})$ is a structure factor, $C(\vec{q})$ is a correction factor and $\vec{q}$ is a scattering vector. The lattice constant along the direction perpendicular to the surface is denoted by $\vec{a}_3$. Instead of determining the structure factor from the intensity profile, which depends on the momentum transfer, an ideal intensity profile was calculated with the correction factors and fitted to the measured intensity profile simply by using a scaling factor. An active area effect on the sample and the footprint of the diffracted beam received by the detector were included in the correction factor. The Lorentz, polarisation (P = 1), and Debye Waller factors as well as the absorption effect of the Cu film were also taken into account using the published data of the bulk values.[2]

2.0 EXPERIMENTS

Two different types of surfaces were prepared. The (1 × 1) unreconstructed surface and the reconstructed ($3\sqrt{3} \times 3\sqrt{3}$) $R30°$ surface. The α-Al_2O_3(00.1) substrates (1 × 1 × 0.2 mm^3) were cut from commercially available substrate (Union Carbide). All substrates were ultrasonically cleaned in distilled water, methanol and chemically etched in boiling trichloroethane for 15 minutes. Substrates were then cleaned with successive rinses of acetone, methanol and distilled water. All substrates were then blown in dry N_2. Cu is deposited on those substrates in a MBE-UHV chamber. Two samples are: (1) A Cu/Al_2O_3 sample without substrate annealing but was flashed at 600°C in the UHV chamber and cooled to room temperature prior to Cu deposition and (2) A Cu/Al_2O_3 sample with substrate annealing in air at 1500°C for 3 hours. Prior to Cu deposition, the annealed substrate was put into the UHV chamber and heated to 1200°C for 30 min. and flashed at 600°C. The sample was allowed to cool to room temperature. Cu was evaporated with a deposition rate of 2 min. at the room temperature. The resulting Cu films were 350 Å thick for both samples #1 and 2. The thickness of the film was monitored by a water-cooled quartz thickness monitor. Before deposition and high temperature annealing, all the substrates showed clean (1 × 1) bulk-like surface periodicity by RHEED. Each sample was held by tantalum clips onto a Ta foil, which was used as a heater. The temperature was monitored by a high-temperature pyrometer with an emissivity setting of 0.1 during annealing.

The intensities of the 10.L crystal truncation rod from Cu/α-Al_2O_3 (00.1) interface were measured by employing ω-rocking curve scans (± 0.2) from L = 0.1 to 10.1 with 0.2 steps in reciprocal space. X-ray wavelength of 1.127 Å was used. Samples were enclosed in an evacuated Be dome to reduce air scattering. The incoming beam size was 1 × 2.5 mm^2 and the receiving slit size was 1 × 4 mm^2. The incident x-ray flux was monitored by a N_2-filled ion chamber. The diffracted beam was detected with a proportional counter at a distance of 610 mm from the sample.

3.0 RESULTS AND DISCUSSION

3.1 Surface Reconstruction

The RHEED patterns showed (1 × 1) structure for the substrate and $(3\sqrt{3} \times 3\sqrt{3})$ $R30°$ for the reconstructed surface. A clean α-Al_2O_3 (00.1) substrate exhibits a complex surface reconstruction after heat treatment at 1200°C. Several experimental investigations have been reported using LEED techniques as a function of the various heat treatment conditions.[3]

3.2 Crystal Truncation Rods from Cu/α-Al_2O_3

3.2.1 Epitaxial Relationships

Figure 1 shows a typical XRD diffraction pattern obtained from Cu grown on unreconstructed α-Al_2O_3 (00.1) substrate at room temperature for determining the texture of the Cu film using conventional rotating anode x-ray. This is the θ-2θ scan i.e. a scan along the specular direction measuring the scattering vector perpendicular to the surface. The predominant reflections from the Cu films were 111 type diffraction peaks with some weaker features (less than 0.1% of the film volume) of 200 and 220 reflections.

For α-Al_2O_3(00.1) substrate a_s = 4.7589 Å and for a (111) plane of Cu, $a_f = \sqrt{6}a_0/2$ where a_0 is the lattice constant of Cu (3.61 Å). Thus, the misfit is -6.9% for Cu(111) || α-Al_2O_3(00.1) system. The crystallographic orientation of an epitaxial film is constrained by energy minimization and kinetic process. This requires the minimization of strain energy. However, different kinetic paths and thermal histories can lead to different orientations. There are two sets of six reflections instead of three reflections expected for 3-fold symmetry along <111> cubic. The six reflections are a common feature for the Cu(200) on α-Al_2O_3(00.1) substrate.[4-6]

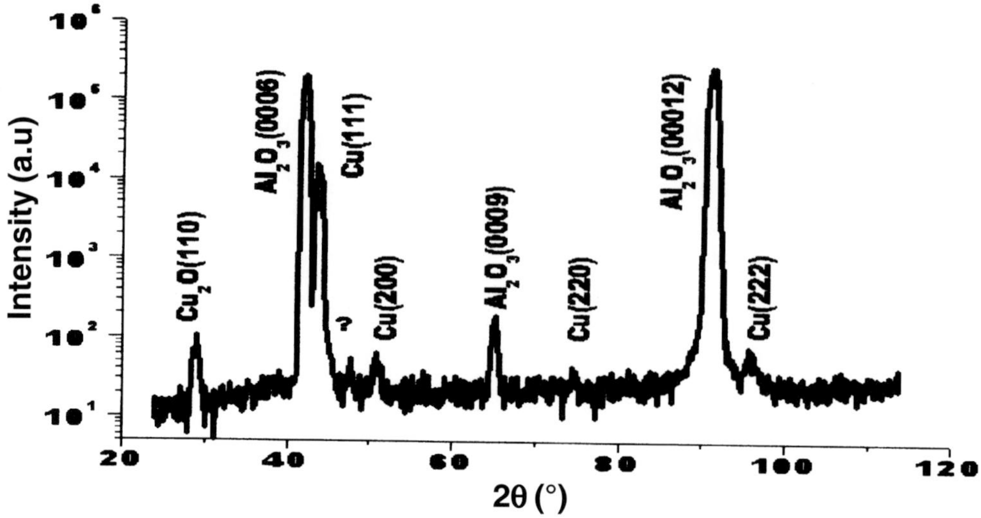

Fig. 1 θ-2θ scans of Cu deposited at room temperature on unteconstructed α-Al_2O_3(00.1) substrate (sample #1) using laboratory X-ray of Cu rotating anode.

There are two possibilities for the six reflections. One is due to the two different sequences of depositing layer for the FCC Cu, that are ...ABCABC...and... ACBABC resulting in pseudo six-fold symmetry which is common in FCC metals. The other possibility is due to the stepped substrate. (There are two types of terraces related 60° from each other on (00.1) surface). This stepped surface result in additional variant in the nucleation growth direction.

The corresponding Cu 200 omega rocking curves are shown in Figure 2 and revealed a smaller mosaic spread for samples #1. There are peak shape differences between unreconstructed surfaces of samples #1 and reconstructed surfaces of samples #2. Samples #1 showed a typical Gaussian peak shape as common in bulk crystals. On the other hand, sample #2 displayed a Lorentzian line shape that can be related to extended structural defects during the heat treatments and the increase of the roughness.[7]

This roughness is determined from the CTR data analysis. From the above x-ray diffraction data, we can conclude that the films are poly-crystal FCC Cu with a preferred orientation along <111> direction and the epitaxial relationship between the film and the substrate is Cu(111)||α-Al_2O_3(00.1).

3.2.2 CTR Analysis

The intensity near the Bragg peak is predominantly effected by the bulk structure factor. The intensity is very weak when L is far away from the Bragg peak. For sample #1, the measured signal is very small relative to the diffuse background.

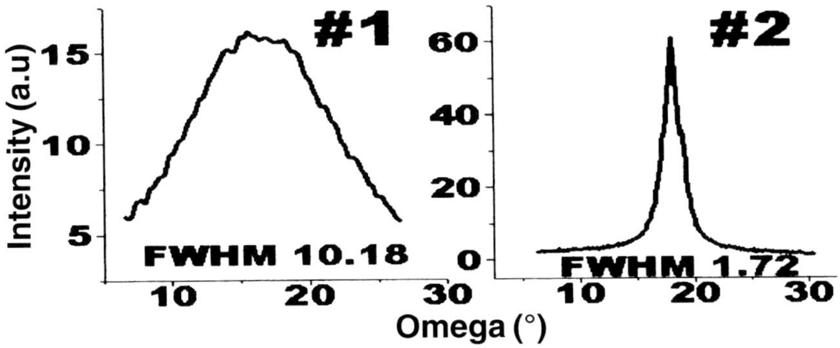

Fig. 2 Synchrotron x-ray diffraction of omega rocking profiles for the Cu (200) bragg peak.

Thus, in order to improve the fitting, the intensity far away from the Bragg peak is weighted proportional to the counting rate. There are total of 8 fitting parameters including 4 atomic displacements, 1 overall scaling factor, 1 roughness, 1 surface fraction of Cu coverage that adopts the structure derived in the analysis and 1 in-plane displacement of first Cu layer. The oxygen fraction in the surface of the substrate is derived from the previous fitting results for samples #1 and 2. The occupancy factor and one more out-of-plane displacement of 4th layer beneath the surface are also adopted as fixed parameters. There are 2 different terminating sequences with 2 types of Cu deposition which require 4 different fitting models. In addition to that, to check the enhancement of epitaxy by heat treatment, a double Cu layer model is also tested. We designate the Al-O-Al termination as T1, O-Al-Al termination as T2, Al-Al-O as T3, Cu (type A)-Al-O as T4, Cu (type B)-Al-O as T5, 2 layer Cu (type A)-Al-O as T6, Cu (type A)-Al-O as T7, Cu (type B)-Al-Al as T8 and 2 layer Cu (type A)-Al-Al as T9. Because there is a possibility of the Cu overlayer's periodicity is far away from the substrate's and produces no scattering intensity along the substrate coordinates based CTR, a model substrate with no Cu (T1, T2 and T3) is also tested. A simple substitution of Cu with first layer of Al is also tested (T10).

The measured CTRs and models that showed the better fittings among 10 different models are shown in Figure 3. The best fitting results are summarised in Table 1. Our model structure consisted of an ideal bulk crystal with additional ideal unit cell. Then we add one more unit cell with the top 5 layers relaxed. The two non-structural parameters, roughness parameter and Cu film coverage, showed physically reasonable values. The rms roughness for sample #1 showed slightly larger values than the bare substrate's.

For samples #1 and 2 a single Cu layer at (h k) = (2/3 1/3), (0 0) and (1/3 2/3) on top of the single Al termination showed reasonable good fits with R factor of 0.21 and 0.24 respectively. Kasowski et al.[8] studied the electronic structure and bonding

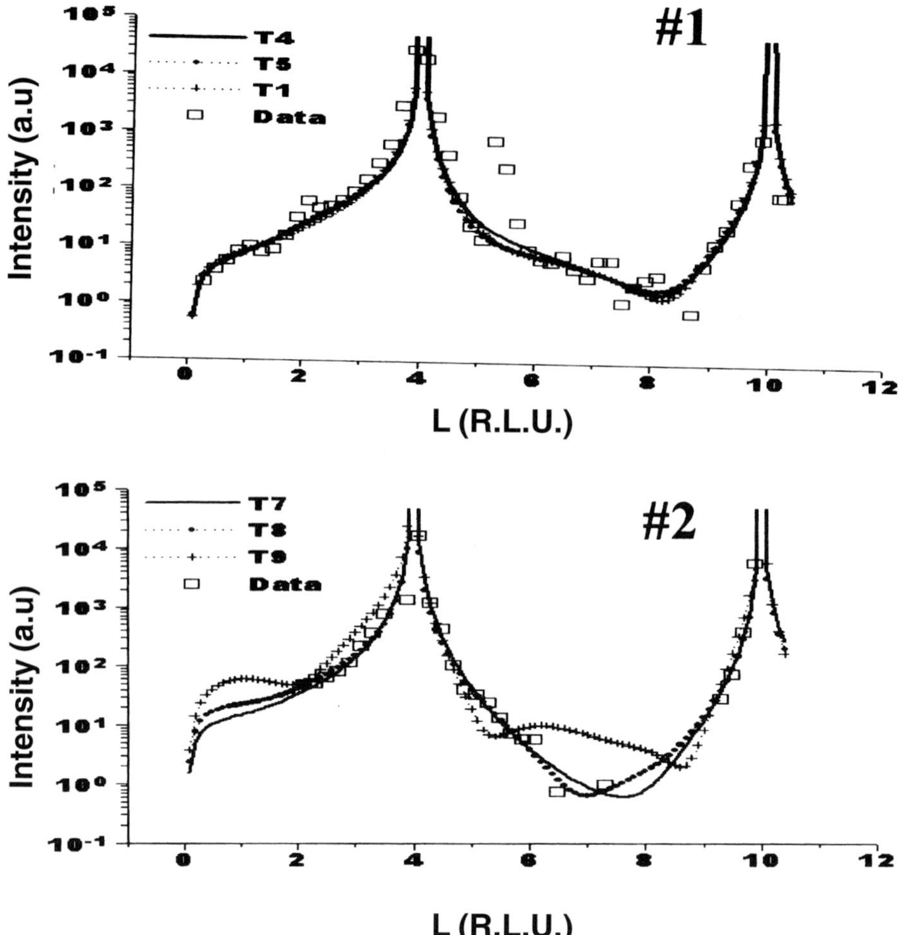

Fig. 3 The measured 10.L CTRs and computed CTRs based upon various surface structures for four samples.

of Cu to Al_2O_3 by using ultraviolet photo-emission spectra (UPS) method with *ab initio* total energy calculation. The suggested bond length of Cu-O is 1.85 Å and bond length of Cu-Al of 2.55 Å. Guo's[5] calculation showed a short Al-O bond length of 1.86 Å and a long bond length of 1.97 Å. Assuming the Cu and all oxygen atoms are located at the ideal position in (h k) in-plane coordinates and varies only along out-of-plane direction, the inter atomic distances from CTR analysis are comparable to the calculated bond length.

From the ϕ and χ scan we found there are two types of surface Cu domains. One is an azimuthal orientation of $\langle 11\bar{2}\rangle$Cu || $\langle 11.0\rangle$ Al_2O_3, (type A) and the other which is 30° away from the former crystallites in ϕ scan with $\langle 2\bar{1}\bar{1}\rangle$ Cu || $\langle 10.0\rangle Al_2O_3$ (type B). Both models (T7 and T8) showed indistinguishable results in our range of precision as shown in Figure 3. It is possible that the actual CTR

Table 1 Fitting parameters observed from the best fitting model.

	Sample #1 (T4)	Sample #2 (T7)
R factor	0.21	0.134
Cu (L)	2.041	2.156
Al (L)	1.966	-
O (L)	1.870	-
Al (L)	1.857	1.677
Al (L)	-	1.682
O (L)	-	1.646
In-plane relax. (H)	± 0.06	± 0.06
Rms roughness (Å)	4.7	4.6
Cu coverage (%)	24	42

Four atoms, including one Cu atom, are relax out-of-plane direction. H and L are r.l.u. of (HKL).

intensity is the average of those two models with a ratio equal to 2:1. It is known that arrangements of alternating domains are favourable since this can minimize the elastic energy associated with the surface stress.[9] Coexistence of two types of FCC grains is favourable in energy consideration.

For sample #1, a single Al termination without any Cu atom on its surface (model T1) also provide a good fitting due to the poor coverage and registry of the Cu. The R factor is 0.22 and rms roughness is 5 Å for this model. In sample #2, there is an abnormally large intensity found near (10.5.3) that is omitted for fitting loop. This position is one of the Cu{200} which is 63° in χ away from the other three sets of Cu{111} peaked 28° away from the surface normal Cu(111). For sample #2, a double layer Cu model (T9) on double Al termination did not showed better fittings. This is expected from the wide FWHM found in ϕ scan for these samples. Both samples showed best fitting with a single Cu layer on double Al termination (T7) model. The fitting results also showed a large inter-atomic distance between surface Cu and surface oxygen for sample #2.

4.0 CONCLUSION

The smaller FWHM in omega rocking curves for Cu film grown on reconstructed surface indicates that it led to better out-of-plane ordering. From the CTR analysis, interatomic distance and roughness of buried interface are obtained. A smaller distance of Cu-O is shown in copper deposited on a bare substrate. The reconstructed surface showed a double Al termination. A large relaxation is found between Al and Al in the bare substrate. A larger separation of Cu-O is shown in copper deposited on the reconstructed surface. This larger distance indicates that the Cu has more metallic nature than the Cu on the bare substrate.

5.0 ACKNOWLEDGEMENTS

This material is based upon work supported by the Division of Materials Sciences, Office of Basic Energy Science, U.S. Department of Energy under Grant No. DEFG02-9IER45439. Acknowledgment is also made to the Illinois Board of Higher Education under Grant No. HECA-NWU-0950-951 0A. The National Synchrotron Light Source is supported by the U.S. Department of Energy under Contract No.DE-AC02-76CH0016.

6.0 REFERENCES

1. K. Chung: Interface Structure Studies of Copper Thin Films on C-Plane Sapphire by Crystal Truncation Rod Analysis using Synchrotron X-ray, Ph.D. Thesis, University of Illinois, 1998.
2. *International Tables for X-Ray Crystallography*, Kynoch Press, Birmingham, England, **3**, 1962, pp.71, 99, 149, **4**, 1974, pp.233.
3. R. H. French, D. J. Jones and S. Loughin: Interband Electronic Structure of α-alumina up to 2167 K, *Journal of American Ceramics Society*, **77**(2), 1994, p.412.
4. D. W. Susnitzky and C. B. Carter: Topotactic Growth of β-Alumina on α-Alumina, *Journal of American Ceramics Society*, 69 C-217, 1986, G. Renaud, B. Villette, I. Vilfan, and A. Bourret, Atomic Structure of the α-Al_2O_3(0001) ($\sqrt{31} \times \sqrt{31}$) R ± 9° Reconstruction, Physics Review Letters, **73**(13), 1994, p.1825.
5. H. Bialas and E. Knoll: Heteroepitaxy of Copper on Sapphire under UHV Conditions, *Vacuum*, **45**(9), 1994, p.959, E. Knoll and H. Bialas, Growth Modes of Epitaxial Copper Films on c-sapphire, *Thin Solid Films*, **250**, 1994, p.42.
6. J. Guo, D. E. Ellis and D. J. Lam: Electronic Structure and Energetics of Sapphire(0001) and (1-102) Surfaces, *Physics Review*, **B45**(23), 1992, p.13647.
7. G. Katz: Ion-Etch Characteristics of Epitaxial Copper on Sapphire, *Journal of Materials Science*, **5**, 1970, p.736.
8. Q. Zhu, A. Botchkarev and W. Kim, et al.: Structural Properties of GaN Films Grown on Sapphire by Molecular Beam Epitaxy, *Applied Physics Letters*, **68**(8), 1996, p.1141.
9. R. V. Kasowski, F. S. Ohuchi and R. H. French: Utraviolet Photoemission and Theoretical Studies of the Electronic Structure of Al_2O_3 up to 1000°C, *Physica B*, **150**, 1988, p.44.
10. O. A. Alerhand, D. Vanderbilt, R. D. Meade and J. D. Joannopoulos, Spontaneous Formation of Stress Domains on Crystal Surfaces, *Physics Review Letters*, **61**, 1988, p.1973.

Author Index

Subject Index